D1596604

THEORY OF MODERN ELECTRONIC SEMICONDUCTOR DEVICES

THEORY OF MODERN ELECTRONIC SEMICONDUCTOR DEVICES

KEVIN F. BRENNAN
APRIL S. BROWN
Georgia Institute of Technology

A Wiley-Interscience Publication
JOHN WILEY & SONS, INC.

This book is printed on acid-free paper. ∞

Published simultaneously in Canada.

For ordering and customer service, call 1-800-CALL-WILEY.

Library of Congress Cataloging-in-Publication Data Is Available

ISBN 0-471-41541-3

Printed in the United States of America

10 9 8 7 6 5 4 3 2 1

To our families,
Lea and Casper
and
Bob, Alex, and John

CONTENTS

†Optional material.

PREFACE

The rapid advancement of the microelectronics industry has continued in nearly exponential fashion for the past 30 years. Continuous progress has been made in miniaturizing integrated circuits, thus increasing circuit density and complexity at reduced cost. These circumstances have fomented the continuous expansion of computing capability that has driven the modern information age. Explosive growth is occurring in computing technology and communications, driven mainly by the advancements in semiconductor hardware. Continued growth in these areas depends on continued progress in microelectronics.

At this writing, critical device dimensions for commercial products are already approaching 0.1 μm. Continued miniaturization much beyond 0.1-μm feature sizes presents myriad problems in device performance, fabrication, and reliability. The question is, then, will microelectronics technology continue in the same manner as in the past? Can continued miniaturization and its concomitant increase in circuit speed and complexity be maintained using current CMOS technology, or will new, radically different device structures need to be invented?

The growth in wireless and optical communications systems has closely followed the exponential growth in computing technology. The need not only to process but also to transfer large packets of electronic data rapidly via the Internet, wireless systems, and telephony is growing at a brisk rate, placing ever increasing demands on the bandwidth of these systems. Hardware used in these systems must thus be able to operate at ever higher frequencies and output power levels. Owing to the inherently higher mobility of many compound semiconductor materials compared to silicon, currently most high-frequency electronics incorporate compound semiconductors such as GaAs and InP. Record-setting frequency performance at high power levels is invariably accomplished using either heterostructure field-effect or heterostructure bipolar transistors. What, though, are the physical features that limit the performance of these devices? What are their limits of performance? What alternatives can be utilized for high-frequency-device operation?

Device dimensions are now well within the range in which quantum mechanical effects become apparent and even in some instances dominant. What quantum mechanical phenomena are important in current and future semiconductor devices? How do these effects alter device performance? Can nanoelectronic devices be constructed that function principally according to quantum mechanical physics that can provide important functionality? How will these devices behave?

The purpose of this book is to examine many of the questions raised above. Specifically, we discuss the behavior of heterostructure devices for communications systems (Chapters 2 to 4), quantum phenomena that appear in miniaturized structures and new nanoelectronic device types that exploit these effects (Chapters 5, 6, and 9), and finally, the challenges faced by continued miniaturization of CMOS devices and futuristic alternatives (Chapters 7 and 8). We believe that this is the first textbook to address these issues in a comprehensive manner. Our aim is to provide an up-to-date and extended discussion of some of the most important emerging devices and trends in semiconductor devices. The book can be used as a textbook for a graduate-level course in electrical engineering, physics, or materials science. Nevertheless, the content will appeal to practicing professionals. It is suggested that the reader be familiar with semiconductor devices at the level of the books by Streetman or Pierret. In addition, much of the basic science that underlies the workings of the devices treated in this text is discussed in detail in the book by Brennan, *The Physics of Semiconductors with Applications to Optoelectronic Devices*, Cambridge University Press, 1999. The reader will find it useful to refer to this book for background material that can supplement his or her knowledge aiding in the comprehension of the current book.

The book contains nine chapters in total. The first chapter provides an overview of emerging trends in compound semiconductors and computing technology. We have tried to focus the book on the three emerging areas discussed above: telecommunications, quantum structures, and challenges and alternatives to CMOS technology. The balance of the book examines these three issues in detail. There are sections throughout that can be omitted without loss of continuity. These sections are marked with a dagger. We end the book with a chapter on magnetic field effects in semiconductors. It is our belief that although few devices currently exploit magnetic field effects, the unusual physical properties of reduced dimensional systems when exposed to magnetic fields are of keen interest and may point out new directions in semiconductor device technology. Again, the instructor may elect to skip Chapter 9 completely without compromising the main focus of the book.

From a pedagogic point of view, we have developed the book from class notes we have written for a one-semester graduate-level course given in the School of Electrical and Computer Engineering at the Georgia Institute of Technology. This course is generally taught in the spring semester following a preparatory course taught in the fall. Most students first study the fall semester course, which is based on the first nine chapters of the book by Brennan, *The Physics of Semiconductors with Applications to Optoelectronic Devices*. Nevertheless, the present book can be used independent of a preparatory course, using the book by Brennan as supplemental reference material. The present book is fully self-contained and refers the reader to Brennan's book only when needed for background material. Typically, we teach Chapters 2 to 8 in the current book, omitting the optional (Sections 2.5, 5.7, 6.3, and 7.1). The students are asked to write a term paper in the course following up in detail on one topic. In addition, homework problems and a midterm and final examinations are given. The reader is invited to visit the book Web site at *www.ece.gatech.edu/research/labs/comp_elec* for updates and supplemental information. At the book Web site a password-protected solutions manual is available for instructors, along with sample examinations and their solutions.

We would like to thank our many colleagues and students at Georgia Tech for their interest and helpful insight. Specifically, we are deeply grateful to Dr. Joe Haralson II, who assisted greatly in the design of the cover and in revising many of the figures used throughout. We are also grateful to Tsung-Hsing Yu, Dr. Maziar Farahmand, Louis Tirino, Mike Weber, and Changhyun Yi for their help on technical and mechanical aspects of manuscript preparation. Additionally, we thank Mike Weber and Louis Tirino for setting up the book Web site. Finally, we thank Dr. Dan Tsui of Princeton University, Dr. Wolfgang Porod of Notre Dame University, Dr. Mark Kastner of MIT, Dr. Stan Williams of Hewlett-Packard Laboratories, and Dr. Paul Ruden of the University of Minnesota at Minneapolis for granting permission to reproduce some of their work in this book and for helpful comments in its construction.

Finally, both of us would like to thank our families and friends for their enduring support and patience.

Atlanta
November 2000

KEVIN F. BRENNAN
APRIL S. BROWN

CHAPTER 1

Overview of Semiconductor Device Trends

The dawn of the third millennium coincides with what has often been referred to as the *information age*. The rapid exchange of information in its various formats has become one of the most important activities of our modern world. Shannon's early recognition that information in its most basic form can be reduced to a series of bits has lead to a vast infrastructure devoted to the rapid and efficient transfer of information in bit form. The technical developments that underlie this infrastructure result from a blending of computing and telecommunications. Basic to these industries is semiconductor hardware, which provides the essential tools for information processing, transfer, and display.

1.1 MOORE'S LAW AND ITS IMPLICATIONS

The integrated circuit is the fundamental building block of modern digital electronics and computing. The rapid expansion of computing capability is derived mainly from successive improvements in device miniaturization and the concomitant increase in device density and circuit complexity on a single chip. Functionality per chip has grown in accordance with Moore's law, an historical observation made by Intel executive Gordon Moore. *Moore's law* states that functionality as measured by the number of transistors and bits doubles every 1.5 to 2 years. As can be seen from Figure 1.1.1, the number of transistors on a silicon chip has followed an exponential dependence since the late 1960s. This in turn has led to dramatic improvements in computing capability, leading the consumer to expect ever better products at reduced cost.

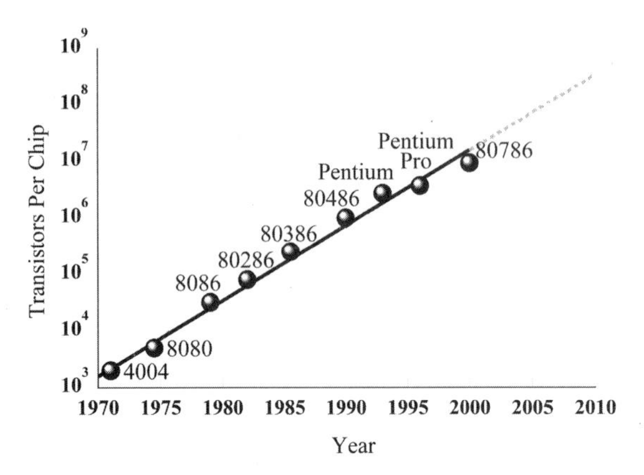

FIGURE 1.1.1 Number of transistors per chip as a function of year, including the names of the processors. The dashed line shows projections to 2010. The exponential growth reflected by this graph is what is commonly referred to as Moore's law. (Data from Birnbaum and Williams, 2000.)

One of the questions that this book addresses is: Is there a limit to Moore's law? Can integrated-circuit complexity continue to grow exponentially into the twenty-first century, or are their insurmountable technical or economic challenges that will derail this progress?

The two prominent technical drivers of the semiconductor industry are dynamic random access memory (DRAM), and microprocessors. Historically, DRAM technology developed at a faster pace than microprocessor technology. However, from the late 1990s microprocessors have become at least an equal partner to DRAMs in driving semiconductor device refinement. In many instances, microprocessor units (MPUs), have become the major driver of semiconductor technology. There are different performance criteria for these two major product families. The major concerns for DRAMs are cost and memory capacity. The usual metric applied to DRAMs is *half-pitch*, which is defined as essentially the separation between adjacent memory cells on the chip. Consequently, minimization of the area of each memory cell to provide greater memory density is the primary development focus for DRAMs. Cost and performance also drive microprocessor development, but the key parameters in this case are gate length and the number of interconnect layers. The maintainence of Moore's law requires, then, aggressive reduction in gate length as well as in DRAM cell area.

The semiconductor industry has tracked the technology trends of both DRAM and microprocessor technology and established technology road maps to project where the technology will be in subsequent years. Such road maps provide the industry with a rough guide to the technology trends expected. Below we discuss some of the implications of these trends and examine how they affect Moore's law.

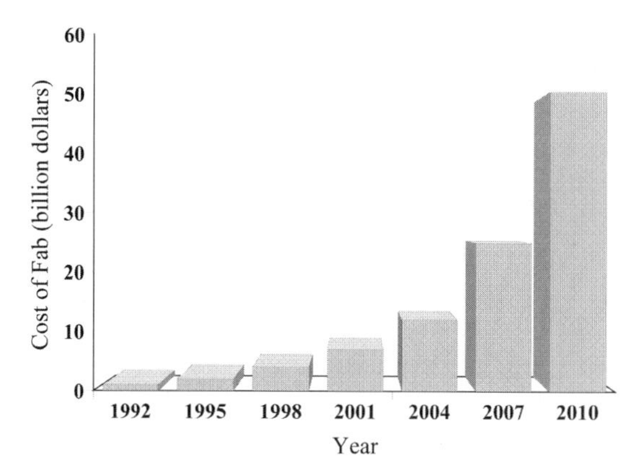

FIGURE 1.1.2 Fabrication plant cost as a function of year. Notice the tremendous cost projected by the year 2010. (Data from Birnbaum and Williams, 2000.)

Let us first consider economic challenges to the continuation of Moore's law. There is a second law attributed to Moore, Moore's second law, which examines the economic issues related to integrated-circuit chip manufacture. According to *Moore's second law*, the cost of fabrication facilities needed to manufacture each new generation of integrated circuits increases by a factor of 2 every three years. Extrapolating from fabrication plant costs of about $1 billion in 1995, Figure 1.1.2 shows that the cost of a fabrication plant in 2010 could reach about $50 billion. That a single company or even a consortium of companies could bear such an enormous cost is highly doubtful. Even if a worldwide consortium of semiconductor industries agreed to share such a cost, continuing to do so for future generations would certainly not be feasible. Therefore, it is likely that the economics of manufacturing will strongly influence the growth of the integrated-circuit industry in the near future.

Aside from the economic issues faced by continued miniaturization, several daunting technical challenges threaten continued progress in miniaturization of integrated circuits in accordance with Moore's law. As discussed in detail in Chapter 7, several technical issues threaten continued exponential growth of integrated-circuit complexity. Generally, these concerns can be classified as either physical or practical. By physical challenges we mean problems encountered in the physical operation of a device under continued miniaturization. Practical challenges arise from the actual fabrication and manufacture of these miniaturized devices. Among the practical challenges are lithography, gate oxide thickness reduction, and forming interconnects to each device. Physical operational problems encountered by continued miniaturization of devices include threshold voltage shifts, random fluctuation in the dopants, short-channel effects, and high-field effects. Any one or a combination of these effects could threaten continued progress of Moore's law. We examine these effects in detail in Chapter 7.

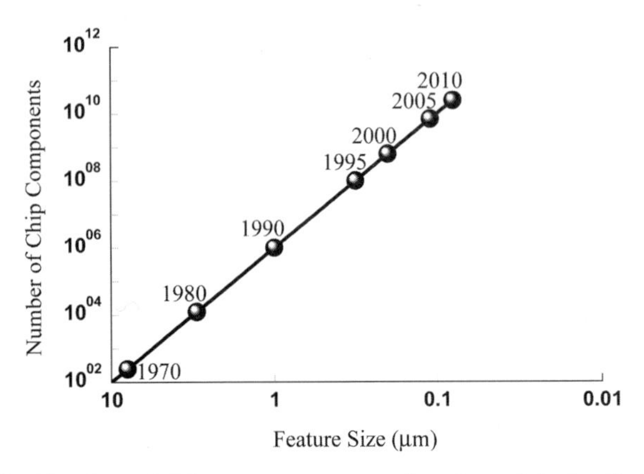

FIGURE 1.1.3 Number of chip components as a function of device feature size. Both the historical trend and projected values according to the Semiconductor Institute of America (SIA) road map are shown. (Data from Birnbaum and Williams, 2000.)

Figure 1.1.3 shows how the number of chip components scales with the on-chip feature size. As can be seen, to maintain the historical trend in the near future, device feature sizes will need to be scaled significantly below 0.1 μm. However, doing so requires overcoming many of the challenges listed above. At the time of this writing, strategies for overcoming the physical and practical challenges to further miniaturization are not known.

In addition to the practical and physical challenges to continued miniaturization, there are other issues that may thwart further progress. Formidable challenges arise in design, testing, and packaging integrated circuits containing billions of transistors. For example, testing a chip containing billions of transistors that operates at gigabit speeds is presently not possible. It is not sufficient that the chip contain billions of devices and operates at gigabit speeds. Just as important is the ability to extract signals from the transistors within the chip at gigahertz frequencies. Therefore, new packaging schemes will need to be developed to ensure that chip output progresses along with the chip itself.

Although there are many practical and physical challenges that may thwart further reduction of integrated circuits, progress may still occur through either evolutionary or revolutionary advancements. By *evolutionary*, we mean continued progress in device reduction through progressive refinements in the main integrated-circuit technology itself, complementary metal-oxide semiconductor (CMOS), technology. In Chapter 7 we examine several evolutionary technologies that could potentially extend CMOS in accordance with Moore's law. Alternatively, *revolutionary* technologies that go well beyond conventional CMOS may be required to continue progress in miniaturization. In Chapter 8 we examine several leading candidate technologies that may form the basis of computing hardware in the future. However, before anyone begins to invest and develop these alternative technologies aggressively, it is first necessary to examine just how much better computing hardware can be made to be. It would be very unwise to pursue an expensive revolutionary technology aggressively if only marginal performance improvements can be made.

To this end, it is important to ask: What are the fundamental limits to computation, if any? Can these limits be quantified? Knowledge of the ultimate limits that the laws of physics place on computation, coupled with where CMOS technology is at present, clearly enable us to assess what more can be done and whether such advancements warrant development. Rolf Landauer and Richard Feynman pondered these questions in the late 1950s. They considered independently what the thermodynamic limits are to computing after recognizing that information could be treated as a physical entity and could thus be quantified.

Work by Lloyd (2000) has examined in detail the physical limits to computation. To that end, it is useful to determine the maximum speed of a logical operation. To perform an elementary logic operation in a certain time, a minimum amount of energy is required. The minimum energy required to perform a logic operation in time Δt is given by Lloyd (2000) as

$$E \geq \frac{\pi\hbar}{2\Delta t} \tag{1.1.1}$$

Therefore, if a computer has energy E available to it for computation, it can perform a maximum number of logical operations N per second given as

$$N = \frac{2E}{\pi\hbar} \tag{1.1.2}$$

It is important to recognize that the foregoing limit applies to both serial and parallel processing. The computational limit is independent of the computer architecture. The total number of computations per second is the same whether one chooses to spend all the available energy to do serial computation faster or spread the energy out to do many calculations at once but with each calculation taking more time.

How do present computers compare to this ultimate speed limit? If one had an ideal computer, one in which all its mass could be transformed into computational energy, the limits imposed by Eqs. 1.1.1 and 1.1.2 would lead to very high computational speeds. Lloyd (2000) has estimated that a 1-kg computer all of whose mass–energy could be used for computation could perform about 10^{50} operations per second. A modern computer falls far short of this rate mainly because (1) very little of its mass–energy is used in performing logic operations, and (2) many electrons are used to encode 1 bit. In modern computers, anywhere from 10^6 to 10^9 electrons are used to encode a "1." In principle, only one electron is needed to encode a "1." As we will see in Chapter 8, single-electron transistors have been developed that require only one electron for charging and storage of a bit.

In addition to the limit placed on the computational speed of a logical operation, thermodynamics limits the number of bits that can be processed using a specific amount of energy in a given volume. The amount of available energy limits the computational rate, while the entropy limits the amount of information that can be stored and manipulated. The classical entropy is related to the multiplicity of states as (Brennan, 1999, Sec. 5.3)

$$S = k_B \sigma = k_B \ln g \qquad 1.1.3$$

where k_B is Boltzmann's constant, σ the entropy, S the conventional entropy, and g the multiplicity function. For a two-level system of m components, the multiplicity function g is equal to 2^m. Such a system can store m bits of information. Thus the number of bits m that can be stored in the system can be related to the entropy as

$$m = \frac{S}{k_B \ln 2} \qquad 1.1.4$$

Combining Eqs. 1.1.2 and 1.1.4, the maximum number of logical operations per second per bit can be found as

$$\frac{\text{operations}}{\text{bit-second}} = \frac{2E}{\pi\hbar}\frac{k_B \ln 2}{S} = \frac{2Ek_B \ln 2}{\pi\hbar S} \qquad 1.1.5$$

Recognizing that the temperature T is given as (Brennan, 1999, Sec. 5.3)

$$\frac{1}{T} = \frac{dS}{dE} \qquad 1.1.6$$

the maximum number of operations per bit per second can be approximated as

$$\frac{\text{operations}}{\text{bit-second}} \sim \frac{k_B T}{\hbar} \qquad 1.1.7$$

Thus the temperature limits the maximum number of operations per bit per second. A very high temperature is needed to maximize the number of operations. High-temperature operation is undesirable in a practical computing system.

In summary, the thermodynamic limit of a nonreversible computer (one in which there exists dissipation of energy during computation) is many orders of magnitude higher than the estimated upper limit of CMOS circuitry. Based on these results, it is clear that there remains the possibility of a huge improvement in computing capability with further miniaturization. Although different paths may need to be taken from the current one for CMOS, there remains the strong possibility that dramatic improvement in computing hardware can be realized with increased miniaturization. Part of the focus of this book is to examine what technologies can be harnessed to provide further miniaturization.

1.2 SEMICONDUCTOR DEVICES FOR TELECOMMUNICATIONS

Although the largest part of the semiconductor industry is devoted to silicon integrated-circuit technology and CMOS in particular, the rapid growth of the telecommunications industry has stimulated significant growth in a different part of the semiconductor industry: compound semiconductors. The total worldwide semiconductor market has exceeded $200 billion in sales in the year 2000. Only about $15 billion of this amount can be attributed to compound semiconductor products. Nevertheless, compound semiconductors have undergone a dramatic increase in sales over the past decade. From the early to late 1990s, shipment of compound semiconductor device products has increased over fourfold. Much of the rapid growth in compound semiconductor products has occurred in telecommunications, owing to expansion in both wireless and wired fiber-optic systems. Projections indicate that due to the rapid growth of the telecommunications industry, the sale of compound semiconductor products will grow more rapidly than that of silicon-based semiconductor products.

The most important compound semiconductor products used in fiber-optic networks are optoelectronic devices, particularly lasers and detectors. Semiconductor lasers are made exclusively from compound semiconductors since these materials have direct bandgaps and can thus be made to lase. In contrast, silicon is an indirect-gap semiconductor and currently cannot lase. In addition, direct-gap materials make more efficient photodetectors, at least for radiation with wavelengths near the bandgap. For these reasons, the optoelectronic component industry utilizes primarily compound semiconductors. Although the overall market for optoelectronic devices is small compared to that for CMOS, it is still substantial. Revenue in 2000 for optical components exceeded $10 billion. Forecasts predict that the worldwide optical component market will exceed $19 billion by 2003, just about doubling in three to four years. In this book we restrict our discussion to electronic devices and do not consider optoelectronic devices. The interested reader is referred to the book by Brennan (1999) for a discussion of optoelectronic devices.

The compound semiconductor industry developed from military electronics needs for microwave and millimeter-wave systems. For these applications, including radar and satellite-based communications, cost was considered a secondary metric to performance. With the advent of microwave commercial electronics, the metrics have changed. Figure 1.2.1 shows the increase in frequency for commercial electronics over the past two decades. Figure 1.2.2 shows the progression of technology drivers and insertion points for microwave electronics. The movement toward commercial electronics has significantly reduced the cycle time, placed a much stronger emphasis on ruggedness or reliability, and broadened the scope of relevant and specific performance criteria.

Advances in epitaxy, the process technique for producing atomically smooth heterojunctions, have, in part, enabled their use in commercial products. Mo-

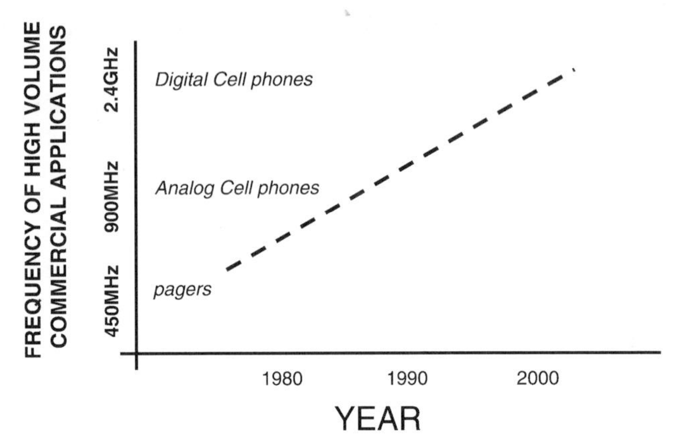

FIGURE 1.2.1 Advancement of frequency for commercial electronics. Progress up in frequency occurs at a rate of approximately 1 octave/decade. (Reprinted with permission from Golio, 2000.)

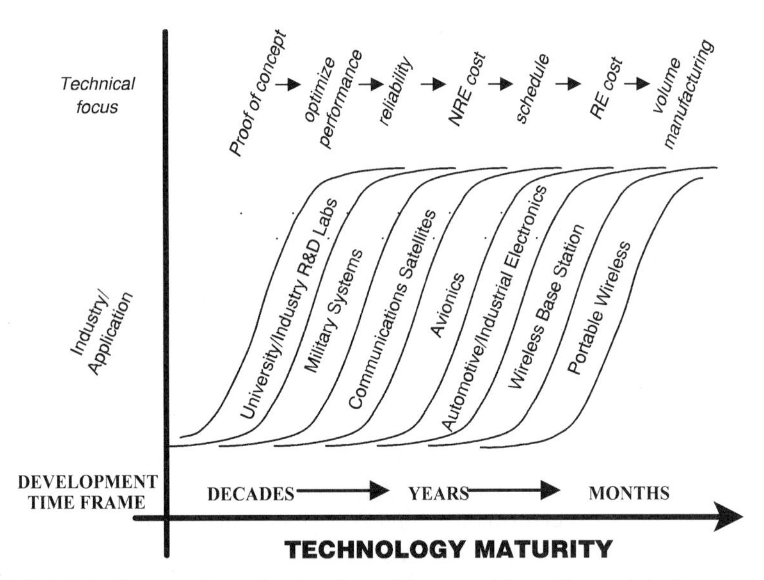

FIGURE 1.2.2 Progression of technology drivers and insertion points for microwave electronics. (Reprinted with permission from Golio, 2000.)

lecular beam epitaxy (MBE) and metal-organic chemical vapor deposition (MOCVD) evolved primarily as laboratory tools for physics experiments and the production of new, exotic materials. Today, both techniques have been scaled for multiwafer growth and are being used for the commercial production of heterojunction-based optical and electronic devices. The size of GaAs and InP wafers has also affected manufacturing and cost, due in part to the

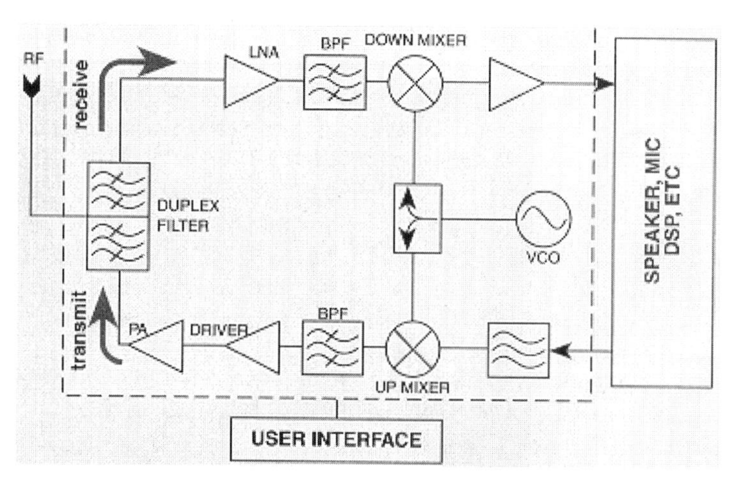

FIGURE 1.2.3 Front end of a radio. (Reprinted with permission from Golio, 1995.)

economies of scale in the numbers of circuits produced per wafer run and the fact that much process equipment, developed originally for silicon, accommodates a minimum wafer size. Today, 6-inch GaAs is used for production and 4-inch InP is close to production.

The front end of a radio (Figure 1.2.3) provides a model system for both military and commercial applications. Device engineering for military applications consisted primarily of pushing the limits of performance with frequency. The power amplifier and low-noise receiver most strongly leveraged the performance advantages of GaAs. Output power and noise figures, as functions of frequency from the C to the W band, were enhanced through device design and the development of new heterojunction systems. Today, these heterojunction devices are being reengineered for commercial applications.

The power amplifier (PA) is the critical portion of the radio-frequency (RF) front end and wireless system as a whole. As mentioned above, the primary applications of PAs are in handsets and base stations. The major trends that influence the device performance for PAs are cost, consumer use, and standards. Cost is an obvious driver in a commercial system. Some consumer-use concerns are weight and talk time. Obviously, a handset must be compact and have little weight and provide ample talk time between battery charging. These constraints require lower operating voltages and greater device efficiency. International standards also influence device performance. The third-generation wireless system, the International Mobile Telecommunication 2000 system (IMT2000), is expected to come into service in 2001. IMT2000 dictates that mobile transmissions around the world have a 2-Mbps (megabits per second) data transmission capability. To achieve these high data rates, power amplifier transistors will necessarily have to operate at about 2 GHz. Although silicon-based transistors can compete in this frequency range, GaAs and particularly heterojunction-based electronics generally offer lower noise and improved linearity.

Heterojunction device engineering requires much stronger coupling to system requirements. An example is in the design for high linearity in power amplifiers. As shown in Chapter 3, linear HFETs can be designed through control of the doping profile or by choosing an enhancement-mode device as opposed to a depletion-mode device. HBT collector thicknesses and doping profiles strongly affect linearity, as discussed in Chapter 4. Specific linearity requirements, in turn, are determined by the specific standard and associated modulation format. Thus, the new device engineer must clearly understand the context of the device in the system specification.

The major challenges that next-generation wireless systems face are propagation loss, shadowing, and multipath fading. Propagation loss arises from reflection and scattering. The loss is proportional to the square of the frequency and the distance to the nth power, where n is 2 for free space, 4.35 for rural open space, and 3 to 4.3 for urban/suburban space. Objects such as autos or a building that temporarily block the base station and the receiver cause shadowing. Multipath fading arises when signals are received after having taken multiple routes. Multipath fading can cause distortion, which can lead to a significant increase in the bit-error rate. The most promising approaches to combating multipath fading are orthogonal frequency-division multiplexing (OFDM) and code-division multiple access (CDMA). OFDM employs multicarriers for one channel in place of a single carrier. The original signal is divided into many narrow bandwidths and sent by a different carrier with a different frequency. CDMA utilizes a spread spectrum to obtain frequency diversity. One of the principal operating requirements of the power amplifiers used in these systems is linearity. Why is linearity of a power amplifier so important? Nonlinearity in a output of the power amplifier leads to power leakage out of the signal channel into adjacent channels. Adjacent channel leakage is caused by third-order intermodulation distortion, which is intermodulation distortion between the fundamental and second harmonic signals.

An additional system constraint results from the current packaging of integrated circuits into modules for RF applications. A mixture of Si circuits and GaAs circuits is common for front-end components. Clearly, a single material and circuit technology will lead to lighter and more compact circuits and therefore systems. However, each specific material and device technology offers specific advantages and disadvantages. SiGe technology offers the advantages of heterojunction design to Si technology but cannot address the RF limitations of the conducting Si substrate. The conducting substrate introduces significant signal loss, as the frequency is increased and therefore limits Si as a monolithic microwave integrated circuit (MMIC). Figure 1.2.4 shows the performance of three technologies—Si BJT, HBT, and HEMT—as a function of metrics. Each metric is relevant to performance in a different block of a front end. An ultimate solution to this problem is heterogeneous integration of dissimilar materials and devices. Advanced materials and process technology

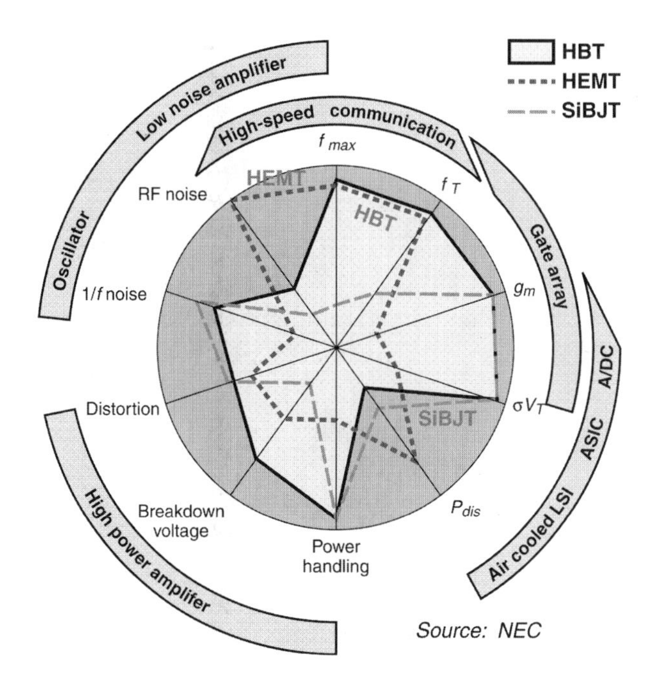

FIGURE 1.2.4 Performance of three device technologies—Si BJT, HBT, and HEMT—as a function of performance metrics. (Reprinted with permission from Honjo, 1997.)

will offer this option. Figure 1.2.5 shows an integrated circuit comprised of two different device types, HEMTs and HBTs, integrated laterally by selective area epitaxy. Other approaches include wafer bonding or device/circuit pick-and-place integration.

1.3 DIGITAL COMMUNICATIONS

The primary driver of the communications industry in the foreseeable future is the ever-growing demand for the rapid, efficient, and accurate transfer of digital information. The recognition that all information can ultimately be digitized makes it possible to convey it through telecommunications systems. Among the many formats of digital information are voice, audio (music, radio, etc.), visual (photos, television, movies, newspapers, etc.), and computer information. All of these formats are either presently being transmitted or planned for transmission via the Internet, wireless, and fiber-optic telephone networks. As a result, telecommunications and computing are blending together.

One of the major issues that confronts telecommunications/computing technology is what conditions will be required of the hardware in fiber-optic and wireless systems that will ensure very high bandwidth communications at a low bit-error rate in the future. Feature size is one of the most important

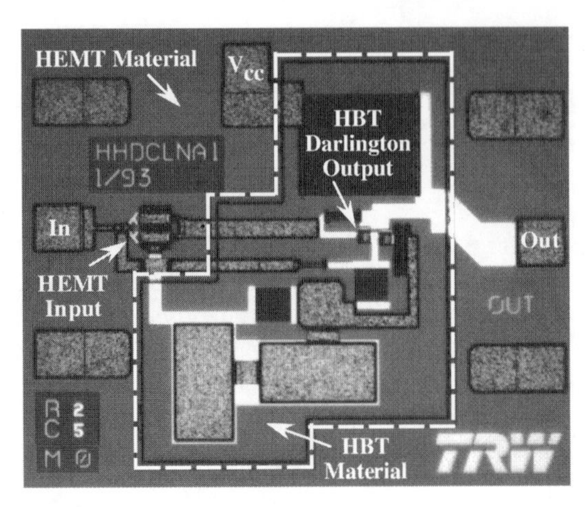

FIGURE 1.2.5 Integration of dissimilar device types: HEMTs and HBTs. (Reprinted with permission from Dwight Streit, TRW.)

parameters that dictates high-speed device performance. In addition, the storage of massive amounts of data for rapid retrieval, transmission, and processing requires very small memory devices. Therefore, in both digital and analog electronics, the major drive is for continued miniaturization, since it provides faster device operation and denser integrated circuits for memory and processing applications. Continued miniaturization will place device performance squarely within the quantum regime, wherein the device physics is governed primarily by quantum mechanical effects.

Quantum devices can be classified as structures with feature sizes comparable to or less than the electron de Broglie wavelength or devices in which charge quantization dominates the device physics. The de Broglie wavelength in semiconductors is less than 100 nm. Devices with feature size dimensions (i.e., gate lengths, well widths, etc.) below about 100 nm, then, will generally exhibit quantum effects. Inspection of Figure 1.1.3 shows that device sizes will be below 100 nm and begin to exhibit quantum effects in the near future if Moore's law continues.

Highly dense memory chips will require very small half-pitch, wherein few electrons will be used to store a bit. Single-electron transistors, devices in which a single-electron represents a bit, have already been made. Memory chips comprised of single-electron transistors offer very high density random access memory that can conceivably store vast amounts of digital information.

Quantum devices need not necessarily be made of semiconductors. New electronic devices have been made using molecules, leading to the new field of moletronics. The particular attraction of molecules for device development is that they are naturally three-dimensional, thus providing massive device densities. In addition, many molecules are self-replicating, leading to the interesting possibility that computers could be self-made. In this book we discuss the workings of various types of quantum devices and examine their possible use in future digital and analog electronics.

The book is organized as follows. In Chapters 2 to 4 we focus on conventional (defined as operating semiclassically), yet state-of-the-art devices based on heterostructures, heterostructure field-effect transistors (HFETs), and heterostructure bipolar transistors (HBTs). The physics of the transferred electron effect, in both k and real space, along with their concomitant devices, is presented in Chapter 5. In Chapter 6 we examine quantum mechanical devices based on resonant tunneling as well as the physics of resonant tunneling. The challenges to CMOS and both evolutionary and revolutionary alternatives to CMOS are examined in Chapters 7 and 8. The revolutionary alternatives we examine include quantum dot cellular automata, single-electron transistors, moletronics, and defect-tolerant computing. The book concludes with a discussion of magnetic field effects in semiconductors, including integer and fractional quantum Hall effects, which may affect futuristic devices.

CHAPTER 2

Semiconductor Heterostructures

In this chapter we discuss the basics of semiconductor heterostructures. Generally, a heterostructure is formed between any two dissimilar materials, examples of which are a metal and a semiconductor, an insulator and a semiconductor, or two different semiconductor materials. In this chapter we restrict our discussion to the formation of a heterostructure between two dissimilar semiconductors. The reader is referred to the book by Brennan (1999) for a detailed discussion of metal–semiconductor and metal–insulator–semiconductor junctions. Semiconductor–semiconductor heterostructures have become of increasing importance in electronic devices since they offer important new dimensions to device engineering. In this chapter we discuss the formation of heterostructures, their physical properties, and aspects of heterostructures that influence device behavior. In later chapters we illustrate how heterostructures can be incorporated into bipolar and field-effect transistors to improve their performance.

2.1 FORMATION OF HETEROSTRUCTURES

As mentioned above, placing two dissimilar semiconductor materials into contact forms a semiconductor–semiconductor heterostructure. Typically, a different semiconductor material is grown on top of another semiconductor using one of several epitaxial crystal growth techniques. Since the two constituent semiconductors within the heterostructure are of different types, many of their properties are distinctly different. The most important properties that influence the behavior of the heterostructure are the material lattice constants, energy gaps, doping concentrations, and affinity differences, among others. Let us examine how differences in these quantities affect the heterostructure formation.

For simplicity, let us refer to the two materials forming the heterostructure as materials 1 and 2, with energy gaps E_{g1} and E_{g2}, respectively. In virtually all cases, the energy gaps of the constituent semiconductors are different. Since the energy gaps are different, the conduction and valence bands of the two materials cannot simultaneously be continuous across the heterointerface. Therefore, at least one of the two, the conduction band or the valence band, must be discontinuous at the interface. Generally, both the conduction band and valence band edges are discontinuous at a heterointerface. The energy differences between the conduction band and valence band edges at the interface are called the *conduction band* and *valence band discontinuities*, respectively.

There are several different ways in which the energy bandgap discontinuity is accounted for at the interface. Generally, there are three different classes of heterojunctions. These three classes are called type I, II, and III heterostructures. The three different types are sketched in Figure 2.1.1. The type I heterostructure is the most common. An important example of a type I heterostructure is the GaAs–AlGaAs materials system. We discuss this system in some detail below. Notice that in a type I heterostructure, the sum of the conduction band and valence band edge discontinuities is equal to the energy gap difference,

$$\Delta E_g = \Delta E_c + \Delta E_v \qquad 2.1.1$$

The type II heterostructure (Figure 2.1.1) is arranged such that the discontinuities have different signs. The bandgap discontinuity in this case is given as the difference between the conduction band and valence band edge discontinuities. A type II heterostructure is formed by $Al_{0.48}In_{0.52}As$ and InP.

In type III heterostructures, the band structure is such that the top of the valence band in one material lies above the conduction band minimum of the other material. An example of this type of heterostructure is the heterostructure formed by GaSb and InAs (Figure 2.1.1). As for the type II case, the bandgap discontinuity is equal to the difference between the conduction band and valence band edge discontinuities.

Certainly, one of the most important heterostructures is that formed between GaAs and AlAs or its related ternary compounds, AlGaAs. As mentioned above, the GaAs–AlGaAs heterojunction forms a type I heterostructure. The energy bandgap of GaAs at 300 K is 1.42 eV, while the gap in $Al_xGa_{1-x}As$ varies with Al composition in accordance with

$$E_g = 1.424 + 1.247x \qquad 0 < x < 0.45 \qquad 2.1.2$$

The GaAs–AlGaAs heterostructure has the additional feature of close lattice matching. Two materials that have nearly identical lattice constants are

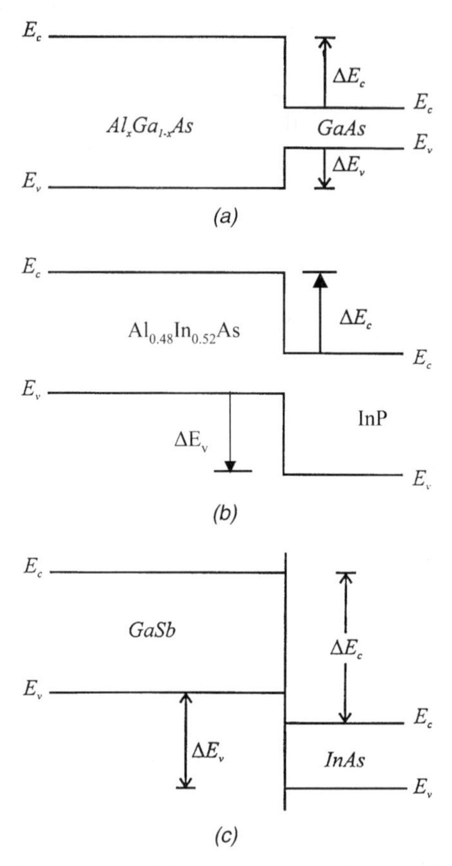

FIGURE 2.1.1 The three types of heterostructures: (*a*) type I; (*b*) type II; (*c*) type III.

said to be *lattice matched* when used to form a heterostructure. As we will see below, if the materials are not lattice matched, the lattice mismatch can be accommodated through strain or by the formation of misfit dislocations.

The doping type and concentration within the constituent materials also affect the heterojunction. Heterojunctions can be formed using two intrinsic materials or by doping one or both materials either n- or p-type. Therefore, a large variety of junctions can be formed (i.e., p-n, n-n, n-i, p-p, etc.). For purposes of illustration, let us consider the formation of an n-i AlGaAs–GaAs heterojunction. As discussed by Brennan (1999), in equilibrium the Fermi level is flat everywhere, and far from the junction the bulklike properties of the materials are recovered. These two conditions are very helpful in constructing the equilibrium energy diagram for the heterostructure. To facilitate construction of the energy band diagram, it is helpful to define a few

EXAMPLE 2.1.1: Energy Band Diagram for a Graded Heterostructure

Let us determine the energy band diagram for a graded heterostructure in equilibrium. In this case, the transition from narrow- to wide-bandgap material is gradual. As is always the case, the Fermi level is flat in equilibrium, and far from the heterojunction the bulklike properties of the constituent materials are recovered. The electron affinity is an intrinsic property of the material. Therefore, in order that the affinity remain the same for the narrow- and wide-gap materials, the vacuum level must bend in equilibrium as shown in Figure 2.1.2. We set E_0 as a reference. As can be seen from the diagram, E_0 is the vacuum level for the wide-gap semiconductor far from the junction. Both the conduction and valence bands can then be described relative to E_0 as follows.

Examination of Figure 2.1.2 shows that $E_c(z)$ and $E_v(z)$ can be expressed as

$$E_c(z) = E_0 - qV(z) - q\chi(z)$$
$$E_v(z) = E_0 - qV(z) - q\chi(z) - E_g(z)$$

quantities:

1. ϕ, *the work function*. The energy $q\phi$ is required to promote an electron from the Fermi level to the vacuum level. In other words, to remove an electron from the material, it must absorb an energy equal to $q\phi$.
2. χ, *the electron affinity*. The energy required to promote an electron from the conduction band edge to the vacuum level is given as $q\chi$.
3. ΔE_c *and* ΔE_v, *the conduction band and valence band edge discontinuities*. Defined above.

The equilibrium band diagrams of the n-i AlGaAs–GaAs heterojunction assuming the materials are apart and placed in contact are shown in Figure 2.1.3*a* and *b*, respectively. The work function, affinity, and band edge discontinuities are shown in the diagram. The equilibrium band diagram of the junction follows from the application of the two rules that the Fermi level is flat everywhere and that far from the junction the bulklike properties of the materials are recovered. In Section 2.2, we discuss formation of the energy bands in this system and present a simplified scheme for determining the carrier concentration within the GaAs layer.

The built-in potential for a type I heterostructure is easily found from the difference between the work functions of the constituent materials as (Brennan,

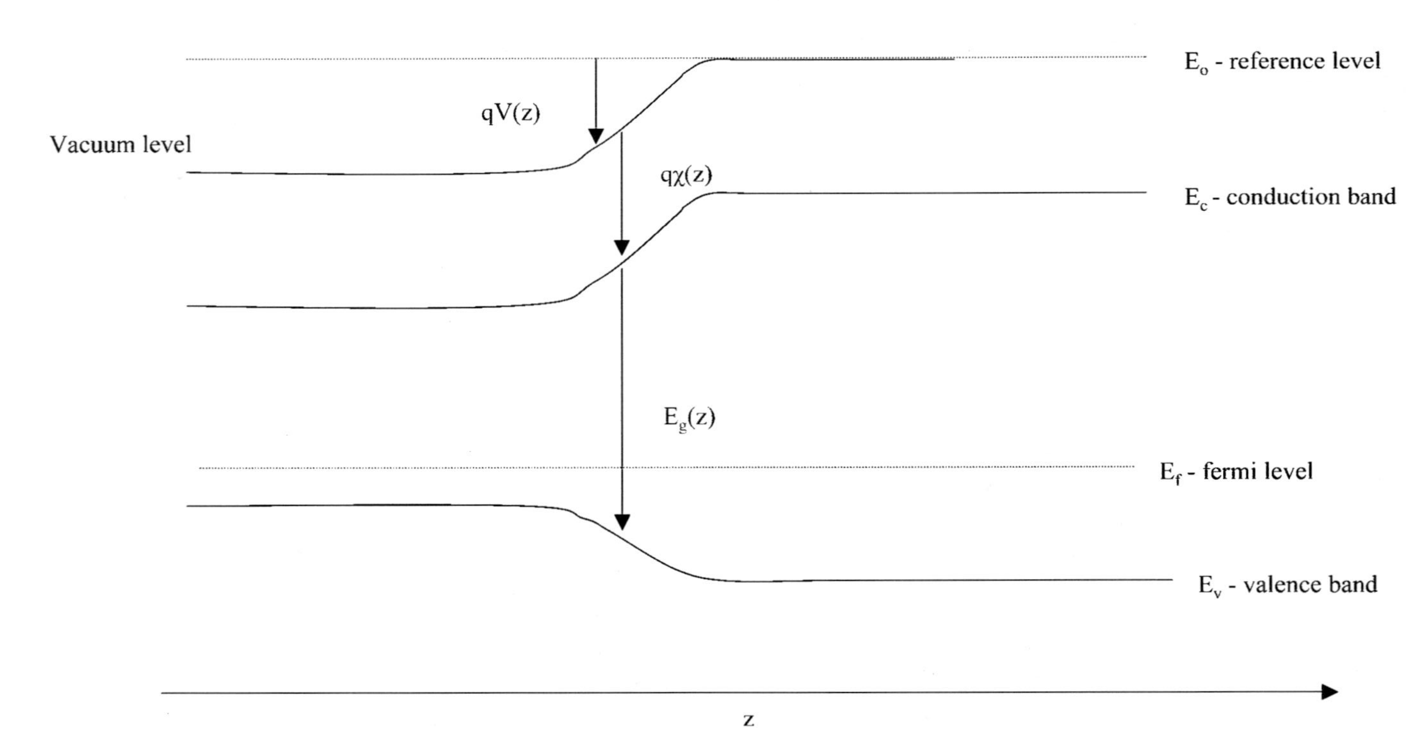

FIGURE 2.1.2 Graded heterojunction band diagram. $V(z)$ is the potential, $\chi(z)$ the electron affinity, and $E_g(z)$ the energy gap as functions of position along the z axis.

FIGURE 2.1.3 n-Type AlGaAs and intrinsic GaAs band structures (*a*) apart and (*b*) placed in contact in equilibrium. Notice that the Fermi level is flat.

1999, Sec. 11.2)

$$V_{bi} = \phi_2 - \phi_1 \tag{2.1.3}$$

where ϕ_2 and ϕ_1 are the work functions for the narrow- and wide-bandgap semiconductors, respectively, as shown in Figure 2.1.3. The built-in voltage for the specific case of a n-type wide-bandgap semiconductor and intrinsic narrow-gap semiconductor heterostructure can be obtained as follows. Using Figure 2.1.3, the difference in the work functions can be found to be

$$V_{bi} = \frac{\Delta E_c}{q} + \frac{k_B T}{q} \ln \frac{n_{10} N_{c2}}{n_{20} N_{c1}} \tag{2.1.4}$$

where n_{10}, and n_{20} are the equilibrium electron concentrations in the wide- and narrow-bandgap semiconductors (in this case, the narrow-gap semiconductor is assumed to be intrinsic). N_{c2} and N_{c1} are the effective density of states for the narrow- and wide-gap semiconductors, respectively (Brennan, 1999, Sec. 10.3).

2.2 MODULATION DOPING

Arguably, one of the most important developments that greatly increased the importance of compound semiconductor materials was the invention of modulation doping (Dingle et al., 1978). Modulation doping offers an important advantage in device engineering since it provides a mechanism by which the free carrier concentration within a semiconductor layer can be increased significantly without the introduction of dopant impurities. Although conventional doping techniques are important for increasing the free carrier concentration and conductivity of a semiconductor, they come at the expense of increased ionized impurity scattering and a concomitant reduction in the carrier mobility. Therefore, conventional doping approaches lead to a trade-off; increased doping concentration is desirable to reduce the resistance and increase the free carrier concentration, but this leads to a serious reduction in the carrier mobility and speed of the device.

Modulation doping provides an extremely attractive alternative. In a modulation-doped heterostructure, the free carriers are spatially separated from the dopants. The spatial separation of the dopants and free carriers reduces the deleterious action of ionized impurity scattering. Therefore, the free carrier concentration can be increased significantly without compromising the mobility.

Modulation doping can be understood as follows. In its simplest implementation, one constructs a heterostructure formed by an n-type wide-bandgap semiconductor with an unintentionally doped, relatively narrow gap semiconductor, as shown in Figure 2.2.1*a*. The most commonly used materials systems for modulation doping are GaAs and AlGaAs. As shown in the di-

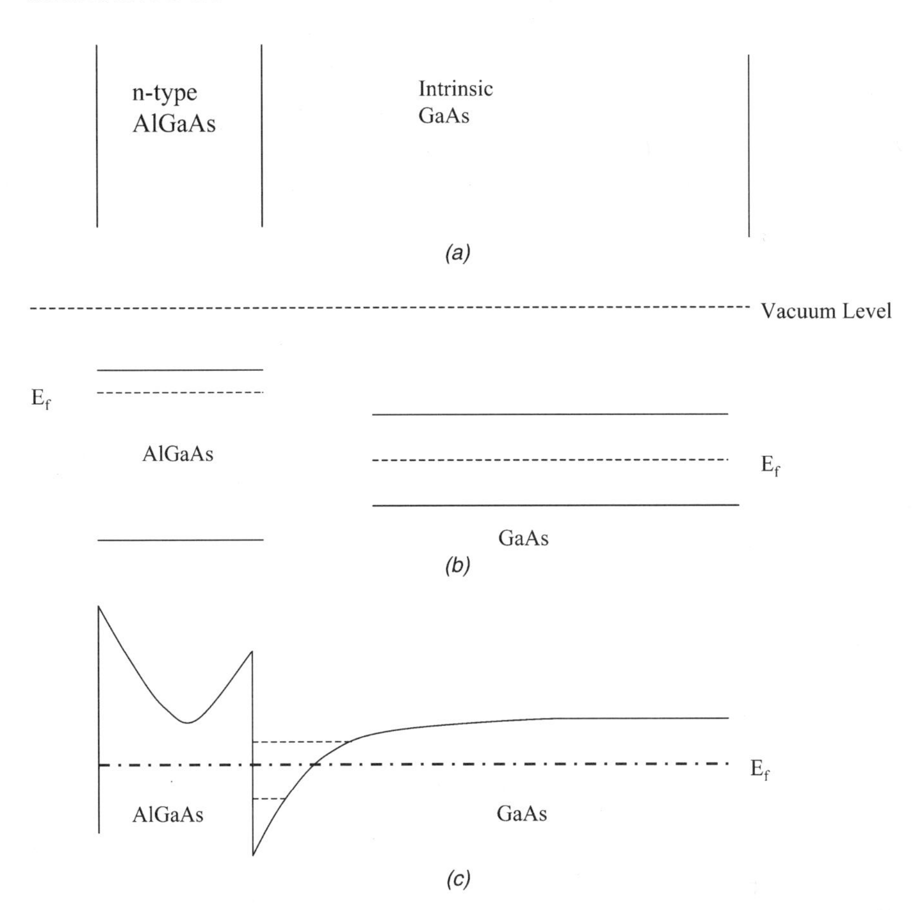

FIGURE 2.2.1 (*a*) Layer structure for a simple modulation-doped heterostructure; (*b*) energy band diagrams of the AlGaAs and GaAs layers when apart and in equilibrium; (*c*) corresponding energy band diagram for the layer structure shown in part (*a*). The two dashed horizontal lines at the heterointerface in the GaAs layer represent energy subbands arising from spatial quantization effects.

agram, the AlGaAs layer, which has the larger bandgap, is doped n-type, while the narrower-gap GaAs layer is unintentionally doped. In equilibrium the Fermi level must align throughout the structure. As can be seen from Figure 2.2.1*b*, when the two materials AlGaAs and GaAs are initially apart, the Fermi level lies closer to the conduction band edge in the AlGaAs than in the GaAs since the AlGaAs is doped n-type. When the two materials are placed into contact, as shown in Figure 2.2.1*c*, electrons must be transferred from the AlGaAs layer into the GaAs layer to align the Fermi level. This results in a sizable increase in the electron concentration within the GaAs layer without the introduction of ionized donor impurities. The ionized donor atoms within the AlGaAs result in a net positive charge, which balances the net negative charge due to the electrons transferred in the GaAs layer. Although the ionized donor atoms in the AlGaAs obviously influence the electrons transferred in the GaAs, the spatial separation between the two charge species mitigates the Coulomb interaction between them. As a result, ionized impurity scattering of the transferred electrons is reduced, resulting in a higher electron mobility. Typically, an undoped AlGaAs spacer layer is formed between the doped AlGaAs and undoped GaAs layers to increase the spatial separation of the electrons from the ionized donors, further reducing the ionized impurity scattering. As a result, the mobility is increased.

Inspection of Figure 2.2.1*c* shows that the conduction band edge in the GaAs layer is strongly bent near the heterointerface. The band bending is a consequence of the electron transfer. The net charge to the left from the heterostructure interface is positive due to the ionized donors. Therefore, a test electron in the GaAs layer near the interface will be attracted to the interface by the action of the positive charge. Electrons roll "downhill" in energy band diagrams (Brennan, 1999, Chap. 11). Therefore, the conduction band must necessarily bend such that a test electron will roll downhill to the interface when placed in the GaAs layer. Therefore, the conduction band must bend as shown in Figure 2.2.1*c*. The sharp bending of the conduction band edge and the presence of the conduction band edge discontinuity forms a potential well within the GaAs layer. In most instances, the band bending is sufficiently strong that the spatial dimensions of the potential well are comparable to the electron de Broglie wavelength. As a result, spatial quantization effects occur (Brennan, 1999, Chap. 2 and Sec. 11.2). Spatial quantization produces discrete energy bands, called *subbands*, in the potential well, as shown by the dashed lines in Figure 2.2.1*c*.

Figure 2.2.2 shows an expanded view of the conduction band edge formed at the interface of n-type AlGaAs and i-GaAs. In the present situation, the direction of quantization is perpendicular to the heterointerface (labeled as the z direction in Figure 2.2.3) and the energy along this direction is quantized. However, in the directions parallel to the interface (x and y), no spatial quantization effects occur since the motion of the electrons is not restricted by the band bending. Therefore, the electrons behave as free particles for motion along the x and y axes (those parallel to the interface) but are quantized in the direction perpendicular to the z direction. Using a parabolic band model approximation, the resulting expression for the electron energies within the potential well in the conduction band is

$$E = \frac{\hbar^2 k_x^2}{2m} + \frac{\hbar^2 k_y^2}{2m} + E_i \qquad 2.2.1$$

where k_x and k_y represent the x and y coordinates of the electron k-vector, respectively. In Eq. 2.2.1, E_i represents the energies due to the spatial quantization in the z direction. The corresponding wavefunction for an electron in

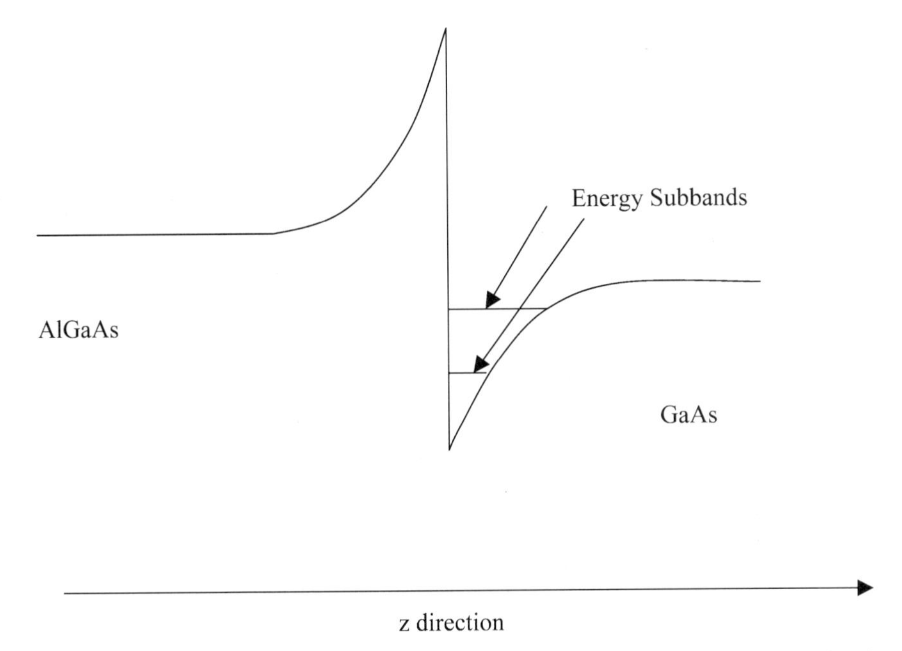

FIGURE 2.2.2 Expanded view of the conduction band edge discontinuity in the n-AlGaAs and i-GaAs heterojunction, showing two energy subbands. The energy subbands form due to spatial quantization effects in the heterostructure.

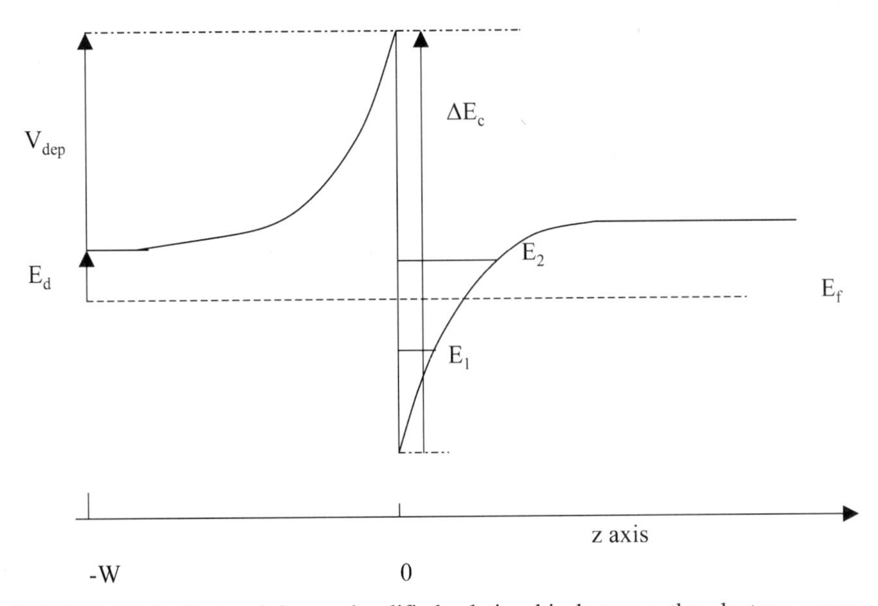

FIGURE 2.2.3 Determining a simplified relationship between the electron concentration within the potential well, N_s, and known quantities.

the potential well is

$$\psi(r,z) = \chi(z)e^{ikr} \qquad 2.2.2$$

where k is the two-dimensional wavevector, r the two-dimensional spatial vector consisting of the x and y coordinates, and $\chi(z)$ the subband wave-function.

The general solution for the eigenenergies of the finite potential well formed in the conduction band requires self-consistent solution of the Schrödinger and Poisson equations, which are discussed briefly in the next section. The eigenenergies can be roughly approximated using the solution for an infinite triangular potential well as

$$E_i = \left(\frac{\hbar^2}{2m}\right)^{1/3} \left(\frac{3}{2}\pi qF\right)^{2/3} \left(i + \frac{3}{4}\right)^{2/3} \qquad 2.2.3$$

where F is the electric field strength corresponding to the slope of the energy band and i is an integer representing the band index. Using Eqs. 2.2.3 and 2.2.1, the energy of an electron in the conduction band well can be roughly approximated.

Before we discuss the general problem of transport in the subbands in a heterostructure, it is useful to formulate a simple yet crude approximate expression for the electron concentration within the potential well. For simplicity we assume that the system is at $T = 0$ K and that only one subband is occupied. In equilibrium the Fermi level aligns throughout the system. On the GaAs side of the junction, at $T = 0$ K, all the electron states are filled up to the Fermi level. Recall that the minimum energy is E_1, the subband energy, and can be approximated using Eq. 2.2.3. The Fermi energy on the GaAs side corresponds to the highest filled energy state above E_1. The location of the Fermi energy can then be determined as follows. Recall that the total electron concentration can be determined by integrating the distribution function times the density-of-states function (Brennan, 1999, Sec. 5.1). This becomes

$$N_s = \int_{E_1}^{E_1+E_f} f(E)D(E)dE \qquad 2.2.4$$

where $f(E)$ is the Fermi–Dirac distribution and $D(E)$ is the density of states. At $T = 0$ K, the Fermi–Dirac distribution function is 1 for $E < E_f$ and zero otherwise. With this simplification, the integral in Eq. 2.2.4 can now be performed easily, using the expression for the two-dimensional density of states, $m^*/\pi\hbar^2$, to yield

$$E_f = \frac{N_s \pi \hbar^2}{m^*} \qquad 2.2.5$$

which is measured relative to E_1. The position of the Fermi level on the GaAs side is then given as

$$E_f = E_1 + \frac{N_s \pi \hbar^2}{m^*} \tag{2.2.6}$$

where the zero of potential energy is taken at the conduction band minimum in the GaAs layer. In the AlGaAs layer we can determine the position of the Fermi level from the diagram. If we assume that the donor levels within the AlGaAs are sufficiently deep that the Fermi level is pinned there, the Fermi level relative to the conduction band minimum in the GaAs layer is given as

$$E_f = \Delta E_c - E_d - qV_{\text{dep}} \tag{2.2.7}$$

where ΔE_c is the conduction band discontinuity, E_d the donor energy, and qV_{dep} the band bending due to depletion of donor atoms in the AlGaAs. V_{dep} can readily be calculated as

$$V_{\text{dep}} = -\int_0^{-W} F\,dz = \int_0^{-W} \frac{qN_D z}{\varepsilon_0 \varepsilon_{\text{AlGaAs}}} dz = \left. \frac{qN_D z^2}{2\varepsilon_0 \varepsilon_{\text{AlGaAs}}} \right|_0^{-W} = \frac{qN_D W^2}{2\varepsilon_0 \varepsilon_{\text{AlGaAs}}} \tag{2.2.8}$$

Equating Eqs. 2.2.6 and 2.2.7 and using Eq. 2.2.8 yields

$$E_1 + \frac{N_s \pi \hbar^2}{m^*} = \Delta E_c - qV_{\text{dep}} - E_d \tag{2.2.9}$$

Using Eq. 2.2.9, an estimate of the two-dimensional electron concentration N_s can be obtained. Recall that this provides only a rough approximation. Since we have assumed that the temperature is 0 K, we use Eq. 2.2.3 for E_1 and assume that the donors pin the Fermi level in the AlGaAs and that the donors are fully ionized. A more exact solution that can be obtained using a numerical approach is discussed in the next section.

2.3 TWO-DIMENSIONAL SUBBAND TRANSPORT AT HETEROINTERFACES

As mentioned above, the electronic structure within the potential well formed in the conduction band at a heterointerface is quantized in one direction, resulting in a two-dimensional system. The transport physics is significantly different in a two-dimensional system from that of a three-dimensional system. Aside from the obvious difference in the allowed energies of a carrier in a two-dimensional versus three-dimensional system, the scattering rates are also different. Let us consider the physics of electronic transport in a two-dimensional system. Although a similar analysis can in principle be applied

EXAMPLE 2.2.1: Determination of the Total Carrier Density in a Two-Dimensional System

Consider a two-dimensional system formed in the GaAs–AlGaAs materials system. With the assumption that the energy levels can be determined from the infinite triangular well approximation, determine the total carrier density in the system at $T = 0$ K if only one subband is occupied. Assume the following information: $m^* = 0.067m$; $\varepsilon_{\text{AlGaAs}} = 13.18 - 3.12x$ (where x is the Al concentration); the donor concentration within the AlGaAs layer, N_D, is 3.0×10^{17} cm^{-3}; and the effective field F in the triangular well is 1.5×10^5 V/cm. The conduction band edge discontinuity in the GaAs–AlGaAs system is usually estimated as 62% of the difference in the energy band gaps. Assume that the Al concentration within the AlGaAs is 40%. The donor energy in AlGaAs is assumed to be 6 meV. The width of the depletion region in the AlGaAs is given as 18.2 nm.

We start with Eq. 2.2.9,

$$E_1 + \frac{N_s \pi \hbar^2}{m^*} = \Delta E_c - V_{\text{dep}} - E_d$$

The first subband energy, E_1, can be calculated using the infinite triangular well approximation with the field F of 3.0×10^5 V/cm:

$$E_i = \left(\frac{\hbar^2}{2m}\right)^{1/3} \left(\frac{3}{2}\pi qF\right)^{2/3} \left(i + \frac{3}{4}\right)^{2/3}$$

Substituting in for i, 1, and E_1 is equal to 0.205 eV. The bandgap discontinuity ΔE_g is found using Eq. 2.1.2 as

$$\Delta E_g = 1.247x = (1.247)(0.40) = 0.50$$

The conduction band edge discontinuity is then

$$\Delta E_c = (0.62)(0.5) = 0.31 \text{ eV}$$

V_{dep} can be calculated from

$$V_{\text{dep}} = \frac{qN_D W^2}{2\varepsilon_0 \varepsilon_{\text{AlGaAs}}}$$

Substituting in the relevant values, V_{dep} is computed to be 0.075 V. N_s, the two-dimensional electron concentration, can now be determined

(*Continued*)

EXAMPLE 2.2.1 (*Continued*)

as

$$\frac{N_s \pi \hbar^2}{m^*} = \Delta E_c - V_{\text{dep}} - E_d - E_1 = 0.31 - 0.075 - 0.006 - 0.205 = 0.024 \text{ eV}$$

Solving for N_s yields

$$N_s = \frac{m^*}{\pi \hbar^2}(0.024) = 6.7 \times 10^{11} \text{ cm}^{-2}$$

to the valence band, the valence band structure is generally more complicated, which results in greater complexity. For simplicity, we limit our discussion here to electron transport within a two-dimensional conduction band system.

The approximation of the electronic energy band structure given by Eqs. 2.2.1 and 2.2.3 is generally very poor. A far more accurate description of the electronic energy structure can be obtained through the self-consistent solution of the Schrödinger and Poisson equations. It is common to approximate the Schrödinger equation using a one-dimensional effective mass model as

$$-\frac{\hbar^2}{2}\frac{\partial}{\partial z}\left[\frac{1}{m^*(z)}\frac{\partial}{\partial z}\right]\chi(z) + V(z)\chi(z) = E\chi(z) \qquad 2.3.1$$

where the wavefunction given by Eq. 2.2.2 has been inserted. The potential includes the electrostatic potential and the conduction band discontinuity at the interface. The Schrödinger equation should be solved numerically along with the Poisson equation. A common approach is to employ the Rayleigh–Ritz method to solve the Schrödinger equation since it determines the eigenenergies and corresponding eigenfunctions for a given set of boundary conditions. The potential $V(z)$, neglecting exchange effects, is given as

$$V(z) = \phi(z) + V_h(z) \qquad 2.3.2$$

where $\phi(z)$ is the electrostatic potential and $V_h(z)$ is the step function describing the interface potential barrier. The electrostatic potential can be determined from the solution of Poisson's equation, given as

$$\frac{d}{dz}\varepsilon_0 \kappa(z)\frac{d\phi(z)}{dz} = q\left[\sum_i N_i \chi_i^2(z) + N_A(z) - N_D(z)\right] \qquad 2.3.3$$

where $\kappa(z)$ is the relative dielectric constant in each layer, $N_A(z)$ and $N_D(z)$ are the acceptor and donor concentrations, respectively, and N_i is the number of

EXAMPLE 2.2.2: Calculation of the Carrier Temperature in a Two-Dimensional System in the Presence of an Electric Field

Consider the system described in Example 2.2.1, but assume that there now exists an applied electric field that heats the carriers in the two-dimensional system to higher energy. What is the carrier temperature if the relative population of the second subband is 15%? Assume that only the first two subbands are occupied and that Boltzmann statistics can be used.

In this problem we need to use elementary statistical mechanics to determine the occupation probability. Generally, the relative occupation probability of the second subband to the first is given as

$$\frac{N_2}{N_1} = \frac{\int_{E_2}^{\infty} f(E)D(E)dE}{\int_{E_1}^{\infty} f(E)D(E)dE}$$

For a two-dimensional system, the density of states, $D(E)$, is constant and divides out of both integrals. Substituting in the Boltzmann factor for the distribution function $f(E)$ yields

$$\frac{N_2}{N_1} = \frac{\int_{E_2}^{\infty} e^{-E/kT} dE}{\int_{E_1}^{\infty} e^{-E/kT} dE} = e^{-(E_2 - E_1)/kT}$$

The second subband energy level, E_2, can readily be found from use of the infinite triangular well approximation as

$$E_i = \left(\frac{\hbar^2}{2m}\right)^{1/3} \left(\frac{3}{2}\pi qF\right)^{2/3} \left(i + \frac{3}{4}\right)^{2/3}$$

Substituting in the appropriate values, E_2 can be determined to be 0.28 eV. The difference in the energy levels is then

$$E_2 - E_1 = 0.28 - 0.205 = 0.075 \text{ eV}$$

The temperature can now be found as

$$\ln(0.15) = -\frac{E_2 - E_1}{kT}$$

The temperature computes to $T = 458$ K. This number may seem very high to the reader, but remember that we are talking about the carrier temperature here, not the lattice temperature. The carrier temperature is due to field heating and simply implies that the distribution is heated well above the equilibrium temperature.

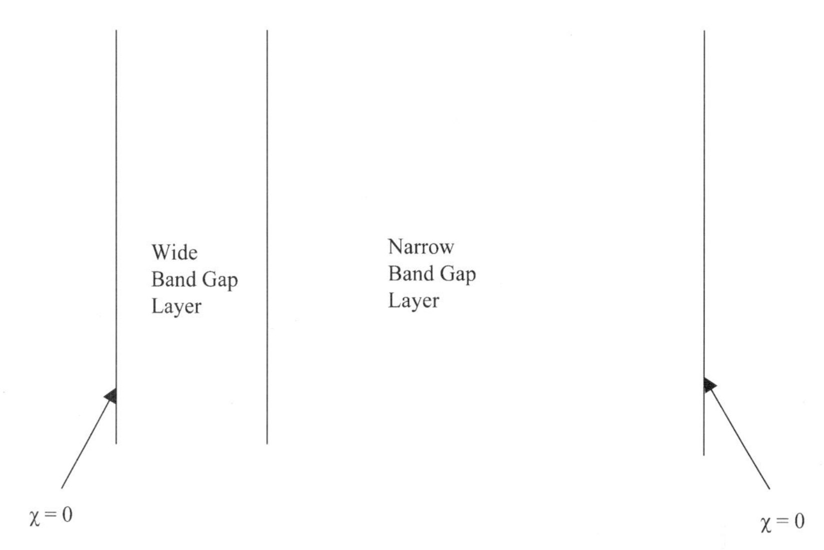

FIGURE 2.3.1 Heterostructure system, showing the boundary conditions on the wavefunction.

electrons within the *i*th subband, given as

$$N_i = \frac{mk_BT}{\pi\hbar^2} \ln[1 + e^{(E_f - E_i)/k_BT}] \tag{2.3.4}$$

Equation 2.3.4 can be determined by using the two-dimensional density-of-states function and the Fermi function (Brennan, 1999, Prob. 8.6).

The boundary conditions for the problem are given as follows. The wavefunction is assumed to vanish as z approaches infinity, that is, deep into the narrow-gap layer, as shown in Figure 2.3.1. Additionally, the normal derivative of the potential vanishes far from the junction. The second condition is that the wavefunction again is assumed to vanish on the far side of the wide-gap material (see Figure 2.3.1). With this choice of boundary conditions, the wavefunctions and energy eigenvalues for the finite potential well formed at the heterointerface can be determined.

The most common heterojunction structure of interest is comprised of several layers as follows. In the ideal situation, the narrow-gap layer, in this case GaAs, is assumed to be undoped. The wide-bandgap AlGaAs layer consists of two parts, a nominally undoped layer called the *spacer* and an intentionally doped region. The undoped spacer layer is formed between the GaAs and the doped AlGaAs. The function of the undoped spacer is to spatially separate the two-dimensional electron gas formed in the GaAs potential well from the ionized donors within the doped AlGaAs. The greater the spatial separation of the electron gas and ionized impurities, the weaker the ionized scattering rate becomes. As a result, the mobility of the two-dimensional electron gas can be

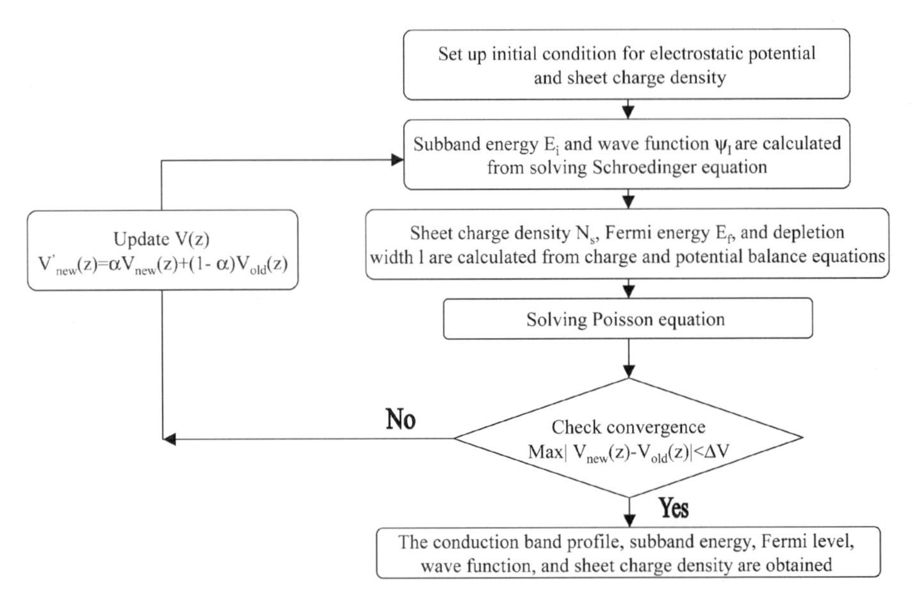

FIGURE 2.3.2 Outline of the numerical solution for the self-consistent calculation.

enhanced. However, there is a limit as to how large the spacer layer can be. As the spacer layer increases in width, the electron concentration formed in the GaAs potential well is reduced. Thus there is a trade-off in selection of the spacer layer width.

The self-consistent solution can be outlined as follows. The procedure is illustrated as a flowchart in Figure 2.3.2. An initial guess of the two-dimensional electron concentration is chosen and the Poisson equation is solved for the potential. The subband energies and wavefunctions are then calculated using this choice of potential from the solution of the Schrödinger equation. The depletion layer width is found in terms of N_s using the charge balance equation,

$$N_d W_d + N_{\text{spacer}} d_s = N_s \qquad 2.3.5$$

where N_d is the depletion layer charge density, N_{spacer} the charge density within the spacer layer, and d_s the spacer layer width. The Fermi energy is determined from

$$\Delta E_c - E_f - E_{DD} = \frac{q^2 N_d W_d^2}{2\varepsilon_s \varepsilon_0} + \frac{q^2 N_d W_d d_s}{\varepsilon_s \varepsilon_0} + \frac{q^2 N_{\text{spacer}} d_s^2}{2\varepsilon_s \varepsilon_0} \qquad 2.3.6$$

where ΔE_c is the conduction band edge discontinuity and E_{DD} is the donor energy. Having found the quantized energy levels, the two-dimensional electron carrier concentration N_s, the depletion layer width within the doped AlGaAs layer W_d, and the Fermi energy E_f are determined using Eqs. 2.3.5, 2.3.6, and the relationship for N_s. The updated total carrier density is calculated from Eq.

2.3.4 by summing over all the energy bands. The electrostatic potential is then recalculated using the new value for the carrier concentration. This procedure continues iteratively until sufficient accuracy is obtained.

For low field transport, arguably the most important transport parameter is the mobility. The mobility is, of course, a strong function of the scattering rates of the carriers. The magnitude and physics of the scattering rates in a two-dimensional system are quite different from those within a three-dimensional system. In the GaAs–AlGaAs heterostructure system, the primary scattering agents within the two-dimensional potential well are acoustic phonons, polar optical phonons, and ionized impurity scattering. For elastic scattering processes, the mobility using the relaxation time approximation is readily determined to be

$$\mu = \frac{q\tau}{m} \tag{2.3.7}$$

where τ is the total relaxation time for all scattering processes. The total relaxation time is given using Matthiessen's rule as

$$\frac{1}{\tau_{\text{total}}} = \sum_i \frac{1}{\tau_i} \tag{2.3.8}$$

where τ_i is the relaxation time for the ith scattering mechanism. The momentum relaxation time approximation is valid when the loss of energy through a scattering process is negligible compared to the carrier energy: in other words, when the process is elastic or can be modeled as elastic (see Box 2.3.1). Unfortunately, this procedure breaks down when inelastic scattering mechanisms, such as polar optical scattering are present since the phonon energy can be comparable to or greater than the carrier energy. In these circumstances the mobility must be calculated either directly from the solution of the Boltzmann equation or by using a variational technique or other methods.

For elastic scattering mechanisms, however, the relaxation time approximation is valid. Before we discuss the more general methodology for treating inelastic scatterings, let us examine the elastic case. The scattering rate in three dimensions is in general given as

$$S(k,k') = \frac{2\pi}{\hbar} \iiint |\langle k'|H_{ep}|k\rangle|^2 \delta_E \frac{\Omega}{(2\pi)^3} dQ\, dQ_z\, d\theta \tag{2.3.9}$$

where Ω is the volume, Q the magnitude of the phonon wavevector, and δ_E a delta function conserving the energy in the process. The squared matrix element in Eq. 2.3.9 for acoustic deformation potential scattering in a two-dimensional system is given as

$$|\langle k'|H_{ep}|k\rangle|^2 = \sum_{q_z} \frac{\hbar D^2 Q^2}{2\omega\rho\Omega}\left(n_Q + \frac{1}{2} \mp \frac{1}{2}\right)|I_{mn}(Q_z)|^2 \tag{2.3.10}$$

BOX 2.3.1: Validity of the Relaxation Time Approximation

The relaxation time approximation is useful only when the relaxation time is independent of the distribution function or the driving force on the distribution. Otherwise, the collision term in the Boltzmann equation cannot be separated from the drift and diffusion terms. From the definition of the Boltzmann equation (Brennan, 1999, Chap. 6), the relaxation time can be defined as

$$\left.\frac{\partial f}{\partial t}\right|_{\text{scattering}} = \int [S(k',k)f' - S(k,k')f]d^3k' = -\frac{f}{\tau}$$

where f' is the nonequilibrium distribution that describes the probability that the state k' is occupied at time t, and f is the nonequilibrium distribution that gives the probability that the state k is occupied at time t. In the above, we have neglected the terms $(1 - f')$ and $(1 - f)$, assuming that vacancies always exist for the final states. In equilibrium, the time rate of change of the nonequilibrium distribution f with respect to time is zero. Therefore,

$$S_0(k',k) = \frac{f_0}{f_0'}S_0(k,k')$$

where the subscript 0 represents equilibrium. Assuming that a similar relationship exists for the nonequilibrium case, $S' = Sf_0/f_0'$, the relaxation time can be written as

$$\frac{1}{\tau} = \frac{\Omega}{(2\pi)^3}\int S\left(1 - \frac{f_0 f'}{f_0' f}\right)d^3k'$$

where Ω is the volume. In general, the nonequilibrium distribution function can be written as the sum of symmetric and asymmetric components as

$$f = f_s + f_a$$

Under low applied fields, the symmetric component is close to the equilibrium value and only the asymmetric term contributes to the current. The symmetric component does not contribute to the current since the weighting of the velocity over a symmetric function will result in an odd function that vanishes when integrated over all space. Therefore, in the expression for the relaxation time above, the only remaining part of f is

(*Continued*)

BOX 2.3.1 (*Continued*)

f_a. The relaxation time equation then becomes

$$\int [S(k',k)f'_a - S(k,k')f_a]d^3k' = -\frac{f_a}{\tau}$$

As discussed by Brennan (1999, Sec. 6.2), the nonequilibrium distribution function under low fields can be written as

$$f = f_0\left(1 - \frac{q\tau\mathbf{v}\cdot\mathbf{F}}{kT}\right)$$

which can now be written using the asymmetric portion of the nonequilibrium distribution as

$$f = f_0 + f_a = f_0 - \frac{q\tau\mathbf{v}\cdot\mathbf{F}}{kT}f_0$$

Clearly, f_a is then given as

$$f_a = -\frac{q\tau\mathbf{v}\cdot\mathbf{F}}{kT}f_0$$

Substituting the expression for f_a above into the expression for $1/\tau$ yields

$$\frac{1}{\tau} = \frac{\Omega}{(2\pi)^3}\int S\left(1 - \frac{f_0 f'_a}{f'_0 f_a}\right)d^3k' = \frac{\Omega}{(2\pi)^3}\int S\left(1 - \frac{\mathbf{v}'\cdot\mathbf{F}}{\mathbf{v}\cdot\mathbf{F}}\right)d^3k'$$

As mentioned above, the relaxation time approximation is truly useful only if the relaxation time does not depend on the distribution function but only on the scattering mechanisms. Inspection of the expression for $1/\tau$ above reveals that there is a dependency on the velocity and hence the distribution function. However, if the scattering mechanisms are elastic, the magnitude of the initial and final velocities, v and v', must be the same. The expression for the relaxation time then becomes

$$\frac{1}{\tau} = \frac{\Omega}{(2\pi)^3}\int S\left(1 - \frac{v'F\cos\theta'}{vF\cos\theta}\right)d^3k' = \frac{\Omega}{(2\pi)^3}\int S\left(1 - \frac{\cos\theta'}{\cos\theta}\right)d^3k'$$

Notice that the relaxation time is now dependent only on the scattering process and the final angles and not on the distribution function itself.

(*Continued*)

BOX 2.3.1 (*Continued*)

Hence, we see that only for elastic scattering and close to equilibrium is the relaxation time approximation valid.

It is useful to further simplify the relaxation time expression by considering the relationship between the angles θ and θ'. Let α be the angle between k and k'. while the angles θ and θ' are the angles between **k** and **F** and **k**$'$ and **F**, respectively, as shown in Figure 2.3.3. The ratio of $\cos\theta'$ to $\cos\theta$ can be obtained from the following. $\text{Cos}\,\theta'$ can be written as

$$\cos\theta' = \frac{\mathbf{k}' \cdot \mathbf{F}}{|k'||F|}$$

The vectors **k**$'$ and **F** can be written using Figure 2.3.3 as

$$\mathbf{k}' = k' \sin\alpha\cos\phi\hat{\mathbf{i}} + k' \sin\alpha\sin\phi\hat{\mathbf{j}} + k'\cos\alpha\hat{\mathbf{k}}$$

$$\mathbf{F} = F\sin\theta\hat{\mathbf{j}} + F\cos\theta\hat{\mathbf{k}}$$

Hence $\cos\theta'$ becomes

$$\cos\theta' = \sin\theta\sin\alpha\sin\phi + \cos\theta\cos\alpha$$

The ratio of $\cos\theta'$ to $\cos\theta$ is then

$$\frac{\cos\theta'}{\cos\theta} = \tan\theta\sin\alpha\sin\phi + \cos\alpha$$

When the expression above is substituted into

$$\frac{1}{\tau} = \frac{\Omega}{(2\pi)^3}\int S\left(1 - \frac{\cos\theta'}{\cos\theta}\right)d^3k'$$

the term containing the angle ϕ will integrate to zero since the integration is taken over all azimuthal angles. Hence the final expression for the relaxation time is

$$\frac{1}{\tau} = \frac{\Omega}{(2\pi)^3}\int S(1 - \cos\alpha)d^3k'$$

where α is the polar angle between the incident and scattered wavevectors.

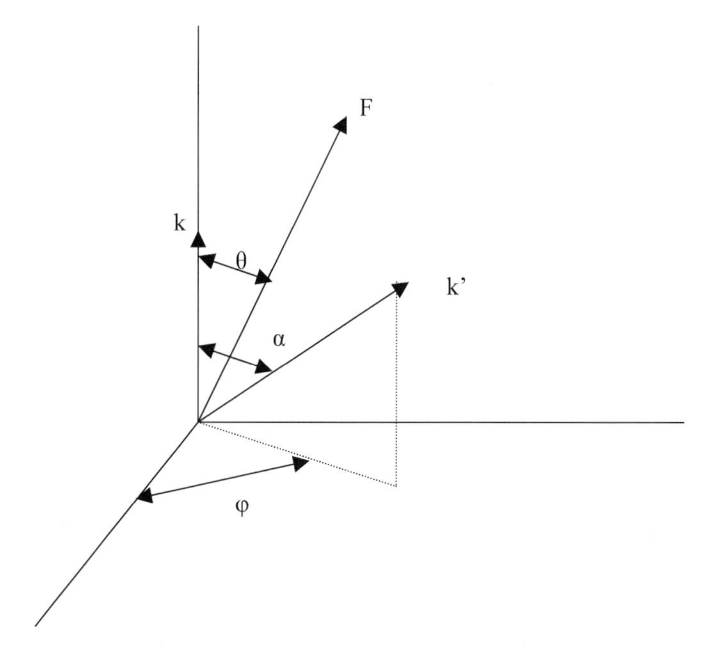

FIGURE 2.3.3 Coordinate system for the scattering events discussed in Box 2.3.1.

where Q_z is the phonon wavevector component perpendicular to the interface, D the deformation potential, ρ the density of the crystal, $\hbar\omega$ the phonon energy, n_Q the phonon occupation function given by the Bose–Einstein distribution (Brennan, 1999, Sec. 5.7), the − sign is for absorption and the + sign for emission, and $I_{mn}(Q_z)$ is the overlap integral of the Bloch periodic functions given as

$$I_{mn}(Q_z) = \int \chi_m(z)\chi_n(z)e^{\mp iQ_z z}dz \tag{2.3.11}$$

For acoustic scattering, the phonon energy is typically small. When the thermal energy is much larger than the phonon energy, the phonon occupation probability function n_Q can be simplified using the equipartition approximation as

$$n_Q \sim n_Q + 1 \sim \frac{kT}{\hbar\omega} \tag{2.3.12}$$

Making the equipartition approximation in Eq. 2.3.10 produces an extra factor of 2 from a combination of the absorption and emission terms. Equation 2.3.10 becomes

$$|\langle k'|H_{ep}|k\rangle|^2 = \sum_{Q_z} \frac{kTD^2}{\rho s_l^2 \Omega}|I_{mn}(Q_z)|^2 \tag{2.3.13}$$

where s_l is the longitudinal sound velocity. Note that for acoustic phonons, $s_l = \omega/Q$. The sum over the variable Q_z can be converted into an integral as

$$\sum_{Q_z} \rightarrow \frac{L}{2\pi} \int e^{iQ_z(z-z')} dQ_z = L\delta(z - z') \qquad 2.3.14$$

Recognizing that the only term that involves Q_z in Eq. 2.3.13 is $I_{mn}(Q_z)$, using Eq. 2.3.14, Eq. 2.3.13 becomes

$$|\langle k'|H_{ep}|k\rangle|^2 = \frac{kTD^2}{\rho s_l^2 \Omega} L \int dz\, \chi_m(z)\chi_n(z) \int dz'\, \chi_m(z')\chi_n(z) e^{iQ_z(z-z')} \delta(z - z') \qquad 2.3.15$$

Performing the integral over z' yields

$$|\langle k'|H_{ep}|k\rangle|^2 = \frac{kTD^2}{\rho s_l^2 A} \int \chi_m^2(z)\chi_n^2(z) dz \qquad 2.3.16$$

where A is the two-dimensional area. The total scattering rate can now be found by inserting the expression for the matrix element given by Eq. 2.3.16 into Fermi's golden rule for a two-dimensional system:

$$S(k,k') = \frac{2\pi}{\hbar} \frac{D^2 kT}{s_l^2 \rho A} \int dz\, \chi_m^2(z)\chi_n^2(z) \int \frac{A}{(2\pi)^2} \delta_E d^2 q \qquad 2.3.17$$

Changing the integral over q in Eq. 2.3.17 into an integral over the energy E and assuming parabolic energy bands yields (see Box 2.3.2)

$$S(k,k') = \frac{m^* kTD^2}{s_l^2 \rho \hbar^3} \int dz\, \chi_m^2(z)\chi_n^2(z) \qquad 2.3.18$$

Equation 2.3.18 is next evaluated numerically using the cell periodic forms of the wavefunctions determined from the self-consistent solution of the Poisson and Schrödinger equations. A similar analysis to the above can be applied to other elastic scattering mechanisms to calculate the scattering rates and hence the relaxation times.

The relaxation time is related to the scattering rate in the following manner. The general expression for the relaxation time is given from Box 2.3.1as

$$\frac{1}{\tau} = \frac{\Omega}{(2\pi)^3} \int S(k,k')(1 - \cos\alpha) d^3 k' \qquad 2.3.19$$

where $S(k,k')$ is the scattering rate and α is the polar angle between k and k'. The inverse scattering time is obtained from Eq. 2.3.19 by replacing the

BOX 2.3.2: Transformation of Integrals over k to over E for a Parabolic Energy Band

The summation over k for a three-dimensional system can be converted into an integral over k using the following prescription:

$$\lim_{V\to 0}\frac{1}{V}f(E(k)) = \int \frac{d^3k}{(2\pi)^3}f(E(k)) = \int \frac{k^2 dk\, d\Omega}{(2\pi)^3}f(E(k))$$

The corresponding integral over k can be transformed into an integral over the energy E in a simple manner if the $E(k)$ relationship is assumed to be parabolic (i.e., $E = \hbar^2k^2/2m$). Using the parabolic relationship, the expressions for k and dk become

$$k = \sqrt{\frac{2mE}{\hbar^2}} \qquad dk = \frac{m}{\hbar^2 k}dE$$

The integral over k can then be transformed into one over E as

$$\int \frac{d^3k}{(2\pi)^3}f(E(k)) \to \int \frac{m^*}{2\pi^2\hbar^2}\sqrt{\frac{2mE}{\hbar^2}}f(E)dE$$

Similarly, the integral over k for a two-dimensional system can be transformed as

$$\int \frac{d^2k}{(2\pi)^2} \to \frac{1}{2}\int \frac{m}{\pi\hbar^2}dE$$

term $(1 - \cos\alpha)$ by 1. Physically, the scattering time represents the mean time between collisions. The momentum relaxation time, on the other hand, gives the mean time it takes to relax the distribution from its forward direction. Notice that if the angle of scattering is very small, α approaches zero; then the momentum relaxation time approaches infinity. An infinite momentum relaxation time implies that the distribution never relaxes in momentum. Physically, this occurs when the electrons are not scattered from their forward direction, and hence they maintain their motion in the field direction. Thus the forward momentum of the distribution is never relaxed.

From the discussion above it is clear that the simple relaxation time approximation cannot be made for inelastic processes. In many important semiconductors even at relatively low electric field strengths, inelastic scattering processes dominate the transport dynamics. This is particularly true in polar semiconductor materials, such as GaAs and InP, where polar optical phonon scattering is very important and can strongly influence the carrier mobility at

EXAMPLE 2.3.1: Momentum Relaxation Time for Ionized Impurity Scattering

Determine an expression for the momentum relaxation time for ionized impurity scattering.

Equation 2.3.19 gives an expression for the momentum relaxation time as

$$\frac{1}{\tau} = \frac{\Omega}{(2\pi)^3} \int S(k,k')[1 - \cos\alpha] d^3k'$$

Therefore, it is necessary to integrate the scattering rate multiplied by $(1 - \cos\alpha)$ over all possible final k states. The expression for the scattering rate for ionized impurity scattering from one impurity in the volume Ω is given by (Brennan, 1999, Eq. 9.4.34)

$$S(k,k') = \frac{2\pi Z^2 q^4}{\hbar\varepsilon^2\Omega^2} \frac{\delta(E_{k'} - E_k)}{[4k^2 \sin^2(\alpha/2) + Q_D^2]^2}$$

where Z is the atomic number, ε the product of the relative and free-space dielectric constants, and Q_D the inverse screening length. For the case of N_I total impurity centers, $S(k,k')$ is multiplied by $N_I\Omega$. Recall that ionized impurity scattering is elastic. Substituting the expression for $S(k,k')$ into the momentum relaxation time yields

$$\frac{1}{\tau} = \frac{N_I\Omega^2}{(2\pi)^3} \frac{2\pi Z^2 q^4}{\hbar\varepsilon^2\Omega^2} \int \frac{[1 - (k'/k)\cos\alpha]\delta(E_{k'} - E_k) k'^2 dk' d(\cos\alpha) d\phi}{[4k^2 \sin^2(\alpha/2) + Q_D^2]^2}$$

Integrating over the azimuthal angle and recognizing that $k = k'$ since the scattering event is elastic yields

$$\frac{1}{\tau} = \frac{N_I}{(2\pi)} \frac{Z^2 q^4}{\hbar\varepsilon^2} \int_0^\infty \int_{-1}^{+1} \frac{[1 - (k'/k)\cos\alpha]\delta(E_{k'} - E_k) k'^2 dk' d(\cos\alpha)}{[4k^2 \sin^2(\alpha/2) + Q_D^2]^2}$$

Using the trigonometric identity

$$\sin^2\frac{\alpha}{2} = \frac{1 - \cos\alpha}{2}$$

the integral over α can be evaluated, leaving

$$\frac{1}{\tau} = \frac{N_I Z^2 q^4}{8\pi\varepsilon^2\hbar} \int_0^\infty \frac{k'^2 dk' \delta(E' - E)}{k^4} \left[\ln(1 + \gamma^2) - \frac{\gamma^2}{1 + \gamma^2}\right]$$

(Continued)

EXAMPLE 2.3.1 *(Continued)*

where γ is defined as

$$\gamma = \frac{4k^2}{Q_D^2}$$

Using the following two important properties of delta functions,

$$\delta(ax) = \frac{1}{|a|}\delta(x) \qquad \delta(k'^2 - k^2) = \frac{1}{2k'}\delta(k' - k)$$

and the assumption of parabolic energy bands, the momentum relaxation time can finally be written as,

$$\frac{1}{\tau} = \frac{N_I Z^2 q^4}{16\pi\varepsilon^2\sqrt{2m}}\left[\ln(1+\gamma^2) - \frac{\gamma^2}{1+\gamma^2}\right]\frac{1}{E^{3/2}}$$

room temperature. Although there are several approaches used to determine the mobility in the presence of inelastic phonon scattering mechanisms, we illustrate only one technique here and refer the interested reader to the literature for discussions of other approaches (Gelmont et al., 1995; Ridley et al., 2000). The technique we consider is based on linearized solution of the Boltzmann equation.

As discussed in Box 2.3.1, the nonequilibrium distribution function can be approximated in terms of the equilibrium distribution function at low fields. A more convenient, yet equivalent form for the distribution function than that used in Box 2.3.1 can be written as

$$f = f_0 - \frac{q\hbar F}{m^*}k\cos\alpha\frac{\partial f_0}{\partial E}\phi(E) \qquad 2.3.20$$

where α is again the scattering angle, but in this case is defined as the angle between k and F, the field direction. $\phi(E)$ is a perturbation distribution, f_0 the equilibrium distribution, and m^* the electron effective mass.

The linearized form of the Boltzmann equation can be obtained using the distribution function given by Eq. 2.3.20 as follows. The collision term in the Boltzmann equation, $I_c(f)$, is given as (Brennan, 1999, Sec. 6.1)

$$I_c(f) = -\int\frac{d^3k'}{(2\pi)^3}\{S(k,k')f(k)[1-f(k')] - S(k',k)f(k')[1-f(k)]\} \qquad 2.3.21$$

In equilibrium, the distribution function cannot change with time. Therefore, the collision integral must vanish. This implies that the scattering rate out of k into k' must be equal to the scattering rate from k' into k. This concept, often referred to as the *principle of detailed balance*, requires then that,

$$S(k,k')f_0(k)[1 - f_0(k')] = S(k',k)f_0(k')[1 - f_0(k)] \qquad 2.3.22$$

where $f_0(k)$ and $f_0(k')$ are the equilibrium distributions, reminding us that Eq. 2.3.22 holds only in equilibrium. Using Eq. 2.3.22, omitting second- and higher-order terms in $\phi(E)$, and assuming that the transition probability depends only on the angle between k and k', defined here as θ, after much algebra we obtain

$$1 = \int \frac{d^3k'}{(2\pi)^3} \frac{[1 - f_0(E')]}{[1 - f_0(E)]} \left[\phi(E) - \frac{k'}{k}\cos\theta\phi(E')\right] S(k,k') \qquad 2.3.23$$

Equation 2.3.23 is the linearized Boltzmann equation for a uniform electric field in a homogeneous system in steady state.

The two-dimensional scattering rate S_{II} is related to the three-dimensional scattering rate S_{III} as

$$S_{\text{II}} = \int |I(Q_z)|^2 S_{\text{III}} dQ_z \qquad 2.3.24$$

where $I(Q_z)$ is the overlap integral for the z components of the wavefunctions, given as

$$I_{ij}(Q_z) = \int \chi_i(z)\chi_j(z)e^{iQ_z z}dz \qquad 2.3.25$$

where the indices ij represent intraband scattering for $i = j$ and interband scattering for $i \neq j$. The two-dimensional differential polar optical phonon scattering rate is then

$$S_{\text{II}}(k,k') = \frac{\pi q^2 \omega_0}{2\varepsilon_0} \left(\frac{1}{\varepsilon_\infty} - \frac{1}{\varepsilon_r}\right) \frac{H_{ij}(Q)}{Q} \delta(E_{k'} - E_k) \qquad 2.3.26$$

where ε_0 is the free-space dielectric constant, ε_r the relative low-frequency dielectric constant, ε_∞ the high-frequency dielectric constant, Q the two-dimensional part of the phonon wavevector, and $H_{ij}(Q)$ is

$$H_{ij}(Q) = \int_0^\infty dz \int_0^\infty dz' \chi^2(z)\chi^2(z')e^{-Q|z-z'|} \qquad 2.3.27$$

Substituting Eq. 2.3.26 into Eq. 2.3.23, the linearized Boltzmann equation for polar optical phonon scattering in two dimensions becomes

$$1 = \frac{q^2\omega_0}{8\pi\varepsilon_0}\left(\frac{1}{\varepsilon_\infty} - \frac{1}{\varepsilon_0}\right)\frac{1}{1 - f_0(E)}\int d^2k'\left[\phi(E) - \frac{k'}{k}\cos\theta\phi(E')\right]$$

$$\times [1 - f_0(E')]\frac{H_{ij}(Q)}{Q}\delta(E_{k'} - E_k) \qquad 2.3.28$$

At this point, the perturbation terms $\phi(E)$ and $\phi(E')$ can be separated into emission and absorption. Equation 2.3.28 and the results for elastic scattering processes can now be combined to determine the two-dimensional mobility.

The two-dimensional mobility can be calculated under two different conditions. If all the scattering processes are elastic or can be treated as elastic, the relaxation time can be found using Eq. 2.3.19 for each process. The total relaxation time can then be determined from Eq. 2.3.8. Once the total relaxation time is determined, the drift mobility is found from

$$\mu = \frac{q\langle\tau\rangle}{m^*} \qquad 2.3.29$$

where $\langle\tau\rangle$ is the mean value of the momentum relaxation time averaged as

$$\langle\tau\rangle = \frac{\int_0^\infty \tau(E)E[\partial f_0(E)/\partial E]dE}{\int_0^\infty E[\partial f_0(E)/\partial E]dE} \qquad 2.3.30$$

Now if inelastic processes are present, the mobility must be calculated differently. In this case we need to put both the elastic and inelastic scattering mechanisms into the linearized Boltzmann equation given by Eq. 2.3.23 and solve it numerically using an iterative technique. Therefore, we substitute in Eq. 2.3.23 for the terms $S(k,k')$ the appropriate transition rates for all the elastic and inelastic processes. We can further define incoming and outgoing scattering rates for scattering out and into the differential volume element dk. The Boltzmann equation can then be written as a difference equation as (Kawamura et al., 1992)

$$1 = S_0(E)\phi(E) - S_a(E)\phi(E + \hbar\omega) - S_e(E)\phi(E - \hbar\omega) \qquad 2.3.31$$

where $S_0(E)$ is the sum of the outgoing scatterings due to inelastic processes and all elastic processes, and $S_a(E)$ and $S_e(E)$ are the in-scattering contributions from inelastic absorption and emission processes, respectively. These terms are given as

$$S_0(E) = \frac{e^2 m^* \omega_0}{4\pi\varepsilon_0 \hbar^2}\left(\frac{1}{\varepsilon_\infty} - \frac{1}{\varepsilon_0}\right)\left\{[N_q + f_0(E + \hbar\omega_0)] \int_0^{\pi} \frac{H_{mn}(q^+)}{q^+} d\theta \right.$$

$$\left. + [N_q + 1 - f_0(E - \hbar\omega_0)] \int_0^{\pi} \frac{H_{mn}(q^-)}{q^-} d\theta \right\} + \frac{1}{\tau_{\text{tot}}^{\text{el}}(E)} \qquad 2.3.32$$

$$S_a(E) = \frac{e^2 m^* \omega_0}{4\pi\varepsilon_0 \hbar^2}\left(\frac{1}{\varepsilon_\infty} - \frac{1}{\varepsilon_0}\right)[N_q + f_0(E + \hbar\omega_0)]\sqrt{\frac{E + \hbar\omega_0}{E}} \int_0^{\pi} \frac{H_{mn}(q^+)}{q^+} \cos\theta d\theta$$

$$S_e(E) = \frac{e^2 m^* \omega_0}{4\pi\varepsilon_0 \hbar^2}\left(\frac{1}{\varepsilon_\infty} - \frac{1}{\varepsilon_0}\right)[N_q + 1 - f_0(E - \hbar\omega_0)]\sqrt{\frac{E - \hbar\omega_0}{E}} \int_0^{\pi} \frac{H_{mn}(q^-)}{q^-} \cos\theta d\theta$$

Using a numerical iteration technique, an expression for $\phi(E)$ can be determined. The phonon wavevector component q parallel to the layer plane is given by

$$q^{\pm} = \frac{\sqrt{2m^*}}{\hbar}\sqrt{E_a^{\pm} - E_b^{\pm}\cos\theta} \qquad 2.3.33$$

with $E_a^{\pm} = 2E(k) \pm \hbar\omega$ and $E_b^{\pm} = 2\sqrt{E(k)[E(k) \pm \hbar\omega]}$. The $\pm$ signs denote the phonon emission and absorption processes. The perturbation term $\phi(E)$ can be obtained by iteratively solving Eq. 2.3.31. First, the starting value ϕ^0 can be obtained by assuming that the S_a and S_e terms both equal zero. Then in the $(n + 1)$th step of the iteration, ϕ^{n+1} is calculated from the value of $\phi^n(E \pm \hbar\omega_0)$. Finally, the iteration will give a converged value of $\phi(E)$ that satisfies Eq. 2.3.31.

Experimentally, the most commonly determined quantity is the Hall mobility rather than the drift mobility (see Chapter 9). The Hall and drift mobilities generally have values that differ by what is commonly called the *Hall factor*. The drift mobility is generally smaller in magnitude than the Hall mobility. The drift and Hall mobilities are equal when the relaxation time from scattering is constant. Typically, this is not the case. Then the drift and Hall mobilities, μ and μ_H, respectively, are related as

$$\mu_H = r\mu \qquad 2.3.34$$

where r is a number that varies between 1 and 2, depending on the type of scattering that dominates. The Hall mobility is defined as

$$\mu_H = \frac{q}{m^*}\frac{\langle\tau^2\rangle}{\langle\tau\rangle} \qquad 2.3.35$$

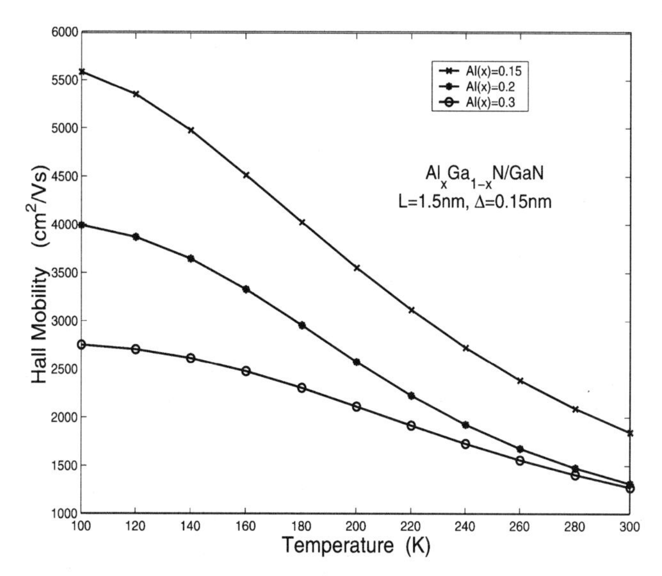

FIGURE 2.3.4 Calculated Hall mobility versus temperature for AlGaN–GaN HEMT structures with Al mole fractions of 15, 20, and 30.

with

$$\langle \tau^n \rangle = \frac{\int_0^\infty \tau^n(E) E[\partial f_0(E)/\partial E] dE}{\int_0^\infty E[\partial f_0(E)/\partial E] dE} \qquad (n = 1, 2) \qquad 2.3.36$$

Once $\phi(E)$ is known, the mobility can be calculated from Eqs. 2.3.35 and 2.3.36, replacing $\tau(E)$ by $\phi(E)$.

It is interesting to examine how the two-dimensional mobility is influenced by the various scattering mechanisms. For purposes of illustration, we examine the two-dimensional mobility in an AlGaN–GaN heterostructure. The calculation is made for zero field and as a function of temperature (see Figure 2.3.4). In the calculation, both elastic and inelastic mechanisms are included. The elastic mechanisms are piezoelectric, acoustic, remote ionized impurity (from donors within the AlGaN layer introduced by modulation doping), and interface roughness scattering. The only inelastic mechanism included in the calculation is that of polar optical phonon scattering. The mobility is calculated following the approach outlined above (i.e., through solution of the linearized Boltzmann equation following a self-consistent solution of the Schrödinger and Poisson equations). Intersubband scattering is included in the calculations, implying that the carriers can scatter between any two subbands subject to both momentum and energy conservation. The calculations are made using a fully numerical approach (i.e., the wavefunctions, subband energies, and carrier concentration are determined numerically). The calculated electron Hall mobility along with an experimentally measured Hall mobility is shown in

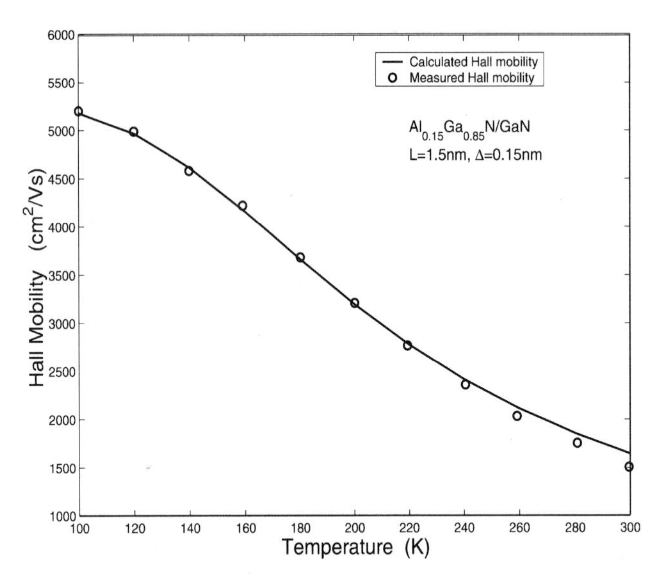

FIGURE 2.3.5 Comparison of calculated and measured Hall mobilities in a $Al_{0.15}Ga_{0.85}N$–GaN HEMT device. The experimental Hall mobility is from Wu et al. (1996).

Figure 2.3.5. The surface roughness parameters, L and Δ, have been selected to ensure good agreement between calculation and experiment. For the specific device structure considered here, the correlation length L and amplitude of the roughness Δ have been chosen as 1.5 nm and 0.15 nm, respectively. Good agreement is obtained between the experimental and theoretical results over the full temperature range. The electron Hall mobility is 5177 $cm^2/V \cdot s$ at 100 K and decreases to 1637 $cm^2/V \cdot s$ at room temperature (Yu and Brennan, 2001).

It is interesting to note that the mobility varies strongly with temperature. To gauge how each scattering mechanism affects mobility, the component mobilities for the AlGaN–GaN system are plotted in Figure 2.3.6 as a function of temperature. As can be seen from the figure, remote impurity scattering is independent of temperature. This is because Coulomb processes are usually approximated as temperature independent. All of the most important scattering mechanisms are included: interface roughness, remote donors within the AlGaN layer, acoustic deformation potential, piezoelectric acoustic phonon, residual impurity, and polar optical phonon scattering. As can readily be observed from the figure, over most of the temperature range examined, interface roughness scattering dominates the mobility until room temperature is reached. The largest effect near room temperature is due to polar optical phonon scattering. At low temperatures, the electron gas has insufficient energy to emit an optical phonon. Therefore, the polar optical phonon scattering rate is suppressed. However, as the temperature increases, the carriers become

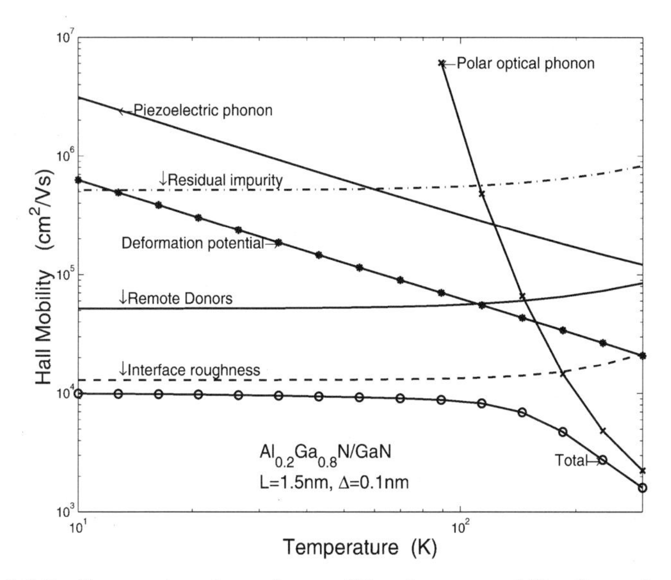

FIGURE 2.3.6 Temperature dependence of the electron mobility for each active scattering mechanism in a $Al_{0.2}Ga_{0.8}N$–GaN HEMT structure, including spontaneous and piezoelectric polarization-induced fields.

sufficiently hot that they can now emit an optical phonon. As such, the mobility becomes a strong function of the optical phonon scattering and decreases accordingly.

2.4 STRAIN AND STRESS AT HETEROINTERFACES

The simplest description of a bulk crystalline semiconductor is that it exhibits perfect or nearly perfect translational symmetry. In other words, suitable translations of the basic unit cell of a crystal restore the crystal back into itself. Implicit in this definition is the assumption that the atoms within the crystal are regularly spaced throughout the entire bulk sample. This assumption is generally true for bulk materials. However, two important exceptions can arise. The first is that a bulk crystal can include impurities and dislocations such that the perfect periodicity of the material is disrupted locally. The crystal can still retain its overall highly ordered structure, yet contain local regions in which perfect periodicity is disrupted by impurities or dislocations. These impurities and dislocations can significantly affect the properties of the material. The second situation arises in multilayered structures. Crystal growth technology has enabled the growth of thin layers of heterogeneous semiconductor material called *heterostructures*. Using exacting crystal growth procedures, heterostructures can be grown with atomic layer precision. A very thin layer of material can be grown on top of or sandwiched between layers grown with a different

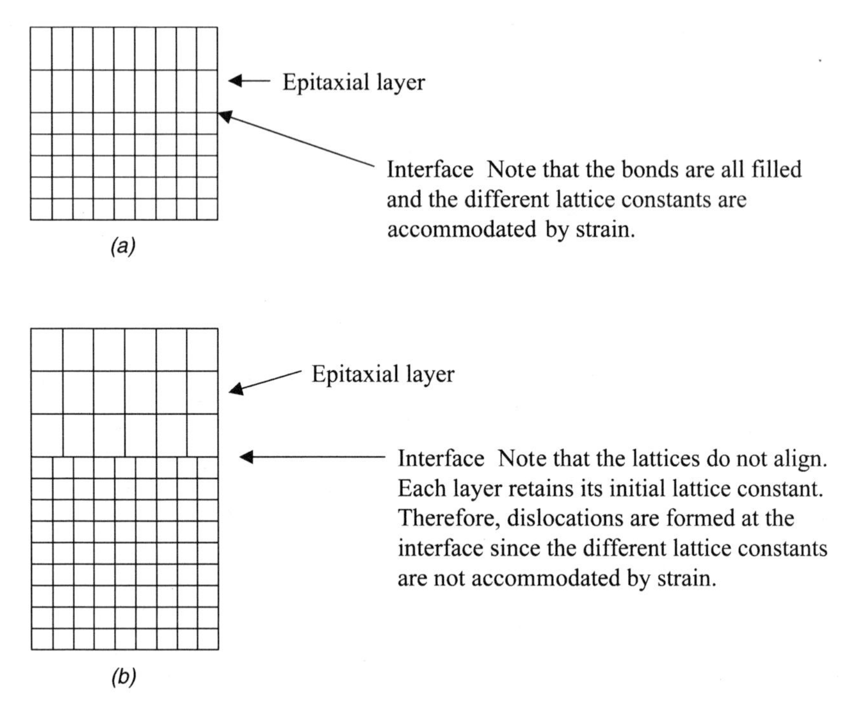

FIGURE 2.4.1 (*a*) Thin epitaxial layer strained to accommodate the various lattice constants of the underlying semiconductor layer and (*b*) a thicker epitaxial layer that has relaxed. In part (*b*) the epitaxial layer is thicker than the critical thickness and dislocations appear at the interface.

type of semiconductor material, even materials in which the lattice constant is different.

When a thin layer of material is grown either on or between layers of a different semiconductor that has a significantly different lattice constant, the thin, epitaxial layer will adopt the lattice constant of the neighboring layers provided that the lattice mismatch is less than about 10%. As can be seen from Figure 2.4.1, when the thin, epitaxial layer adopts the lattice constant of the surrounding layers, it becomes strained, i.e., it is either compressed or expanded from its usual bulk crystal shape. There exists a maximum thickness of the thin layer below which the lattice mismatch can be accommodated through strain. For layer thickness above the critical thickness, the lattice mismatch cannot be accommodated through strain, dislocations are produced and the strain relaxes as is seen in Figure 2.4.1*b*. The strain within the layer is homogeneous. The strained layer can be in either compressive or tensile strain. If the lattice constant of the strained layer is less than that of the surrounding layers the system is in tension. Conversely, if the lattice constant of the strained layer is greater than that of the surrounding layers, the strained layer is in compression.

BOX 2.4.1: Tensor Properties

Generally, a tensor can be of any integer rank. A scalar is simply a tensor of zero rank, a vector is a tensor of rank 1, and a matrix is a tensor of rank 2. Let us illustrate tensors through a few examples.

Consider two vectors, **a** and **b**, that are related as follows. Let the components of vectors **a** and **b** be **a** = $[a_1, a_2, a_3]$ and **b** = $[b_1, b_2, b_3]$. Let vectors **a** and **b** be related by the tensor **T** as

$$\begin{bmatrix} a_1 \\ a_2 \\ a_3 \end{bmatrix} = \begin{bmatrix} T_{11} & T_{12} & T_{13} \\ T_{21} & T_{22} & T_{23} \\ T_{31} & T_{32} & T_{33} \end{bmatrix} \begin{bmatrix} b_1 \\ b_2 \\ b_3 \end{bmatrix}$$

The relationship above can be expanded out following the usual rules of matrix multiplication, to give

$$a_1 = T_{11}b_1 + T_{12}b_2 + T_{13}b_3$$
$$a_2 = T_{21}b_1 + T_{22}b_2 + T_{23}b_3$$
$$a_3 = T_{31}b_1 + T_{32}b_2 + T_{33}b_3$$

The relationship between a and b can be represented as

$$[a] = [T][b]$$

or in indicial notation as

$$a_j = T_{ij}b_i = \sum_i T_{ij}b_i$$

where the index i is called a *dummy index* and j is called a *free index*. The indicial equations can be expanded out as

$$a_1 = T_{11}b_1 + T_{12}b_2 + T_{13}b_3$$
$$a_2 = T_{21}b_1 + T_{22}b_2 + T_{23}b_3$$
$$a_3 = T_{31}b_1 + T_{32}b_2 + T_{33}b_3$$

which are identical to the expansions for the matrix multiplication given above.

We begin our study of the effects of strain on the electrical properties of a semiconductor with the definitions of strain and stress (Nye, 1985). Strain is a measure of the change in length of a material undergoing small deformations arising from an applied force per unit area. This force per unit area is called the stress. The stress and strain are generally represented as tensor quantities. (See Box 2.4.1 for a description of tensor properties.) These tensors have nine components that describe the relative stress and strain along the principal directions of the crystal on planes that are described with normal vectors lying along these principal directions. The stress tensor can be defined as follows. The stress vector $\mathbf{p}_n$ acting at a point P on a surface with unit normal vector $\mathbf{n}$ is defined as

$$\mathbf{p}_n = \lim_{A \to 0} \frac{F}{A} \tag{2.4.1}$$

where F is the force and A is the area of the surface. The stress vector can be written as

$$\mathbf{p}_n = \sigma \mathbf{n} \tag{2.4.2}$$

where σ is the stress tensor and $\mathbf{n}$ the normal vector. Expanding Eq. 2.4.2 yields

$$\begin{aligned} \mathbf{p}_1 &= \sigma_{11}\hat{\mathbf{e}}_1 + \sigma_{12}\hat{\mathbf{e}}_2 + \sigma_{13}\hat{\mathbf{e}}_3 \\ \mathbf{p}_2 &= \sigma_{21}\hat{\mathbf{e}}_1 + \sigma_{22}\hat{\mathbf{e}}_2 + \sigma_{23}\hat{\mathbf{e}}_3 \\ \mathbf{p}_3 &= \sigma_{31}\hat{\mathbf{e}}_1 + \sigma_{32}\hat{\mathbf{e}}_2 + \sigma_{33}\hat{\mathbf{e}}_3 \end{aligned} \tag{2.4.3}$$

where $\mathbf{e}_1$, $\mathbf{e}_2$, and $\mathbf{e}_3$ are the three normal vectors to the ordinate planes. We choose to use the more general coordinates $\mathbf{e}_1$, $\mathbf{e}_2$, and $\mathbf{e}_3$ here since only for cubic symmetry are the normal vectors the more usual $\mathbf{x}$, $\mathbf{y}$, and $\mathbf{z}$ vectors. In words, $\mathbf{p}_1$ represents the stress vector acting on the surface whose outward normal vector is $\mathbf{e}_1$. Notice that $\mathbf{p}_1$ has three components, σ_{11}, σ_{12}, and σ_{13}. σ_{11} is the normal component, and σ_{12} and σ_{13} are the tangential components of the stress vector. If we represent the stress tensor as σ_{ij}, the index i indicates the direction of the component of the force per unit area on a plane whose normal is parallel to the $\mathbf{e}_j$ direction. For example, if we consider σ_{12}, the stress is along the $\mathbf{e}_1$ direction acting on the plane with a normal vector parallel to the $\mathbf{e}_2$ axis. Therefore, in this case, σ_{12}, the stress is tangential to the surface of the plane. Such a stress is called a *shear stress*. Alternatively, the stress component σ_{11} is the normal component of the stress on the 23 plane (that with a normal vector parallel to the $\mathbf{e}_1$ axis). These stresses are called *normal stresses*. If the normal stress is positive, acting outward from the surface, it is called a *tensile stress*. If the normal stress is negative, acting inward toward the surface, it is called a *compressive stress*.

In many circumstances the relationship between the stress and strain is linear. Under this condition, the deformation of the solid is called *elastic*, meaning that upon removal of the load, all deformations disappear. In other words, once the stress is removed, the system returns to its original shape and form. Such behavior is, of course, linear. The general relationship between the stress tensor σ_{ij} and the strain tensor ε_{kl} is then given as

$$\sigma_{ij} = C_{ijkl}\varepsilon_{kl} \tag{2.4.4}$$

where C_{ijkl} is the elasticity tensor. The elasticity tensor can be simplified by recognizing that there are some inherent symmetries that reduce the number of independent components of C. The first simplification arises from the fact that there can be no net torque from external forces on an infinitesimal volume element. Therefore, the stress components must obey (assuming that there are no internal torques present)

$$\sigma_{ij} = \sigma_{ji} \tag{2.4.5}$$

Similarly, $\varepsilon_{kl} = \varepsilon_{lk}$. Therefore, the components of the elasticity tensor must obey

$$C_{ijkl} = C_{jikl} = C_{ijlk} = C_{jilk} \tag{2.4.6}$$

The elasticity tensor can be further simplified as follows. The differential work done, dW, when a unit volume element is reversibly deformed by differential strain increments $d\varepsilon_{ij}$ is

$$dW = \sigma_{ij}d\varepsilon_{ij} = C_{ijkl}\varepsilon_{kl}d\varepsilon_{ij} \tag{2.4.7}$$

The Helmholtz free energy F is defined as (Brennan, 1999, Sec. 5.5)

$$F = U - TS \tag{2.4.8}$$

and the associated Maxwell relation is (Brennan, 1999, Table 5.5.1)

$$dF = -S\,dT - P\,dV \tag{2.4.9}$$

For an isothermal system, $dF = -P\,dV$, which is directly related to $\sigma_{ij}d\varepsilon_{ij}$ as

$$dF = C_{ijkl}\varepsilon_{kl}d\varepsilon_{ij} \tag{2.4.10}$$

Taking the second derivative of F with respect to ε, Eq. 2.4.10 becomes

$$\frac{\partial^2 F}{\partial\varepsilon_{ij}\partial\varepsilon_{kl}} = C_{ijkl} \qquad \frac{\partial^2 F}{\partial\varepsilon_{kl}\partial\varepsilon_{ij}} = C_{klij} \tag{2.4.11}$$

But the fact that dF is an exact differential implies that

$$\frac{\partial^2 F}{\partial \varepsilon_{ij} \partial \varepsilon_{kl}} = \frac{\partial^2 F}{\partial \varepsilon_{kl} \partial \varepsilon_{ij}} \qquad 2.4.12$$

since the order of differentiation is unimportant. Using Eqs. 2.4.12 and 2.4.11, we obtain

$$C_{ijkl} = C_{klij} \qquad 2.4.13$$

Equation 2.4.13 implies that C is symmetric along the diagonal.

At this point it is important to examine how C_{ijkl} acts on the tensor ε_{ij}. From Eq. 2.4.4 it is clear that for the multiplication of C and ε to be valid, the number of elements in the rows of C must match the number of elements in the columns of ε. It is further evident that since σ_{ij} and ε_{ij} have nine elements each, that C_{ijkl} must then be 9×9. The multiplication is then between C, a 9×9 matrix, and ε, a 9×1 matrix to form σ, a 9×1 matrix. Notice that the number of elements in a row of C, 9, matches the number of elements in the single column of ε, 9, as required. From Eq. 2.4.13, the matrix C is obviously symmetric about the diagonal. The multiplication can now be written in matrix form as

$$\begin{bmatrix} \sigma_{11} \\ \sigma_{22} \\ \sigma_{33} \\ \sigma_{23} \\ \sigma_{31} \\ \sigma_{12} \\ \sigma_{32} \\ \sigma_{13} \\ \sigma_{21} \end{bmatrix} = \begin{bmatrix} c_{11} & c_{12} & c_{13} & c_{14} & c_{15} & c_{16} & c_{14} & c_{15} & c_{16} \\ c_{12} & c_{22} & c_{23} & c_{24} & c_{25} & c_{26} & c_{24} & c_{25} & c_{26} \\ c_{13} & c_{23} & c_{33} & c_{34} & c_{35} & c_{36} & c_{34} & c_{35} & c_{36} \\ c_{14} & c_{24} & c_{34} & c_{44} & c_{45} & c_{46} & c_{44} & c_{45} & c_{46} \\ c_{15} & c_{25} & c_{35} & c_{45} & c_{55} & c_{56} & c_{45} & c_{55} & c_{56} \\ c_{16} & c_{26} & c_{36} & c_{46} & c_{56} & c_{66} & c_{46} & c_{56} & c_{66} \\ c_{14} & c_{24} & c_{34} & c_{44} & c_{45} & c_{46} & c_{44} & c_{45} & c_{46} \\ c_{15} & c_{25} & c_{35} & c_{45} & c_{55} & c_{56} & c_{45} & c_{55} & c_{56} \\ c_{16} & c_{26} & c_{36} & c_{46} & c_{56} & c_{66} & c_{46} & c_{56} & c_{66} \end{bmatrix} \begin{bmatrix} \varepsilon_{11} \\ \varepsilon_{22} \\ \varepsilon_{33} \\ \varepsilon_{23} \\ \varepsilon_{31} \\ \varepsilon_{12} \\ \varepsilon_{32} \\ \varepsilon_{13} \\ \varepsilon_{21} \end{bmatrix} \qquad 2.4.14$$

where we have used Eq. 2.4.6 to further simplify the result. The contracted notation used in Eq. 2.4.14 can be understood as follows. As mentioned above, the matrix is symmetric about the diagonal. The indices ij increase downward in the matrix for C, while the indices kl increase across in the matrix for C. Since $ij = kl$ by Eq. 2.4.13, we can replace ij by m and kl by n. The correspondence between the indices is given in Table 2.4.1. Notice that there are only 21 independent constants in the 81 elements of C (see Problem 2.2). For a material with cubic symmetry, the system is invariant under a rotation by 90° about the cube axes. As a result, only three independent components remain of the original 81 elements of C for the cubic case. These are c_{11}, c_{12},

TABLE 2.4.1 Correspondence Between the Indices ij, kl and m, n Used in Constructing the Matrix C

ij or kl	11	22	33	23	31	12	32	13	21
m or n	1	2	3	4	5	6	7	8	9

and c_{44}, which are equal to

$$\begin{aligned} c_{11} &= c_{1111} = c_{2222} = c_{3333} \\ c_{12} &= c_{1122} = c_{1133} = c_{2233} \\ c_{44} &= c_{1212} = c_{2323} = c_{3131} \end{aligned} \tag{2.4.15}$$

With these simplifications, the elastic constants matrix C for a cubic material becomes

$$c_{mn} = \begin{bmatrix} c_{11} & c_{12} & c_{12} & 0 & 0 & 0 \\ c_{12} & c_{11} & c_{12} & 0 & 0 & 0 \\ c_{12} & c_{12} & c_{11} & 0 & 0 & 0 \\ 0 & 0 & 0 & c_{44} & 0 & 0 \\ 0 & 0 & 0 & 0 & c_{44} & 0 \\ 0 & 0 & 0 & 0 & 0 & c_{44} \end{bmatrix} \tag{2.4.16}$$

With the following introduction to stress and strain, we are now equipped to study piezoelectric effects in strained heterostructure systems. A complete discussion of this topic is well beyond the scope of this book. Instead, we focus on the origin of piezoelectrically induced strain fields in heterostructures and how these fields can potentially be utilized in device structures. The first question is: How does strain occur in a heterostructure? As discussed above, the lattice mismatch between two semiconductor layers can be accommodated either through the formation of misfit dislocations or by strain. If the layers are grown sufficiently thin that they do not relax through the formation of misfit dislocations, the mismatch is accommodated by internal strains. These strains can, in turn, generate strain-induced polarization fields (Smith, 1986). Generally, it is important to consider the effects of the substrate on which the layers are grown. For simplicity, we confine our discussion here to thin semiconductor layers, assuming that all the strain arises from the lattice mismatch of only these layers. The strain-induced polarization fields can be used to alter the carrier concentration in the immediate vicinity of a heterostructure layer by changing the local electric field profile. One important application of these fields is to accumulate a large free carrier concentration at the heterointerface without introducing modulation doping (Kuech et al., 1990). (See Section 2.5 for a discussion of modulation doping.) Experimental work has shown that very high electron concentrations can be induced in this way at certain heterointerfaces (Yu et al., 1997; Smorchkova et al., 1999).

The piezoelectric effect can readily be understood as follows. A polarization field can be produced within a material when a stress is applied to it. This is known as the *direct piezoelectric effect* and can be expressed mathematically as

$$P_i = d_{ijk}\sigma_{jk} \qquad 2.4.17$$

where the components of d are called the *piezoelectric moduli*. Equation 2.4.17 implies that each component of P depends upon all the components of σ. P_1 given by Eq. 2.4.17 can be expanded out as

$$\begin{aligned} P_1 = {} & d_{111}\sigma_{11} + d_{112}\sigma_{12} + d_{113}\sigma_{13} + d_{121}\sigma_{21} + d_{122}\sigma_{22} \\ & + d_{123}\sigma_{23} + d_{131}\sigma_{31} + d_{132}\sigma_{32} + d_{133}\sigma_{33} \end{aligned} \qquad 2.4.18$$

Similar equations can be written for P_2 and P_3. Using the repeated index notation, Eq. 2.4.18 can be written in a much simpler and compact form as

$$P_1 = d_{1jk}\sigma_{jk} \qquad 2.4.19$$

where, as is usual, there is an implied sum over both indices j and k on the right-hand side. It is useful to relate the polarization field P to the strain ε using Eqs. 2.4.4 and 2.4.17. In matrix form, P can be written as

$$P = d\sigma = d(C\varepsilon) = e\varepsilon \qquad 2.4.20$$

which in component form is

$$P_i = d_{ijk}C_{jklm}\varepsilon_{lm} = e_{ilm}\varepsilon_{lm} \qquad 2.4.21$$

Depending on the symmetry of the crystal, only some components are nonzero in Eq. 2.4.21. In this way, an expression for the polarization vector can be obtained from the piezoelectric tensor e and the strain tensor ε.

Alternatively, a piezoelectric crystal becomes deformed when an external electric field is applied to it, an effect often referred to as the *converse effect*. In either case, deforming the crystal to produce an electric field or applying an electric field to deform the crystal, there exists a "one-wayness" to the process. By this we mean that if the crystal is under pressure or stretched, the sign of the electric field changes appropriately. Similarly, depending on the sign of the applied electric field, the crystal will be either compressed or stretched. The internal structure of the crystal determines how it will deform.

The group III-nitride semiconductor materials are often used to illustrate the strain-induced polarization field effect since these materials form highly strained heterolayers and are strongly piezoelectric. However, these materials typically crystallize in the wurtzite phase, which has hexagonal symmetry, as

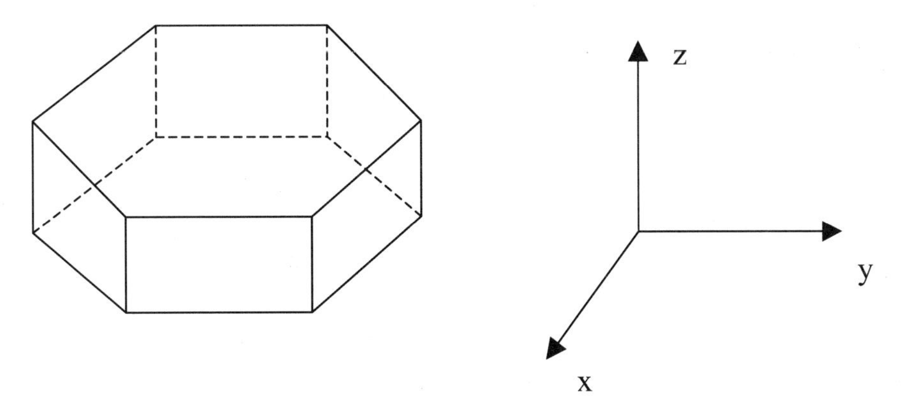

FIGURE 2.4.2 Wurtzite unit cell, showing the hexagonal symmetry of the system. The z axis direction is often referred to as the c axis; the in-plane directions (x and y) are also referred to as the *basal plane directions*.

shown in Figure 2.4.2. From the figure, it is clear that the crystal is different along the z axis then along the in-plane axes, x and y. The lattice constants along the z axis, c, and in-plane, a, are generally different. The strains along the z axis and in-plane are defined as (Ambacher et al., 2000)

$$\varepsilon_{zz} = \frac{c - c_0}{c_0} \qquad \varepsilon_{xx} = \varepsilon_{yy} = \frac{a - a_0}{a_0} \qquad 2.4.22$$

where c_0 and a_0 are the lattice constants in the unstrained system and c and a are the lattice constants under strain conditions, and we have used the xyz coordinate system shown in Figure 2.4.2 for the strains. The piezoelectrically induced polarization field in the z direction is given as (Ambacher et al., 2000)

$$P_z = e_{33}\varepsilon_{zz} + e_{31}(\varepsilon_{xx} + \varepsilon_{yy}) \qquad 2.4.23$$

Inspection of Eq. 2.4.23 shows that the polarization field is induced by variations in the lattice constants in the basal plane and c axis directions. Physically, the strain, caused by compression or expansion of the crystalline lattice from its unstrained condition, pushes the ions together or pulls them apart. In either case, the local dipole moments are different, resulting in a change in the macroscopic polarization vector, **P**.

Let us consider the situation shown in Figure 2.4.3, wherein a thin AlN layer is sandwiched between two relatively thick GaN layers. The lattice mismatch of the system is accommodated by straining the thin AlN layer, provided that this layer is grown less than the critical thickness, which for AlN is about 3.0 nm. From Eq. 2.4.23 a polarization field **P** is produced in the longitudinal

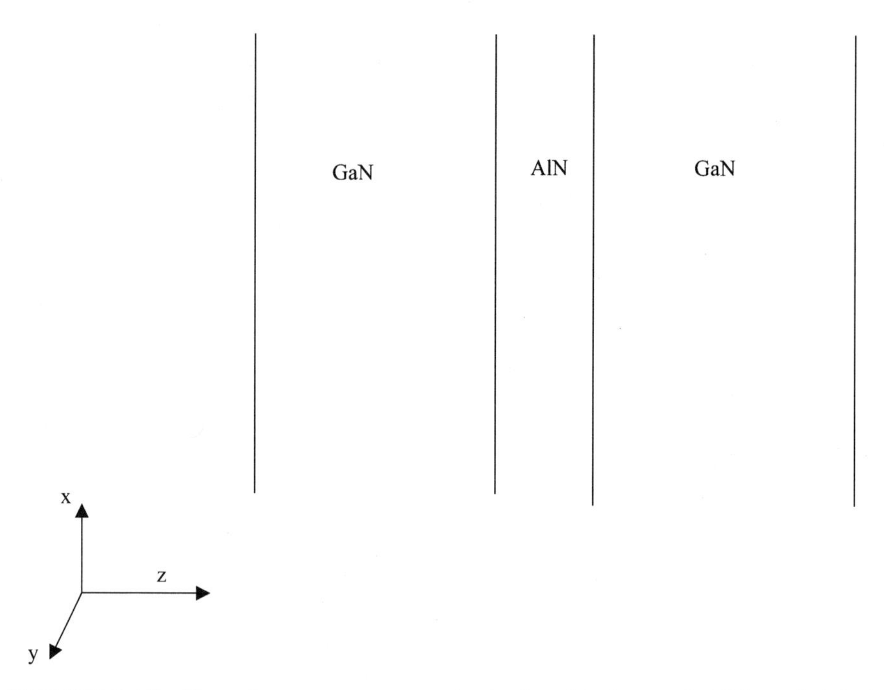

FIGURE 2.4.3 Three-layer system comprised of a thin layer of AlN sandwiched between two GaN layers. The AlN layer is strained, resulting in the production of strain-induced polarization fields pointing along the z direction.

direction (z axis as shown in Figure 2.4.3). As is well known from electromagnetics, one can define the electric displacement vector **D** within a macroscopic media, which contains a polarization field **P** as

$$\mathbf{D} = \kappa_0 \mathbf{E} + \mathbf{P} \qquad 2.4.24$$

where κ_0 is the free-space dielectric constant. One of the four Maxwell's equations,

$$\nabla \cdot \mathbf{D} = \rho \qquad 2.4.25$$

where ρ is the charge concentration, can be modified by substituting in the expression given by Eq. 2.4.24 for **D** to yield

$$\nabla \cdot (\kappa_0 \mathbf{E} + \mathbf{P}) = \rho \qquad 2.4.26$$

Rearranging Eq. 2.4.26 and defining the polarization charge ρ_{pz} yields

$$\nabla \cdot \kappa_0 \mathbf{E} = \rho - \nabla \cdot \mathbf{P} = \rho + \rho_{pz} \qquad 2.4.27$$

Thus, the polarization vector can be written as

$$\nabla \cdot \mathbf{P} = -\rho_{pz} \tag{2.4.28}$$

The strain condition changes abruptly at the two heterointerfaces. As a result, Eq. 2.4.23 implies that the polarization vector changes across the interface. From Eq. 2.4.28, this implies that there exists a polarization charge at the interface due to the nonzero divergence. One side of the heterointerface has a positive charge while an equal but opposite negative charge is produced on the other side.

As mentioned earlier, the strain-induced polarization field can be used to change the local carrier concentration near the heterointerface. An example system is shown in Figure 2.4.4. In this system a Schottky metal gate contact is formed on top of a thin AlGaN layer that is in turn grown on top of a GaN layer. The thin AlGaN layer is strained, producing a polarization field. This field in turn creates a polarization charge density, σ_{pz}, at both interfaces. Notice that the polarization charge density at each interface induces in turn a compensating charge density in the top metal Schottky barrier and in the bottom GaN layer. The charge density induced in the bottom GaN layer is due to free electrons accumulated at the interface. Thus the local electron concentration at the heterointerface within the GaN layer is increased significantly. The overall effect is that the electron concentration within the GaN layer at the heterointerface is drastically increased without the introduction of dopants. As a result, a high carrier concentration at low impurity concentration with a concomitant high carrier mobility is obtained.

In addition to the piezoelectrically induced polarization fields, materials such as GaN, InN, and AlN exhibit *spontaneous polarization*, which arises even in the absence of strain. It is a property of low-symmetry materials in their ground state. Spontaneous polarization can be thought of as arising from the nonideality of the crystalline structure of materials such as wurtzite GaN. In these materials the c/a ratio (ratio of the lattice constants in the c axis direction to that in the basal plane) deviates from its ideal value. This results in a difference in the bond lengths and a concomitant change in the dipole moment. The net effect is that the material exhibits a built-in polarization that in many cases is quite significant in magnitude. The spontaneous and piezoelectric polarizations can be either aligned or antialigned, depending on the properties of the material and whether the layer is in tensile or compressive strain. When the spontaneous and piezoelectric polarizations are aligned, the net polarization is given simply as

$$\mathbf{P} = \mathbf{P}_{pz} + \mathbf{P}_{\text{spon}} \tag{2.4.29}$$

The combined action of the piezoelectric and spontaneous polarization fields can be used to change the free carrier concentration in a heterolayer. For further details about this effect, the reader is referred to the references.

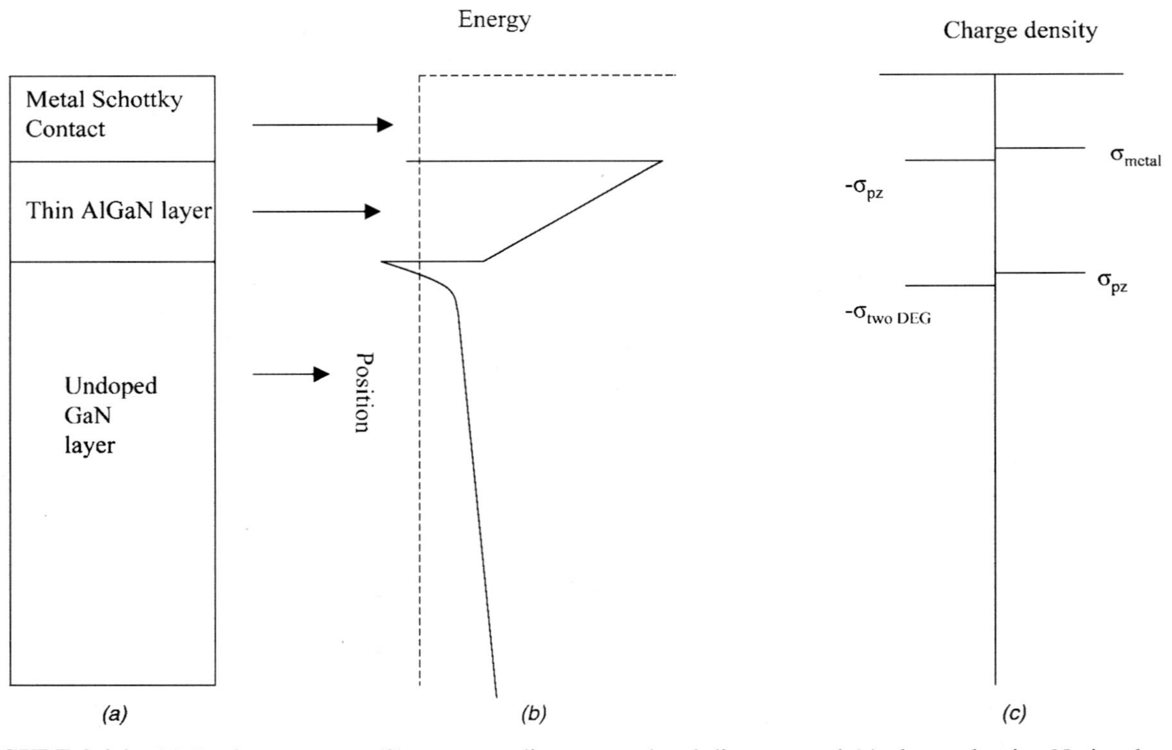

FIGURE 2.4.4 (*a*) Device structure, (*b*) corresponding energy band diagram, and (*c*) charge density. Notice that there is a compensating negative and positive charge density at both interfaces, metal–AlGaN layer and AlGaN–GaN layer.

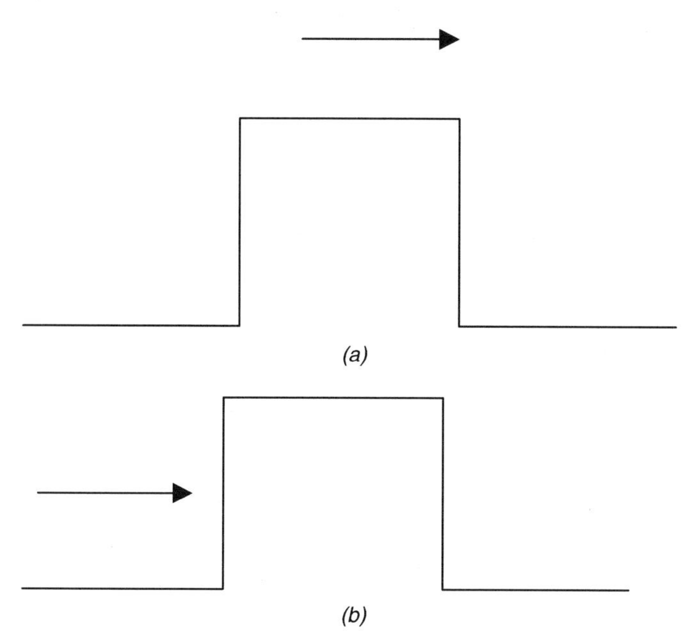

FIGURE 2.5.1 Potential diagrams for two separate cases: (*a*) carrier energy greater than the potential barrier height; (*b*) carrier energy less than the barrier height.

2.5 PERPENDICULAR TRANSPORT IN HETEROSTRUCTURES AND SUPERLATTICES

We next consider the physics of transport perpendicular to the heterojunction. Two conditions are of interest: when the carrier energy is greater than or less than the conduction band minimum within the barrier region formed by the wide-gap semiconductor layer. These two conditions are shown in Figure 2.5.1. Classically, the situation is quite simple. When the electron has energy greater than the potential barrier, as shown in Figure 2.5.1*a*, it simply moves from one region into the other with no chance of reflection. However, if the electron energy is less than the potential barrier in the second region, as in Figure 2.5.1*b*, the electron is simply reflected from the barrier back into the incident layer. The picture is far more complicated and interesting, however, when treated quantum mechanically. Before we discuss the quantum mechanical treatment, let us first consider the classical case when the carrier energy exceeds the potential barrier. For simplicity, we restrict our discussion to electrons.

The case when the electron energy exceeds the conduction band edge in the potential barrier corresponds to the situation encountered in a heterostructure bipolar transistor (HBT). In an n-p-n HBT, electrons are injected from a wide-bandgap emitter into a narrower-gap base region. Therefore, the electron

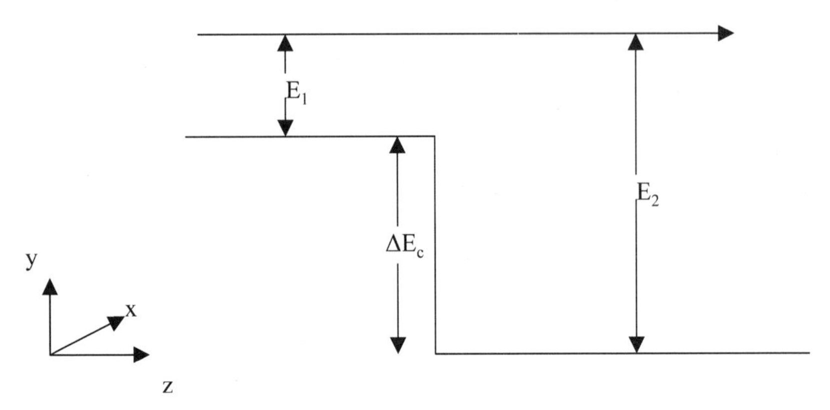

FIGURE 2.5.2 Heterostructure system for the above-the-barrier injection condition.

gains kinetic energy from the potential energy difference between the emitter and the base in a manner similar to that shown in Figure 2.5.1*a*. The simplest formulation of the transport physics in this case is to treat the electron classically, and as such, it suffers no reflection at the heterointerface. With this assumption, it is relatively straightforward to calculate the final electron state after crossing the interface.

The electron energy is, of course, conserved upon crossing the heterobarrier. The electron transmission at a potential energy step can be assessed quantitatively as follows. The system of interest is sketched in Figure 2.5.2. Assuming parabolic energy bands, the conservation of energy requirement demands that

$$\frac{\hbar^2 k_2^2}{2m_2} = \frac{\hbar^2 k_1^2}{2m_1} + \Delta E_c \qquad 2.5.1$$

where k_2 and k_1 are the magnitudes of the k-vectors, and m_2 and m_1 are the effective masses in the narrow- and wide-gap semiconductor layers, respectively. If the z direction is assumed to be the direction perpendicular to the heterostructure, conservation of linear momentum implies that

$$k_{1x} = k_{2x} \qquad k_{1y} = k_{2y} \qquad 2.5.2$$

For simplicity, let us assume that the electron is confined completely to the x–z plane. The condition on y can then be neglected. Using Eqs. 2.5.2 and 2.5.1 yields

$$\frac{\hbar^2}{2m_1}(k_{1x}^2 + k_{1z}^2) + \Delta E_c = \frac{\hbar^2}{2m_2}(k_{1x}^2 + k_{2z}^2) \qquad 2.5.3$$

If it is assumed that the incident condition is known, both k_{1x} and k_{1z} are known. The only unknown in Eq. 2.5.3 then is k_{2z}. Solving for k_{2z} yields

$$k_{2z}^2 = \left[\frac{m_2}{m_1}(k_{1z}^2 + k_{1x}^2) - k_{1x}^2\right] + \frac{2m_2 \Delta E_c}{\hbar^2} \qquad 2.5.4$$

Hence, the final state of the electron in the second (narrow-gap) semiconductor layer is given as

$$\begin{aligned} E_2 &= E_1 + \Delta E_c \\ k_{2x} &= k_{1x} \\ k_{2z} &= \sqrt{\left[\frac{m_2}{m_1}(k_{1z}^2 + k_{1x}^2) - k_{1x}^2\right] + \frac{2m_2 \Delta E_c}{\hbar^2}} \end{aligned} \qquad 2.5.5$$

The formulation above applies to the case when the electron can be treated as a classical particle. The classical picture is most appropriate when the carrier traverses a single potential barrier and if reflective loss at the barrier is safe to neglect. As mentioned above, typically this is the case in an HBT structure. We discuss HBTs in detail in Chapter 4.

Generally, an electron must be considered as a quantum mechanical entity. Hence, at a heterostructure interface there is some probability that the electron will be reflected as well as transmitted, even when its energy exceeds the potential barrier height. Whether the electron is incident from the wide-gap semiconductor into the narrow-gap semiconductor, or vice versa, there is a nonzero probability that it will undergo reflection at the interface. Although the reflection coefficient is typically very small, in some heterostructure systems with very small dimensions, the reflection coefficient can become quite appreciable. The reflection and transmission coefficients for a one-dimensional potential barrier are derived in detail in Brennan (1999, Sec. 2.2) for the special case of constant mass, when the electron energy exceeds the potential barrier height. Following the same procedure as in Brennan (1999), when the mass is different between the two semiconductor layers, the transmission and reflection coefficients become

$$T = \frac{4(k_{2z}/k_{1z})(m_1/m_2)}{[1 + (k_{2z}/k_{1z})(m_1/m_2)]^2} \qquad 2.5.6$$

$$R = \frac{[1 - (k_{2z}/k_{1z})(m_1/m_2)]^2}{[1 + (k_{2z}/k_{1z})(m_1/m_2)]^2} \qquad 2.5.7$$

T and R in Eqs. 2.5.6 and 2.5.7 correspond to the transmission and reflection coefficients for particle flux at the heterointerface, respectively.

It is also relatively easy to consider the situation where the electron is incident from the narrow-gap layer into the wide-gap layer such that its energy is less than the potential barrier height formed by the wide-gap semiconductor material. Classically, the electron would, of course, be reflected at the interface and there would be no chance of it emerging on the other side of the potential barrier. However, quantum mechanically, if the potential barrier is finite in both width and height, there is a nonzero probability of the electron emerging on the other side of the barrier even when its energy is less than the barrier height. This is called *tunneling*. The mathematical details of one-dimensional tunneling through a potential barrier are presented in Brennan (1999, Sec. 2.4). The salient features of the problem can be summarized as follows. When the carrier energy is less than the potential barrier height, the transmission and reflection coefficients are given as

$$T = \left[1 + \frac{V_0^2 \sinh^2 k_2 b}{4E(V_0 - E)}\right]^{-1} \qquad R = \left[1 + \frac{4E(V_0 - E)}{V_0^2 \sinh^2 k_2 b}\right]^{-1} \tag{2.5.8}$$

Notice that, as expected, there is a nonzero probability of the electron being transmitted through the barrier when its energy is less than the barrier height. An interesting condition occurs when two barriers are used to enclose a small potential well. At certain energies, the electron wave "resonates" with the well and the transmission coefficient approaches unity. This effect is called *resonant tunneling*. In Chapter 6 we discuss this effect and devices that utilize resonant tunneling, resonant tunneling diodes and transistors, and discuss their application in high-frequency oscillators and digital logic circuits.

The results above can be extended to the case of a repeated periodic potential. Two different situations can again be considered. These are the cases when the carrier has energy greater than the potential barrier height, and when it has energy less than the potential barrier height. In either situation, the Schrödinger equation is solved throughout the structure subject to the boundary conditions at each interface. The solution for the transmission coefficient of the entire system is then obtained using a transfer matrix approach. The full details of the technique for the special case of equal masses and transport for energies less than the potential barrier height are discussed in Brennan (1999, Sec. 2.5). We discuss this result in Chapter 6 and defer further discussion of this case until then. Here, we consider the case when the electron energies exceed the potential barrier in a multiquantum-well system.

The steady-state Schrödinger equation for a heterostructure assuming that the energy everywhere exceeds the potential barrier height can readily be solved for the incident, transmitted, and reflected electron wavefunctions as

$$\begin{aligned} \psi_i &= A e^{i(k_{ix}x + k_{iy}y)} \\ \psi_t &= B e^{i(k_{tx}x + k_{ty}y)} \\ \psi_r &= C e^{i(k_{rx}x + k_{ry}y)} \end{aligned} \tag{2.5.9}$$

respectively, where A, B, and C are the amplitude constants and x is the direction of propagation. Note that $k_{rx} < 0$ implies that the reflected wave moves in the negative x direction of propagation. The continuity of the wavefunction across the heterointerface implies that

$$\psi_i(0, y) + \psi_r(0, y) = \psi_t(0, y) \tag{2.5.10}$$

Since the y components of each k vector are equal (i.e., $k_{iy} = k_{ty} = k_{ry}$), Eq. 2.5.10 implies that

$$A + C = B \tag{2.5.11}$$

or, equivalently,

$$1 + r = t \tag{2.5.12}$$

where $r = C/A$ is the electron amplitude reflectivity and $t = B/A$ is the electron amplitude transmissivity. The probability current density normal to the interface, j_x, is conserved, which yields

$$j_{ix} + j_{rx} = j_{tx} \tag{2.5.13}$$

Using the result for the quantum mechanical probability current density given by Brennan (1999, Sec. 1.5), j_{ix}, j_{tx}, and j_{rx} are given as

$$\begin{aligned} j_{ix} &= \frac{\hbar}{m_1} AA^* k_{ix} \\ j_{tx} &= \frac{\hbar}{m_2} BB^* k_{tx} \\ j_{rx} &= \frac{\hbar}{m_1} CC^* k_{rx} \end{aligned} \tag{2.5.14}$$

Substituting Eqs. 2.5.14 into Eq. 2.5.13 and recognizing that $k_{rx} = -k_{ix}$ yields

$$\frac{m_1}{k_{ix}} \frac{k_{tx}}{m_2} t^2 + r^2 = 1 \tag{2.5.15}$$

Combining Eqs. 2.5.12 and 2.5.15, the electron amplitude transmissivity becomes

$$t = \frac{2(m_2/k_{tx})}{(m_2/k_{tx}) + (m_1/k_{ix})} \tag{2.5.16}$$

and the electron reflectivity is

$$r = \frac{(m_2/k_{tx}) - (m_1/k_{ix})}{(m_2/k_{tx}) + (m_1/k_{ix})} \tag{2.5.17}$$

Finally, recognizing that $k_{ix} = k_i \cos\theta_i$ and $k_{tx} = k_t \cos\theta_t$, Eqs. 2.5.16 and 2.5.17 give for the transmission and reflection amplitudes,

$$t = \frac{2\sqrt{E/m_1}\cos\theta_i}{\sqrt{E/m_1}\cos\theta_i + \sqrt{(E-V)/m_2}\cos\theta_t} \qquad 2.5.18$$

$$r = \frac{\sqrt{E/m_1}\cos\theta_i - \sqrt{(E-V)/m_2}\cos\theta_t}{\sqrt{E/m_1}\cos\theta_i + \sqrt{(E-V)/m_2}\cos\theta_t} \qquad 2.5.19$$

Using the results above it is now relatively straightforward to determine the transmission coefficient for a multiquantum-well stack under the condition that the incident electron energy is greater than the potential barrier height. Consider a system of M layers comprising a multiquantum-well structure. We assume that an electron is incident onto the $m = 0$ layer at an angle θ_0 with respect to the normal direction and that the direction of propagation is the $+x$ direction. The solution is obtained by solving the Schrödinger equation in each layer and matching the appropriate boundary conditions at the interface. The procedure is similar to that discussed in detail by Brennan (1999, Sec. 2.5) and is not repeated here. Additionally, the similarities of electron wave propagation to that of optics enables using optical techniques for calculating electron wave propagation in a multiquantum-well system. The full details of this technique are given by Gaylord and Brennan (1989). The general expression for the probability amplitudes for the m layer in terms of the $(m + 1)$th layer is then given as

$$\begin{bmatrix} \psi_{i,m} \\ \psi_{r,m} \end{bmatrix} = \frac{1}{t_m}\begin{bmatrix} 1 & r_m \\ r_m & 1 \end{bmatrix}\begin{bmatrix} e^{ik_m d_m \cos\theta_m} & 0 \\ 0 & e^{-ik_m d_m \cos\theta_m} \end{bmatrix}\begin{bmatrix} \psi_{i,m+1} \\ \psi_{r,m+1} \end{bmatrix} \qquad 2.5.20$$

where t_m and r_m are the amplitude transmissivity and reflectivity at the $m - 1$ and m interfaces as described by Eqs. 2.5.18 and 2.5.19, k_m is the magnitude of the electron wavevector in layer m, d_m is the thickness of layer m, and θ_m is the angle of the wavevector direction in layer m. For the entire stack of M layers, the total normalized transmitted electron wave amplitude, $\Psi_{t,M+1}$ (the final region is $M + 1$ after the Mth stack), and the total normalized reflected electron wave amplitude, $\Psi_{r,0}$ (in region 0), are obtained through chain multiplication of the total $M + 1$ versions of Eq. 2.5.20. The resultant expression is

$$\begin{bmatrix} 1 \\ \psi_{r,0} \end{bmatrix} = \prod_{m=1}^{M} \frac{1}{t_m}\begin{bmatrix} 1 & r_m \\ r_m & 1 \end{bmatrix}\begin{bmatrix} e^{ik_m d_m \cos\theta_m} & 0 \\ 0 & e^{-ik_m d_m \cos\theta_m} \end{bmatrix}$$

$$\times \frac{1}{t_{M+1}}\begin{bmatrix} 1 & r_{M+1} \\ r_{M+1} & 1 \end{bmatrix}\begin{bmatrix} \psi_{t,M+1} \\ 0 \end{bmatrix} \qquad 2.5.21$$

which can be solved directly for amplitude transmissivity and reflectivity.

EXAMPLE 2.5.1: Calculation of the Transmitted Angle at an Interface

Consider an electron of wavevector k_1, with effective mass m_1, incident from region I into region II, where it has an effective mass m_2, as shown in Figure 2.5.3. Assume that the motion of the electron is completely confined within the x–y plane and that the energy versus wavevector relationship is given as

$$E = \frac{\hbar^2 k^2}{2m}$$

(a) Determine the angle that the transmitted wave makes with the normal if the angle of incidence is θ_i. Let the potential in region I be zero and in region II be V.

(b) Determine a numerical value for the transmitted angle if the potential height is 0.2 eV, the incident energy $E_i = 0.3$ eV, $m_1 = 0.067$ m, $m_2 = 0.1087$ m, and the angle of incidence is 30°.

Let us consider the solution to part a first. From Eq. 2.5.5, the parallel k-vector components are conserved across the interface. Therefore, $k_{1y} = k_{2y}$. In addition, $k_{1y} = k_{ry}$. This implies that the angle of incidence is equal to the angle of reflection,

$$\theta_i = \theta_r$$

Resolving the incident and transmitted wavevectors into their components (see Figure 2.5.4) the y components can easily be related as

$$k_1 \sin\theta_i = k_2 \sin\theta_2 \qquad \sqrt{m_1 E}\sin\theta_i = \sqrt{m_2(E-V)}\sin\theta_2$$

which is a Snell's law for electrons. The quantities equivalent to the refractive indices, n_1 and n_2, are

$$n_1 = \sqrt{m_1 E} \qquad \text{and} \qquad n_2 = \sqrt{m_2(E-V)}$$

The angle transmitted, θ_2, is then

$$\theta_2 = \sin^{-1}\left(\frac{k_1}{k_2}\sin\theta_i\right) = \sin^{-1}\left\{\left[\frac{m_1 E}{m_2(E-V)}\right]^{1/2}\sin\theta_i\right\}$$

(*Continued*)

EXAMPLE 2.5.1 (*Continued*)

where we have recognized that k_2 and k_1 are given as

$$k_2 = \sqrt{\frac{2m_2(E-V)}{\hbar^2}}$$

$$k_1 = \sqrt{\frac{2m_1E}{\hbar^2}}$$

For part (b), we substitute in numerical values to obtain

$$k_1 = 7.26 \times 10^6 \text{ cm}^{-1} \qquad k_2 = 5.34 \times 10^6 \text{ cm}^{-1}$$

With an angle of incidence of 30°, the angle transmitted, θ_2, is calculated to be 42.8°.

Using Eq. 2.5.21, it is possible to calculate what the transmissivity of an electron at an arbitrary angle of incidence is on a multiquantum-well stack under the condition that the electron energy exceeds the potential barrier height. One can design multilayered heterostructure stacks, often referred to as *superlattices*, such that the transmissivity can be maximimized for certain electron energies. Alternatively, superlattice filter designs can be made that only pass electrons with certain specific incident energies, rejecting all others. These superlattice filters have direct analogy to thin-film optical interference filters. Therefore, many different superlattice structures can be made that mimic optical thin-film interference filters and perform for electrons functions similar to those performed for light (Gaylord and Brennan, 1988). An important application of these types of devices is a multiquantum-well barrier or superlattice that is designed to reflect incident electrons. Through a judicious choice of layer widths and barrier heights, following the design methodology discussed above, a multiquantum-well stack can be designed such that the electron reflectivity is high for a range of energies near the potential barrier height. The multiquantum well barrier (MQB) acts effectively to increase the potential barrier height, thus enhancing electron confinement in the region bounded by the barrier. Consequently, using an MQB, the electron leakage out of a region, such as the active region of a laser or light-emitting diode, can be greatly suppressed, enhancing the carrier confinement and the efficiency of the device.

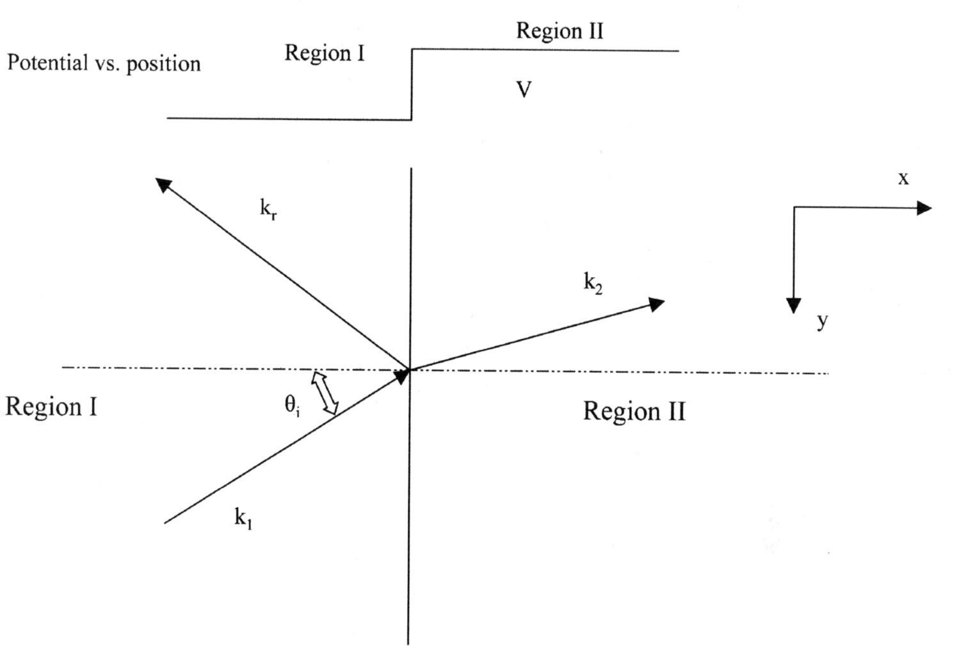

FIGURE 2.5.3 Incident, k_1, reflected, k_r, and transmitted, k_2, wavevectors. The potential in region II is equal to V, and the potential in region I is zero.

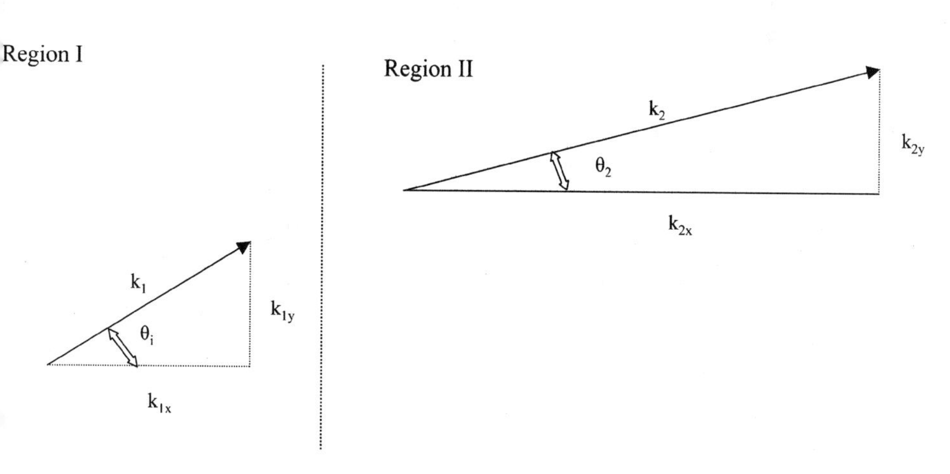

FIGURE 2.5.4 Phase-matching conditions. Notice that $k_{1_y} = k_{2y}$.

2.6 HETEROJUNCTION MATERIALS SYSTEMS: INTRINSIC AND EXTRINSIC PROPERTIES

Heterojunctions are generally formed from materials that can be grown upon each other epitaxially with low defect densities. Useful heterojunction systems are therefore comprised of materials that are relatively closely lattice matched. Figure 2.6.1 shows bandgaps versus lattice constants of common (*a*) cubic III–V and Si-based materials, and (*b* and *c*) GaN and related materials. The AlGaAs–GaAs system possesses only a small lattice mismatch over the entire composition range from GaAs to AlAs and thus was one of the first heterojunction systems to be developed and exploited in device structures.

Heterojunctions may be comprised of *pseudomorphic* layers in which one of the layers is lattice mismatched and coherent, or elastically strained. As mentioned in Section 2.4, for a given strain, determined by the lattice mismatch, a maximum thickness exists, called the *critical thickness* h_c, for the strained layer to be completely coherent with the substrate. The critical thickness has a minimum under conditions of thermodynamic equilibrium, but can be enhanced by the epitaxial process. This is discussed in more detail below. For thicknesses greater than the critical thickness, dislocations are created to reduce the energy of the system. These dislocations can significantly degrade device performance and reliability.

The velocity–field characteristics for relevant heterojunction materials are shown in Figure 2.6.2. These characteristics determine the usefulness of a material in a given application, for example as the channel for a HEMT or as a collector in an HBT. Generally speaking, the higher the frequency, or speed, required for an application, the higher the desired velocity. Thus the InGaAs alloys have shown excellent performance for millimeter-wave applications. A trade-off, however, in materials properties often exists. For example, the mobility of carriers in semiconductors tends to be greater for small-bandgap materials, as shown in Figure 2.6.3. However, the critical electric field for breakdown tends to be greater for wider-bandgap materials. Thus, numerous material parameters must be considered for an application. A high velocity of electrons is desirable for the collector of an HBT, but the collector must also be able to support a high electric field resulting from a large output voltage. Thus InP is often used for an HBT collector for microwave or millimeter-wave power amplifier applications.

The properties of heterojunctions discussed thus far in this chapter have not been related to the specific materials systems used for advanced devices. In Appendix C we list the heterojunction type (I or II, for example, as discussed in Section 2.1) and the heterojunction conduction and valence band offsets for a range of materials systems. As discussed in Chapters 3 and 4, the values of these heterojunction offsets are extremely important in determining the usefulness of a given materials system for a specific device type. HEMTs, for example, benefit from a large conduction band discontinuity, leading to a high 2DEG concentration from modulation doping, as shown in Section 2.2—

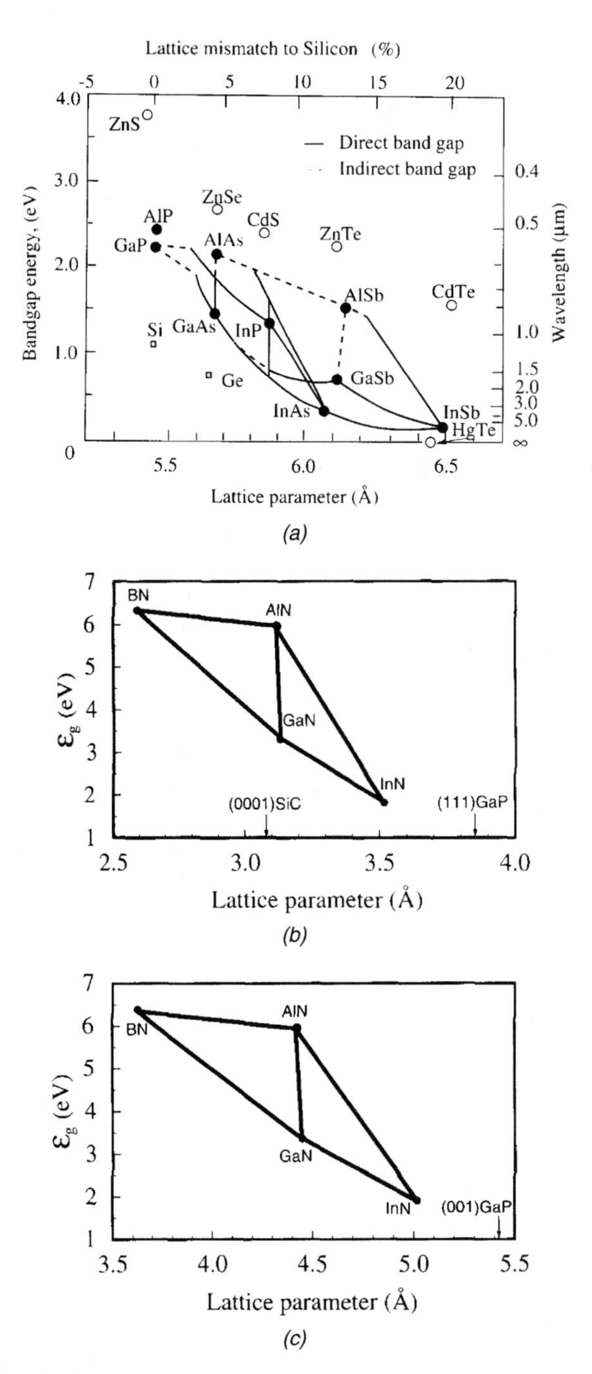

FIGURE 2.6.1 Lattice constant versus bandgap for common semiconductor alloys: (*a*) cubic III–V and Si-based alloys; (*b*) GaN and related materials (wurtzite); (*c*) GaN and related materials (cubic). (Reprinted with permission from Bhattacharya, 1997.)

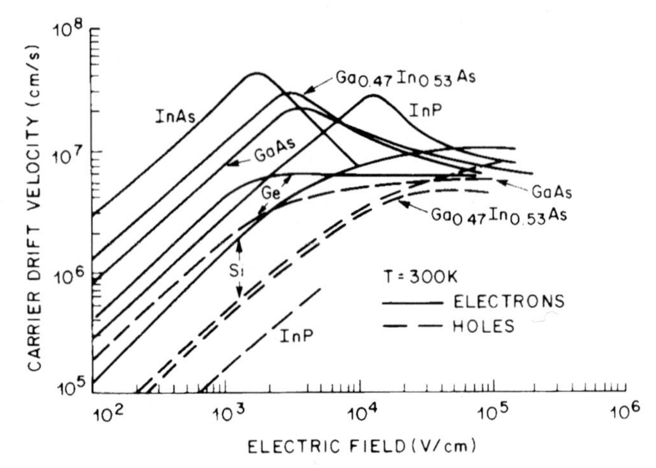

FIGURE 2.6.2 Velocity–field characteristics of common semiconductors. (Reprinted with permission from Bean, 1990.)

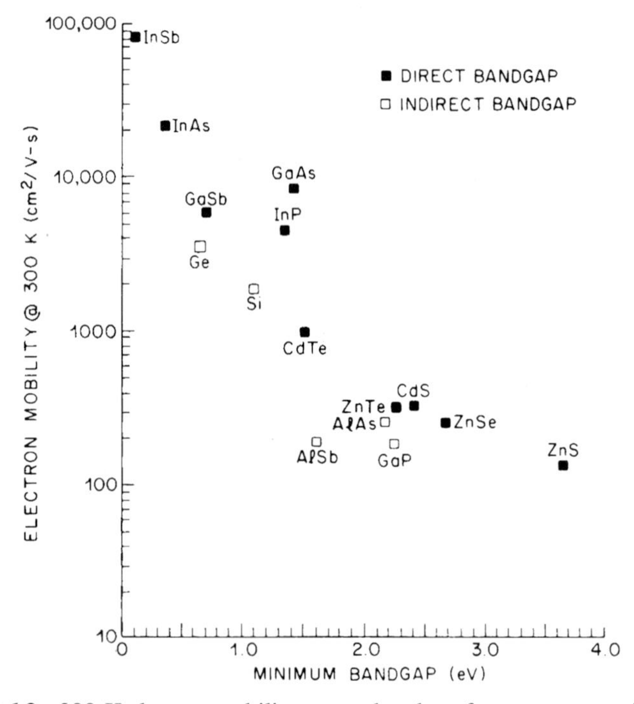

FIGURE 2.6.3 300-K electron mobility versus bandgap for common semiconductors. (Reprinted with permission from Bean, 1990.)

specifically, Eq. 2.2.9. HBTs, on the other hand, benefit from a large valence band discontinuity. This is discussed in detail in Chapter 4.

The materials properties discussed above represent the *intrinsic* properties of individual semiconductor alloys and their heterojunctions. Materials also possess *extrinsic* properties that result from the specific process used for their formation and subsequent processing for devices and circuits. The common production and research epitaxial techniques, molecular beam epitaxy (MBE) and metal-organic vapor-phase epitaxy (MOCVD), yield similar extrinsic properties. It is important to develop a general understanding of some of these properties to design effective devices based on heterojunctions. The subsequent device process steps are also relevant but are beyond the scope of this book.

The designer will choose materials based on the intrinsic properties of the materials and the device requirements. In addition to these considerations, other factors should be considered, such as the interface perfection, the lateral and vertical compositional uniformity of any alloy material used in the structure, the abruptness of the doping layer and control of the dopant activity, and the strain in the material.

To consider these issues, a brief introduction to the two most common epitaxial techniques is given below. [More detailed descriptions are given in Liu (1999).] MBE will be used as the focus of discussions relating extrinsic properties to process conditions.

Molecular Beam Epitaxy (MBE). A diagram of an MBE growth chamber is shown in Figure 2.6.4. MBE occurs through deposition of the desired alloy constituents in the form of molecular beams onto a heated crystal substrate. The substrate is produced by bulk growth techniques, sawed into wafers, and then polished to yield a highly perfect surface. MBE growth occurs in an ultrahigh-vacuum (UHV) environment with a base pressure of approximately 10^{-10} T. The UHV nature of the growth environment minimizes the concentration of impurities that can be incorporated into the growing film, and maintains the beam nature of the impinging fluxes of constituent materials onto the substrate. The beam fluxes are low enough to yield commonly used growth rates of 0.5 to 2.0 μm/h. Typically, solid sources of the constituent elements are used and held in furnaces covered by mechanical shutters to block the beams, or fluxes, of the constituent atoms or molecules that impinge on the substrate. The slow growth rate and mechanical blocking of the beams lead to easy control of deposited materials thicknesses to an atomic level.

The processes governing the interactions of the impinging atoms and molecules and the semiconductor surface leading to epitaxial growth are fairly complex and are not discussed here in detail (see the references). However, a simplified view will aid in understanding certain limitations relevant to devices. Let's take the growth of GaAs as an example. The group V element As and the group III element Ga are produced by the heating of solid sources of Ga and As within separate *effusion* cells. Upon commencement of growth,

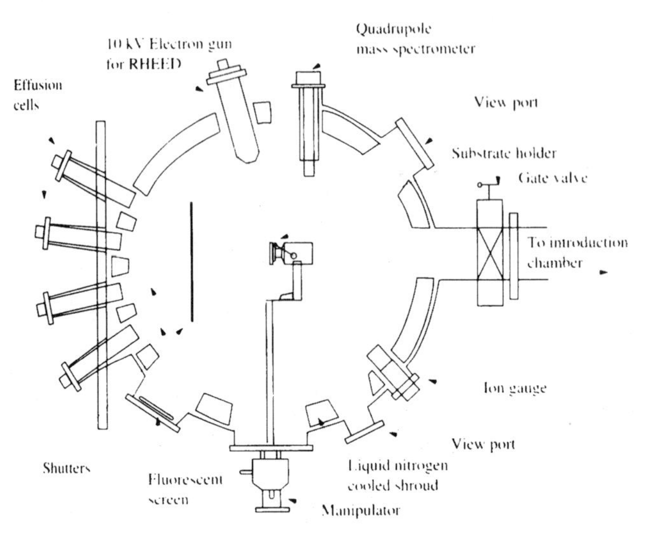

FIGURE 2.6.4 Diagram of MBE growth chamber. (Reprinted with permission from Bhattacharya, 1997.)

the shutters blocking the beams are opened, and Ga atoms and As_4 molecules impinge on a GaAs substrate heated to approximately 600°C. Generally, all the Ga atoms incident onto the film "stick" to the surface and are incorporated into the growing film, and only those As atoms required to bond to Ga are incorporated. The As from the As_4 is incorporated through a fairly complex process of breaking apart into As_2 and then finding the proper configuration of Ga atoms on the surface with which to bond. Excess As is desorbed, primarily as As_2. Thus, two important facts develop from this picture: An excess flux of As should be maintained during growth, and the growth rate is determined solely by the flux of group III, in this case, Ga atoms.

Metal-Organic Chemical Vapor Deposition (MOCVD). MOCVD is also commonly used for the growth of heterojunction devices. The MOCVD process is much more complex than MBE. MOCVD involves the chemical reactions of various gas sources or molecules in carrier gases to produce the constituents. A basic reaction is that between a metal-organic compound such as trimethylgallium [$(CH_3)_3Ga$] and a hydride such as arsine (AsH_3). GaAs is formed during this reaction over a heated substrate, releasing methane:

$$(CH_3)_3Ga + AsH_3GaAs + 3CH_4 \qquad 2.6.1$$

Mass flow controllers are used to regulate the gases entering the chamber and therefore control the growth rate. The metal-organic sources are typically liquids held near room temperature in stainless steel bubblers in tempera-

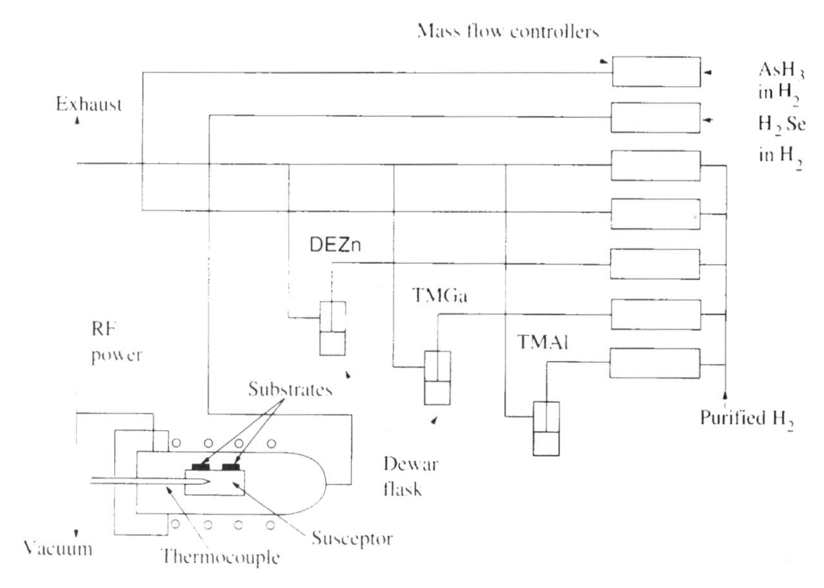

FIGURE 2.6.5 Schematic diagram of MOCVD process. (Reprinted with permission from Bhattacharya, 1997.)

ture baths to control the temperature and therefore the vapor pressures of the sources. A diagram of the process is shown in Figure 2.6.5. Hydrogen gas is used to carry the vapors into the growth chamber. The reaction in Eq. 2.6.1 actually represents the end result of several subprocesses leading the crystal growth. These processes occur in a gas flow environment over the substrate surface, and therefore the flow rates and reactor geometry are critical to achieving uniform deposition.

MOCVD growth rates are comparable to MBE, on the order of 3 μm/h instead of 1 μm/h. Generally, MBE- and MOCVD-produced devices yield similar performance characteristics, with specific advantages and disadvantages for both.

Growth Modes and Heterojunction Smoothness. Various models exist for epitaxial processes that address the perfection of materials at both the atomistic scale and the mesoscopic scale. A brief discussion of the two models is given below. They are chosen to be illustrative of the limits to achieving heterojunction perfection and provide insight to processes that may be exploited increasingly for advanced quantum effect devices such as those discussed in Chapters 6 and 8.

On the mesoscopic scale, three growth modes have been categorized and are shown in Figure 2.6.6 as a function of deposited layer thickness θ in monolayers (MLs) or the thickness of a bilayer (Ga plus As) of the material deposited (Venables, 2000).

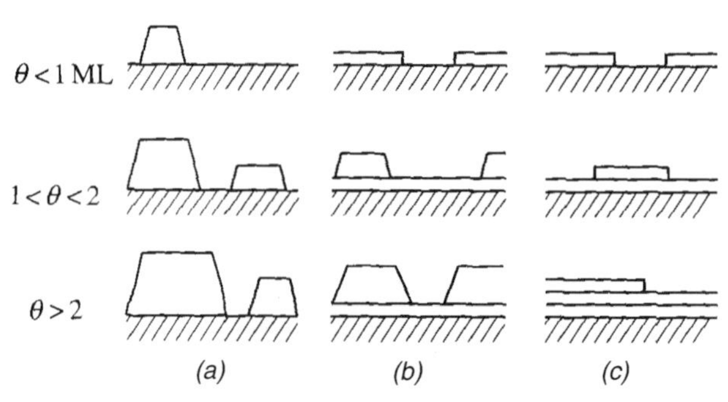

FIGURE 2.6.6 Schematic model of three growth modes as a function of layer coverage in ML (θ): (*a*) Volmer–Webb; (*b*) Stranski–Krastanov; (*c*) Frank–van der Merwe. (Reprinted with permission from Venables, 2000.)

1. The Frank–van der Merwe mode, in which the atoms of the deposited material are more strongly bound to the substrate than to each other, thus yielding an atomically smooth two-dimensional growth mode.
2. The Volmer–Weber mode, in which the atoms of the deposited constituents are more strongly bound to each other than the substrate, thus forming three-dimensional islands.
3. The Stranski–Krastinov mode, intermediate to the two described above, in which initial layer-by-layer growth is followed after a few monolayers by island growth.

For the production of ideal interfaces, clearly the Frank–van der Merwe mode is desirable. An ideal heterojunction is formed by the completion of an atomically smooth two-dimensional layer of one material, GaAs for example, and then followed by the two-dimensional deposition of another material, such as AlGaAs. Figure 2.6.7 shows a high-resolution transmission electron microscope (TEM) image of an atomically abrupt GaAs–GaInP heterojunction interface formed by MBE. Such interfaces are desirable for device applications.

Examination of the two-dimensional layer-by-layer growth mode in more detail leads to a better understanding of device limitations. How are atoms added to the two-dimensional layer during the growth process?

The Burton, Cabrera, and Frank (BCF) theory is a classic description of this process. Figure 2.6.8 shows a diagram of a Kossel crystal surface with common microscopic features such as terraces, ledges, kinks, steps, and adatoms (Tsao, 1993). In the BCF theory, two-dimensional growth occurs through the lateral movement of steps—step flow growth—across the surface, due to the incorporation of adsorbed atoms (adatoms on the crystal surface) or molecules at these edges. Surface processes, such as desorption or diffusion. are governed by Arrhenius-type temperature-activated processes. The ability to maintain the two-dimensional nature of the surface rests, in part, on the ability of incoming atoms to diffuse to such appropriate sites rather than nucleating new

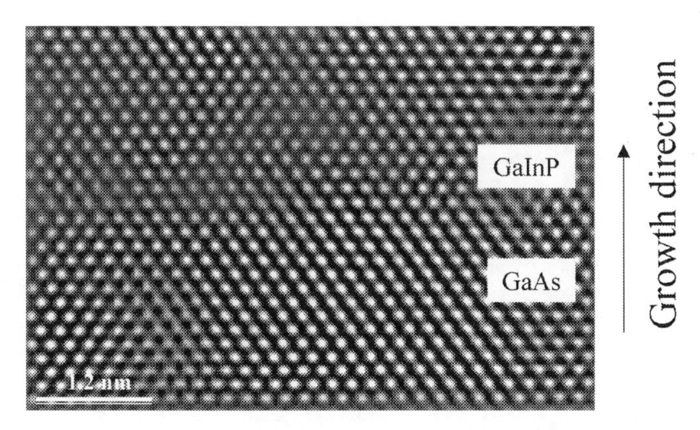

FIGURE 2.6.7 High-resolution TEM image of GaInP–GaAs heterojunction.

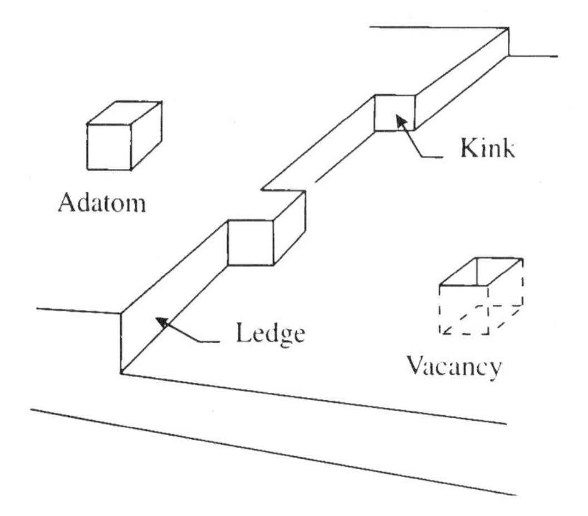

FIGURE 2.6.8 Perspective drawing of a Kossel crystal, showing terraces, steps, kinks, adatoms, and vacancies. (Reprinted with permission from Venables, 2000.)

clusters or islands on terraces, leading to a high concentration of islands that must coalesce presumably with imperfect boundaries. Thus relatively large adatom diffusion lengths are desirable, and therefore growth conditions are often chosen to enhance adatom diffusion. These growth conditions include higher substrate temperature and lower group V flux. The substrate temperature provides thermal energy for the motion of adatoms on the surface. A high group V flux may inhibit diffusion due to a higher probability of bonding at a higher surface As coverage.

Steps may be observed either in plan view using atomic force microscopy (AFM) or in cross section using TEM. Figure 2.6.9 shows AFM images of

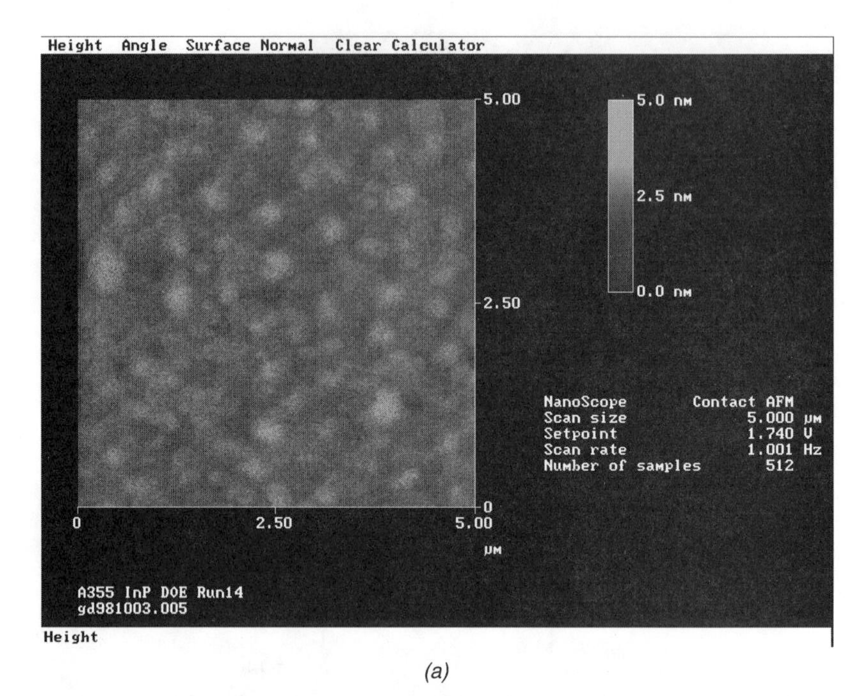

(a)

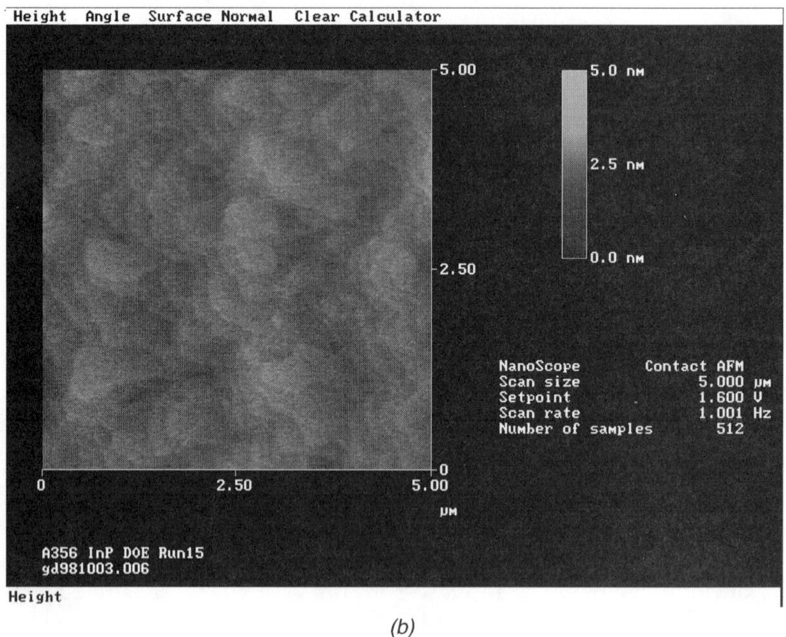

(b)

FIGURE 2.6.9 AFM images of InP grown at (*a*) a temperature of 420°C, exhibiting three-dimensional island growth; and (*b*) a temperature of 500°C exhibiting step flow growth. (From Dagnall, 2000.)

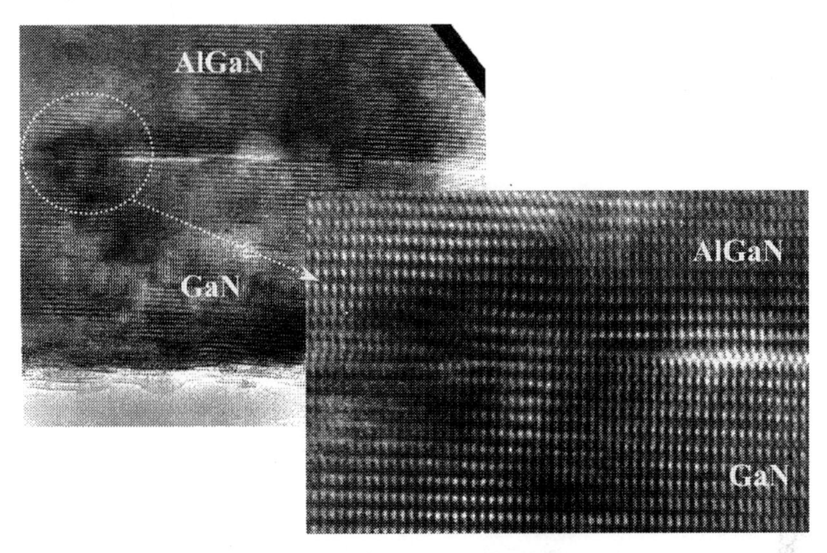

FIGURE 2.6.10 High-resolution TEM image of AlGaN–GaN heterojunction showing a step at the interface. (Reprinted with permission from Kang et al., 2000.)

InP grown on InP exhibiting (*a*) island growth at lower substrate temperature (420°C) and (*b*) step flow growth at higher temperature (500°C). Figure 2.6.10 shows a cross-sectional high-resolution image in an AlGaN–GaN heterostructure. A step can be observed at the heterojunction interface. Smooth step edges are produced by diffusion along a step to a kink site, whereas rough steps may result from atom attachment at the step with little diffusion along the step. Growth of alloys, such as AlGaAs or InAsP, is more complex. In the former case, one must depend on optimizing substrate temperature and fluxes to control the diffusion of two cations, or group III atoms. This is more difficult than for the simple binary compound. Generally, these alloys are atomically rougher, and one can expect that the "inverted" heterojunction interface, that in which AlGaAs is beneath the GaAs, will be rougher than the "normal" interface, that in which the GaAs is beneath the AlGaAs. Even though these ternary alloys exhibit "rougher" atomic structure, they are relatively easy to grow by MBE since the composition, x, is determined by the ratios of the group III fluxes and independent of temperature or group V overpressure. In the latter case, InAsP, the control of the composition is very difficult since the incorporation of the group V elements depends sensitively on the substrate temperature. Figure 2.6.11 shows a cross-sectional scanning tunneling microscope image of an InAs–GaInSb multilayer structure. In part (*b*) the interface profiles are extracted in the two perpendicular (110) and $(1\bar{1}0)$ directions. An edge detection algorithm is used to extract the profile from the STM data. A Fourier analysis of the profiles can be used to quantify the roughness. For the images presented in Figure 2.6.11, in the $(1\bar{1}0)$ direction

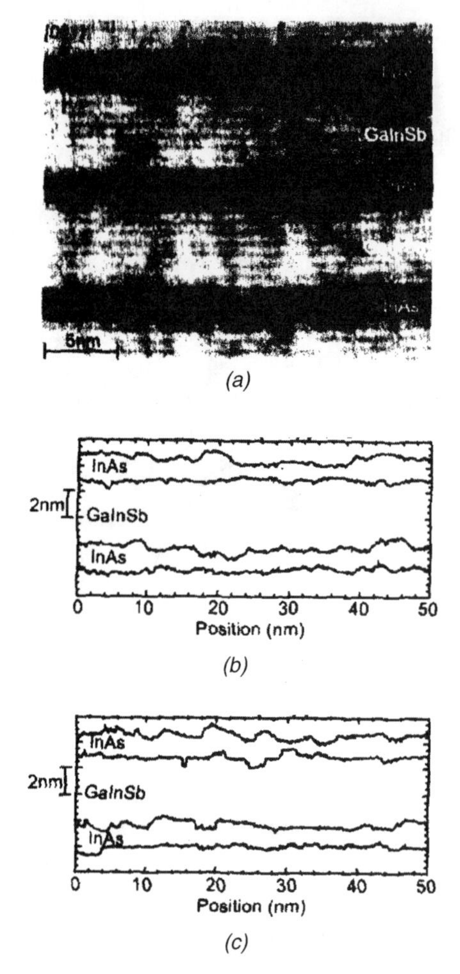

FIGURE 2.6.11 (*a*) STM image of InAsP–InP interface; (*b*) interfaces in (110) plane; (*c*) interfaces in $(1\bar{1}0)$ plane. (Reprinted with permission from Lew et al., 1998.)

the InAs–on–GaInSb interface exhibits a roughness amplitude of 2.8 ± 0.2 A with a correlation length of 174 ± 21 A (Lew et al., 1998). This quantification process can be used to assess the effects on transport, as discussed in Section 2.3.

Pseudomorphic Growth and the Critical Thickness. As mentioned at the beginning of this section, lattice-mismatched, coherent (or elastically strained or *pseudomorphic*) layers are limited in thickness to a critical thickness h_c. Beyond the critical thickness, the strain is relieved at least in part by the introduction and movement of dislocations in the material structure. The specific type of dislocation is dependent on the material system.

A number of theories have developed to specify h_c quantitatively. Following Tsao (1993), we develop expressions for the critical thickness as determined for mechanical equilibrium, as introduced by Matthews and Blakeslee (1974). A discussion of the limits of this model is provided at the end of this section.

Hooke's law, described in Eq. 2.4.14, can be written for cubic materials in terms of the Poisson's ratio ν and the shear modulus μ. ν is the negative of the ratio between lateral and longitudinal strains under a uniaxial longitudinal stress, and μ is the ratio between the applied shear stress and shear strain under pure shear. We can therefore write

$$\begin{bmatrix} \varepsilon_x \\ \varepsilon_y \\ \varepsilon_z \end{bmatrix} = \frac{1}{2\mu(1+\nu)} \begin{bmatrix} 1 & -\nu & -\nu \\ -\nu & 1 & -\nu \\ -\nu & -\nu & 1 \end{bmatrix} \begin{bmatrix} \sigma_x \\ \sigma_y \\ \sigma_z \end{bmatrix} \qquad 2.6.2$$

and

$$C_{11} = 2\mu\frac{1-\nu}{1-2\nu} \qquad C_{12} = 2\mu\frac{\nu}{1-2\nu} \qquad 2.6.3$$

We can simplify Eq. 2.6.2 by considering a cubic semiconductor oriented in a $\langle 100 \rangle$ direction, assuming the in-plane strains to be symmetric and along the x and y axes. We describe these as the in-plane, $\varepsilon_{\parallel}$, or *parallel strains*, and the out-of-plane strain, $\varepsilon_{\perp}$ or *perpendicular strains*.

$$\begin{bmatrix} \varepsilon_{\parallel} \\ \varepsilon_{\perp} \end{bmatrix} = \frac{1}{2\mu(1+\nu)} \begin{bmatrix} 1-\nu & -\nu \\ -2\nu & 1 \end{bmatrix} \begin{bmatrix} \sigma_{\parallel} \\ \sigma_{\perp} \end{bmatrix} \qquad 2.6.4$$

The parallel strain is simply determined by the lattice mismatch between two materials (also denoted as f, or the lattice parameter misfit). The perpendicular stress vanishes since the film is free to expand vertically. Thus, the two unknown quantities can be determined in terms of the lattice mismatch, or $\varepsilon_{\parallel}$.

$$\sigma_{\parallel} = 2\mu\left(\frac{1+\nu}{1-\nu}\right)\varepsilon_{\parallel} \qquad 2.6.5$$

$$\varepsilon_{\perp} = \frac{-2\nu}{1-\nu}\varepsilon_{\parallel} \qquad 2.6.6$$

We now determine the energy per unit area associated with the strain:

$$u_{\text{coh}} = \frac{1}{2}h(2\sigma_{\parallel}\varepsilon_{\parallel} + \sigma_{\perp}\varepsilon_{\perp}) = 2\mu\frac{1+\nu}{1-\nu}h\varepsilon_{\parallel}^2 \qquad 2.6.7$$

where h is the thickness of the film. For multilayered structures with different lattice constants, we can calculate the entire coherency energy by summing over the contribution from each layer:

$$u_{\text{coh}} = 2\mu \frac{1+\nu}{1-\nu} \sum_i h_i \varepsilon_{i,\parallel}^2 \qquad 2.6.8$$

To determine the critical thickness, we must determine the strain energy associated with epitaxial films that are semicoherent, or containing *disregistry* in the lattice through the presence of dislocations. The disregistry can be described by the Burgers vector $\bar{b}$. Different types of dislocations—mixed, edge, and screw, for example—are described by their Burgers vectors. Only certain types (or components) of dislocations actually relieve strain. We describe the relevant component as $b_{\text{edg},\parallel}$. This component of the Burgers vector describing the dislocation is "edgelike" and in the plane of the interface. This is the only component of a general mixed-character dislocation that relieves strain. [See Tsao (1993) for more details.]

In calculating this energy we assume that the interface is composed of a crossed grid of two identical arrays of dislocations, having a linear density of ρ_{md}. We simply assume that the misfit taken up by dislocations, f_{dis}, is equal to $b_{\text{edg},\parallel}$, the lattice displacement parallel to the interface per dislocation, divided by the spacing between the dislocations $(1/\rho_{md})$:

$$f_{\text{dis}} = \rho_{md} b_{\text{edg},\parallel} \qquad 2.6.9$$

If we assume that the coherency strain in the epitaxial film decreases linearly as the dislocation density increases, then

$$\varepsilon_{\parallel} \approx f - f_{\text{dis}} = f - \rho_{md} b_{\text{edg},\parallel} \qquad 2.6.10$$

This simplification does not take into account dislocation interaction, which is important during relaxation. Inserting Eq. 2.6.10 into Eq. 2.6.7, we have

$$u_{\text{coh}} = 2\mu \frac{1+\nu}{1-\nu} h(f - \rho_{md} b_{\text{edg},\parallel})^2 \qquad 2.6.11$$

As the dislocation density increases, the total energy of the dislocation array increases. This is given by [see Tsao (1993, Sec. 5.1.2) for derivation]

$$u_{\text{dis}} \approx \rho_{md} \frac{\mu b^2}{4\pi} \frac{1-\nu \cos^2 \beta}{1-\nu} \ln \frac{4h}{b} \qquad 2.6.12$$

where h is the distance from the interface to the surface of the film (assumed to be the film thickness for our derivation) and β the angle between the Burgers vector and the dislocation line (the edge component is $b \sin \beta$). The total areal energy density is the sum of the coherency strain and the strain of both arrays of dislocations, or

$$
\begin{aligned}
u_{\text{tot}} &= u_{\text{coh}} + 2u_{\text{dis}} \\
&= 2\mu\left(\frac{1+\nu}{1-\nu}\right) h(f - \rho_{md} b_{\text{edg},\|})^2 \\
&\quad + \rho_{md}\frac{\mu b^2}{2\pi}\frac{1-\nu\cos^2\beta}{1-\nu}\ln\frac{4h}{b}
\end{aligned} \tag{2.6.13}
$$

Two regions associated with strain accommodation are revealed from Eq. 2.6.13. For thin, low-misfit epitaxial films, the total energy is minimized for a film with no dislocations. The energy of the dislocation array is greater than the coherency strain energy reduction. For thick, high-misfit films, the energy is minimized for $\rho_{md} > 0$.

Expressions for the critical misfit f_c for a given epitaxial layer thickness and the critical thickness, h_c, for a given misfit can be found by minimizing the total energy as a function of dislocation density. These expressions are

$$
\begin{aligned}
f_c &= \frac{b}{8\pi h\cos\lambda}\left(\frac{1-\nu\cos^2\beta}{1+\nu}\right)\ln\frac{4h_c}{b} \\
h_c &= \frac{b}{8\pi h\cos\lambda}\left(\frac{1-\nu\cos^2\beta}{1+\nu}\right)\ln\frac{4f_c}{b}
\end{aligned} \tag{2.6.14}
$$

In these expressions we have replaced $b_{\text{edg},\|}$ with $b\cos\lambda$. Typically, for face-centered-cubic (fcc)-lattice-matched diamond and zincblende crystals, we can assume that $\beta = \lambda = 60°$. Figure 2.6.12 shows linear and log plots of the critical thickness as a function of strain. Experimental data are shown for samples (InGaAs on GaAs and SiGe on Si) that are grown and then annealed. The filled points are for coherent structures, while the open ones are for relaxed structures. Thus the curve represents the boundary for relaxation onset. The fact that they were annealed is critically important. This mechanical equilibrium model is often not applicable to real structures grown under conditions that are away from thermodynamic equilibrium (i.e., kinetically-limited) and therefore metastable. Such structures may yield critical thicknesses much greater than the equilibrium h_c but may be unstable to the high-temperature processing used for fabricating devices. Different materials systems may exhibit widely varying behavior with respect to the "real" critical thickness that can be exploited in device structures.

Quantum Dot Structures. The Stranski–Krastinov (S-K) growth mode described above can provide an intermediary means of strain relaxation prior to the introduction of dislocations. InAs films of 1 to 4 monolayers in thickness can be deposited onto GaAs (approximately 7% lattice mismatch) and

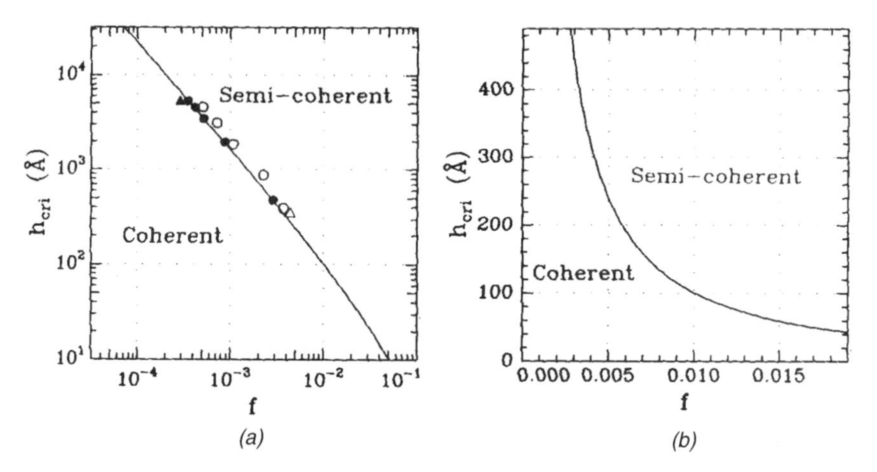

FIGURE 2.6.12 Critical thickness versus misfit f: logarithmic (a) and linear (b) plots. (Reprinted with permission from Tsao, 1993.)

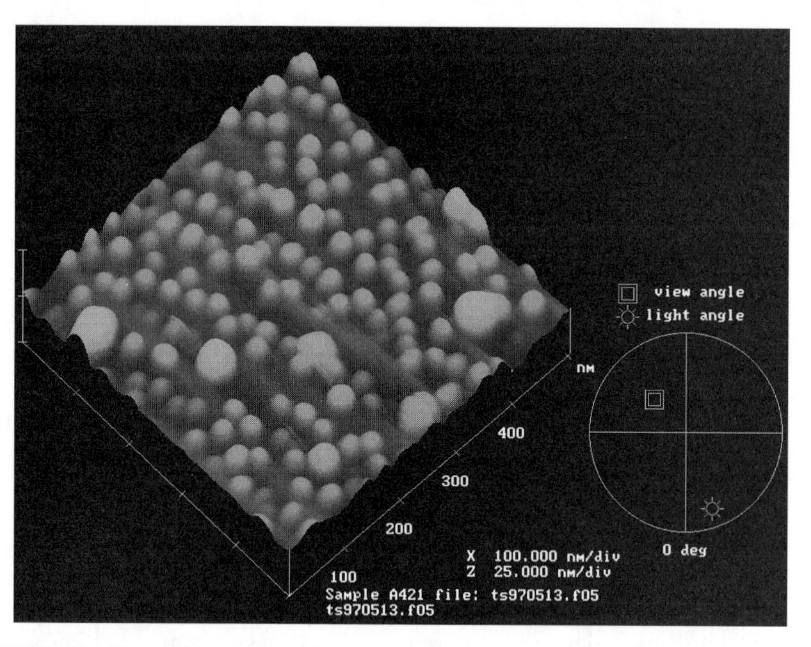

FIGURE 2.6.13 AFM image of InAs quantum dots on GaAs. (Reprinted with permission from Shen et al., 1998.)

S-K growth is observed. This process can be used to form "self-assembled" quantum dots. Figure 2.6.13 shows an AFM image of InAs dots. While the specific structural characteristics of the dots depend on the growth conditions, they typically exhibit diameters on the order of 20 nm and heights of roughly

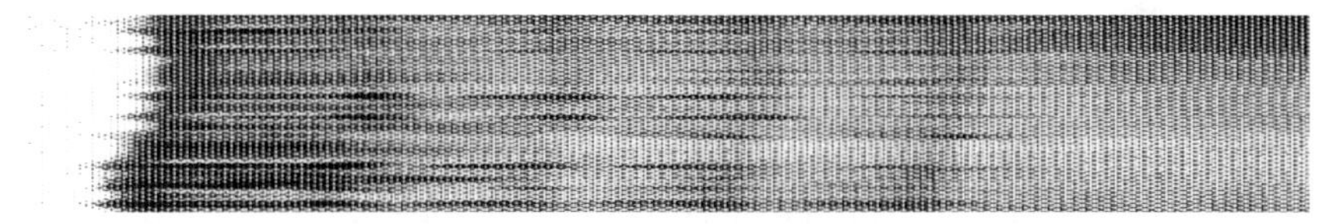

FIGURE 2.6.14 Cross-sectional TEM showing the vertical self-assembly of InAs quantum dots in GaAs.

3 nm. To first order, they are randomly oriented on the surface, although they do exhibit lateral interactions through their substrate strain fields. Remarkable ordering is observed to occur in the growth direction, however, as shown in Figure 2.6.14. The strain fields resulting from a buried quantum dot (i.e., GaAs covering InAs) biases the nucleation of the subsequent dot layers, thus leading to self-assembly.

Compositional Uniformity. Alloy compositions must be vertically and laterally uniform. In the growth (or vertical) direction, some alloy constituents have a tendency to surface segregate, or "ride" the growth front. This can occur during MBE of InGaAs. During MBE growth, the more weakly bound In atom tends to remain on the surface during growth, creating an In-rich surface. This effect can be minimized by reducing the substrate temperature and/or increasing the As flux during growth. As described above, however, these growth conditions are inconsistent with those for the most atomically smooth interface.

Doping Control. Heterojunction-based devices require the ability to abruptly introduce n- and p-type dopants of a range of composition from roughly 10^{15} to 10^{20} cm^{-3}. To control the actual electron and hole concentrations, compensating defects or background impurities must be negligible. While the MBE and MOCVD processes are inherently of high purity, background impurities are introduced during growth. Epitaxial GaAs is typically p-type (10^{14} cm^{-3} range or below), due to the incorporation of carbon, a shallow acceptor. The Al-containing materials are typically semi-insulating and can exhibit high concentrations of deep traps, resulting, presumably, from oxygen incorporation. In-containing materials tend to have increasing residual electron concentrations with increasing In concentration. This results from both native defects and the incorporation of shallow donors from In sources during growth. It is necessary to be aware of common compensation issues in epitaxial materials during device design.

Both HEMTs and HBTs require high concentrations ($> 10^{18}$ cm^{-3}) of dopants with abrupt interfaces. Diffusion and surface segregation of dopants can introduce nonideal doping profiles. Diffusion of dopants can occur either during growth or after growth during device operation due to self-heating effects. These effects are discussed in more detail in Chapters 3 and 4.

PROBLEMS

2.1. Find an expression for the electron current density in a heterostructure in terms of the conduction band edges E_c, the conduction band effective density of states N_c, and the electron carrier concentrations. Start with the expression for the electron current density given in terms of the quasi-Fermi level ϕ_n as (Brennan, 1999, p. 573)

$$J_n = -q\mu_n n \nabla \phi_n$$

Assume that the current flows only in the z direction, that the effective density of states is a function of z, and the conduction band edge is a function of z.

2.2. In the text it is argued that the number of elements in C_{ijkl} can be reduced from 81 to 21. Show that (**a**) the number of independent elements along the main diagonal of C reduces to six, c_{11}, c_{22}, c_{33}, c_{44}, c_{55}, and c_{66}, as given by Eq. 2.4.14; and (**b**) the number of independent off-diagonal elements reduces to 15.

2.3. Determine the relative carrier concentration within the highest level of a two-level system if the temperature of the electron gas is 25,000 K. Use the infinite triangular well approximation and assume that the effective mass is $0.067m_e$, and the effective electric field is 1.0×10^5 V/cm.

2.4. Determine the average relaxation time $\langle\tau\rangle$ in a semiconductor using

$$\langle\tau\rangle = \frac{\int \tau E f(E) D(E) dE}{\int E f(E) D(E) dE}$$

where $f(E)$ is the distribution function, $D(E)$ the density of states, and τ the relaxation time, has an energy dependence given as

$$\tau = aE^{-s}$$

where a is a constant and s is a fraction. Assume that $D(E)$ corresponds to a three-dimensional system, and approximate $f(E)$ by a Boltzmann distribution. Use the Γ function to simplify your result.

2.5. Based on Figure 2.1.3 and the results in Example 2.1.1, develop expressions for the quasielectric fields for electrons and holes in a graded heterojunction due to the change in the affinity, and the energy gap. Show the distinction between the quasielectric fields and the total electric field acting on the carriers.

2.6. Consider a p-n heterojunction in which the p-side is formed in the wider-gap material compared to the n-side. Let ϕ_p and ϕ_n represent the voltage drop across the p and n sides of the junction, respectively. Determine an expression for the ratio of the voltage drop across the p and n sides of the junction, ϕ_p/ϕ_n. This expression holds independent of whether or not the junction is biased. Assume that the depletion approximation holds and that the field and potential vanish at the edge of the depletion regions on either side of the junction.

2.7. Using the result of Problem 2.6, determine an expression for the depletion region width in the heterojunction in terms of the acceptor and donor concentrations N_a and N_d and the built-in voltage.

CHAPTER 3

Heterostructure Field-Effect Transistors

In this chapter we discuss heterostructure field-effect transistors (HFETs). The field-effect transistor (FET) is the most commonly used transistor in digital electronics, owing to its relative ease of fabrication, planar geometry, reliability, reproducibility, and miniaturization capability. FETs all operate by exploiting the field effect: that the underlying conductivity of a semiconductor can be modulated through the application of a bias applied to a surface gate. For a full discussion of the field effect and the basics of field-effect transistors, the reader is referred to Brennan (1999, Chap. 14). It is the purpose of this chapter to build on the initial conceptual ideas on FETs to discuss HFETs and their operation.

3.1 MOTIVATION

The most ubiquitous field-effect transistor is the metal-oxide semiconductor field-effect transistor MOSFET. The semiconductor constituent material in a MOSFET is silicon. The particular advantage of silicon MOSFETs is that silicon can readily be oxidized to form SiO_2, in a highly controllable and reproducible manner, which in turn is used to form the gate of the device. The Si–SiO_2 interface can be formed with very high regularity such that few defects are produced. The end result is that nearly perfect Si MOSFETs can be made in vast quantities, in a highly reproducible manner, and can readily be integrated to form large-scale circuits. For this reason, virtually all digital logic circuits, at least for most commercial products, are made with MOSFETs.

Although silicon MOSFETs are the most important devices for digital electronics, in high-frequency electronics, other device structures are often employed. The major limitation of silicon MOSFETs is that silicon itself is an inherently low-mobility material. Most compound semiconductors (GaAs, InP, etc.) have significantly higher electron mobilities than silicon. As a result, devices made using compound semiconductors can exhibit higher frequencies of operation than comparable silicon MOSFETs. Although compound semiconductor materials are inherently faster than silicon, the failure to identify suitable insulators makes the fabrication of metal–insulator–semiconductor FETs using these materials unattractive.

There are many emerging applications that require high-frequency, high-power transistor amplifiers. Among these are space applications such as satellite-to-satellite cross-links and satellite-to-terrestrial communications. Further applications of emerging importance are in automobile collision avoidance radar systems, wireless communications systems, and high-speed communications systems in general, both wireless and optical fiber. To meet the performance requirements imposed by these systems, the most promising materials are the compound semiconductors, mainly InP and InGaAs. Devices made from these materials have already demonstrated record performance in cutoff frequency f_t and maximum frequency of oscillation $f_{\max}$ for any transistor. Thus, it is important to develop compound semiconductor–based FETs to meet future demands in high-frequency electronics.

Instead of MIS structures, the gating action in a FET can be accomplished using Schottky barriers. Field-effect transistor devices that utilize Schottky barrier gates are called MESFETs (Brennan, 1999, Chap. 14). MESFETs can readily be made using compound semiconductors since no insulating layer is needed. The only requirements are a Schottky barrier and relatively high quality semiconductor material. In a typical MESFET, the channel layer is composed of a uniformly doped bulk semiconductor layer grown on top of a semi-insulating layer as shown in Figure 3.1.1. The source and drain regions, marked S and D in Figure 3.1.1, are comprised of highly doped n-type regions. Application of a positive voltage onto the drain leads to electron flow from the source to the drain. The positive drain voltage acts to reverse bias the Schottky barrier gate near the drain end as shown in Figure 3.1.2. The depletion region beneath the Schottky barrier is largest near the drain side. Of course, the electron flux from the source to the drain moves primarily through the conducting region between the depletion region of the Schottky barrier and the semi-insulating substrate. As can be seen from Figure 3.1.2, the conducting channel is narrowest near the drain end of the device. Under these operating conditions, the current is essentially linear with respect to the voltage as shown in Figure 3.1.2. If a reverse gate voltage is applied, the conducting channel between the source and drain can be choked off completely, as shown in Figure 3.1.3. Under these conditions, the device is said to be in *pinch-off* and the current saturates as shown in the current–voltage characteristic in Figure 3.1.3.

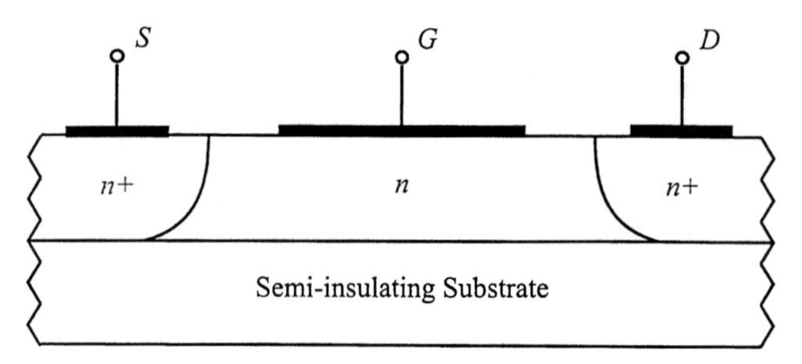

FIGURE 3.1.1 Simple MESFET device. The top layer forms the active region of the device and is typically doped n-type. The bottom layer is a semi-insulating substrate. Current flow is mainly within the top conducting layer and is modulated by the depletion region formed underneath the gate by the action of the gate and drain biases.

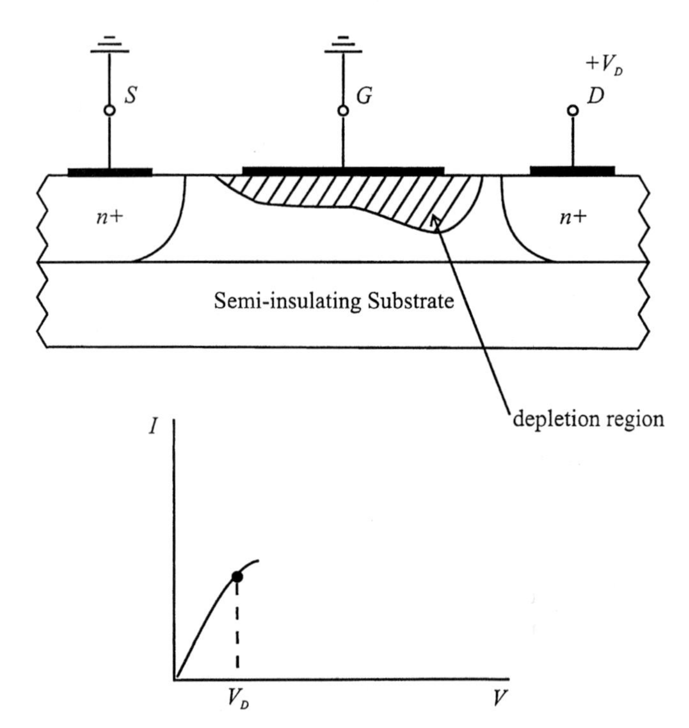

FIGURE 3.1.2 MESFET under a drain bias of V_D below pinch-off and its corresponding current–voltage characteristic. Notice that the current has not yet saturated.

In a MOSFET, the current flow is at the surface in the inversion layer formed between the Si and SiO_2 layers. Alternatively, in a MESFET the current flow is in the bulk semiconductor. Since the current flow is in the bulk in a MESFET, the charge carriers that contribute to the current arise from

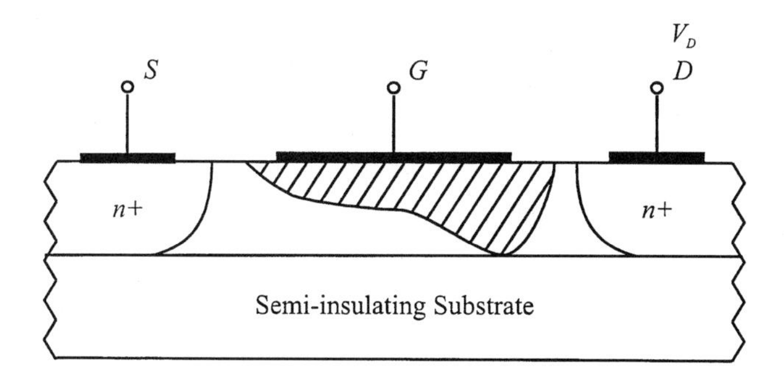

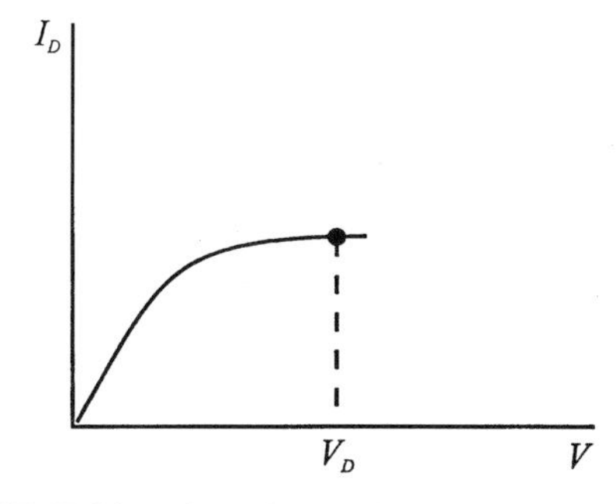

FIGURE 3.1.3 MESFET biased at pinch-off along with its corresponding current–voltage characteristic. In this case, the current has saturated.

the bulk doping concentration. In many applications a high current density is desired, requiring a concomitant high doping concentration in the MESFET. For example, in digital applications large current densities are necessary to ensure that the device can drive subsequent stages and overcome capacitances. In high-frequency applications, a high output power is generally desired, requiring a high drive current. Since the current is in the bulk in a MESFET, the current and donor atoms share the same space. This results in a substantially large Coulomb scattering of the current-carrying electrons by the ionized donor atoms in the channel. This Coulomb scattering is called *ionized impurity scattering* (Brennan, 1999, Sec. 9.4). The strength of Coulomb scattering depends on the spatial separation of the scattering centers (in this case the ionized donor atoms) and the current-carrying electrons. If the spatial separation is on average large, the Coulomb scattering is reduced. However, when the two species exist in the same spatial domain, the Coulomb scattering is substantial. Ionized impurity scattering increases with increasing donor concentration and acts to reduce the electron mobility in the device. Therefore, a clear trade-off occurs in a MESFET (i.e., an increased donor concentration is desirable for increased current drive but at the expense of a decrease in the electron mobility and hence speed of the device). In addition, the device becomes noisier with decreasing electron mobility and saturation velocity.

If the donor atoms could be spatially separated from the current-carrying electrons, the limitations placed on the donor concentration could be alleviated since the Coulomb scattering would be reduced. This can be accomplished using the modulation doping technique discussed in Section 2.2. Modulation doping provides a convenient way in which the ionized donor atoms can be spatially separated from the current-carrying electrons. Heterostructure field-effect transistors (HFETs) exploit this technique to achieve high current densities while retaining high carrier mobility. Thus, HFETs can achieve optimal high-frequency performance since they utilize high-mobility semiconductor materials, and through the use of modulation doping, retain the inherent high mobility without sacrificing current-carrying capability. For these reasons, HFETs currently hold the record for highest-frequency field-effect transistors and are the device type of choice for high-frequency electronics. The basics of HFET operation are discussed in the next section.

3.2 BASICS OF HETEROSTRUCTURE FIELD-EFFECT TRANSISTORS

The most pervasive heterostructure FET has more than one name. It is sometimes called the *modulation-doped-field-effect transistor* (MODFET), or alternatively, a *high-electron mobility transistor* (HEMT). Regardless of the name, the device utilizes the modulation doping technique within a field-effect transistor structure. Until recently, all HFETs were MODFETs, and the names HFETs, MODFETs, and HEMTs could be used interchangeably. However, a new HFET has been developed that does not exploit modulation doping. Therefore, we make the distinction that an HFET represents the general class of heterostructure FETs, with MODFETs (or equivalently, HEMTs) being a type of HFET. The other type of HFET that has emerged is one that relies on spontaneous and piezoelectrically induced charge at the interface, as discussed in Section 2.4. In these structures, modulation doping is not necessarily included; thus the devices are not truly MODFETs.

As discussed in Section 2.2, modulation doping (Dingle et al., 1978) provides a useful means of spatially separating the ionized donor atoms from the current-carrying electrons. The basic structure of a GaAs–AlGaAs MODFET is sketched in Figure 3.2.1. The device structure shown in Figure 3.2.1 is a representative structure illustrating the simplest implementation of a GaAs-based MODFET. As can be seen from the figure, an AlGaAs layer is grown epitaxially onto the GaAs forming a heterostructure. The AlGaAs layer is doped n-type, while the GaAs layer is unintentionally doped. To equilibrate the Fermi level, electrons are transferred from the AlGaAs layer to the GaAs layer, where they are confined by the potential barrier formed by the heterojunction. Thus the surface of the GaAs layer has a high electron concentration that is spatially separated from the ionized donors within the AlGaAs layer. The corresponding band diagram is also sketched in Figure 3.2.1. As can be seen from the diagram, the electrons reside in the region close to the hetero-

EXAMPLE 3.1.1: Threshold and Pinch-off Voltages of a MESFET

The depletion layer width of a Schottky barrier can be found from the expression for a p^+-n junction. A Schottky barrier is, in the limit, equivalent to a highly doped p^+-n junction in the sense that all of the depletion layer width lies within the semiconductor layer. For an asymmetrically doped homojunction, the depletion layer width, W, is given as (Brennan, 1999, Chap. 11)

$$W = \sqrt{\frac{2\varepsilon_s(V_{bi} - V_a)}{qN_B}}$$

where $N_B = N_d$ (the donor concentration) if N_a (the acceptor concentration) $\gg N_d$ or $N_B = N_a$ if $N_d \gg N_a$, Va is the applied voltage, and V_{bi} is the built-in voltage of the junction. Generally, V_a is the voltage that is applied from the metal to the semiconductor and is a function of both the gate and drain voltages. If the drain voltage is zero, then the pinch-off is due to the gate voltage. The pinch-off voltage is typically defined to include the built-in voltage and is the voltage necessary to completely pinch-off the channel of the MESFET. If we defined the thickness of the active layer of the MESFET as a, then the pinch-off voltage for a n-type semiconductor is given as

$$V_p = \frac{qN_d a^2}{2\varepsilon_s}$$

The threshold voltage, V_T, is defined as the voltage that has to be applied in addition to the built-in voltage of the gate to completely close the channel when the drain voltage, $V_D = 0$. Clearly, if the channel is not closed in equilibrium a reverse bias must be applied to the Schottky barrier gate to increase the depletion layer width and thus close the channel. This applied voltage is called the threshold voltage, V_T, and must be negative. Thus V_T is given as

$$V_p = V_{bi} - V_T$$

junction. Depending on the degree of band bending within the GaAs layer, spatial quantization can occur. Relatively strong band bending at the interface can form a potential well of dimensions comparable to the electron de Broglie wavelength. Hence, spatial quantization levels will be produced and the system will behave as a two-dimensional electron gas (see Chapter 2).

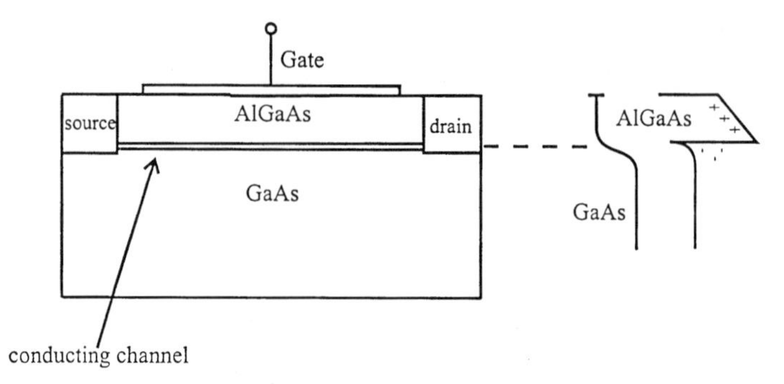

FIGURE 3.2.1 GaAs–AlGaAs MODFET device, showing the band structure. Notice that through modulation doping, electrons are transferred to the GaAs layer, leaving the AlGaAs layer depleted. A conducting channel is formed between the GaAs and AlGaAs layers at the heterointerface.

There are several different MODFET designs than that shown in Figure 3.2.1. The most common variation from the simple structure shown in Figure 3.2.1 is to introduce an undoped AlGaAs layer between the doped AlGaAs and GaAs layers. The additional undoped AlGaAs layer serves to further spatially separate the electrons within the GaAs layer from the ionized donors in the AlGaAs. This has the obvious benefit of further reducing the ionized impurity scattering between the ionized donors and electrons, thus further increasing the electron mobility. The undoped AlGaAs layer is called the *spacer layer*. Typical doping level and thicknesses are given in Section 2.6.

Current flow in a MODFET can be understood as follows. Through modulation doping, charge carriers can be produced at the heterolayer without doping the GaAs side. The total charge concentration is dependent on the gate voltage and, of course, whether the structure is an enhancement- or depletion-mode device. Enhancement-mode devices are normally off with zero gate voltage (i.e., the conducting channel is not present in equilibrium). In an n-channel enhancement-mode device, a sufficient positive gate voltage is required to induce the channel. Alternatively, a depletion-mode device is normally on with zero gate voltage. In an n-type depletion-mode device, a negative gate voltage is necessary to deplete the channel and thus turn the device off. In any event, once the two-dimensional electron gas is induced at the heterointerface, a conducting path is opened between the source and drain contacts. Application of a positive voltage at the drain results in a current flow in the device, arising mainly from the motion of electrons from the source to the drain in the two-dimensional system. If electrons moving within the two-dimensional system deliver most of the current, this condition yields very high mobilities and large electron velocities at a very small applied drain voltage. This results in extremely fast charging of capacitances at low power consumption.

In Chapter 2 we discussed the transport physics of electrons in a two-dimensional system, including calculations of the electron mobility. The situation in a MODFET is somewhat more complicated by the action of the drain voltage in addition to the gate voltage. When a drain voltage is also applied, the band bending is different near the drain than near the source region. For an enhancement-mode device, the drain and gate voltages are of the same sign. Therefore, near the drain the surface potential is less positive than near the source. Thus the band bending is stronger near the source region than near the drain, implying that the well is narrower near the source than near the drain. This leads, in turn, to a difference in the subband energies between the two regions such that the electron energy structure changes continuously from the source to the drain. This effect greatly complicates theoretical models of MODFETs. However, to a reasonable approximation, it can be assumed that the subband structure remains the same from the source to the drain regions. With this assumption, the carrier mobility can be calculated using the techniques discussed in Chapter 2.

A more important consequence of the applied drain voltage is field heating of the carriers. At the drain side of the gate, the electric field increases sharply depending on the magnitudes of the gate and drain voltages and the channel length. In submicron-gate-length devices, significant field strengths can occur near the drain end of the channel. In this region, the average electron energy increases dramatically, leading to significant velocity overshoot. Velocity overshoot arises from the difference in the energy and momentum relaxation times and occurs whenever the electron distribution is heated rapidly to energies much greater than the steady-state average energy. Since the energy relaxation rate of the carrier distribution is not instantaneous, the electrons maintain a high average energy for a period of time during which their velocities can well exceed their steady-state value. Rapid carrier heating can occur either from the injection of cold electrons into a relatively high field region, as occurs near the drain of a short-channel MODFET, or through high-energy injection into a low-field region, as occurs in a heterostructure bipolar transistor (see Chapter 4). In either case, the electron distribution is initially much hotter than its corresponding steady-state value, with a concomitant higher velocity. Velocity overshoot has consistently been predicted to occur in short-channel MODFETs based on numerical models (Park and Brennan, 1989, 1990). Recent experimental measurements have confirmed the effect (Passlack et al., 2000). Velocity overshoot results in a significant increase in the average velocity within the channel. As a consequence, devices that exhibit velocity overshoot have higher frequencies of operation.

As in a MOSFET, the current flow in a MODFET saturates at relatively high drain voltage. The origin of the current saturation can be understood as follows. First, it is useful to review current saturation in a MOSFET. Simple models of MOSFETs include only the drift current, neglecting any contribution from diffusion. This assumption is valid provided that the FET is not operated within the saturation region. At saturation, the simple drift current model encounters difficulties due to the depletion of carriers in the pinch-off region. Unless there is an extremely high electric field produced near the pinch-off point, the drift current alone cannot sustain the observed saturation current (Pao and Sah, 1966). A very high electric field is expected to exist only in short-channel MOSFETs. In long-channel devices, such a high field is improbable. Therefore, current saturation in a long-channel MOSFET cannot be explained solely with the drift current. It is precisely at the pinch-off point that the diffusion current becomes important. Although the carrier concentration is small, the gradient of the carrier concentration is extremely large in this region. Therefore, the diffusion current can be quite large, and as a result it plays an important role in describing the saturation current in a MOSFET.

What mechanism is responsible for current saturation in a MODFET? In long-channel MODFETs the drain saturation current has a origin similar to that in a MOSFET, and inclusion of the diffusion component is necessary to properly describe the saturation condition. In short-channel MODFETs the situation is slightly different. Under this condition, the drain current saturation is determined primarily by velocity saturation at the drain end of the channel.

3.3 SIMPLIFIED LONG-CHANNEL MODEL OF A MODFET

It is highly desirable to develop a simple, analytical model for a device since such a model can be used in circuit simulation. As we will see in Section 3.4, most state-of-the-art devices have very short channel lengths, and simple analytical models can no longer be used. In these instances, numerical approaches are needed to fully capture the physics of the device operation. However, in longer-channel devices and when the applied voltages are relatively small, analytical formulas can be used to describe device behavior. It is the purpose of this section to present a simple analytical model of a MODFET that can be used for circuit simulation. In the following section, we examine the physical effects that appear in short-channel MODFETs and under high-voltage operation.

The fundamental assumption we make in deriving our model is the gradual channel approximation. In the gradual channel approximation (sometimes referred to as the long-channel approximation) it is assumed that the field along the channel direction changes slowly with position compared to that for the field perpendicular to the channel. This assumption enables us to treat the field in only one dimension. The model we present is based on the work by Ruden (1990) and includes gate leakage and velocity saturation effects.

The two-dimensional sheet charge concentration, $n(x)$, can be expressed in terms of the sheet carrier concentration at the source end of the channel, n_{so}, using the results derived in Box 3.3.1, as

$$n(x) = n_{so} - \frac{\varepsilon}{qd} V(x) \qquad 3.3.1$$

BOX 3.3.1: Two-Dimensional Electron Concentration in a HEMT Channel

The electron concentration as a function of x given by Eq. 3.3.1 can be determined as follows. Consider the band diagram shown in Figure 3.3.1. The distance d is the thickness of the AlGaAs layer, d_1 the thickness of the undoped AlGaAs spacer layer, E_f the Fermi level, ΔE_c the conduction band discontinuity, V_G the gate voltage, and ϕ_B the barrier height. The doping concentration within the AlGaAs layer is assumed to be

$$N_D - N_A = 0 \qquad -d_1 < x < 0$$

$$N_D \qquad -d < x < -d_1$$

Within the doped region, $-d < x < -d_1$, the Poisson equation is

$$\frac{d^2V(x)}{dx^2} = -\frac{qN_D}{\varepsilon}$$

From Gauss's law, the field at $x = 0$ must be equal to the field at $x = -d_1$. Calling the field at $x = 0$ the surface field F_s yields

$$F(0) = F(-d_1) = -\left.\frac{dV(x)}{dx}\right|_{x=-d_1} = F_s \qquad V(-d_1) = -F_s d_1$$

Integrating Poisson's equation once and applying the boundary conditions for the field yields

$$\frac{dV(x)}{dx} = -\frac{q}{\varepsilon}\int_{-d_1}^{x} N_d(x)dx - F_s = -\frac{q}{\varepsilon}N_d(x + d_1) - F_s$$

Integrating again, the potential $V(x)$ is

$$V(x) = -F_s x - \frac{q}{\varepsilon}N_D\frac{(x + d_1)^2}{2}$$

Therefore, the potential at $x = -d$ is

$$V(-d) = F_s d - \frac{qN_D}{2\varepsilon}(d - d_1)^2$$

(*Continued*)

BOX 3.3.1 (*Continued*)

From examination of Figure 3.3.1, the potential at $-d$, $V(-d)$, is equal to the voltage difference between $x = 0$ and $x = -d$. This is given as

$$-V(-d) = \phi_B - V_G - \left[\frac{\Delta E_c}{q} - \left(\psi_s - \frac{E_f}{q}\right)\right]$$

which is simply

$$-V(-d) = \phi_B - V_G - \frac{\Delta E_c}{q} + \psi_s - \frac{E_f}{q}$$

Substituting in for $V(-d)$ yields

$$\phi_B - V_G - \frac{\Delta E_c}{q} + \psi_s - \frac{E_f}{q} = -F_s d + \frac{qN_D(d - d_1)^2}{2\varepsilon}$$

If we define the threshold voltage V_T below which there is no charge in the channel as

$$V_T = \phi_B - \frac{\Delta E_c}{q} - \frac{qN_D(d - d_1)^2}{2\varepsilon}$$

the surface electric field F_s can be expressed as

$$F_s d = V_G - V_T + \frac{E_f}{q} - \psi_s$$

Multiplying by ε, the expression for $F_s\, d$ becomes

$$\varepsilon F_s d = \varepsilon\left(V_G - V_T + \frac{E_f}{q} - \psi_s\right) = qn(x)d$$

where we have used Gauss's law and recognized that $n(x)$ is the sheet charge within the two-dimensional electron gas formed at the heterointerface. The surface potential ψ_s at $x = 0$ and arbitrary x is then given as

$$\psi_s(0) = -\frac{qn(x)d}{\varepsilon} + V_G - V_T + \frac{E_f}{q}$$

$$\psi_s(0) = -\frac{qn_{\mathrm{so}}d}{\varepsilon} + V_G - V_T + \frac{E_f}{q}$$

(*Continued*)

BOX 3.3.1 (*Continued*)

where n_{so} is the sheet charge carrier concentration at the source end of the channel. The potential drop along the channel is now easily calculated as the difference between $\psi_s(x)$ and $\psi(0)$ as

$$V(x) = \psi_s(x) - \psi_s(0) = \frac{qd}{\varepsilon}[n_{so} - n(x)]$$

The surface concentration $n(x)$ is then given as

$$n(x) = n_{so} - \frac{\varepsilon}{qd}V(x)$$

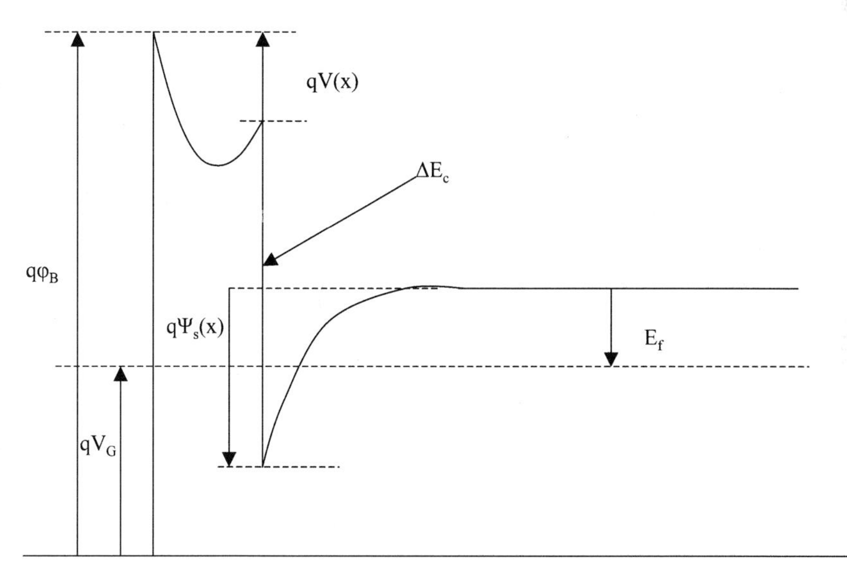

FIGURE 3.3.1 Energy band bending diagram for a HFET used in calculating the surface sheet charge concentration in Box 3.3.1.

The local channel current $I_c(x)$ assuming drift only is given simply as

$$I_c(x) = qWn(x)v(x) \qquad 3.3.2$$

where $v(x)$ is the velocity of the electrons within the channel and W is the gate width. The velocity can be written as a function of the field instead of position. A useful model for the velocity field relation that includes velocity saturation effects is given as

$$v(F) = \begin{cases} \dfrac{\mu|F|}{1 + |F|/F_1} & F \leq F_c \\ v_{\text{sat}} & F > F_c \end{cases} \qquad 3.3.3$$

where v_{sat} is the saturation velocity, μ the mobility, F_c the critical field after which negative differential resistance occurs in the velocity–field relation (see Chapter 5), and F_1 is

$$F_1 = \frac{F_c}{(\mu F_c/v_{\text{sat}}) - 1} \qquad 3.3.4$$

Substituting the expression for $v(F)$ into Eq. 3.3.2 and recognizing that $|F| = dV/dx$ yields

$$I_c(x) = qWn(x)\frac{\mu(dV/dx)}{1 + (dV/dx)/F_1} \qquad 3.3.5$$

When the gate leakage current is significant, the channel current is not conserved. Under these conditions the channel current is equal to the difference between the source current and the gate leakage current:

$$I_s = W\int_0^x j_g(x)dx + I_c(x) \qquad 3.3.6$$

The gate leakage current density $j_g(x)$ is the local gate current density. For simplicity, we assume that the integral in Eq. 3.3.6 can be approximated using the average value of $j_g(x)$, and $\langle j_g \rangle$ as $W\langle j_g \rangle x$. Substituting in for $I_c(x)$ in Eq. 3.3.5, Eq. 3.3.6 then becomes

$$I_s - W\langle j_g \rangle x = \frac{qWn(x)\mu(dV/dx)}{1 + (dV/dx)/F_1} \qquad 3.3.7$$

Integrating the expression for the channel current given by Eq. 3.3.7 over the channel length yields

$$\int_0^L (I_s - W\langle j_g \rangle x)\left(1 + \frac{dV/dx}{F_1}\right) dx = qW\mu \int_0^{V_D} n(V)dV \qquad 3.3.8$$

The left-hand side of Eq. 3.3.8 integrates to

$$L\left(I_s - \frac{\langle j_g \rangle WL}{2} + \frac{V_D}{LF_1}I_s\right) - \frac{\langle j_g \rangle W}{F_1}\int_0^L x\left(\frac{dV}{dx}\right)dx \qquad 3.3.9$$

But the product of the average gate current density $\langle j_g \rangle$ and the area WL is I_G, where I_G is the gate current. The drain current I_D is equal to the difference between the source current I_s and the gate leakage current I_G as

$$I_s = I_D + I_G \tag{3.3.10}$$

Using Eq. 3.3.10, Eq. 3.3.9 can be simplified to

$$L\left[I_D + \frac{I_G}{2} + \frac{V_D}{LF_1}(I_D + I_G)\right] - \frac{\langle j_g \rangle W}{F_1}\int_0^L x\left(\frac{dV}{dx}\right)dx \tag{3.3.11}$$

The right-hand side of Eq. 3.3.8 can be simplified as

$$qW\mu\int_0^{V_D} n(V)dV = qW\mu\int_0^{V_D}\left(n_{so} - \frac{\varepsilon}{qd}V\right)dV = qW\mu\left(n_{so}V_D - \frac{\varepsilon}{2qd}V_D^2\right) \tag{3.3.12}$$

Substituting the results from Eqs. 3.3.11 and 3.3.12 into Eq. 3.3.8 yields

$$L\left[I_D + \frac{I_G}{2} + \frac{V_D}{LF_1}(I_D + I_G)\right] - \frac{\langle j_g \rangle W}{F_1}\int_0^L x\left(\frac{dV}{dx}\right)dx = qW\mu\left(n_{so}V_D - \frac{\varepsilon}{2qd}V_D^2\right) \tag{3.3.13}$$

We can further evaluate the integral on the left-hand side of Eq. 3.3.13:

$$\frac{\langle j_g \rangle W}{F_1}\int_0^L x\left(\frac{dV}{dx}\right)dx \tag{3.3.14}$$

using an integration by parts with the following assignments:

$$\begin{aligned} u &= x \qquad & v &= V \\ du &= dx \qquad & dv &= \left(\frac{dV}{dx}\right)dx \end{aligned} \tag{3.3.15}$$

The integral in Eq. 3.3.14 then becomes

$$\frac{\langle j_g \rangle W}{F_1}\int_0^L x\left(\frac{dV}{dx}\right)dx = \frac{\langle j_g \rangle W}{F_1}LV_D - \frac{\langle j_g \rangle W}{F_1}\int_0^L V(x)dx \tag{3.3.16}$$

Using the definition of I_G given above, Eq. 3.3.16 can be rewritten as

$$\frac{I_G}{F_1}\left[V_D - \frac{1}{L}\int_0^L V(x)dx\right] \tag{3.3.17}$$

Equation 3.3.13 becomes, after substituting Eq. 3.3.17,

$$L\left[I_D + \frac{I_G}{2} + \frac{V_D}{LF_1}(I_D + I_G)\right] - \frac{\langle j_g \rangle W}{F_1} LV_D - \frac{\langle j_g \rangle W}{F_1}\int_0^L V(x)dx$$
$$= qW\mu\left(n_{so}V_D - \frac{\varepsilon}{2qd}V_D^2\right) \qquad 3.3.18$$

The integral of $V(x)$ over the channel length that appears in Eq. 3.3.18 can be approximated by assuming that the voltage varies regularly with x. This is, of course, an approximation, since most of the voltage drop generally occurs closer to the drain than to the source. We approximate the last two terms on the left-hand side of Eq. 3.3.18 as

$$\frac{\langle j_g \rangle W}{F_1}\left[LV_D - \int_0^L V(x)dx\right] = \frac{I_G}{F_1}\alpha V_D \qquad 3.3.19$$

With the approximation of Eq. 3.3.19, Eq. 3.3.18 becomes

$$L\left[I_D + \frac{I_G}{2} + \frac{V_D}{LF_1}(I_D + I_G)\right] - \frac{I_G}{F_1}V_D\alpha = qW\mu\left(n_{so}V_D - \frac{\varepsilon}{2qd}V_D^2\right) \qquad 3.3.20$$

Finally, we solve Eq. 3.3.20 for I_D, the drain current, as

$$I_D = \frac{1}{1 + V_D/LF_1}\left\{\left[-I_G\left(\frac{1}{2} + \frac{V_D}{LF_1}(1-\alpha)\right)\right] + \frac{qW\mu}{L}\left(n_{so}V_D - \frac{\varepsilon}{2qd}V_D^2\right)\right\}$$
$$3.3.21$$

The expression for I_D given by Eq. 3.3.21 describes the drain current in the linear region of a MODFET. Notice that the two-dimensional electron gas enters the result in the charge concentration at the source end of the channel, n_{so}, and the low-field mobility for the two-dimensional system. These quantities can be calculated using the techniques described in Chapter 2. One must be careful to be sure that the applied gate voltage is greater than the threshold voltage such that a two-dimensional electron gas is induced. Once n_{so} and μ are known, one can determine the drain current in the linear regime from knowledge of the dimensions of the device and the velocity field relation. For simplicity the parameter α can be taken as $\frac{1}{2}$.

The drain current at saturation can be found as follows. Once the field at L reaches F_c, saturation occurs. The channel current as a function of x is now given as

$$I_c(x) = qWn(x)v_{sat} \qquad 3.3.22$$

since the velocity is now equal to the constant value v_{sat}. At $x = L$, the channel current is equal to the drain current, which in this case is the drain saturation current $I_{D,sat}$. Thus the drain saturation current is given as

$$I_c(L) = I_{D,sat} = qW\left[n_{so} - \frac{\varepsilon}{qd}V(L)\right]v_{sat} = qW\left(n_{so} - \frac{\varepsilon}{qd}V_{D,sat}\right)v_{sat} \qquad 3.3.23$$

At this point one needs to determine the drain saturation voltage, $V_{D,sat}$. Alternatively, it is possible to solve Eq. 3.3.23 for $V_{D,sat}$ in terms of $I_{D,sat}$. Then this expression for $V_{D,sat}$ can be substituted into Eq. 3.3.21 at saturation and solved for $I_{D,sat}$. The result, using the notation of Ruden (1990), follows from some lengthy but straightforward algebra (see Problem 3.1) as

$$I_{D,sat} = -I_{00} \pm \sqrt{I_{00}^2 - I_{0A}^2} \qquad 3.3.24$$

where I_{00} and I_{0A} are defined as

$$\begin{aligned} I_{00} &= \frac{\varepsilon LWv_{sat}^2F_1}{d(\mu F_1 - 2v_{sat})}\left[1 + \frac{d}{\varepsilon LF_1}\left(qn_{so} - \frac{I_G}{2Wv_{sat}}\right)\right] \\ I_{0A}^2 &= \frac{2\varepsilon LWv_{sat}^2F_1}{d(\mu F_1 - 2v_{sat})}\left[\frac{I_G}{2}\left(1 + \frac{qdn_{so}}{\varepsilon LF_1}\right) - \frac{q^2Wd\mu n_{so}^2}{2\varepsilon L}\right] \end{aligned} \qquad 3.3.25$$

Using Eqs. 3.3.24 and 3.3.25, the drain saturation current can be calculated from known quantities.

Finally, to complete the model, we need to specify the gate leakage current I_G. The gate leakage current is a function of the applied gate voltage and is generally device dependent. It is further found that the gate leakage current has different dependencies on the applied gate voltage (i.e., one expression can be applied at low gate voltage while another holds for high gate voltage). Approximate semiempirical expressions for I_G have been determined of the form

$$\begin{aligned} I_{G1} &= I_{s1}(e^{qV_G/n_1kT} - 1) \\ I_{G2} &= I_{s2}e^{qV_G/n_2kT} \end{aligned} \qquad 3.3.26$$

where n_1, n_2, I_{s1}, and I_{s2} are parameters that are determined by matching to experimental data for the gate leakage current.

Often, it is useful to utilize a simplified model for the current–voltage (I–V) characteristic of a MODFET. Equation 3.3.21 can be simplified if it is assumed that $I_G = 0$, the gate current vanishes, and that $V_D/LF_1 \ll 1$. Under this condition, the first term on the right-hand side of Eq. 3.3.21 vanishes and the prefactor in Eq. 3.3.21 can be approximated as unity. With these assumptions, the I–V characteristic for the linear regime of a MODFET simplifies to

$$I_D = \frac{qW\mu}{L}\left(n_{so}V_D - \frac{\varepsilon}{2qd}V_D^2\right) \tag{3.3.27}$$

The simplified result given by Eq. 3.3.27 shows clearly how the unique characteristics of the MODFET influence the device performance. Notice that I_D depends on μ and n_{so}, the two-dimensional electron mobility and concentration, respectively. These quantities can be determined using techniques developed in Chapter 2. The drain current in saturation $I_{D,sat}$ can also be determined with the assumptions leading to Eq. 3.3.27. Starting from Eq. 3.3.27, $I_{D,sat}$ can be determined as (see Problem 3.6)

$$I_{D,sat} = \frac{q\mu W n_{so}F_s}{d}(\sqrt{1+a^2} - a) \tag{3.3.28}$$

where a is defined as

$$a \equiv \frac{LF_s\varepsilon}{qdn_{so}} \tag{3.3.29}$$

and F_s, the field at the drain end of the device in saturation, is given as

$$F_s = \frac{1}{Ln_d}\left(n_{so}V_{D,sat} - \frac{\varepsilon}{2qd}V_{D,sat}^2\right) \tag{3.3.30}$$

In Eq. 3.3.30, n_d is the two-dimensional electron concentration near the drain.

The simplified form of the MODFET I–V characteristic is generally not applicable to GaAs since the value of F_1 is too small to satisfy the condition $V_D/LF_1 \ll 1$. GaN is closer to satisfying this condition but only for devices of relatively long channel length. Therefore, Eq. 3.3.27 provides only a very rough approximation to I_D for a MODFET under most conditions. Nevertheless, as we will see below, the simplified expression for I_D is informative since it relates closely to that for MOSFETs.

It is interesting to compare Eq. 3.3.27 to the simplified I–V characteristic for a MOSFET. The simplified expression for the drain current in a MOSFET below pinch-off is given by Brennan (1999, Eq. 14.6.10) as

$$I_D = \frac{\mu_n WC_i}{L}\left[(V_G - V_T)V_D - \frac{V_D^2}{2}\right] \tag{3.3.31}$$

Comparing Eq. 3.3.31 to Eq. 3.3.27, it is clear that the functional form of the drain currents is similar between the MOSFET and the MODFET. In fact, Eqs. 3.3.27 and 3.3.31 are essentially the same. This can be seen from the

EXAMPLE 3.3.1: Threshold Voltage of a GaAs–AlGaAs MODFET

Determine the threshold voltage of a GaAs–AlGaAs MODFET whose geometry and doping concentrations are given as follows. The Al composition is 25%. Assume a conduction band to valence band discontinuity ratio of 60%/40%, that the Schottky barrier height is 1.0 V, and that the AlGaAs layer is 33.0 nm thick with an undoped spacer layer of 3.0 nm. The dielectric constant of the AlGaAs layer is 12.4.

The threshold voltage V_T is given in Box 3.3.1 as

$$V_T = \phi_B - \frac{\Delta E_c}{q} - \frac{qN_D(d - d_1)^2}{2\varepsilon}$$

where d_1 is the undoped space layer width, d the AlGaAs layer width, and ϕ_B the Schottky barrier height. The conduction band edge discontinuity ΔE_c can be determined as follows. The energy bandgap of the AlGaAs layer is found from Eq. 2.1.2 as

$$E_g = 1.424 + 1.247x = 1.424 + 1.247 \times 0.25 = 1.736 \text{ eV}$$

The conduction band edge discontinuity between the AlGaAs and GaAs is then found by taking 60% of the difference in the energy bandgaps:

$$\Delta E_c = 0.60(1.736 - 1.424) \text{ eV} = 0.187 \text{ eV}$$

The threshold voltage can now be found as

$$V_T = 1.0 \text{ V} - \frac{0.187}{q} \text{ eV} - \frac{(1.6 \times 10^{-19} \text{ C})(10^{18} \text{ cm}^{-3})[(33.0 - 3.0) \times 10^{-7} \text{ cm}]^2}{2(12.4)(8.85 \times 10^{-14} \text{ C/V} \cdot \text{cm})}$$

Evaluating the expression for V_T above yields a threshold voltage of 0.157 V.

following. The gate capacitance per unit area C_i is given as

$$C_i = \frac{\varepsilon}{d} \qquad 3.3.32$$

where ε is the dielectric constant of the semiconductor and d is the thickness of either the insulator in the MOSFET or the AlGaAs layer in the MODFET.

EXAMPLE 3.3.2: Determination of the Transconductance of a MESFET Using the Saturated Velocity Model

In the text we derived an expression for the drain current in a MODFET assuming the field-dependent mobility model given by Eq. 3.3.3. Let us now consider what happens in the limiting case where current saturation is assumed to occur everywhere under the gate for a uniformly doped device. This saturated velocity model reasonably approximates the performance of a device in the limit of very small gate lengths. For simplicity we treat a MESFET structure since the current in the saturated velocity model is particularly easy to describe. The current in the device under these conditions can be written as

$$I = qv_{\text{sat}}W(a - h)N_D$$

where W is the device width, v_{sat} the saturation velocity, N_D the uniform doping concentration within the active layer, a the depth of the doped active region of the MESFET, and h the depth of the depletion region. The relationship for I above assumes that the current flow is completely confined within the channel formed between the edge of the depletion region, h, and the edge of the active region, a. In general, when the doping density is nonuniform, the current is given as

$$I = v_{\text{sat}}W \int_h^a \rho(y)dy$$

The voltage drop across the depletion layer is obtained through solution of the Poisson equation. The voltage at the edge of the depletion region, $V(h)$, can be written as (see Problem 3.2)

$$V(h) = \frac{1}{\varepsilon}\int_0^h y\rho(y)dy$$

Using the rule for differentiation of a definite integral,

$$\frac{d}{dx}\int_{u(x)}^{v(x)} f(t)dt = f[v(x)]\frac{dv}{dx} - f[u(x)]\frac{du}{dx}$$

dI/dh and dV/dh can be found as

$$\frac{dI}{dh} = -v_{\text{sat}}W\rho(h) \qquad \frac{dV}{dh} = \frac{h\rho(h)}{\varepsilon}$$

(*Continued*)

EXAMPLE 3.3.2 (*Continued*)

The transconductance can now be expressed as the absolute value of the ratio of dI/dh to dV/dh,

$$g_m = \frac{dI}{dV} = \frac{dI}{dh}\frac{dh}{dV} = -\frac{v_{\text{sat}} W \rho(h)\varepsilon}{h\rho(h)} = -\frac{v_{\text{sat}} W \varepsilon}{h}$$

Notice that the transconductance depends on the width of the depletion region, h. If h changes slowly with the gate voltage, the device will behave linearly.

The product of C_i and $(V_G - V_T)$ is simply equal to the channel charge Q_n:

$$Q_n = C_i(V_G - V_T) \tag{3.3.33}$$

Using Eqs. 3.3.32 and 3.3.33, the expression for the drain current in a MOSFET given by Eq. 3.3.31 becomes

$$I_D = \frac{\mu_n W}{L}\left(Q_n V_D - \frac{\varepsilon}{2d} V_D^2\right) \tag{3.3.34}$$

If we further factor out q from Eq. 3.3.34 and call the remaining electron concentration n_{so}, Eq. 3.3.34 becomes

$$I_D = \frac{qW\mu_n}{L}\left(n_{\text{so}} V_D - \frac{\varepsilon}{2qd} V_D^2\right) \tag{3.3.35}$$

which is essentially the same expression for the MODFET as given by Eq. 3.3.27. It is not very surprising that to first order at least, the I–V characteristics for the MOSFET and the MODFET are essentially the same since the transport physics is similar. In both structures the transport is of two-dimensional electrons trapped at an interface, the insulator–semiconductor interface in the MOSFET and the heterointerface in the MODFET. The mobility in the MOSFET is the two-dimensional electron mobility in the silicon inversion layer. Typically, this mobility is substantially lower than the two-dimensional electron mobility within the GaAs quantum well. As a result, we see an approximate explanation of the origin of why the speed of response of the MODFET is substantially higher than that of a MOSFET.

3.4 PHYSICAL FEATURES OF ADVANCED STATE-OF-THE-ART MODFETS

The analytical model presented in Section 3.3 is based on the gradual channel approximation and thus is best applied to relatively long channel devices. Although it is highly useful in this regime, its extension to submicron-gate-length structures, as well as to devices that have relatively high voltage swings, is questionable. The highest-frequency MODFET devices used in practice have very short channel lengths and typically operate at high voltages. Use of a simple analytical model such as the one presented in Section 3.3 under these conditions is questionable.

In practice, several different types of device models are needed. These models range from relatively simple compact models of great use in computer-aided design of digital and microwave circuits to highly sophisticated, detailed microscopic models that illustrate the fundamental physics of device behavior and check the validity of the elementary models. The compact models are of a form similar to the model presented in Section 3.3. The more sophisticated models are generally numerical and can range from advanced drift-diffusion studies to self-consistent Monte Carlo simulations. The actual details of how these simulations are made are beyond the level of this book. We refer the interested reader to a book by Snowden (1988) and the article by Smith and Brennan (1998) for an introduction to this subject. In this section we focus instead on the physical features that arise in advanced structures that necessitate numerical modeling techniques. It is our goal to discuss the physical features that affect MODFET device performance that occur in short-channel structures with concomitant high electric field strengths.

The first question the reader might ask is why we have to dispense with analytical models and adopt numerical techniques in modeling short-channel structures. There are several reasons why simple analytical models fail in short-channel devices which hold true whether one is modeling a MODFET, a MOSFET, or a MESFET. The first problem that is encountered in short-channel devices is the fact that the potential profile becomes two-dimensional. Most analytical models adopt the long-channel approximation such that the field can safely be assumed to be one-dimensional. As the channel length decreases, the validity of the gradual channel approximation becomes suspect since the field lines under the gate can no longer be assumed to be one-dimensional (see Chapter 7). Thus, it is necessary to solve the problem using two dimensions, making an analytical formulation much more difficult. A more important limitation arises from the fact that as the device active region shrinks, the applicability of any transport theory based on the average behavior of the carriers having undergone a relatively large number of scatterings becomes questionable. Virtually all analytical models and some numerical models assume that the carriers undergo sufficient scatterings such that they are thermalized into a steady-state distribution. Steady state occurs, of course, when the input power from the field is balanced by the output power lost to the lattice through phonon scatterings. In this case, the average energy of the distribution no longer changes with time. However, in a device with dimensions comparable to the mean free path for electron scattering, the carriers can completely transit the structure without suffering any collisions. Even if the device is somewhat larger than the mean free path for electron scattering, many of the carriers will not suffer sufficient scatterings in order to reach steady-state conditions prior to their exiting the device. In either case, use of a theoretical model that assumes thermalization of the carriers can produce inaccurate results. This assumption underlies not only most analytical theories but also the most commonly used numerical model, the drift-diffusion technique. [See Brennan (1999, Sec. 6.3) for a derivation of the drift-diffusion equations.]

The most compelling reason, then, to turn to advanced numerical simulation tools is that in the current state-of-the-art very short channel devices, the carriers experience few, if any, collisions during their transit, and thus their dynamics are radically different from those predicted using steady-state formulations. Non-steady-state transport is often referred to as nonstationary transport. The two most important elements of nonstationary transport are ballistic transport and velocity overshoot. Ballistic transport occurs when the carriers transit the device without suffering any collisions. Obviously, ballistic transport can be appreciable only if the device dimensions are equal to or smaller than the scattering mean free path for electrons. In devices with longer dimensions than the mean free path, some but not many collisions can occur. Recognizing that it takes many inelastic collisions (phonon scattering is the dominant inelastic relaxation mechanism in semiconductors) to relax the carrier's energy, if insufficient scatterings occur during the carrier's transit, it will "overshoot" and not settle into steady state. The principal consequence of this "overshoot" is that the carrier's energy will be higher than its corresponding steady-state value, with a concomitant increase in its velocity. Both ballistic transport and velocity overshoot can dominate device behavior in very short channel structures. It should be further emphasized that nonstationary transport can occur in any short-channel device. Thus the discussion below of ballistic transport and velocity overshoot holds for short-channel MOSFETs, MESFETs, MODFETs, and so on.

The momentum balance equation (Brennan, 1999, Prob. 6.2) can be determined by taking moments of the Boltzmann equation. The momentum balance equation for the total momentum density is given as

$$\frac{dP}{dt} = -qFn - \frac{P}{\tau_m} \qquad 3.4.1$$

where P is the total momentum density, F the applied field, q the charge, n the carrier concentration, and τ_m is the momentum relaxation time. The momentum relaxation time is defined as the average time needed to relax the electron's momentum. In practice, the momentum relaxation time can be quite fast since it can take only one large angle scattering event to redirect the electron's momentum. The force term is negative here since we assume that electrons are the carriers. The average momentum is given simply by the ratio of the total momentum density, P, to the concentration, n. With this substitution, Eq. 3.4.1 can be solved for the mean momentum to obtain

$$\bar{p} = qF\tau_m(e^{-t/\tau_m} - 1) \tag{3.4.2}$$

If one examines Eq. 3.4.2, it is clear that the system initially has a mean momentum of zero at $t = 0$. After a time on the order of the momentum relaxation time, the mean momentum approaches $-qF\tau_m$. A similar relationship is easily found for the average carrier velocity as

$$\bar{v} = \frac{qF\tau_m}{m^*}(e^{-t/\tau_m} - 1) \tag{3.4.3}$$

We can further estimate the mean distance traveled before steady-state conditions are achieved as follows. The mean distance traveled as the system evolves to steady state is obtained from integrating Eq. 3.4.3 with respect to the time t as

$$d = \frac{1}{m^*}\int_0^{\tau_m} qF\tau_m(e^{-t/\tau_m} - 1)dt \tag{3.4.4}$$

Evaluating the integral in Eq. 3.4.4 yields, for the distance d,

$$d = -\frac{qF\tau_m^2}{m^*}e^{-1} \tag{3.4.5}$$

The mean distance can be estimated for GaAs and Si at an applied field of 10 kV/cm to be $d \sim 11.0$ nm for Si and $d \sim 100$ nm for GaAs. In these calculations, the momentum relaxation time can be estimated from the low-field mobility and we have used 1450 and 8500 $\text{cm}^2/\text{V}\cdot\text{s}$ for Si and GaAs, respectively. Therefore, if the active device dimensions are less than d, the electrons will not, on average, achieve a steady state distribution. Instead, they will exit the device before they have suffered sufficient scatterings to produce steady state. Under this condition, steady state is not achieved. It is important to notice that the mean distance required for the system to reach steady state is much larger in GaAs than in Si. Therefore, velocity overshoot effects are expected to be much more pronounced in GaAs and related compounds than in silicon, implying potentially faster device operation (see Problem 3.9).

The question becomes, then, why does velocity overshoot occur? Velocity overshoot is a consequence of the difference in the momentum and energy relaxation times. To see this, it is useful to consider the energy balance equation (Brennan, 1999, Prob. 6.3), given as

$$\frac{d\bar{E}}{dt} = -qF\bar{v} - \frac{\bar{E} - \bar{E}_0}{\tau_E} \tag{3.4.6}$$

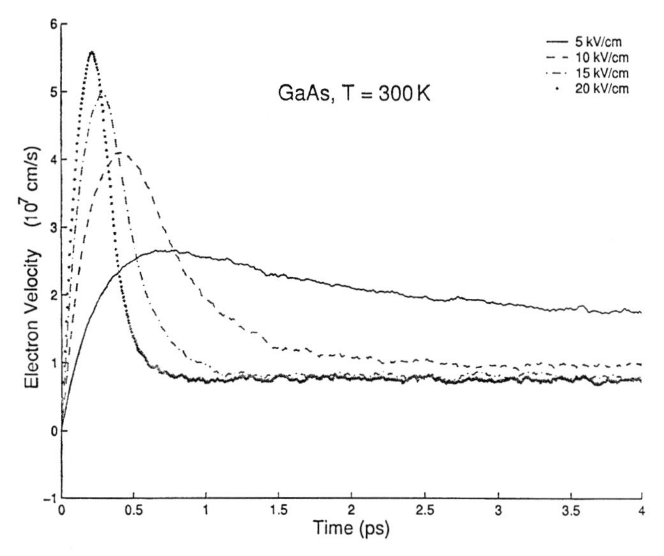

FIGURE 3.4.1 Electron drift velocity as a function of time for various electric field strengths to show the overshoot effect. Notice that at low electric field strengths, the velocity overshoot is less dramatic than at high applied field strengths.

where $\bar{v}$ is the mean velocity, $\bar{E}_0$ the average energy in equilibrium, and τ_E the energy relaxation time. Again we have assumed that electrons are the carriers, making the first term on the right-hand side of Eq. 3.4.6 negative. Solving for the carrier energy as a function of time yields

$$\bar{E} = (qF\bar{v}\tau_E - \bar{E}_0)(e^{-t/\tau_E} - 1) \quad 3.4.7$$

The steady-state average energy is approached in a time on the order of the energy relaxation time, just as before for the momentum. The energy relaxation time is defined in a similar manner as the momentum relaxation time (i.e., as the mean time needed to relax the electron's energy). The most effective mechanism for energy relaxation (at least at relatively low energies below the impact ionization threshold) is phonon scattering. Generally, many phonon scatterings have to occur to relax the carrier energy, whereas momentum relaxation can occur after a single scattering event. Hence, the energy relaxation time τ_E is larger than the momentum relaxation time τ_m. The momentum relaxation time is not constant but varies with the carrier energy. As the carrier accelerates and reaches higher energies, the scattering rate increases and thus τ_m decreases. When τ_m adjusts slowly, the transient velocity overshoots its steady-state value, as shown in Figure 3.4.1. As can be seen in Figure 3.4.1, the velocity increases well above its steady-state value for a short duration. After some time, the carrier velocities relax back to steady state.

In both GaAs and silicon, the velocity overshoots its saturation value at high applied field strengths. As mentioned above, the momentum relaxation time τ_m depends on the average carrier energy or temperature. At low energy, τ_m is largest, implying that the carrier travels with relatively few scatterings. As the carrier heats up in energy, τ_m decreases due to an increase in the scattering rate. Since the momentum relaxation time is smaller than the energy relaxation time in both Si and GaAs, the average electron energy changes slowly, on the time scale of the momentum relaxation. The momentum balance equation becomes

$$\frac{dP}{dt} = -qFn - \frac{P}{\tau_m(T_e)} \tag{3.4.8}$$

where we have rewritten τ_m as a function of the electron temperature T_e. T_e changes slowly as a function of time, depending on the energy relaxation time τ_E. At short times, $t < \tau_E$, during transient transport, the momentum gain from the field exceeds that lost via collisions and the net momentum increases, leading to a higher drift velocity. Consequently, the electron drifts with its low-field mobility since insufficient scatterings occur during the transient to reduce the mobility. As a result, the carrier velocity can significantly overshoot its steady-state value. With time, the energy of the system relaxes from repeated scattering events, leading to a broadening of the distribution function. As a result, the drift velocity decreases and the velocity of the electrons approach their steady-state value. Since the phonon scattering rate is much higher in silicon than in GaAs at low energies, the transient effects persist longer in GaAs, leading to more extensive velocity overshoot.

Figure 3.4.1 shows the electron drift velocity as a function of time for various applied electric field strengths in GaAs. At low applied electric field strengths, 5.0 kV/cm, the velocity overshoot is less dramatic than at higher field strengths. Not much is gained in the average speed of the carriers over that for the steady state at low applied field strengths. As the field strength is increased, the velocity overshoots its steady-state value significantly, as seen in Figure 3.4.1. If the device length can be traversed in the short time during which the velocity overshoot is significant, in principle there should be a substantial reduction in the transit time of the carriers. However, as the applied field is increased further, up to 20 kV/cm and above, the overshoot decreases dramatically in a relatively short period of time, due to the transferred electron effect. Owing to the large density of states within the satellite valleys, upon transferring, the electron drift velocity decreases sharply. Clearly, there is only a limited range of applied field strengths that will lead to substantial velocity overshoot effects over a significant portion of the device.

The variation of the carrier energy with time is shown in Figure 3.4.2. As can be seen from the figure, the carrier energy also increases above its steady-state value at high applied field strengths and then relaxes. Although the calculations presented in Figure 3.4.1 apply to GaAs, as mentioned above, velocity overshoot also occurs in silicon. In silicon, velocity overshoot arises

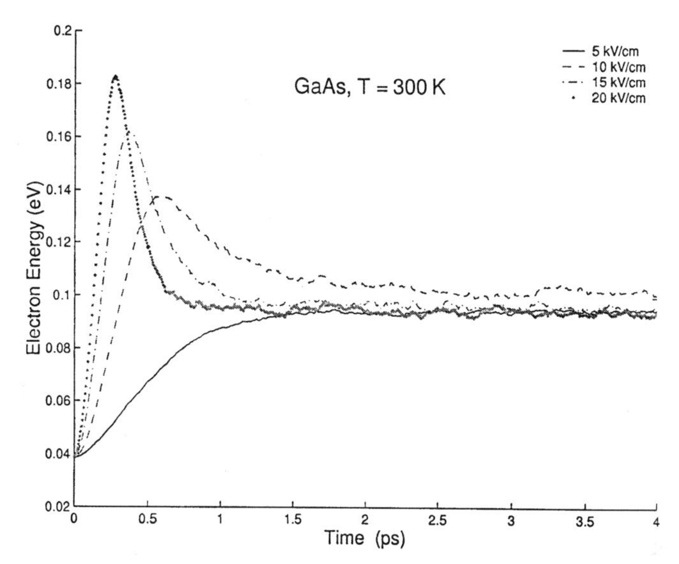

FIGURE 3.4.2 Average electron energy as a function of time in GaAs. Notice that at high electric field strengths, the energy overshoots its steady-state value significantly.

because $\tau_E > \tau_m$. However, in GaAs additional effects arise due to intervalley transfer, which is discussed in more detail in Chapter 5.

Velocity overshoot can influence device performance. Although velocity overshoot is expected in simple structures, its observance in MODFET devices has been somewhat elusive. Part of the problem has been that the experimental verification of velocity overshoot effects is made indirectly. Additionally, parasitics in the extrinsic device and circuit further complicate the experiment. Recent experimental measurements made on InGaAs MODFETs show convincing signs of velocity overshoot (Passlack et al., 2000). In these experiments it has been shown that pronounced velocity overshoot effects do occur and that the average velocity under the gate is increased above its steady-state value. The measurements show that the average carrier velocity increases monotonically with decreasing gate length, indicating the clear presence of velocity overshoot effects. Velocity overshoot acts to increase the speed of response of the device, thus leading to a higher cutoff frequency, as discussed in Section 3.5. It is for this reason that it is important to account for velocity overshoot when modeling very short channel FET devices. Thus, the simple analytical and drift-diffusion models typically become unreliable at these dimensions and one must utilize more comprehensive approaches, such as the ensemble Monte Carlo method, to achieve accuracy.

The steady-state electronic velocity–field curve in GaAs is somewhat more complicated than in silicon. In silicon, the electron velocity increases essentially monotonically with increasing applied electric field until it levels off roughly at its saturated value. In GaAs the situation is quite different. As the

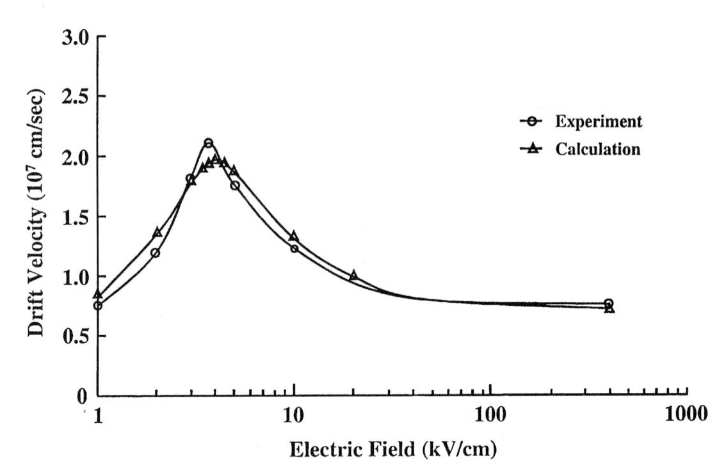

FIGURE 3.4.3 Calculated and experimental steady-state electron drift velocity in bulk GaAs as a function of applied electric field strength.

electric field increases, the electron velocity increases, reaches a maximum, and then decreases with increasing field strength, as can be seen in Figure 3.4.3. In Figure 3.4.3, both the experimental and calculated steady-state electron drift velocity are shown. As we discuss in detail in Chapter 5, the cause of this behavior is electron transfer from the low effective mass central valley to the higher-energy, higher-effective-mass satellite valleys. At low field, electrons in GaAs occupy the central valley, in which they have a low effective mass. Consequently, the electron mobility is relatively high. Under the application of an applied field, the electrons are readily accelerated and gain energy rapidly since their mass is relatively low. If the applied field is sufficiently high, after some time the electron's gain and energy equal that of the satellite valleys, 0.28 eV in GaAs. Once the electrons reach an energy equal to that of the satellite valleys, they can transfer into them via phonon scattering. Upon transferring into satellite valleys, the electrons effective mass increases and their concomitant velocity decreases. Thus once the electrons gain sufficient energy to transfer to the satellite valleys, the velocity of the carriers drops substantially. This leads to a negative differential resistance in the velocity–field curve.

The key to maintaining a very high velocity in GaAs-based devices, then, is to ensure that the electrons remain within the low effective mass central valley [also known as the *gamma* (Γ) *valley*] during their flight through the device. The presence of a very high electric field within the device can, of course, lead to intervalley transfer and thus a reduction in the velocity of the carriers within the device. The high electric field alone is not fully responsible for carrier heating. In a MODFET the carrier concentration within the two-dimensional system can be very high. As a result, the electron gas can be degenerate, meaning that the Fermi level is close to or within the conduction

EXAMPLE 3.4.1: Current–Voltage Characteristic for an n^+-i-n^+ Diode Exhibiting Ballistic Transport

A simple device that is often used as a theoretical tool to study nonstationary transport is an n^+-i-n^+ diode, often referred to as a *ballistic diode*. The particular advantage of a ballistic diode is that its field roughly approximates the field within a channel of a FET but is one-dimensional. Thus the complexities that accompany the two-dimensional electric field profile within a FET are removed, yet the basics of channel transport are preserved within the ballistic diode. If the carriers are injected with an initial energy close to zero, the average velocity of the carriers within the device can be computed by recognizing that the change in kinetic energy is equal to the work done by the potential $V(z)$ as

$$\tfrac{1}{2}mv^2(z) = qV(z)$$

Solving for the velocity $v(z)$ yields

$$v(z) = \sqrt{\frac{2qV(z)}{m}}$$

Assuming only a drift component to the current density $J(z)$, it can be written in terms of the velocity as

$$J(z) = -qn(z)v(z)$$

where $n(z)$ is the carrier concentration within the ballistic diode. The middle region of the ballistic diode is intrinsic, so the background doping concentration is generally negligible with respect to the free carrier concentration. Therefore, Poisson's equation becomes

$$\frac{d^2V}{dz^2} = \frac{J\sqrt{m/2q}}{\varepsilon}\frac{1}{\sqrt{V}}$$

The solution of the equation above can be found assuming that the current density is constant. Let

$$\frac{du}{dz} = KV^{-1/2} \qquad \frac{dV}{dz} = u$$

where K is

$$K = \frac{J\sqrt{m/2q}}{\varepsilon}$$

(*Continued*)

EXAMPLE 3.4.1 (*Continued*)

Then dividing du/dz by dV/dz and integrating the result with the boundary condition that $V(0) = 0$ and $u(0) = 0$ yields

$$u = \frac{dV}{dz} = 2\sqrt{K}V^{1/4}$$

Integrating to find V gives us

$$\int_0^{V_A} \frac{dV}{2\sqrt{K}V^{1/4}} = \int_0^L dz$$

which simplifies to

$$J = \frac{4}{9}\frac{\varepsilon}{L^2}\sqrt{\frac{2q}{m}}V_A^{3/2} .$$

It is interesting to compare the result above to the opposite case, that where the transport is collision-dominated. In the collision dominated case the drift velocity is given simply as

$$v(z) = \mu F_z$$

where F_z is the component of the field in the z direction. Under this condition the concentration is

$$n = -\frac{J}{q\mu F_z}$$

Again assuming that current density and the mobility are constant, Poisson's equation becomes

$$\frac{dF_z}{dz} = -\frac{qn}{\varepsilon} = \frac{J}{\varepsilon\mu F_z}$$

Integrating from 0 to z and noting that $F_z(0) = 0$ (the field at the edge of the depletion region is assumed to be zero) yields

$$F_z^2 = \frac{2Jz}{\varepsilon\mu}$$

(*Continued*)

EXAMPLE 3.4.1 (*Continued*)

Recognizing that $-dV/dz = F_z$, we have

$$\int_0^{V_a} dV = -\int_0^L \sqrt{\frac{2Jz}{\varepsilon\mu}}\,dz$$

Integrating and simplifying obtains an expression for the current density as a function of the applied voltage V_a:

$$J = \frac{9\varepsilon\mu V_a^2}{8L^3}$$

Notice that collision-dominated transport has a different dependency on the applied voltage than that of the ballistic transport case.

band. In a degenerate system the equilibrium distribution can no longer be approximated by the Boltzmann distribution but must be described using the Fermi–Dirac distribution. Since many of the allowed states are occupied within a degenerate semiconductor, the high electron concentration near the source forces carriers to relatively high energies with respect to the conduction band edge in accordance with the Pauli exclusion principle (Brennan, 1999, Chap. 3). Thus the carriers begin their flight through the device at relatively high energy even before they experience the electric field. As a result, intervalley transfer occurs more readily, resulting in a loss in the average velocity of the carriers.

Materials such as InP and InGaAs, in which the satellite valleys lie at substantially higher energies than in GaAs, offer obvious advantages. In these materials the satellite valleys lie at much higher energies, such that few carriers suffer intervalley transfer during their transit through the device. In this way, the electrons remain confined within the central (Γ) valley and maintain a high velocity during their transit through the device. For this reason, InP- and InGaAs-based MODFET devices have shown very high frequency performance and typically are the materials of choice for ultrahigh-frequency operation.

Collective effects also affect FET performance. Collective effects incorporate electron–electron interactions. In a degenerate system the high concentration of electrons results in a substantial increase in electron–electron scatterings. Electron–electron scattering involves collisions between electrons only and as such should provide no net energy or momentum loss to the electron system. Nevertheless, the electron–electron interaction within a solid is highly complicated by screening and the presence of positively charged background atoms. As a result, the electron gas exhibits both collective and individual particle behaviors. The collective response, called the *plasma oscillation* (Brennan, 1999, Sec. 9.3), is derived from the collective oscillation of the system, owing to the long-range nature of the Coulomb interaction. Plasma oscillations are manifest over distances greater than the characteristic screening length; within smaller distances the electron gas behaves more as a collection of individual particles interacting through a screened Coulomb interaction. Thus it is natural to separate the electron–electron interaction into long- and short-range components.

The long-range component of the electron–electron interaction involves the collective response of the electron gas, called a *plasmon oscillation.* Through plasmon scattering the net momentum and energy of the electron gas is altered since the collective excitation is ultimately coupled to the phonons. Hence, both electron momentum and energy are relaxed through the plasmon interaction. In contrast, short-range electron–electron scattering affects primarily the electron momentum and acts typically to "smear out" the electron distribution in momentum space. The combined influence of the short- and long-range components of the electron–electron interaction is to randomize the fast electrons, leading to a smearing of the electron distribution in k space. Consequently, the path length of the electrons is lengthened, and enhanced energy-loss collisions occur. The result is that the distribution is lowered in energy with a concomitant lowered average electron velocity near the drain region of the device. Thus, it can be concluded that collective effects can reduce nonstationary transport effects in short-channel devices.

The physical features discussed above that appear in short-channel devices are generally common to all FET device types. We discuss short-channel effects in the context of MOSFETs further in Chapter 7. However, before we end this section it is useful to point out a feature that can influence device performance that is mainly important only to MODFET structures. As mentioned, in short-channel devices, the electric field strength can be quite high. The high electric field strength within the channel has some important consequences. As discussed in Chapter 7, the high electric field within the channel can lead to impact ionization near the drain, carrier velocity saturation, and transfer of the carriers out of the channel into the neighboring material. In the case of MODFETs, the carriers can be transferred out of the GaAs channel into the AlGaAs layer. The conduction band edge discontinuity formed separating the GaAs channel and the AlGaAs layer is relatively low, about 0.3 eV, depending on the Al concentration present in the AlGaAs. Therefore, the electrons within the channel can readily be heated to energies sufficiently high such that they can be scattered from the GaAs layer into the AlGaAs layer. Once within the AlGaAs, the electron mobility is significantly lower than in the GaAs, and the concomitant electron drift velocity is lower. Thus a parallel conduction path is formed in the AlGaAs layer but with a lower drift velocity. This results in a slower speed of response of the device. It is thought that the ultimate limits of MODFET performance are defined by the onset of this parallel conduction within the AlGaAs layer (Quay et al., 2001).

A similar effect occurs within a MOSFET (see Chapter 7). The high electric field within the channel can heat the electrons to sufficiently high energy that they can be transferred into the SiO_2 layer. Called *gate oxide charging*, this can result in a significant change in the threshold voltage of the device, as we will see in Chapter 7. It should be noted that in a MOSFET, upon transferring into the SiO_2 layer, the electrons are localized and no longer can contribute to the current. In the MODFET device, the electrons may avoid capture and still contribute to the drain current, albeit with a lower drift velocity.

3.5 HIGH-FREQUENCY PERFORMANCE OF MODFETS

Perhaps the most important application of MODFETs is in high-frequency amplifiers. Therefore, it is useful to discuss the high-frequency performance of MODFETs. Here we present a brief introduction to the subject and refer the reader to the references for a more thorough discussion. We assume that the ac model for a MODFET can be approximated by that for a MESFET. Further we examine only quasistatic small-signal equivalent-circuit models of the device, referring the reader to the references for a full discussion of the small-signal analysis (Tiwari, 1992; Liu, 1999) of a MODFET.

The small-signal drain current i_D can be written in terms of the small-signal drain and gate voltages, v_D and v_G, respectively, as (Brennan, 1999, Sec. 14.8)

$$i_D = g_D v_D + g_m v_G \tag{3.5.1}$$

where g_D is the channel or drain conductance and g_m is the transconductance. The drain conductance and the transconductance are defined as

$$g_D = \left.\frac{\partial I_D}{\partial V_D}\right|_{V_G} \qquad g_m = \left.\frac{\partial I_D}{\partial V_G}\right|_{V_D} \tag{3.5.2}$$

The transconductance is particularly important in describing the RF performance of the transistor. In most RF applications, it is desirable that the transistor exhibit linear behavior. Linearity implies that the output frequency is precisely the same as the input frequency and that no higher harmonics are generated. In a FET, the RF input is the gate-to-source voltage, while the output is the drain current. Thus the transconductance is a measure of how the output drain current varies with respect to the input gate-to-source voltage. If the transconductance is constant, the device is linear.

The cutoff frequency of the device is defined as the frequency at which the device can no longer amplify the input signal. This can be determined by calculating the frequency corresponding to unity gain with the output short-circuited. In other words, the cutoff frequency is the frequency at which the

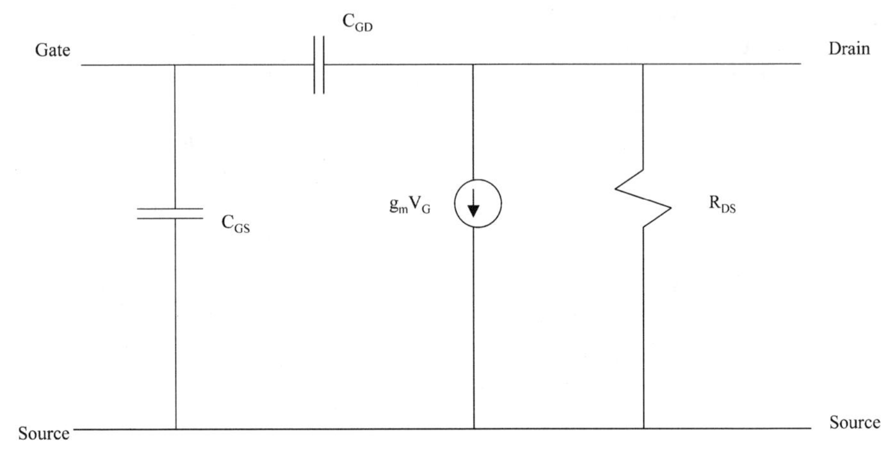

FIGURE 3.5.1 Simplified small-signal equivalent circuit of a MODFET used in determining the cutoff frequency of the device.

magnitude of the forward current gain is unity with the output short-circuited. Using Figure 3.5.1, the cutoff frequency f_t can be calculated as follows. The magnitude of the small-signal input current i_{in} is given as

$$i_{\text{in}} = 2\pi f_t (C_{GD} + C_{GS}) v_G = 2\pi f_t C_G v_G \quad 3.5.3$$

where $C_G = C_{GD} + C_{GS}$. The output current i_{out} is

$$i_{\text{out}} = g_m v_G \quad 3.5.4$$

For unity gain, the input current is equal to the output current. Equating Eqs. 3.5.3 and 3.5.4 and solving for f_t yields

$$f_t = \frac{g_m}{2\pi C_G} \quad 3.5.5$$

It is interesting to examine the functional dependence of f_t on the device parameters. As an illustrative example, it is useful to determine the cutoff frequency for a JFET or a MESFET. The transconductance of a JFET in saturation is given from Box 3.5.1 as

$$g_m \leq G_0 = \frac{2qW\mu N_d a}{L} \quad 3.5.6$$

The gate capacitance of a JFET is somewhat complicated. Due to the reverse bias on the p-n junctions, the junction capacitance is distributed unevenly, due to the differing shape of the depletion layer, along the channel. The gate

BOX 3.5.1: Calculation of the Transconductance of a JFET

The transconductance above pinch-off in the saturation region can be determined as follows. The saturation drain current of a simple JFET is given by Brennan (1999, Sec. 14.3) as

$$I_{D,\text{sat}} = G_0 \left\{ V_G - V_p - \frac{2}{3}(V_{bi} - V_p) \left[1 - \left(\frac{V_{bi} - V_G}{V_{bi} - V_p} \right)^{3/2} \right] \right\}$$

where G_0 is given as

$$G_0 = \frac{2qW\mu N_d a}{L}$$

V_p is the pinch-off voltage of the JFET, L the channel length, a the half-width of the channel in equilibrium, W the width of the device, N_d the doping concentration, V_{bi} the built-in voltage of the p$^+$-n junctions forming the gates, and μ the electron mobility. The transconductance is obtained by taking the derivative of $I_{D,\text{sat}}$ with respect to V_G as

$$g_m = G_0 \left[1 - \left(\frac{V_{bi} - V_G}{V_{bi} - V_p} \right)^{1/2} \right]$$

Clearly, from the expression above, g_m is certainly less than or equal to G_0. Therefore, we can approximate g_m by G_0 to give an upper bound on the transconductance. Therefore,

$$g_m \leq G_0 = \frac{2qW\mu N_d a}{L}$$

capacitance can be approximated using an average depletion layer width x as

$$C_G = \frac{2WL\varepsilon_0\varepsilon_s}{x} \qquad 3.5.7$$

where the extra factor of 2 arises from the two gates in the device. WL gives the area of the structure. If the gate voltage is assumed to be zero and the device operated at pinch-off, the average depletion layer width is $a/2$. With this assumption, the gate capacitance can be approximated as

$$C_G = \frac{4\varepsilon_0\varepsilon_s WL}{a} \qquad 3.5.8$$

Substituting Eqs. 3.5.6 and 3.5.8 into Eq. 3.5.5 yields for the cutoff frequency

$$f_t \leq \frac{q\mu N_d a^2}{4\pi\varepsilon_0\varepsilon_s L^2} \qquad 3.5.9$$

Equation 3.5.9 provides an upper bound for the cutoff frequency of a JFET. A similar relationship to that of Eq. 3.5.9 holds for MESFETs. Notice that the cutoff frequency is directly proportional to the carrier mobility and inversely proportional to the square of the channel length. High-frequency operation thus requires a very high carrier mobility as well as a very short channel length.

Before we discuss the high-frequency properties of MODFETs, it is useful to further elaborate on the characteristics of JFET devices that influence high-frequency behavior. The high-frequency performance of JFETs depends on both the intrinsic transport properties of the constituent materials, the device geometry, and the circuit loading. Inspection of Eq. 3.5.9 shows that the carrier mobility plays a dominant role in determining the cutoff frequency. Since the electron mobility is usually much higher than the hole mobility, only n-channel high-frequency FETs are used. The electron mobility of GaAs and of similar group III–V compounds, such as InP and InGaAs, is substantially higher than that of silicon. For this reason, most high-frequency FETs are made from compound semiconductors. The most important parameter in the device geometry is channel length. As can be seen from Eq. 3.5.9, the cutoff frequency of a JFET is inversely proportional to the square of the channel length. High-speed operation is achieved in general by using very short channel FETs.

In addition to the material parameters and device geometry, the frequency operation of a FET depends to some extent on circuit loading. A more complete small-signal model of the JFET–MESFET device is shown in Figure 3.5.2. The intrinsic model of the device consists of the elements found within the dotted-line boundary in the sketch. The elements R_i and $R_{DS} = 1/g_{DS}$ in Figure 3.5.2 represent the effects of the channel resistance, and C_{GD} and C_{GS} represent the total gate-to-channel capacitance, The extrinsic elements, which essentially represent the parasitics, are the source resistance R_S, the gate-metal resistance R_G, the substrate capacitance C_{DS}, and the drain resistance R_D. Although the cutoff frequency has some dependency on the parasitics, the maximum frequency of oscillation $f_{\max}$ is much more strongly affected. The maximum frequency of oscillation $f_{\max}$ is defined as the frequency at which the unilateral power gain of the transistor rolls off to unity. The unilateral power gain is the maximum power gain achievable by the transistor. A transistor is said to be *unilateral* when its reverse transmission parameter is zero. In other words, for a unilateral transistor the output is completely isolated from the input; there is no reverse transmission. Knowledge of both the cutoff frequency and the maximum frequency of operation are important in characterizing the device. The cutoff frequency characterizes the switching speed of the device,

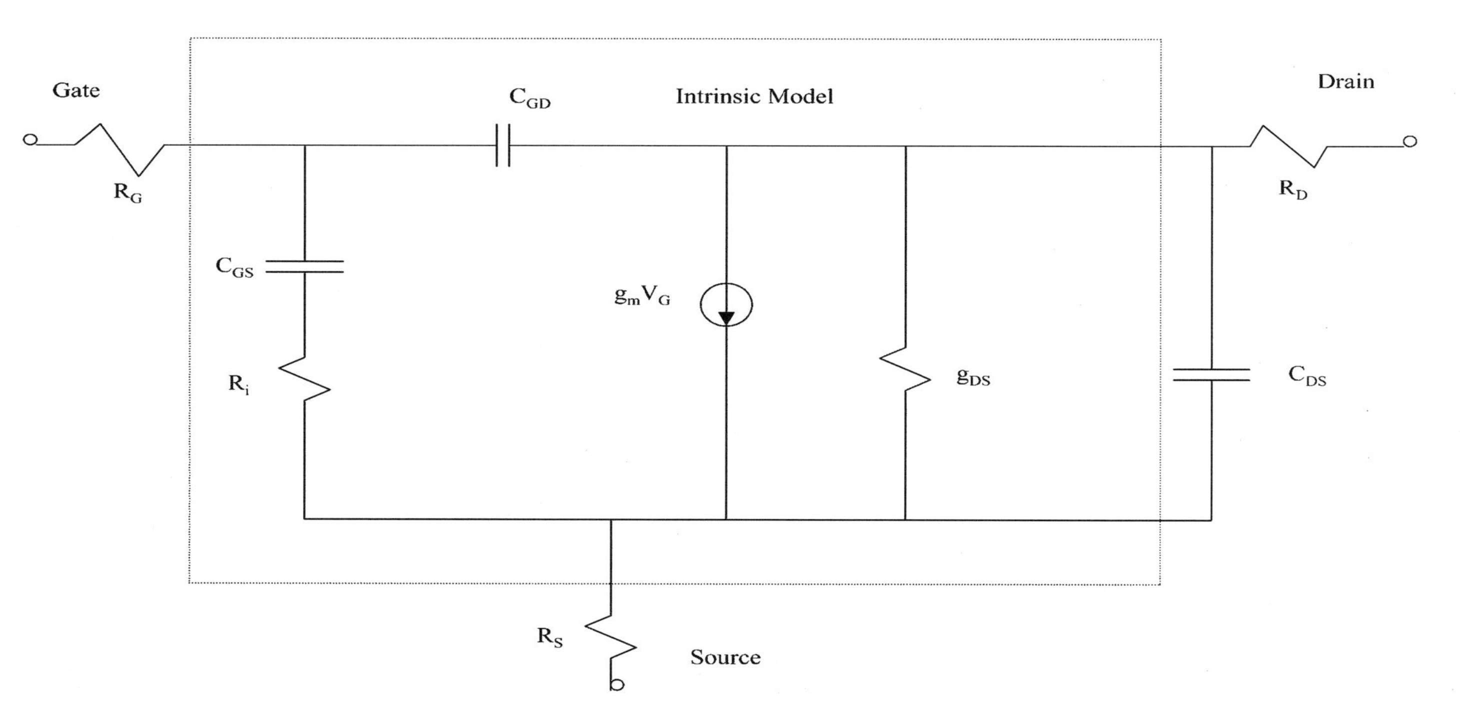

FIGURE 3.5.2 Quasistatic small-signal equivalent circuit model for a MODFET. The intrinsic model of the device is the part shown within the dotted-line boundary. The extrinsic model encompasses the remaining part of the diagram. R_i is the intrinsic channel resistance; C_{GD}, C_{DS}, and C_{GS} are the gate–drain, drain–source, and gate–source capacitances, respectively; g_m is the transconductance; and g_{DS} is the drain–source conductance.

while the maximum frequency of operation is more important in defining the RF performance.

The maximum frequency of oscillation $f_{\max}$ can be determined from the y parameters of the device. The details of this approach are beyond the level of this book and the interested reader is directed to the book by Liu (1999) for a thorough discussion. From knowledge of the y parameters, $f_{\max}$ is determined under saturation conditions when $g_D = 0$ to be

$$f_{\max} = \sqrt{\frac{f_t}{8\pi R_G C_{GD}}} \qquad 3.5.10$$

Notice that $f_{\max}$ depends strongly on the gate resistance and the gate-to-drain capacitance. It is interesting to consider the dependency of $f_{\max}$ on the gate length L. As discussed above, f_t increases strongly with decreasing L. However, it is not always guaranteed that $f_{\max}$ will increase as L decreases. This is because $f_{\max}$ is much more strongly dependent on the parasitics than f_t, which do not necessarily scale with L.

The cutoff frequency can be reformulated in a different way that illustrates its dependence on the carrier velocity. This approach is particularly useful in describing MODFETs. The two-dimensional electron gas concentration within a MODFET can be approximated in terms of the gate capacitance C_G as

$$n_s \sim C_G \frac{V_G - V_T}{qLW} \qquad 3.5.11$$

where V_G is the gate voltage, V_T the threshold voltage, W the width of the MODFET, and L the channel length. The drain current, assuming drift only, is given by Eq. 3.3.2 as

$$I_D = qn_s vW \qquad 3.5.12$$

Substituting Eq. 3.5.11 into Eq. 3.5.12 and taking the derivative with respect to V_G yields the approximate transconductance of a MODFET as

$$g_m = \frac{\partial I_D}{\partial V_G} = \frac{vC_G}{L} \qquad 3.5.13$$

The cutoff frequency of a MODFET can then be found by substituting Eq. 3.5.13 into Eq. 3.5.5 to yield

$$f_t = \frac{g_m}{2\pi C_G} = \frac{v}{2\pi L} = \frac{1}{2\pi \tau_{\text{tr}}} \qquad 3.5.14$$

where τ_{tr} is the carrier transit time under the gate. Thus the faster the electrons transit the device, the greater the cutoff frequency of the device.

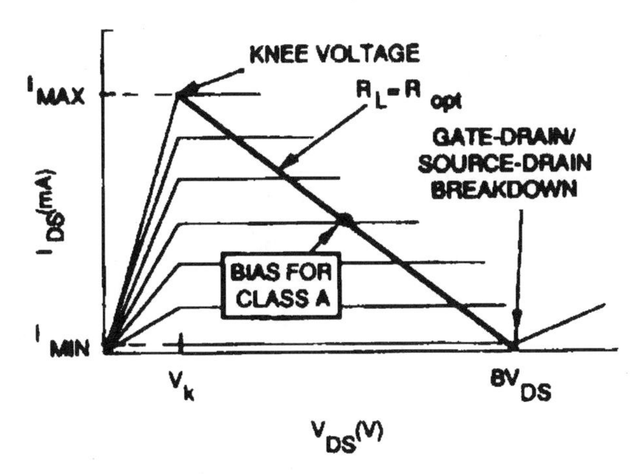

FIGURE 3.5.3 Representative *I–V* characteristic for class A amplifier operation. (Reprinted with permission from IEEE.)

From the discussion above it is clear that high-frequency performance of a FET requires a high carrier velocity and short channel lengths. Clearly, if the electron velocity overshoots its steady-state value during its transit through the device, the cutoff frequency of the device will be higher than if the carriers move at their steady-state velocity. Thus velocity overshoot effects clearly improve the high-frequency operation of a FET. Specific high-frequency applications are for low noise and power amplification. The most important figures of merit related to these applications are defined below.

From Nguyen et al. (1992a), the power performance can be characterized in terms of the maximum output power P_o, the associated gain G_a, and the maximum power added efficiency (PAE). For class A operation, the maximum output power is given by

$$P_o = \tfrac{1}{8}(I_{\max} - I_{\min})(\mathrm{BV}_{gd} - V_{po} - V_k) \qquad 3.5.15$$

where $I_{\max}$ is the maximum channel current, $I_{\min}$ the minimum channel leakage current, BV_{gd} the gate-to-drain breakdown voltage, V_{po} the pinch-off voltage, and V_k the knee voltage. Figure 3.5.3 shows a schematic representing the device characteristics and the load line, given by

$$R_L = \frac{\mathrm{BV}_{gd} - V_{po} - V_k}{I_{\max} - I_{\min}} \qquad 3.5.16$$

The power added efficiency PAE is related to device gain by

$$\mathrm{PAE} = \frac{P_o - P_i}{P_{\mathrm{dc}}} \qquad 3.5.17$$

where the dc power, P_{dc}, is given by

$$P_{dc} = \tfrac{1}{4}(I_{max} + I_{min})(BV_{dc} + V_k) \qquad 3.5.18$$

Assuming that the gate–drain capacitance C_{gd} is negligible, the small-signal gain G_a is

$$\left(\frac{f_t}{f}\right)^2 \frac{R_{opt}}{4R_{in}} < G_a < \left(\frac{f_{max}}{f}\right)^2 = \left(\frac{f_t}{f}\right)^2 \left(\frac{R_{ds}}{4R_{in}}\right)^2 \qquad 3.5.19$$

where

$$R_{opt} = \frac{BV_{ds} - V_k}{I_{max} - I_{min}} \qquad 3.5.20$$

is the load resistance for optimal output power and R_{ds} and R_i are the load resistance for maximum gain and the total input resistance. Most power amplifiers today operate in class AB operation, which is a compromise between the maximum power output, efficiency, and linearity requirements.

For low-noise amplifiers, the minimum noise figure of the device is of critical importance. The minimum noise figure F_{min} as a function of frequency is given by the Fukui equation:

$$F_{min} = 1 + K_f \frac{f}{f_t}\sqrt{g_m(R_g + R_s)} \qquad 3.5.21$$

where K_f is a fitting factor that relates to the material parameters of the devices, g_m the tranconductance, and R_g and R_s are the gate and source resistances, respectively.

Linearity is increasingly important for high-frequency amplification applications. One figure of merit associated with linearity for analog applications is the ratio of the input third-order intercept point (IP3) to the dc power dissipation (Larsen, 1997). For digital applications power amplifier applications, linearity is usually specified as the adjacent channel power ratio (ACPR) in dBc. ACPR measures the spectral signal "spillover" due to nonlinearities of the amplifier into the adjacent frequency bands. The required IP3 for a specified ACPR for a CDMA system is

$$IP3 = -5\log\left[\frac{P_{IM3}(f_l, f_2)B^3}{P_o[(3b - f_1)^3 - (3B - f_2)^3]}\right] + 22.2 \qquad 3.5.22$$

where IP3 is the required output third-order intercept point in dBm, B is one-half of the signal bandwidth, f_1 and f_2 are the out-of-band frequency boundaries, P_o is the amplifier output power, and $P_{IM3}(f_1, f_2)$ is the out-of-band specified power. FETs exhibit good linearity due to the nearly square-law relationship between current and voltage.

TABLE 3.6.1 Representative HFET Layer Structure (InP-Based)

Layer	Material	Thickness (nm)	Doping (cm^{-3})
Cap	InGaAs	5	$N^{+} = 2 \times 10^{19}$
Schottky layer	AlInAs	20	Undoped
Donor layer	AlInAs	10	$N = 5 \times 10^{18}$
Spacer	AlInAs	2	Undoped
Channel	InGaAs	30	Undoped
Buffer	InGaAs	500	Undoped
Substrate	InP		

Most of the discussion above centered on MESFETs and JFETs. The high-frequency performance of MODFETs is slightly more complicated by the modulation doping present in the device. Some issues specific to MODFET performance are discussed in the next section.

3.6 MATERIALS PROPERTIES AND STRUCTURE OPTIMIZATION FOR HFETS

We have quantified relationships and identified important design concerns required for achieving good device performance for a range of applications. Many of the design requirements are dependent on the intrinsic materials properties of the heterojunction materials, as discussed in Section 2.6. The focus of this section is to highlight some of the more important materials issues that affect HFET performance and to discuss optimization of device structures.

A representative layer structure for an InP-based HFET is shown in Table 3.6.1. The electron velocity and mobility of the channel material are key materials figures of merit that relate to high-frequency performance. Figure 2.6.2 compares the velocity–field characteristics of a range of semiconductor materials. In addition to transport characteristics, the two-dimensional sheet charge concentration n_s is important to achieving high current drive as well as high g_m and f_t. Typically, a higher sheet charge is associated with a higher *modulation efficiency* (Nguyen, 1991). Although a full discussion of modulation efficiency is beyond the scope of this book, this concept describes the efficiency of charge modulation by the gate in achieving incremental changes in drain current. The total charge that is modulated by the gate potential is the charge in the channel (n_s), and all other parasitic charges affected by the gate fields, such as that charge which may reside in the wider-bandgap donor supply layer above the channel, as described at the end of Section 3.5. For heterojunction material systems with smaller conduction band discontinuities, and consequently smaller n_s, the parasitic charge components are larger and the modulation efficiency is decreased. The requirements of high electron velocity and mobility, as well as high sheet charge, point to specific materials

TABLE 3.6.2 Representative Charge Concentrations and Mobilities in Modulation-Doped Structures

Heterojunction	Two-Dimensional Charge (cm^{-2})	Mobility ($cm^2/V \cdot s$)
$Al_{0.3}Ga_{0.7}As$–GaAs	1×10^{12}	7,000
$Al_{0.3}Ga_{0.7}As$–$In_{0.2}Ga_{0.8}As$	2.5×10^{12}	7,000
$Al_{0.48}In_{0.53}As$–$Ga_{0.47}In_{0.53}As$	3.0×10^{12}	10,000
AlGaSb–InAs	2×10^{12}	20,000
$Al_{0.3}Ga_{0.7}N$–GaN	1×10^{13}	1,500
$Si_{0.2}Ge_{0.8}$	p-type: 2×10^{12}	1,000
Si (strained)	n-type: 1×10^{12}	2,000

systems and design trends for improved HFET performance. Table 3.6.2 shows a comparison of representative 300-K sheet charges and electron mobilities achieved in the dominant HFET materials systems.

Use of the $In_xGa_{1-x}As$ ternary alloy to improve device performance has significantly advanced HFET performance at high frequencies. We discuss the insertion of InGaAs in device structures as a model of HFET design evolution. Taking the binary semiconductor GaAs as the starting point, the addition of InAs to create the ternary alloy, $In_xGa_{1-x}As$, increases both the electron mobility and peak velocity. Thus, devices that can accommodate InGaAs channels have very high f_t values, approaching 350 GHz for pseudomorphic InP-based structures (Nguyen et al., 1992b). Trade-offs in performance and manufacturing constraints must be considered for HFET devices containing InGaAs active layers.

First, InGaAs substrates do not exist, and therefore the commonly available GaAs and InP substrates must be used for epitaxial growth of the device structures. These substrates are generally advantageous for microwave and millimeter-wave applications, in comparison to Si, due to their high resistivity and resultant low signal loss in integrated circuits. The limits of the $In_xGa_{1-x}As$ alloy composition x that can be accommodated are determined by the critical thickness of the materials comprising the structure, as described in Section 2.6. For InGaAs channels on GaAs with a reasonable thickness of 15 nm, the indium concentration is limited to roughly 30%. Although the *unstrained* electron mobility and velocity increase with indium composition, the electron transport properties do not generally improve with increasing x for strained systems. Due to increased alloy scattering and the effects of strain and quantum confinement on the band structure, the low-temperature mobility of electrons in InGaAs on GaAs *decreases* with the addition of In from 0 to 25% (Nguyen et al., 1992a). However, due to the increase in conduction band discontinuity with increasing indium concentration, the sheet charge can increase from 1.2×10^{12} to 2.9×10^{12} cm^{-2} as x increases from 0 to 30%. The improved f_t value of pseudomorphic InGaAs on GaAs HFETs (or PHEMTs,

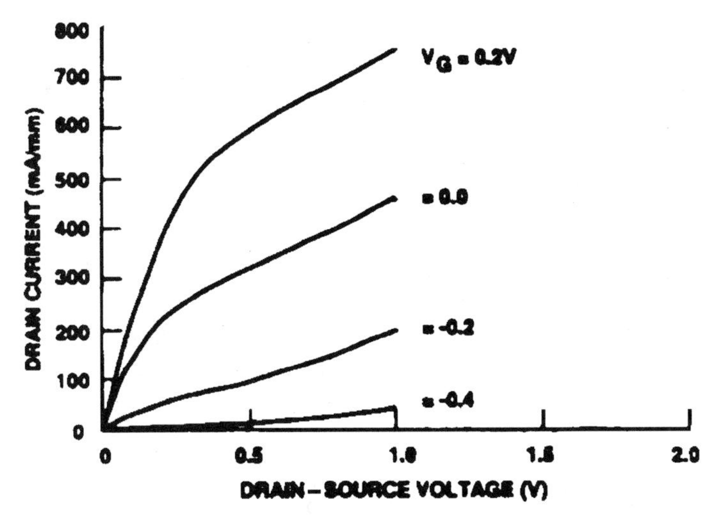

FIGURE 3.6.1 *I–V* characteristics of pseudomorphic InP-based HFET. (Reprinted with permission from IEEE.)

as they are often referred to) results primarily from improved modulation efficiency and higher n_s.

The InP substrate-based lattice-matched $Al_{48}In_{0.52}As–Ga_{0.47}In_{0.53}As$ HFET structure provides enhanced f_t performance beyond that of the PHEMT, due to its' high conduction band discontinuity and small effective mass of electrons. With increasing indium composition x beyond the lattice-matched limit, the electron mobility increases in these structures, unlike the pseudomorphic HEMT devices. With increasing indium concentration above 50%, the alloy scattering decreases in these structures. Figure 3.6.1 shows a dc current–voltage characteristic of an InGaAs–AlInAs HFET structure with an $In_{0.8}Ga_{0.2}As$ channel and a 50-nm gate length. This device exhibited a transconductance of 1740 mS/mm and an f_t of 340 GHz. The 300-K n_s and mobility of electrons in the channel was 3.6×10^{12} cm^{-2} and 12,800 cm^2/V $\cdot$ s, respectively (Nguyen et al., 1992b).

A relatively high output conductance is observed due to short-channel effects. Short-channel effects are minimized by the appropriate vertical scaling of HFETs. As the gate length is reduced, the vertical material dimensions should also be reduced to maintain a high aspect ratio between the gate length and the gate-to-channel separation. A recommended aspect ratio of greater than 5 has been determined for proper control of the field effect (Nguyen et al., 1992). Appropriate scaling is also used to minimize parasitic delays, such as charging of capacitances and extension of the drain-depletion region in the devices.

Decreasing the thickness of the doped region (supplying n_s) can decrease the gate-to-channel separation beneath the gate. Dopant atoms can be con-

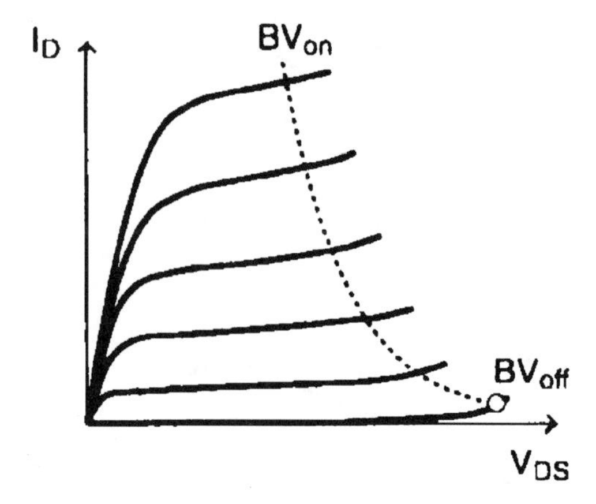

FIGURE 3.6.2 On- and off-state breakdown voltages. (Reprinted with permission from Del Alamo and Somerville, 1999.)

fined to a few atomic planes to minimize the thickness of the doped layer. This approach, called *planar doping*, can be used to improve scaling. Some spreading of the dopant does occur, however, due to diffusion or segregation of the impurities during growth.

Despite the improved high-frequency performance of HFETs as indium is added to the channel, such structures have limited use for power applications, due to their low breakdown voltage. Figure 3.6.2 shows the off- and on-state breakdown voltages of typical HFETs (Del Alamo and Somerville, 1999). The off-state voltage refers to the maximum drain-to-source (or drain-to-gate) voltage that can be applied in the off-condition (below threshold). The off-state must be defined by a specific gate current (1 mA/mm of gate width is typically used). The on-state breakdown voltage is defined by the locus in the I–V characteristics for V_{gs} above a specified threshold.

Two physical effects dominate the breakdown voltage: tunneling or the thermionic emission of gate electrons, and impact ionization of channel electrons. InP-based devices suffer from a relatively low Schottky barrier height of metals to AlInAs, and relatively high impact ionization in InGaAs due to its low bandgap. Consequently, InP-based HFETs have an off-state breakdown voltage of 2 to 3 V lower than GaAs pseudomorphic structures and an on-state breakdown voltage far lower than that of GaAs structures. Design approaches to overcoming these limits include increasing the Al concentration in the AlInAs Schottky layer (see Table 3.6.1) beneath the gate to increase the Schottky barrier height, and the introduction of InP in the InGaAs channel to increase the on-state breakdown, as shown in Figure 3.6.3 (Chen et al., 1999).

Devices with even higher velocities may be used. As shown in Table 3.6.2, the AlGaSb–InAs heterostructure yields the highest room-temperature electron mobility. The design of such structures is difficult, however, due to the low bandgap of InAs and the resultant low breakdown voltage of the devices.

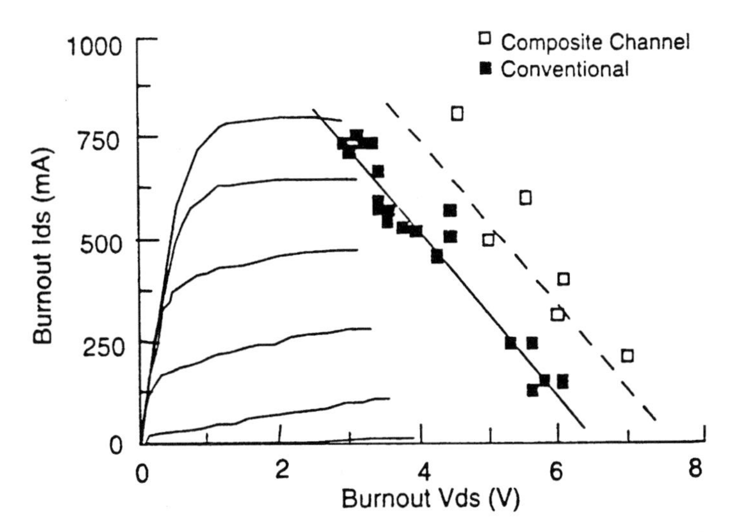

FIGURE 3.6.3 *I–V* characteristic and breakdown voltages for InGaAs and InGaAs–InP composite channel FETs. (From Kim, 2000.)

Advances in materials technology are enabling a wider range of HFET devices for specific applications. Some materials systems are limited by the lack of an appropriate substrate. For high electron and hole mobilities in Si, strain is used to engineer improvements in the band structure. The required strains are not achievable on Si substrates. Si-based HFETs are enabled by the creation of *pseudosubstrates* in which a relaxed SiGe template can be used to grow a tensilely strained Si channel for n-type HFETs and a compressively strained SiGe channels for p-type HFETs.

GaN is emerging as an important material for future high-power, high-frequency amplifiers, due to its very large bandgap and relatively high saturated drift velocity. However, GaN device technology is presently challenged because GaN bulk substrates are not available. Identification of a suitable substrate is presently still a challenge. Instead, GaN devices are produced by decoupling the AlGaN–GaN HFET structure from a highly lattice-mismatched substrate through appropriate growth conditions.

PROBLEMS

3.1. Derive Eq. 3.3.24 starting from Eq. 3.3.21 and using Eq. 3.3.23 to find $V_{D,\text{sat}}$ in terms of $I_{D,\text{sat}}$.

3.2. Determine the expression for the voltage at the edge of the depletion region, $V(h)$, in a MESFET device.

3.3. Consider an n-type GaAs MESFET with a barrier height of 0.9 V and a channel doping of 5.0×10^{16} cm^{-3} doping concentration. Assume that the active layer width $a = 0.3$ μm. The relative dielectric constant for GaAs is 13.1. Determine (**a**) the pinch-off voltage and (**b**) the threshold voltage.

3.4. Calculate the channel conductance (defined by Eq. 3.5.2) in saturation for the MODFET using the dc model developed in Section 3.3.

3.5. Start with the expression for $n(x)$ derived in Box 3.3.1 and determine an expression for the drain saturation voltage $V_{D,\text{sat}}$. Use the simplified expression for the drain current given by Eq. 3.3.27.

3.6. Using the expression derived in Problem 3.5, determine an expression for the saturation drain current, $I_{D,\text{sat}}$. Use the same simplified expression for I_D given by Eq. 3.3.27.

3.7. Consider bulk semiconductor material. As was found in Section 3.4, the velocity saturates at high electric field strengths. We can use a simple approach to show physically why the velocity saturates at high fields in bulk semiconductors. Start with the energy balance (Eq. 3.4.6) in steady state. If we assume that

$$\tau_E = \frac{E}{\hbar\omega}\tau_m$$

where $\hbar\omega$ is the phonon energy, show that the saturation velocity v_{sat} is equal to

$$|v_z| = v_{\text{sat}} = \sqrt{\frac{\hbar\omega}{m^*}}$$

Assume further that the field is along the z direction, that the equilibrium average energy E_0 can be neglected compared to the average energy E, and that the simple mobility formula

$$\mu = \frac{q\tau_m}{m^*}$$

can be used.

3.8. Determine the drain current for the following MODFET device if it is assumed that $I_G = 0$, and that $V_D = 1.0$ V, gate length = 1.0 μm, device width = 10.0 μm, $\mu = 7400$ cm^2/V $\cdot$ s, $d = 8.0 \times 10^{-6}$ cm, $\varepsilon_s = 13.1\varepsilon_0$. Assume that the two-dimensional electron concentration can be calculated at $T = 0$ K. The donor energy is 6 meV, the Al concentration is 25%, the equivalent gate field forming the channel is 0.5×10^5 V/cm, $\varepsilon_{\text{AlGaAs}} = 13.18 - 3.12x$ (where x is the Al concentration), $N_d = 3.0 \times 10^{17}$ cm^{-3}, depletion layer width = 18.0 nm, and the effective mass in GaAs is 0.067 m. Assume that only one subband is occupied. Also assume that the electron concentration at the source can be found by using the analytical technique given in Chapter 2.

3.9. Compare the mean distance traveled in GaAs to that in InP, diamond, p-type silicon, and InGaAs if the low field mobilities at 300 K are: GaAs $= 8500\ \mathrm{cm^2/V \cdot s}$; InP $= 4600\ \mathrm{cm^2/V \cdot s}$; diamond $= 2200\ \mathrm{cm^2/V \cdot s}$, InGaAs $= 13{,}800\ \mathrm{cm^2/V \cdot s}$, and p-type silicon $= 500\ \mathrm{cm^2/V \cdot s}$. The effective masses are: GaAs $= 0.067$, InP $= 0.077$, diamond $= 1.15$, InGaAs $= 0.041$, and p-type silicon $= 0.81$. Evaluate the mean distance at a field of 10 kV/cm, and assume that the momentum relaxation time can be determined from the low field mobility.

3.10. Determine an expression for the channel conductance of a MODFET below pinch-off assuming the simplified I–V characteristic given by Eq. 3.3.27.

CHAPTER 4

Heterostructure Bipolar Transistors

In this chapter we discuss the operation of heterostructure bipolar transistors (HBTs). HBTs are promising devices for many important device applications that require ultrahigh speed, high frequency, large current drive, and low noise performance. HBTs have three important attributes that make them a competitive technology: high power-handling capability, high current drive capability, and low $1/f$ noise while operating at high frequency and high power. The high power capability of HBTs is of great utility in power transistor applications, while the low $1/f$ noise is important in oscillator applications. The high current drive capability of HBTs is useful for high-speed digital applications which require charging and discharging load capacitances that arise from wiring capacitances.

4.1 REVIEW OF BIPOLAR JUNCTION TRANSISTORS

In this section we present a brief review of the basics of bipolar junction transistor (BJT) behavior. For a full discussion at an introductory level, the reader is referred to the books by Streetman and Banerjee (2000) or Pierret (1996). A BJT can be used in several circuit configurations. In most applications the input signal is across two of the BJT leads, and the output signal is extracted from a second pair of leads. Since there are only three leads for a BJT, one of the leads must be common to both the input and output circuitry. Hence we call the various circuit configurations, common emitter, common base, and common collector, to identify the lead common to both the input and output. These configurations are shown in Figure 4.1.1. The most commonly employed configuration is the common-emitter configuration.

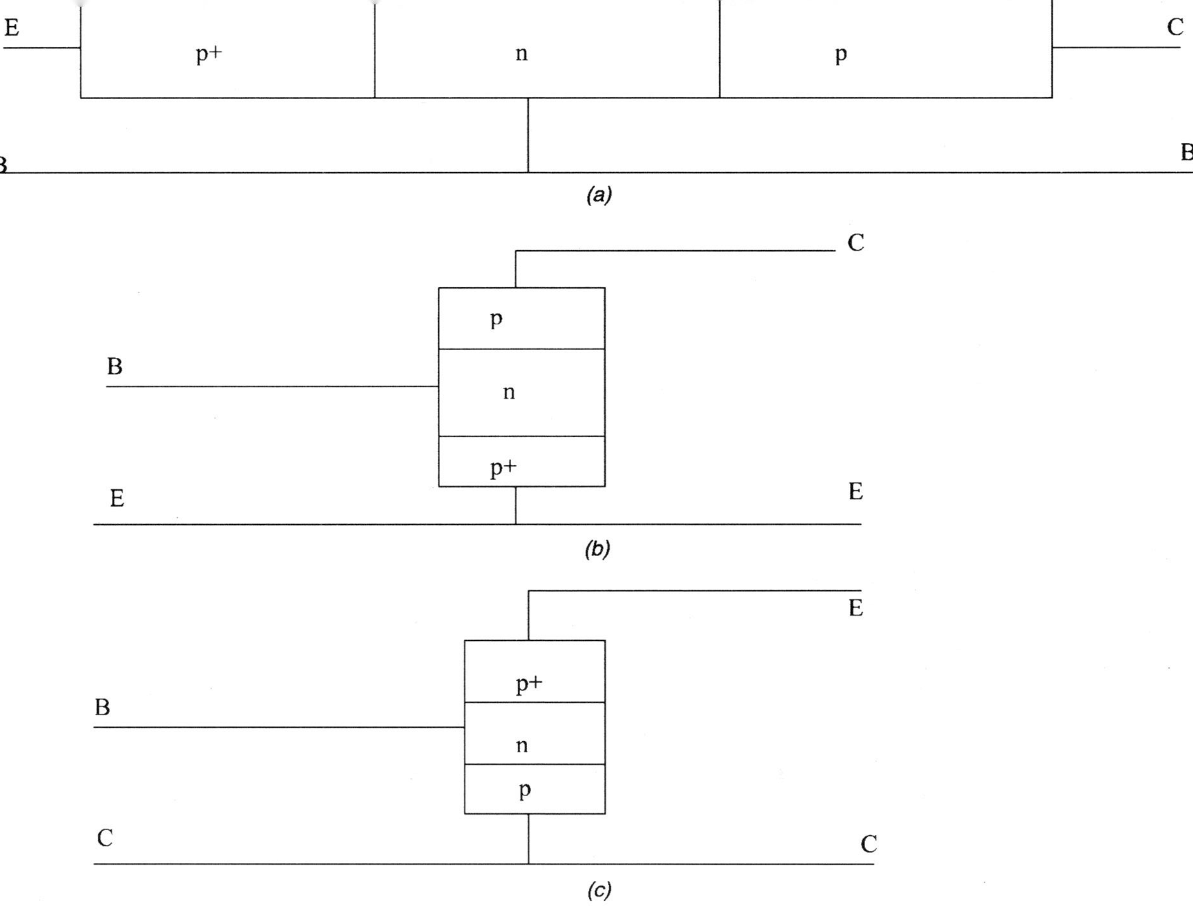

FIGURE 4.1.1 Three configurations of a BJT: (*a*) common base; (*b*) common-emitter; (*c*) common collector.

A BJT can be biased into one of four possible modes: saturation, active, inverted, and cutoff. In the saturation mode both the emitter–base and collector–base junctions are forward biased. The active mode consists of a forward-biased emitter–base junction and a reverse-biased collector–base junction. The inverted mode has the emitter–base junction reverse biased, while the collector–base junction is forward biased. Finally, in the cutoff mode, both the emitter–base and collector–base junctions are reverse biased.

The active mode is used for amplification. Under active mode biasing, the output collector current I_C is approximately equal to the input emitter current I_E. However, a small portion of the emitter current is not collected at the output. The base current, I_B, consists of three components which arise from the following:

1. Recombination of injected holes from the emitter with electrons stored in the base. The electrons lost through recombination must be restored due to space-charge neutrality requirements. I_B restores the electrons lost in the base from recombination through the base contact.
2. Electron injection into the emitter from the base in the forward-biased p-n emitter–base junction. I_B must supply the electrons lost.
3. The reverse-biased base–collector junction sweeps electrons into the base. These electrons reduce the magnitude of the base current I_B.

In the common-emitter configuration, the input base current is amplified by a factor β, which is the ratio of the collector current to the base current. The forward-biased emitter–base junction of a p-n-p BJT produces hole injection into the base from the emitter. The base must remain space-charge neutral. Therefore, the injected excess holes from the emitter that enter the base require compensating excess electrons to enter the base. These electrons enter the base mainly from the base contact, although a small portion enters from the reverse-biased collector–base junction. The injected holes spend less time in the base than the electrons since they are injected at higher velocity and energy. The holes spend on average a time called the base transit time τ_B in the base. The base transit time is essentially the time it takes the holes to transit the neutral base region and enter the collector–base depletion region. Typically, the base width W_B is small compared to the hole diffusion length L_p. Hence the base transit time is much less than the hole recombination lifetime in the base, τ_p. Recall that the hole recombination lifetime is the average time before a hole recombines with an electron within the base. Since the base transit time is much less than the hole recombination lifetime, there are many holes that pass through the base for each electron that enters the base. Thus τ_p/τ_B holes pass through the base for each electron that enters the base. The device then has a gain of

$$\frac{i_C}{i_B} = \frac{\tau_p}{\tau_B} = \beta_{\mathrm{dc}} \tag{4.1.1}$$

Equation 4.1.1 can be derived in an alternative way under steady-state conditions (Brennan, 1999, Sec. 12.8).

The currents in a BJT can be determined from solution of the continuity equation. The simplest formulations of the currents in a BJT are made using the following assumptions.

1. Drift is negligible in the base region. Holes diffuse from the emitter to the collector.
2. The emitter current is comprised entirely of holes. This implies that the emitter injection efficiency is unity.
3. The collector saturation current is negligible.
4. Current flow in the base can be treated as one-dimensional from the emitter to the collector.
5. All currents and voltages are in steady state.

For simplicity we continue to work with a p-n-p transistor, since in this case the hole current flows in the same direction as the hole flux. We assume further that the BJT is biased in the active mode with the emitter–base (EB) junction strongly forward biased, and the collector–base (CB) junction strongly reverse biased. Generally, the excess hole carrier concentrations on the n-side of the EB junction and on the n-side of the CB junction are given as (Brennan, 1999, Sec. 11.5)

$$\Delta p_E = p_B(e^{qV_{EB}/kT} - 1)$$
$$\Delta p_C = p_B(e^{qV_{CB}/kT} - 1) \qquad 4.1.2$$

where p_B is the equilibrium hole concentration within the n-type base region. The value of p_B can be found in terms of the base donor doping concentration N_{dB} as $p_B = n_i^2/N_{dB}$. If the junctions are strongly forward and reverse biased, Eqs. 4.1.2 become

$$\Delta p_E \sim p_B e^{qV_{EB}/kT}$$
$$\Delta p_C \sim -p_B \qquad 4.1.3$$

The one-dimensional continuity equation assuming no drift component is given as

$$\frac{d}{dt}\delta p = D_B \frac{\partial^2}{\partial x^2}\delta p - \frac{\delta p}{\tau_p} + G_L \qquad 4.1.4$$

where D_B is the hole diffusion constant within the n-type base, δp the excess hole concentration, τ_p the hole lifetime, and G_L the generation rate. In our notation we use for the diffusion coefficients and the diffusion lengths D and L, respectively, subscripted by E, B, or C to denote the emitter, base, or collector.

It is further understood that D and L correspond to the minority carrier within each of these regions. For example, D_B is the diffusion coefficient within the base of the minority carrier. For a p-n-p device, the minority carriers within the base are holes. Thus, D_B for the p-n-p structure corresponds to the diffusion coefficient for holes within the base. Under steady-state conditions, with no illumination, Eq. 4.1.4 becomes

$$\frac{d^2\delta p}{dx^2} = \frac{\delta p}{L_B^2} \qquad 4.1.5$$

where L_B is the hole diffusion length within the base. The general solution of Eq. 4.1.5 is

$$\delta p = C_1 e^{x/L_B} + C_2 e^{-x/L_B} \qquad 4.1.6$$

The boundary conditions are applied at the edge of the depletion region of the EB junction within the base, called $x = 0$, and at the end of the quasineutral base region W_B, as shown in Figure 4.1.2. The end of the quasineutral base region occurs at the edge of the CB depletion region within the base. The boundary conditions are then

$$\begin{aligned} \delta p(x=0) &= \Delta p_E = C_1 + C_2 \\ \delta p(x=W_B) &= \Delta p_C = C_1 e^{W_B/L_B} + C_2 e^{-W_B/L_B} \end{aligned} \qquad 4.1.7$$

Solving for the excess hole concentration as a function of x yields

$$\begin{aligned} C_1 &= \frac{\Delta p_C - \Delta p_E e^{-W_B/L_B}}{e^{W_B/L_B} - e^{-W_B/L_B}} \\ C_2 &= \frac{\Delta p_E e^{W_B/L_B} - \Delta p_C}{e^{W_B/L_B} - e^{-W_B/L_B}} \end{aligned} \qquad 4.1.8$$

The emitter hole current can now be determined from the relation

$$I_p = -qD_B A \frac{d}{dx}\delta p \qquad 4.1.9$$

using the expression for δp given by the second of Eqs. 4.1.7. The emitter hole current is found to be

$$I_{Ep} = \frac{qAD_B}{L_B}\left(\Delta p_E \coth\frac{W_B}{L_B} - \Delta p_C \operatorname{csch}\frac{W_B}{L_B}\right) \qquad 4.1.10$$

The collector hole current can be found as follows. If the current component corresponding to electron injection from the collector back into the base can

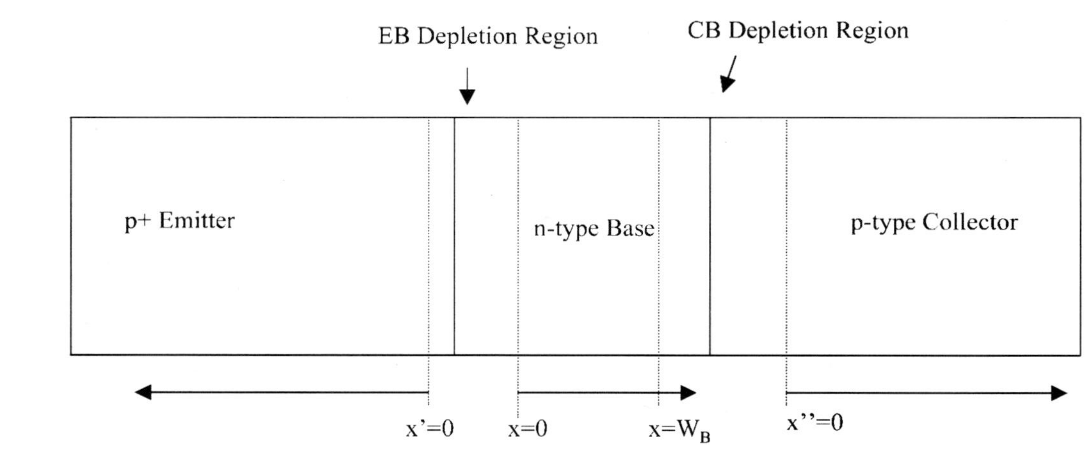

FIGURE 4.1.2 BJT showing the depletion regions and coordinate systems used in determining the device currents.

be neglected, the total collector current is given by the hole current as

$$I_{Cp} = I_p(x = W_B) = -qAD_B \frac{d}{dx}\delta p(x)\bigg|_{x=W_B} \tag{4.1.11}$$

Substituting into Eq. 4.1.11 for δp yields

$$I_{Cp} = -qAD_B \frac{d}{dx}(C_1 e^{x/L_B} + C_2 e^{-x/L_B})\bigg|_{x=W_B} \tag{4.1.12}$$

which reduces to

$$I_{Cp} = \frac{qAD_B}{L_B}(C_2 e^{-W_B/L_B} - C_1 e^{W_B/L_B}) \tag{4.1.13}$$

Substituting in for C_2 and C_1 the expressions given by Eq. 4.1.8 and simplifying the result Eq. 4.1.13 becomes

$$I_{Cp} = \frac{qAD_B}{L_B}\left(\Delta p_E \operatorname{csch}\frac{W_B}{L_B} - \Delta p_C \coth\frac{W_B}{L_B}\right) \tag{4.1.14}$$

The base current can readily be found from the node law. If the electron currents are neglected and the total emitter and collector currents are assumed to be those corresponding to the hole currents, the base current is given as

$$I_B = I_{Ep} - I_{Cp} \tag{4.1.15}$$

Substituting Eqs. 4.1.10 and 4.1.14 into Eq. 4.1.15 yields

$$I_B = \frac{qAD_B}{L_B}\left[(\Delta p_E - \Delta p_c)\left(\coth\frac{W_B}{L_B} - \operatorname{csch}\frac{W_B}{L_B}\right)\right] \tag{4.1.16}$$

Equation 4.1.16 can be simplified using the relationship between the hyperbolic functions,

$$\coth u - \operatorname{csch} u = \frac{\cosh u - 1}{\sinh u} = \tanh\frac{u}{2} \tag{4.1.17}$$

to be

$$I_B = \frac{qAD_B}{L_B}\left[(\Delta p_E - \Delta p_C)\tanh\frac{W_B}{2L_B}\right] \tag{4.1.18}$$

It is useful to determine the total currents within the device, including the electron component. The total emitter and collector currents are found simply by adding the electron components to the hole components as

$$I_E = I_{En} + I_{Ep} \qquad I_C = I_{Cn} + I_{Cp} \tag{4.1.19}$$

Let us first consider the electron component to the emitter current. Generally, the excess electron concentration in the emitter (recall that this is a p-n-p device) quasineutral region, $\delta n_E(x')$ is

$$\delta n_E(x') = A_1 e^{-x'/L_E} \tag{4.1.20}$$

where x' is the coordinate within the emitter, as shown in Figure 4.1.2 ($x' = 0$ defines the edge of the EB depletion region on the emitter side of the junction) and L_E is the electron diffusion length within the p-type emitter. Notice that only the exponentially decaying solution exists, since it is assumed that the emitter region width is sufficiently large that the excess electron concentration decays to zero. At $x' = 0$ the excess electron concentration is given by (Brennan, 1999, Sec. 11.5)

$$\Delta n_E = n_E(e^{qV_{EB}/kT} - 1) \tag{4.1.21}$$

where n_E is the equilibrium electron concentration within the emitter. Recall for this example that the structure is a pnp transistor. Therefore, n_E can be related to the emitter acceptor doping concentration N_{aE} as $n_E = n_i^2/N_{aE}$, where n_i is the intrinsic carrier concentration within the emitter. The excess electron concentration in the emitter as a function of position x' is then

$$\delta n_E = \Delta n_E e^{-x'/L_E} = n_E(e^{qV_{EB}/kT} - 1)e^{-x'/L_E} \tag{4.1.22}$$

The electron component of the emitter current is given as

$$I_{En} = -qAD_E \frac{d}{dx'} \delta n_E \bigg|_{x'=0} \tag{4.1.23}$$

where D_E is the electron diffusion coefficient within the emitter. Substituting into Eq. 4.1.23 the expression for δn_E and performing the derivative yields

$$I_{En} = \frac{qAD_E}{L_E} n_E(e^{qV_{EB}/kT} - 1) \tag{4.1.24}$$

A similar analysis can be performed to find the electron component of the collector current. In the collector region, we utilize the x'' coordinate axis, as shown in Figure 4.1.2. The origin, $x'' = 0$ occurs at the edge of the CB depletion region on the collector side of the junction. The excess electron concentration within the collector at $x'' = 0$, Δn_C, is

$$\Delta n_C = n_C(e^{qV_{CB}/kT} - 1) \tag{4.1.25}$$

where n_C is the equilibrium electron concentration within the p-type collector. The electron collector current I_{Cn} is then

$$I_{Cn} = -\frac{qAD_C}{L_C} n_C(e^{qV_{CB}/kT} - 1) \tag{4.1.26}$$

where L_C and D_C are the electron diffusion length and electron diffusion coefficient within the collector, respectively. The total emitter and collector currents are determined by adding the electron components to the hole components. The total emitter current is then

$$I_E = qA\left[\left(\frac{D_E}{L_E}n_E + \frac{D_B}{L_B}p_B\coth\frac{W_B}{L_B}\right)(e^{qV_{EB}/kT} - 1) - \left(\frac{D_B}{L_B}p_B\operatorname{csch}\frac{W_B}{L_B}\right)(e^{qV_{CB}/kT} - 1)\right] \tag{4.1.27}$$

The total collector current is given as

$$I_C = qA\left[\left(\frac{D_B}{L_B}p_B\operatorname{csch}\frac{W_B}{L_B}\right)(e^{qV_{EB}/kT-1}) - \left(\frac{D_C}{L_C}n_C + \frac{D_B}{L_B}p_B\coth\frac{W_B}{L_B}\right)(e^{qV_{CB}/kT} - 1)\right] \tag{4.1.28}$$

From knowledge of the currents, the performance parameters that characterize the device operation can be determined. Four useful quantities are typically used to describe the static characteristics of a BJT:

1. The emitter injection efficiency γ, defined as the ratio of the emitter hole current to the total emitter current,

$$\gamma = \frac{I_{Ep}}{I_{Ep} + I_{En}} \tag{4.1.29}$$

2. The base transport factor α_T, defined as the ratio of the collector hole current to the injected emitter hole current,

$$\alpha_T = \frac{I_{Cp}}{I_{Ep}} \tag{4.1.30}$$

3. The common-base current gain α_{dc},

$$\alpha_{dc} = \gamma \alpha_T \qquad 4.1.31$$

4. The common-emitter current gain β_{dc},

$$\beta_{dc} = \frac{\alpha_{dc}}{1 - \alpha_{dc}} \qquad 4.1.32$$

Eq. 4.1.32 is equivalent to Eq. 4.1.1. Therefore, the common-emitter current gain can be determined from α_{dc} or from the ratio of the collector and base currents.

Let us consider each performance parameter in turn. Typically, these parameters are defined for the device while operating in the active mode. Under these conditions it is acceptable to approximate the currents by assuming that the EB voltage is sufficiently large that the term $e^{qV_{EB}/kT} \gg 1$. Additionally, the CB junction is assumed to be sufficiently reverse biased such that the second term in Eqs. 4.1.10 and 4.1.14 can be neglected. With these assumptions I_{Ep}, I_{Cp}, and I_{En} are

$$\begin{aligned} I_{Ep} &\sim \frac{qAD_B}{L_B} p_B e^{qV_{EB}/kT} \coth \frac{W_B}{L_B} \\ I_{Cp} &= \frac{qAD_B}{L_B} p_B e^{qV_{EB}/kT} \operatorname{csch} \frac{W_B}{L_B} \qquad 4.1.33 \\ I_{En} &= \frac{qAD_E}{L_E} n_E e^{qV_{EB}/kT} \end{aligned}$$

The emitter injection efficiency is then

$$\begin{aligned} \gamma &= \frac{I_{Ep}}{I_{Ep} + I_{En}} = \frac{(p_B D_B/L_B)\coth(W_B/L_B)}{(p_B D_B/L_B)\coth(W_B/L_B) + n_E D_E/L_E} \\ &= \frac{1}{1 + D_E n_E L_B / L_E p_B D_B \tanh(W_B/L_B)} \qquad 4.1.34 \end{aligned}$$

The base transport factor is

$$\alpha_T = \frac{1}{\cosh(W_B/L_B)} \qquad 4.1.35$$

Inspection of Eq. 4.1.35 indicates that the base transport factor approaches 1 with decreasing base width. This is as expected. As the base width decreases, most of the injected holes survive their flight through the base prior to recombining. Similarly, if the hole diffusion length within the base L_B increases, the holes diffuse further on average before recombining. Provided that the base width is small with respect to L_B, most of the injected holes are collected at the collector, implying a high base transport factor. The common base current gain is

$$\alpha_{\rm dc} = \gamma\alpha_T = \frac{1}{\cosh(W_B/L_B) + D_E L_B n_E / D_B L_E p_B \sinh(W_B/L_B)} \qquad 4.1.36$$

Finally, the common-emitter current gain is found using Eq. 4.1.36 as

$$\beta_{\rm dc} = \frac{1}{\cosh(W_B/L_B) + D_E L_B n_E / D_B L_E p_B \sinh(W_B/L_B) - 1} \qquad 4.1.37$$

The common-emitter current gain can be simplified further if it is assumed that W_B is much less than L_B. Under this assumption, the cosh and sinh functions can be approximated as

$$\cosh\frac{W_B}{L_B} \sim 1 + \frac{1}{2}\left(\frac{W_B}{L_B}\right)^2$$
$$\sinh\frac{W_B}{L_B} \sim \frac{W_B}{L_B} \qquad 4.1.38$$

Substituting the relations given by Eq. 4.1.38 into Eq. 4.1.37 yields

$$\beta_{\rm dc} = \frac{1}{(D_E n_E W_B / D_B p_B L_E) + \frac{1}{2}(W_B/L_B)^2} \qquad 4.1.39$$

Notice that the common-emitter dc current gain depends strongly on the base width W_B. As the base width decreases, the common-emitter gain increases. The common-emitter dc current gain can be rewritten in terms of the emitter and base doping concentrations, N_{aE} and N_{dB}, as follows. Recognizing that n_E, the electron concentration within the p-type emitter, and p_B, the hole concentration within the n-type base, can be expressed as

$$n_E = \frac{n_{iE}^2}{N_{aE}} \qquad p_B = \frac{n_{iB}^2}{N_{dB}} \qquad 4.1.40$$

where n_{iE} and n_{iB} are the intrinsic concentrations within the emitter and base regions, respectively. In an ordinary BJT, $n_{iE} = n_{iB}$, since the emitter and base are made of the same semiconductor material. Using Eq. 4.1.40, $\beta_{\rm dc}$ becomes

$$\beta_{\rm dc} = \frac{1}{(D_E N_{dB} W_B / D_B N_{aE} L_E) + \frac{1}{2}(W_B/L_B)^2} \qquad 4.1.41$$

If W_B is further assumed to be small with respect to L_B, the last term in the denominator of Eq. 4.1.41 can be neglected. Under this assumption, Eq. 4.1.41 becomes

$$\beta_{\mathrm{dc}} \sim \frac{1}{D_E N_{dB} W_B / D_B N_{aE} L_E} \tag{4.1.42}$$

Inspection of Eq. 4.1.42 indicates that β_{dc} is inversely proportional to the product of the base doping concentration and the base width. Thus, β_{dc} is inversely proportional to the total number of donors within the base region, which is often called the *Gummel number*.

The performance of the BJT can be summarized using the Ebers–Moll model. The details of the model are developed in Box 4.1.1. The Ebers–Moll model is often used to describe the dc operation of the device and can be used for all four biasing modes.

4.2 EMITTER–BASE HETEROJUNCTION BIPOLAR TRANSISTORS

It is important to consider the factors that influence the common emitter current gain β_{dc}. From Eq. 4.1.1, β_{dc} is simply equal to the ratio of the collector to base currents. Therefore, to have a common emitter current gain of about 100, it is necessary to have the collector current about 100 times greater than the base current. To see how this can be obtained in practice, let us consider the simplified formulation of β_{dc} given by Eq. 4.1.42. Notice that β_{dc} depends on the ratio of the base-to-emitter diffusion constants, emitter diffusion length to base width, and the emitter-to-base doping concentrations. To the first order of approximation, the ratios of the diffusion constants and emitter diffusion length to base width are about 1. Therefore, the factors that most influence the value of β_{dc} that can be engineered are the doping concentrations within the emitter and base regions. If the emitter doping concentration N_{aE} is significantly larger than the base doping concentration N_{dB}, β_{dc} is increased. Thus it is desirable to dope the emitter more highly than the base, and to achieve a gain of about 100 the emitter would need to be doped about 100 times more heavily than the base. However, if the base doping is relatively low, the base resistance increases. This has the unfortunate effect of lowering the frequency performance of the device. In addition, a high emitter doping concentration results in an increased emitter–base capacitance, further lowering the frequency performance of the device. Thus, an engineering trade-off occurs in the design of a BJT; it is desirable to achieve a high current gain to dope the emitter more highly than the base, but this results in a reduction in the frequency performance of the device.

As discussed above, in an ordinary BJT, an engineering trade-off occurs in its design since a high current gain requires a high ratio of emitter-to-base doping concentrations but results in a reduction in frequency performance. This trade-off can be seen clearly by examining the frequency dependence

BOX 4.1.1: Ebers–Moll Model for a BJT

As discussed in the text, there are four biasing modes for a BJT. Therefore, it is useful to describe the behavior of the device under general biasing conditions. To do so it is useful to examine the emitter current in the device. Let us consider an n-p-n BJT. The emitter current I_E is expressed by Eq. 4.1.19 in terms of two components, I_{En} and I_{Ep}. I_{En} is the forward diffusion current, while I_{Ep} is comprised of two components, which we will refer to as I_{Ep1} and I_{Ep2}. The emitter current can then be written as

$$I_E = I_{En} + I_{Ep1} - I_{Ep2}$$

I_{Ep1} arises from hole injection into the emitter from the base. In the active biasing mode, I_{Ep1} constitutes most of the base current. I_{Ep2} represents the current flow in the emitter from holes injected from the reverse-biased collector–base junction that diffuse through the base and are collected at the base–emitter junction. Since the collector–base junction is reverse biased, the sign of the current is negative. Under active biasing, it is reasonable to further assume that the collector current is equal to the forward electron diffusion current, $I_C \sim I_{Cn} \sim I_{En}$. Therefore, if we make the assumptions that $I_{Ep1} \sim I_B$, and $I_C \sim I_{En}$, we can relate I_{Ep1} to I_{En} as

$$I_{Ep1} = \frac{I_{En}}{\beta_F}$$

where we have added the subscript F to β to distinguish that this is the gain for the forward active bias condition. The emitter current is then

$$I_E = I_{En} + \frac{I_{En}}{\beta_F} - I_{Ep2} = \left(1 + \frac{1}{\beta_F}\right) I_{En} - I_{Ep2}$$

But

$$1 + \frac{1}{\beta_F} = \frac{1}{\alpha_F}$$

where the subscript F on β and α again represents the forward active biasing condition. Therefore, the emitter current becomes

$$I_E = \frac{I_{En}}{\alpha_F} - I_{Ep2}$$

(Continued)

BOX 4.1.1 (*Continued*)

Currents I_{En} and I_{Ep2} can both be expressed using the simple diode equations. For simplicity, we define I_S as

$$I_S = \frac{qAD_{nB}}{L_{nB}} n_B$$

where D_{nB}, L_{nB}, and N_B are the electron diffusion coefficient, electron diffusion length, and equilibrium electron concentration in the base region, respectively. With this substitution, I_{En} and I_{Ep2} are

$$I_{En} = I_S(e^{qV_{EB}/kT} - 1) \qquad I_{Ep2} = I_S(e^{qV_{CB}/kT} - 1)$$

The emitter current is then

$$I_E = \frac{I_S}{\alpha_F}(e^{qV_{EB}/kT} - 1) - I_S(e^{qV_{CB}/kT} - 1)$$

A similar relationship holds for I_C which is given as

$$I_C = I_S(e^{qV_{EB}/kT} - 1) - \frac{I_S}{\alpha_R}(e^{qV_{CB}/kT} - 1)$$

where α_R holds for the reverse active bias. The relationships above are the Ebers–Moll relationships in transport form. The standard form of the Ebers–Moll equations are given using the definitions

$$I_{ES} = \frac{I_S}{\alpha_F} \qquad I_{CS} = \frac{I_S}{\alpha_R}$$

as,

$$I_E = I_{ES}(e^{qV_{EB}/kT} - 1) - \alpha_R I_{CS}(e^{qV_{CB}/kT} - 1)$$

$$I_C = \alpha_F I_{ES}(e^{qV_{EB}/kT} - 1) - I_{CS}(e^{qV_{CB}/kT} - 1)$$

The corresponding equivalent circuit that these equations represent is shown in Figure 4.1.3.

Basically, the model shows that the emitter current is comprised of two components. These are that due to a simple diode, marked as I_F in the diagram, and a current source due to the action of the collector–base junction, marked as $\alpha_R I_R$ in the diagram, where we have made the

(*Continued*)

BOX 4.1.1 (*Continued*)

definitions,

$$I_F = I_{ES}(e^{qV_{EB}/kT} - 1) \qquad I_R = I_{CS}(e^{qV_{CB}/kT} - 1)$$

Similarly, the collector current is comprised of two components. Again, one component is due to a simple diode to account for the collector–base junction, and the other is a current source that accounts for the action of the emitter–base junction. From the relationship above for I_S, we obtain an equation that relates I_{ES} to I_{CS} as

$$\alpha_F I_{ES} = \alpha_R I_{CS}$$

of the device. The maximum frequency of operation f_{max} is one of the most important figures of merit that characterizes RF performance. As we stated in Section 3.5, f_{max} is defined as the frequency at which the unilateral (condition under which there is no reverse transmission) power gain of the transistor goes to 1. The most common formulation for f_{max} for a BJT is given as

$$f_{max} = \sqrt{\frac{f_t}{8\pi r_{bb} C_C}} \qquad 4.2.1$$

where f_t is the cutoff frequency, r_{bb} the base resistance, and C_C the collector–base junction capacitance. Inspection of Eq. 4.2.1 indicates that f_{max} is inversely proportional to the square root of the base resistance. Therefore, as the base resistance increases, which results from lowering the base doping concentration, the maximum frequency of operation is reduced. Clearly, to maintain a high frequency of operation it is important to reduce the base resistance, which in turn requires a high base doping concentration.

The insertion of a wide-bandgap semiconductor material for the emitter enables the use of a higher base doping concentration without compromising the dc common-emitter current gain. Thus, high gain and high-frequency performance can be obtained simultaneously. Bipolar junction transistors that utilize a wider-bandgap material to form the emitter region are called *heterostructure bipolar transistors* (HBTs). Let us now consider how the device performance is affected by the presence of a wide-bandgap emitter.

The energy band diagram of an HBT in equilibrium and under bias is shown in Figure 4.2.1. The heterojunction is assumed to be abrupt and the device type is n-p-n, implying that the emitter is doped n-type, the base p-type, and the

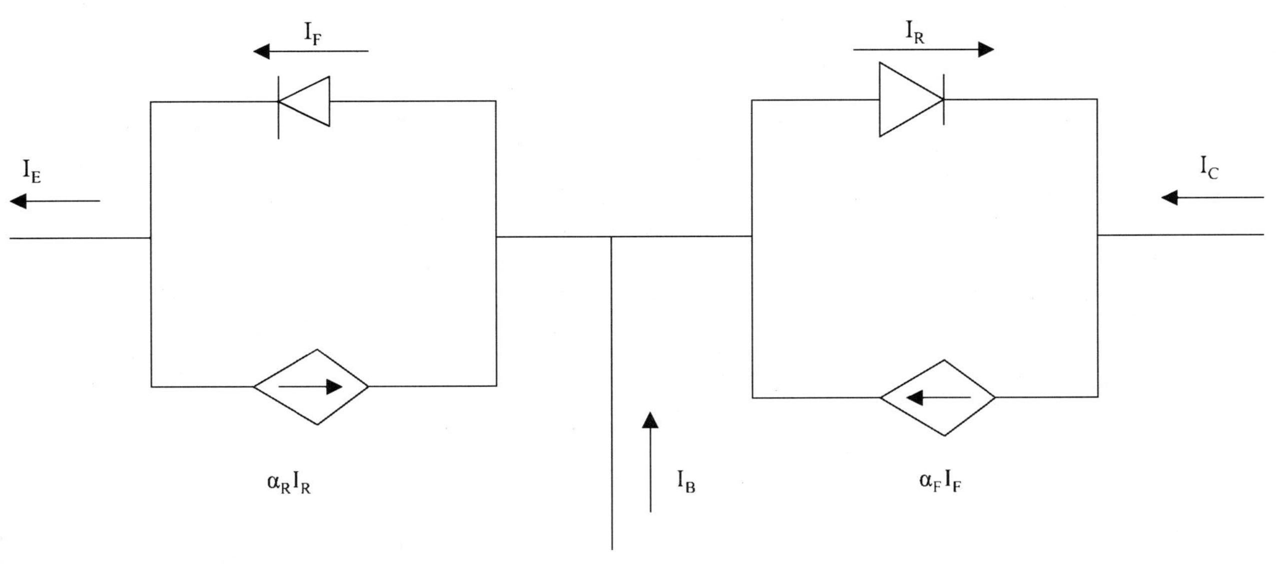

FIGURE 4.1.3 Ebers–Moll model equivalent circuit for an n-p-n device.

EXAMPLE 4.1.1: Application of the Ebers–Moll Model

Using the Ebers–Moll model, determine an expression for the emitter–collector voltage for a silicon n-p-n BJT biased into saturation.

In saturation both junctions are forward biased. The emitter and collector currents are given in the Ebers–Moll model as

$$I_E = I_{ES}(e^{qV_{EB}/kT} - 1) - \alpha_R I_{CS}(e^{qV_{CB}/kT} - 1)$$
$$I_C = \alpha_F I_{ES}(e^{qV_{EB}/kT} - 1) - I_{CS}(e^{qV_{CB}/kT} - 1)$$

In these equations, under forward bias the exponential terms are very much larger than unity. These equations can be solved to obtain V_{CB} and V_{EB} as

$$V_{EB} = \frac{kT}{q} \ln \frac{I_E - \alpha_R I_C}{I_{ES}(1 - \alpha_R \alpha_F)}$$
$$V_{CB} = \frac{kT}{q} \ln \frac{I_C - \alpha_F I_E}{I_{CS}(\alpha_R \alpha_F - 1)}$$

The voltages can be related as $V_{EC} = V_{EB} - V_{CB}$. V_{EC} can then be found as

$$V_{EC} = \frac{kT}{q} \ln \left[\frac{I_E - \alpha_R I_C}{I_{ES}(1 - \alpha_R \alpha_F)} \frac{I_{CS}(\alpha_R \alpha_F - 1)}{I_C - \alpha_F I_F} \right]$$

The above can be simplified using the following relationships:

$$I_E = I_B + I_C$$
$$\beta_F = \frac{\alpha_F}{1 - \alpha_F}$$
$$\beta_R = \frac{\alpha_R}{1 - \alpha_R}$$
$$\frac{I_{CS}(\alpha_F \alpha_R - 1)}{I_{ES}(1 - \alpha_F \alpha_R)} = -\frac{\alpha_F}{\alpha_R}$$

V_{EC} becomes

$$V_{EC} = \frac{kT}{q} \ln \left[-\frac{\alpha_F}{\alpha_R} \frac{I_B + (1 - \alpha_R) I_C}{-\alpha_F I_B + (1 - \alpha_F) I_C} \right]$$

(Continued)

EXAMPLE 4.1.1 (*Continued*)

The final expression for V_{EC} after some algebra is

$$V_{EC} = \frac{kT}{q} \ln \frac{1/\alpha_R + \beta_s/\beta_R}{1 - \beta_s/\beta_R}$$

where β_s is defined as I_C/I_B.

collector n-type. An n-p-n transistor is used here instead of the p-n-p discussed in Section 4.1, since n-p-n HBTs are far more commonly employed. This is due to the fact that n-p-n HBTs offer far superior frequency performance than p-n-p devices due to the much higher velocity of the electrons. Examination of Figure 4.2.1 shows that the energy barrier for electron injection from the emitter into the base is significantly less than the corresponding energy barrier for hole injection from the base into the emitter. Notice that the band bending is the same for both the conduction and valence bands within the emitter, but the valence band edge discontinuity adds to the energy barrier for the holes. Thus hole injection from the base into the emitter requires a higher energy to surmount the potential barrier at the emitter–base junction than electron injection from the emitter into the base. Under active biasing conditions, electrons are more readily injected from the emitter into the base than holes from the base into the emitter. The base current due to hole injection is reduced significantly without compromising the emitter current. Therefore, the base doping concentration can be increased without altering the common-emitter current gain, so high-frequency performance can be maintained at high β_{dc}.

The actual common-emitter current gains for a HBT and a BJT can be compared as follows. For an npn BJT the simplified value of β_{dc} is given as

$$\beta_{\mathrm{dc}} \sim \frac{D_B L_E N_{dE}}{D_E W_B N_{aB}} \tag{4.2.2}$$

where N_{dE} is the emitter donor concentration and N_{aB} is the base acceptor concentration. It should be noted that Eq. 4.2.2 applies mainly to a BJT but can be applied to an HBT in which diffusion processes dominate the current flow. As we will see from the discussion below, if the emitter–base junction is graded, diffusion dominates the current flow. In contrast within an abrupt HBT device, similar to that shown in Figure 4.2.1, the current flow has a strong thermionic emission component and Eq. 4.2.2 no longer applies. Therefore, Eq. 4.2.2 applies only to graded HBT structures, and a different relationship holds for abrupt HBT designs. Below we discuss in more detail the differences between the abrupt and graded HBT structures and provide expressions for

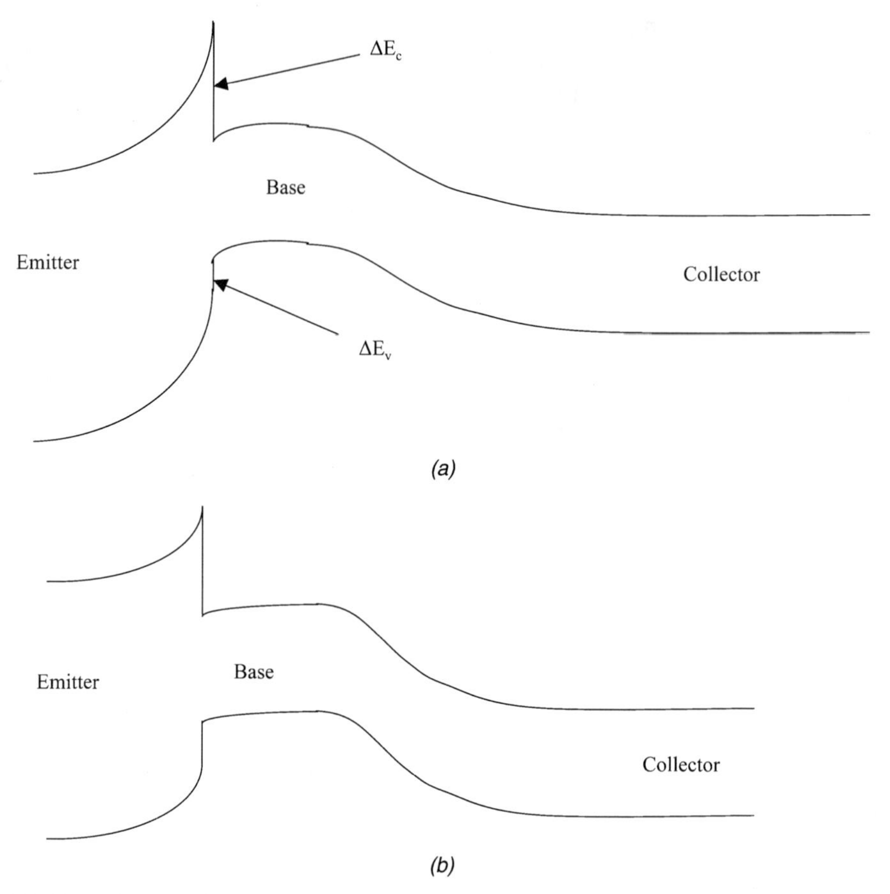

FIGURE 4.2.1 Abrupt HBT: (*a*) in equilibrium; (*b*) under active biasing conditions.

β_{dc} for both graded and abrupt heterojunction designs. In Eq. 4.2.2, D_B is the electron diffusion coefficient in the base and D_E is the hole diffusion coefficient in the emitter. Equation 4.2.2 can be rewritten in terms of the intrinsic carrier concentrations for the emitter and base, n_{iE} and n_{iB}, respectively, as follows. For an n-p-n transistor, Eq. 4.1.39 becomes

$$\beta_{dc} = \frac{1}{(D_E W_B p_E / D_B L_E n_B) + \frac{1}{2}(W_B/L_B)^2} \qquad 4.2.3$$

As in Section 4.1, if W_B is much less than L_B, the second term in the denominator of Eq. 4.2.3 can be neglected. With this assumption, Eq. 4.2.3 simplifies to

$$\beta_{dc} \sim \frac{D_B L_E n_B}{D_E W_B p_E} \qquad 4.2.4$$

But n_B, the electron concentration within the base, and p_E, the hole concentration within the emitter, are given as

$$n_B = \frac{n_{iB}^2}{N_{aB}} \qquad p_E = \frac{n_{iE}^2}{N_{dE}} \tag{4.2.5}$$

Using Eq. 4.2.5, Eq. 4.2.4 becomes

$$\beta_{\mathrm{dc}} \sim \frac{D_B L_E N_{dE}}{D_E W_B N_{aB}} \frac{n_{iB}^2}{n_{iE}^2} \tag{4.2.6}$$

In the case considered in Section 4.1, the intrinsic concentrations within the emitter and base regions are the same. However, in an HBT, n_{iB} and n_{iE} are different since the constituent semiconductor materials and their corresponding bandgaps are different. In an HBT the bandgap of the emitter is larger than that of the base. The difference in bandgaps is ΔE_g. Clearly, n_{iE} is smaller than n_{iB} since the gap of the emitter is larger than that of the base. As a result, β_{dc} is larger for a HBT than a BJT by the square of the ratio of the intrinsic concentrations in the base and emitter. The important result given by Eq. 4.2.6 is that due to the difference in the intrinsic concentrations, the base doping concentration can be larger than the emitter doping concentration, and the gain can still be relatively large. For example, in an HBT with an $Al_{0.3}Ga_{0.7}As$ emitter and a GaAs base, the ratio n_{iB}/n_{iE} squared is about 10^5. Therefore, the base can be doped 100-fold larger than the emitter and β_{dc} is still about 1000, larger than is typically necessary. Thus, use of a wide-bandgap emitter provides an excellent means of maintaining a high current gain at a high base doping concentration and concomitant high-frequency performance.

HBT devices can be made using either an abrupt or graded heterojunction to form the emitter–base junction. In an abrupt HBT the wide-bandgap emitter is grown immediately in contact with the narrow-gap base, forming an abrupt transition from one material to the other. In the graded HBT the transition from the emitter to the base is made gradually using compositional grading. Thus no sharp potential discontinuity is present at the heterointerface in the graded HBT device. The behavior of the devices is somewhat different. To understand the essential difference between the abrupt and graded HBT devices, it is important to compare their equilibrium band diagrams for the emitter and base regions as shown in Figure 4.2.2. The potential energy barriers for electron and hole injection from the emitter and base, respectively, are qV_n and qV_p. Inspection of Figure 4.2.2 shows that $q(V_p - V_n)$ is different between the two device types. This implies that the potential energy barriers are different depending on whether the device is made with an abrupt or a graded heterojunction. Since the current flow in the heterojunction depends on the potential energy barrier, the emitter and base currents and

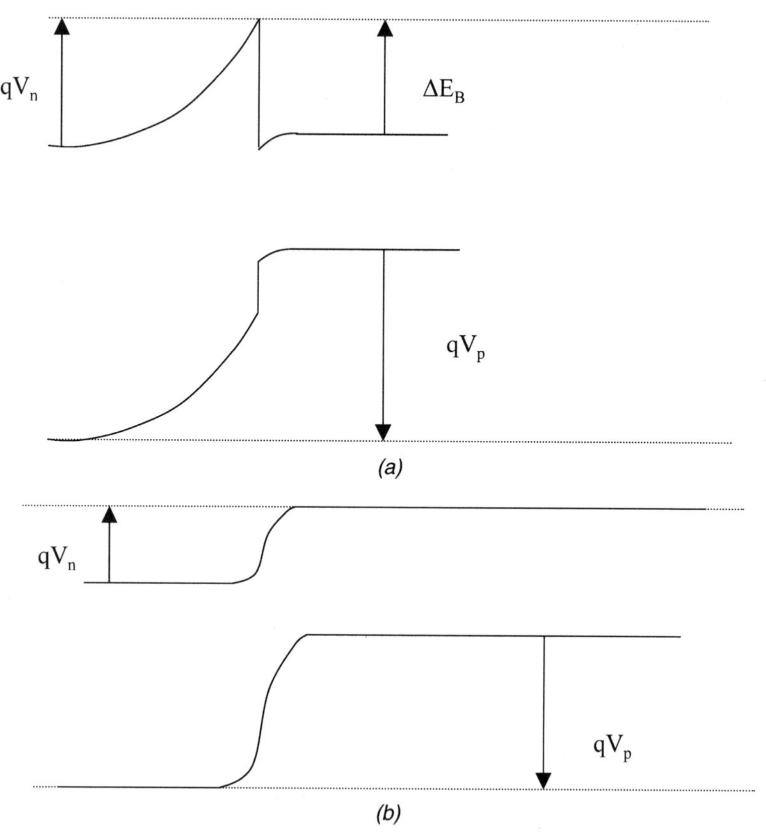

FIGURE 4.2.2 Equilibrium band diagrams for (*a*) abrupt and (*b*) graded HBT devices. Notice that $V_p - V_n$ is different in each case.

their ratio will be different between the two devices. Hence, the value of the common-emitter current gain β_{dc} is different for the two devices. For the abrupt HBT device, $V_p - V_n$ can be found as follows. Summing up the energy differences between the two layout lines on each side of the junction yields

$$qV_p + E_{g2} + \Delta E_B = qV_n + E_{g1} \qquad 4.2.7$$

where E_{g1} is the bandgap energy of the wide-bandgap material shown on the left-hand side of the figure and E_{g2} is the bandgap energy of the narrow-bandgap material shown on the right-hand side of the figure. Recognizing that the difference in the bandgaps is simply ΔE_g, Eq. 4.2.7 becomes

$$q(V_p - V_n) = \Delta E_g - \Delta E_B \qquad 4.2.8$$

From Figure 4.2.2*a*, to a good approximation, $\Delta E_B \sim \Delta E_c$. Using this assumption, the difference between V_p and V_n for an abrupt HBT becomes

$$q(V_p - V_n) = \Delta E_g - \Delta E_C = \Delta E_v \qquad \text{(abrupt HBT)} \tag{4.2.9}$$

where ΔE_v is the valence band edge discontinuity. For the graded HBT device, $(V_p - V_n)$ is different. Using Figure 4.2.2*b*, summing up the energy differences between the two layout lines on either side of the junction yields

$$qV_n + E_{g1} = qV_p + E_{g2} \tag{4.2.10}$$

Collecting terms and again using ΔE_g for the difference in the energy gaps, eq. 4.2.10 for the graded HBT becomes

$$q(V_p - V_n) = \Delta E_g \qquad \text{(graded HBT)} \tag{4.2.11}$$

As mentioned above, the current flow within the junction depends on the magnitude of the potential energy barriers for the electrons and holes at the heterointerface. Following the approach by Kroemer (1982), the electron and hole current densities can be expressed in terms of the potential energy barriers as

$$J_n = N_{dE} v_n e^{-qV_n/kT} \qquad J_p = N_{aB} v_p e^{-qV_p/kT} \tag{4.2.12}$$

where v_n and v_p are the mean electron and hole velocities due to the combined effects of drift and diffusion. The common-emitter current gain, β_{dc}, is given by the ratio of the collector to base currents. If we assume further that the base transport factor is 1, the collector and emitter currents are equal. With this assumption we obtain an estimate of the maximum value of β_{dc} (denoted as β_{max} below) as

$$\beta_{max} = \frac{J_n}{J_p} = \frac{N_{dE} v_n}{N_{aB} v_p} e^{(q/kT)(V_p - V_n)} \tag{4.2.13}$$

Using the results given by Eqs. 4.2.9 and 4.2.11, β_{max} for the abrupt and graded HBTs is given as

$$\beta_{max} = \begin{cases} \dfrac{N_{dE} v_n}{N_{aB} v_p} e^{\Delta E_v/kT} & \text{(abrupt)} \qquad 4.2.14 \\[2ex] \dfrac{N_{dE} v_n}{N_{aB} v_p} e^{\Delta E_g/kT} & \text{(graded)} \qquad 4.2.15 \end{cases}$$

Notice that β_{max} is greater in the graded HBT device than in the abrupt structure. The important consequence is that the graded device can have a substantially higher base doping concentration than the abrupt device without

EXAMPLE 4.2.1: Gain Calculation in HBTs

Estimate the improvement in gain between a graded and an abrupt AlGaAs–GaAs HBT if the Al composition in the emitter is 25%.

From Eq. 2.1.2, the energy bandgap discontinuity between the AlGaAs and GaAs is given as

$$\Delta E_g = 1.247x$$

Using $x = 0.25$, $\Delta E_g = 0.31$ eV. The conduction band edge discontinuity, ΔE_c, is 62% of ΔE_g, which is equal to 0.19 eV. The improvement in the gain can be estimated by taking the ratio of $\beta_{\max}$ for each case as

$$\frac{\beta_{\max}(\text{graded})}{\beta_{\max}(\text{abrupt})} = \frac{e^{\Delta E_g/kT}}{e^{\Delta E_v/kT}}$$

Since AlGaAs–GaAs forms a type I heterostructure, ΔE_v is found simply as 0.12 eV. The ratio of the maximum dc common-emitter gains is then

$$\frac{\beta_{\max}(\text{graded})}{\beta_{\max}(\text{abrupt})} = 103$$

We see that the use of the graded HBT structure provides roughly two orders of magnitude higher gain than that of the abrupt HBT structure.

sacrificing gain. Therefore, graded emitter–base junction HBTs are generally more attractive than abrupt HBT designs.

Abrupt junction HBT structures do, however, offer one advantage over graded devices. As we discuss below, the presence of the conduction band edge discontinuity at the emitter–base junction provides a "launching ramp" for the electrons. In other words, the electrons are injected from the emitter into the base with a minimum kinetic energy equal to the conduction band edge discontinuity. Such high-energy injection can be important in enhancing the electron velocity within the base provided that intervalley transfer does not occur. Consequently, the electron transit time within the base is reduced, leading to a higher-frequency performance. We return to this point below.

4.3 BASE TRANSPORT DYNAMICS

The frequency performance of a BJT or an HBT is characterized by both the cutoff frequency f_t and $f_{\max}$. The cutoff frequency is defined as the frequency at which the common-emitter short-circuit current gain, $h_{fe} \equiv \partial I_C/\partial I_B$, is unity. The cutoff frequency can be written in terms of the emitter–collector delay time τ_{EC} as (Sze, 1981)

$$f_t = \frac{1}{2\pi\tau_{EC}} \tag{4.3.1}$$

The emitter–collector delay time depends on several factors. It is common to express τ_{EC} in terms of four quantities: the emitter junction charging time τ_E, the base transit time τ_B, the collector–base junction depletion layer transit time, τ_{dC}, and the collector junction charging time τ_C as

$$\tau_{EC} = \tau_E + \tau_B + \tau_{dC} + \tau_C \tag{4.3.2}$$

Therefore, the cutoff frequency depends directly on the base transport time τ_B. Generally, the most limiting parameter that influences the transistor frequency response is the base transit time. Its value and methods by which it can be reduced are the main topics of this section.

There are multiple ways to calculate the base transit time. We present a simplified approach for an n-p-n BJT. We start by determining the excess minority carrier concentration within the base region as a function of position, $\delta n_B(x)$. In most situations, the base doping concentration is sufficiently high that the transport is dominated by diffusion and the effects of drift can be neglected. Under this assumption, using the solution of the diffusion equation in the base (see Problem 4.2) the excess electron concentration can be approximated as

$$\delta n_B(x) = \Delta n_E \frac{e^{W_B/L_n} e^{-x/L_n} - e^{-W_B/L_n} e^{x/L_n}}{e^{W_B/L_n} - e^{-W_B/L_n}} \tag{4.3.3}$$

where $\Delta n_E = n_B(e^{qV_{EB}/kT} - 1)$ and n_B is the equilibrium electron concentration within the base. Plotting Eq. 4.3.3 for narrow-basewidth devices produces a nearly straight line with only some slight bowing. Therefore, a straight line can closely approximate Eq. 4.3.3. Evaluating Eq. 4.3.3 at $x = 0$ and $x = W_B$ yields

$$\delta n_B(x = 0) = \Delta n_E \qquad \text{and} \qquad \delta n_B(x = W_B) = 0 \tag{4.3.4}$$

The straight-line fit that satisfies these boundary conditions is

$$\delta n_B = \Delta n_E \left(1 - \frac{x}{W_B}\right) \tag{4.3.5}$$

The base transit time can now be determined as follows. The time required for an electron to traverse the base is

$$\tau_B = \int_0^{W_B} \frac{dx}{v(x)} \tag{4.3.6}$$

The velocity $v(x)$ can be determined from the electron current within the base I_{nB} as

$$I_{nB} = -qA\delta n_B(x)v(x) \tag{4.3.7}$$

Note that the electron current moves in the opposite direction to the velocity, so the sign of I_{nB} is negative. Substituting Eq. 4.3.7 for $v(x)$ into Eq. 4.3.6 yields

$$\tau_B = -\int_0^{W_B} \frac{qA\delta n_B(x)}{I_{nB}} dx = -\int_0^{W_B} \frac{qA}{I_{nB}} \Delta n_E \left(1 - \frac{x}{W_B}\right) dx \tag{4.3.8}$$

If the base width is small such that little recombination occurs, the electron current within the base region I_{nB} is constant. With this assumption, the integral within Eq. 4.3.8 is evaluated to be

$$\tau_B = -\frac{qA\Delta n_E}{I_{nB}} \frac{W_B}{2} \tag{4.3.9}$$

Since the electron current within the base region I_{nB} is constant, it can be approximated by its value at the edge of the emitter–base depletion region at $x = 0$. I_{nB} at $x = 0$ is given as

$$I_{nB} = qAD_{nB} \frac{d\delta n_B}{dx} = -\frac{qAD_{nB}\Delta n_E}{W_B} \tag{4.3.10}$$

Substituting Eq. 4.3.10 for I_{nB} into Eq. 4.3.9, the base transit time becomes

$$\tau_B = \frac{W_B^2}{2D_{nB}} \tag{4.3.11}$$

Equation 4.3.11 shows that the base transit time is directly proportional to the square of the base width. It is important to recognize that Eq. 4.3.11 describes the base transit time assuming that diffusion dominates the electron transport through the base. To achieve high-frequency operation, it is important to design a bipolar transistor with a small base width. In addition, the base must be highly doped to reduce the base resistance.

The base transit time can be reduced further through drift-aided diffusion. Through the introduction of compositional grading, a quasielectric field can be produced within the base region that provides for field-aided diffusion. The quasielectric field can persist even for very high doping concentrations, and thereby transistors incorporating a heavily doped graded base have both a reduced base transit time and base resistance, resulting in improved high-frequency performance. Drift-aided diffusion can also be accomplished using a spatially varying doping concentration. In this case, the nonconstant doping profile gives rise to a built-in electric field similar to the quasielectric field

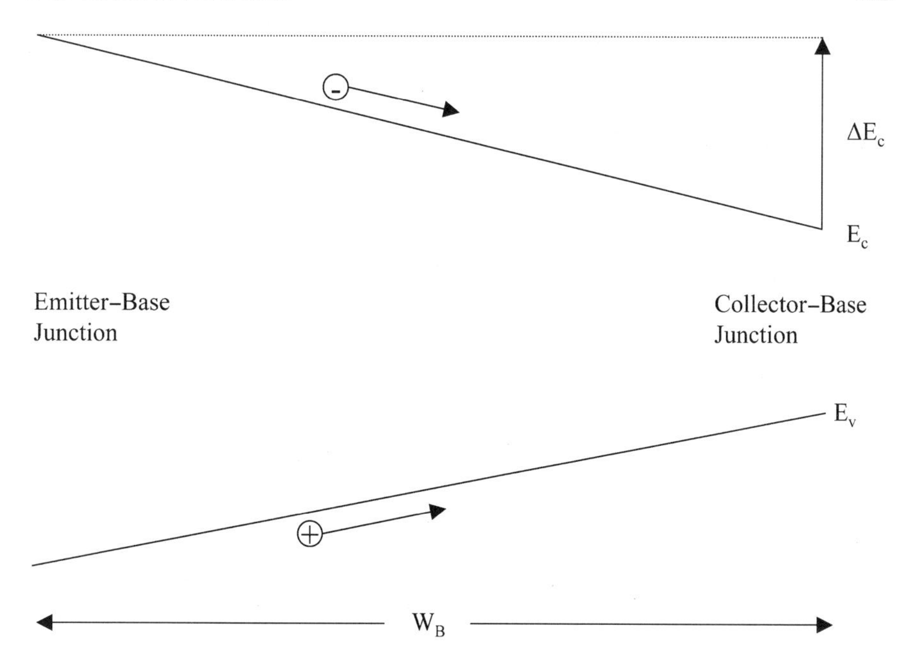

FIGURE 4.3.1 Conduction and valence bands within a compositionally graded semiconductor. The composition near the emitter–base junction has a wider bandgap than that near the collector–base junction as shown in the diagram. Electrons "roll downhill" within the potential energy diagram as shown and holes "roll uphill," thus resulting in drift aided diffusion within the base.

produced by compositional grading. Either grading method, compositional, or nonconstant doping concentration can be used. The particular advantage of a graded base device is that for a given base transit time a significantly thicker base region can be utilized with a concomitant lower base resistance due to the action of the quasielectric field.

Let us first consider the compositional grading technique. Recall that electrons "roll downhill" in potential energy diagrams. Therefore, it is necessary to grade compositionally from a higher-bandgap material at the emitter–base junction to a lower-bandgap material at the collector–base junction, as shown in Figure 4.3.1, to provide drift-aided diffusion. The quasielectric fields for electrons and holes due to compositional grading along the x direction are given as

$$F_e = -\frac{1}{q}\frac{dE_c}{dx} \qquad F_h = +\frac{1}{q}\frac{dE_v}{dx} \qquad 4.3.12$$

where E_c and E_v are the conduction and valence band edges. Notice that the quasielectric fields act to accelerate both the electrons and holes in the same direction (see Figure 4.3.1). For the graded base structure shown in Figure

4.3.1, the quasielectric fields for the electrons and holes become

$$F_e = -\frac{1}{q}\frac{\Delta E_c}{W_B} \qquad F_h = +\frac{1}{q}\frac{\Delta E_v}{W_B} \tag{4.3.13}$$

It is interesting to compare the base transit time with and without drift-aided diffusion in the base. The base transit time in the graded base device can be estimated as follows. The velocity of the electrons within the graded base can be closely approximated by the product of the electron mobility and the quasielectric field. Neglecting diffusion effects, the base transit time can be estimated for the graded base device, referred to as τ_B' as

$$\tau_B' = \frac{W_B}{-\mu_{nB}F_e} = \frac{qW_B^2}{\mu_{nB}\Delta E_c} \tag{4.3.14}$$

The ratio of the base transit times for the compositionally graded to conventional devices is given from Eqs. 4.3.14 and 4.3.11 as

$$\frac{\tau_B'}{\tau_B} = \frac{2kT}{\Delta E_c} \tag{4.3.15}$$

where D_{nB} has been replaced using the Einstein relation, $D_{nB} = \mu_{nB}kT/q$. Notice that Eq. 4.3.15 implies that the base transit time within the graded device is less than that in the conventional device by the factor $2kT/\Delta E_c$. Clearly, by choosing ΔE_c to be large, the base transit time can be greatly reduced. However, it is important to note that the conduction band edge difference should not exceed the intervalley separation energy since k-space transfer results in a dramatic decrease in the electron velocity and, consequently, an increase in the base transit time.

The base can also be graded using a nonuniform doping concentration. A spatially varying doping concentration gives rise to a built-in electric field. In equilibrium the Fermi level must be constant throughout the system. If there is a doping gradient, then in order to align the Fermi level everywhere within the semiconductor, there must be band bending, resulting in a built-in electric field. The electric field resulting from a doping concentration gradient can be determined as follows. In equilibrium the hole current within the base is zero, implying that

$$J_p = q\mu_p pF - qD_p\frac{dp}{dx} = 0 \tag{4.3.16}$$

Assuming that the hole concentration can be replaced by the acceptor concentration within the base, N_{aB}, the built-in field in the base becomes

$$F = \frac{kT}{q}\frac{dN_{aB}/dx}{N_{aB}} \tag{4.3.17}$$

EXAMPLE 4.3.1: Punch-Through Voltage in a Silicon n-p-n Transistor

Consider a silicon n-p-n transistor with the following parameters: $N_{aB} = 10^{17}$ cm^{-3}, $N_{dC} = 10^{16}$ cm^{-3}, and $W_B = 0.25$ μm. Determine the punch-through voltage and the average electric field strength at punch-through.

Under punch-through conditions, the base–collector depletion region enlarges until it reaches the emitter–base depletion region fully depleting the base of free charge carriers. Therefore, the depletion-region width on the p-side of the collector–base junction, x_p (the base side) must extend over the full base width, W_B. Thus $W_B = x_p$ and x_p is given as (Brennan, 1999, Sec. 11.1)

$$W_B = x_p = \sqrt{\frac{2\varepsilon_s}{q}\frac{N_{dC}(V_{bi} + V_{pt})}{N_{aB}(N_{aB} + N_{dC})}}$$

Solving for the punch-through voltage V_{pt} by neglecting the built-in voltage V_{bi} with respect to the punch-through voltage yields

$$V_{pt} = \frac{qW_B^2 N_{aB}(N_{aB} + N_{dC})}{2\varepsilon_s N_{dC}}$$

Substituting in the given values, the punch-through voltage becomes $V_{pt} = 47.5$ V.

The average electric field is given by the ratio of the change in voltage over the base width as

$$\langle F \rangle = \frac{\Delta V}{\Delta W} = \frac{47.5 \text{ V}}{0.25 \times 10^{-4} \text{ cm}} = 1.9x10^6 \text{ V/cm}$$

Once the doping profile within the base is known, the built-in field can be determined from Eq. 4.3.17. As an example, consider an exponential doping profile given as

$$N_{aB}(x) = N_{aB}(0)e^{-ax/W_B} \tag{4.3.18}$$

where a is a parameter equal to

$$a = \ln\frac{N_{aB}(0)}{N_{aB}(W_B)} \tag{4.3.19}$$

Using Eq. 4.3.17, the built-in field in the base for the exponential doping concentration given by Eq. 4.3.18 is

$$F = -\frac{kT}{q}\frac{a}{W_B} \tag{4.3.20}$$

If we again assume that the velocity of the electrons within the base can be approximated by the product of the mobility and the field, the base transit time (referred to here as τ_B'') becomes

$$\tau_B'' = \frac{W_B^2}{aD_{nB}} \tag{4.3.21}$$

The ratio of the base transit times for the nonuniform doping concentration graded structure to the conventional transistor is found from Eqs. 4.3.21 and 4.3.11 as

$$\frac{\tau_B''}{\tau_B} = \frac{2}{a} \tag{4.3.22}$$

From inspection of Eq. 4.3.22 it is clear that to obtain improved performance, as measured by a reduced base transit time, the parameter a must be greater than 2. The important issue here is that in using grading through a doping concentration gradient, the variation in the doping concentration within the base must be significant. Typically, the doping concentration within the base must vary by at least one or more orders of magnitude to provide a substantial improvement in the base transit time.

One of the key assumptions we have made to this point is that the base width does not vary with bias. The width of the depletion regions formed by the emitter–base and collector–base junctions change with bias. The collector–base depletion region length can change substantially since this junction is typically reverse biased. Therefore, the effective base width within the device is generally not constant under biased conditions. The action of the reverse-biased collector–base junction shortens the effective base width. This behavior is often referred to as *base narrowing* or the *Early effect.* The Early effect acts principally to alter β_{dc}. As can be seen from inspection of Eq. 4.1.41, β_{dc} increases systematically with decreasing base width under increasing collector–base bias. Therefore, the collector current for a given base current increases with increasing bias giving rise to a slope in the I_C versus V_{CE} characteristic.

Under extreme biasing conditions, the effective base width can be completely reduced to zero by the encroaching collector–base depletion region. This condition is called *punch-through.* Once punch-through is reached, the device breaks down since a large injection of carriers from the emitter directly into the collector can now occur. However, in most transistors, before punch-through can occur, avalanche multiplication within the collector–base depletion region becomes sufficiently large to trigger breakdown. Avalanche breakdown is discussed in detail in the following section.

4.4 NONSTATIONARY TRANSPORT EFFECTS AND BREAKDOWN

As we discussed in Section 3.4, nonstationary transport occurs in devices with very short active regions. These nonstationary effects are expected to appear in any structure in which the dimensions of the active regions are comparable to the electron mean free path. Carriers can traverse these short active regions without suffering sufficient collisions to relax into steady state. In an HBT, the base region is typically very small, and again nonstationary transport effects are expected to appear. As in FETs, the presence of nonstationary transport effects can greatly alter the carrier dynamics and, as such, the frequency performance of the device.

In an abrupt emitter–base junction npn HBT, the presence of the conduction band edge discontinuity provides a "launching ramp" for injected electrons (Tang and Hess, 1982; Brennan et al., 1983). Electrons entering the base from the emitter are injected with a minimum kinetic energy equal to the potential energy of the conduction band edge discontinuity, ΔE_c. Consequently, the injected electrons enter the base at a relatively high energy and a concomitant high drift velocity. It is possible, then, that the transit time within the base can be diminished. However, the actual conditions under which significant improvement in the carrier transit time can be realized depend on several parameters. Intervalley transfer in direct-gap semiconductors leads to a significant decrease in the electron drift velocity. For example, in GaAs, electrons occupy the low-effective-mass, high-mobility Γ valley under low applied electric field strengths. Even under steady-state conditions the electron drift velocity can be relatively high. As the electron energy increases to the threshold energy for intervalley transfer, the energy relaxation rate increases dramatically due to the presence of very strong deformation potential scattering. The strong deformation potential scattering acts to transfer the electrons into the secondary minima (L) valleys, in which the effective mass is substantially higher than in the Γ valley. The sudden increase in the carrier effective mass results in a sharp drop in the electron drift velocity. Therefore, for high-speed transport it is important that intervalley transfer be avoided.

To understand the influence of nonstationary transport in HBTs it is necessary to examine transient electronic transport following high-energy injection. Transient transport is defined as the dynamical regime prior to achieving steady-state conditions. Following high-energy injection from the emitter, the electrons typically traverse the narrow base region before they relax to steady state. Therefore, the transient electron dynamics within the narrow base of an HBT strongly influence the device performance. The transient can be measured either in time or distance. In Chapter 3 we presented an expression for the mean distance d an electron travels before steady-state conditions are achieved (Eq. 3.4.4). Inspection of Eq. 3.4.4 shows that the distance d depends on the applied electric field strength. The higher the electric field strength, the longer d is. One could conclude, then, that to achieve high-speed operation it is essential to use very high strength applied electric fields. However, this is incorrect. Equation 3.4.4 does not include any effects from intervalley transfer. If the applied field is too high, the electrons are heated quickly to the threshold energy for intervalley transfer. As a result, intervalley transfer occurs and the electron drift velocity decreases abruptly. Hence there is a "window" of electric field strengths under which high transient speed transport can be achieved.

Tang and Hess (1982) investigated in detail the transient electron dynamics following high-energy injection in GaAs. They concluded that a small collision-free window exists in GaAs with respect to parameters such as electric field, injection energy, external voltage, and device dimensions. Tang and Hess (1982) employed a full band ensemble Monte Carlo simulation in their research. This technique provides a full solution of the Boltzmann transport equation and naturally includes the effects of intervalley transfer (Jacoboni and Lugli, 1989). They have shown that a high transient electron velocity persists in GaAs over a typical length of 100 nm or more, provided that the external applied electric field does not accelerate the electrons to energies far above the threshold energy for intervalley transfer.

It is interesting to examine the transient electron velocity in bulk GaAs and InP. InP has similar electronic transport properties to GaAs but has a greater intervalley separation energy. The lowest-lying satellite valleys in GaAs and InP are the L valleys that lie along the $\langle 111 \rangle$ crystallographic directions. In GaAs the Γ–L separation energy is about 0.28 eV. In contrast, the Γ–L intervalley separation energy in InP is 0.54 eV, almost twice as high. Therefore, intervalley transfer and its concomitant reduction in the electron drift velocity is delayed in InP as compared to GaAs. The transient electron velocity as a function of position calculated using the ensemble Monte Carlo method for GaAs and InP are presented in Figures 4.4.1 and 4.4.2, respectively. The calculations are made for constant applied electric fields of 10 and 30 kV/cm applied along the $\langle 100 \rangle$ crystallographic direction for GaAs and InP, respectively. The electrons are launched with different initial momentum as shown in the diagram. Each value of momentum corresponds to a different launching energy. For example, in GaAs (Figure 4.4.1) an initial momentum of $0.04(2\pi/a)$ corresponds to an initial energy of 0.11 eV. Physically, this would correspond to launching an electron from the emitter into the base with a conduction band edge discontinuity of 0.11 eV. In this way, the electrons start with velocities much larger than their steady-state velocity. As can be seen in Figure 4.4.1, the initial launching energy and hence velocity greatly affect the velocity as a function of distance. If the electron is launched initially at energy close to or greater than the intervalley separation energy of 0.28 eV (cases g and h in Figure 4.4.1), the velocity drops precipitously with distance; this is due, as mentioned above, to electron transfer to the larger effective mass satellite valleys. On the other hand, low-energy injection fails to produce very high drift velocities, as seen from curves a and b in Figure 4.4.1. Intermediate choices of injection energy result in substantial velocity overshoot over the full 150 nm.

The physical explanation for the results shown in Figure 4.4.1 is simple. The electrons initially assume the small effective mass of the central (Γ) valley, whereupon they are accelerated by the electric field in the forward direction. For modest injection energies and field strengths, electrons suffer few intervalley scattering events. Most electrons move up in energy with little or modest

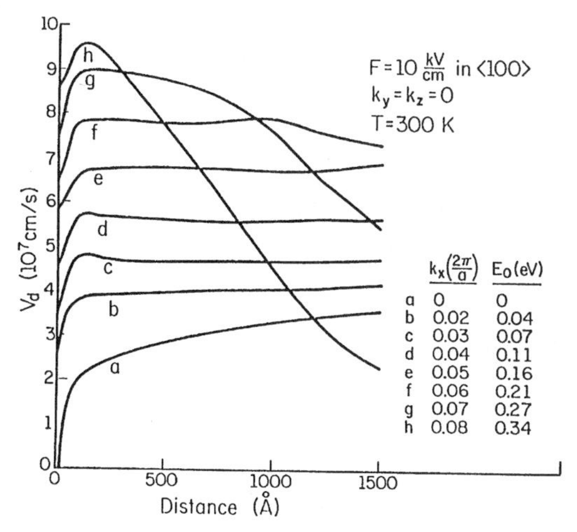

FIGURE 4.4.1 Calculated transient electron drift velocity as a function of distance in GaAs at an applied field of 10 kV/cm along the ⟨100⟩ direction at various launching energies. As can be seen, a high velocity is achieved for a window of launching energies, below the intervalley threshold energy but above the band minimum.

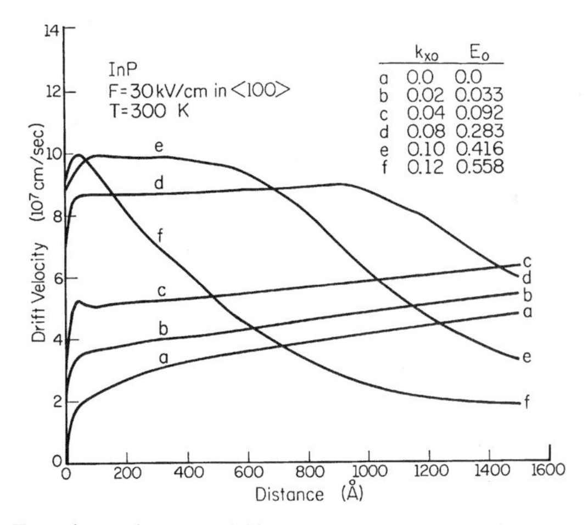

FIGURE 4.4.2 Transient electron drift velocity as a function of distance in InP for an applied field of 30 kV/cm along the ⟨100⟩ direction. Due to the larger intervalley separation energy in InP, there exists a larger range of injection energies for which velocity overshoot persists than in GaAs.

polar optical scattering in the central valley, thereby raising the ensemble average of the electron drift velocity. However, for injection energies approaching the intervalley separation energy, the electrons are promoted to the high effective mass satellite valleys. Strong intervalley scattering occurs, which reduces the drift velocity of the electrons after a relatively short transit distance.

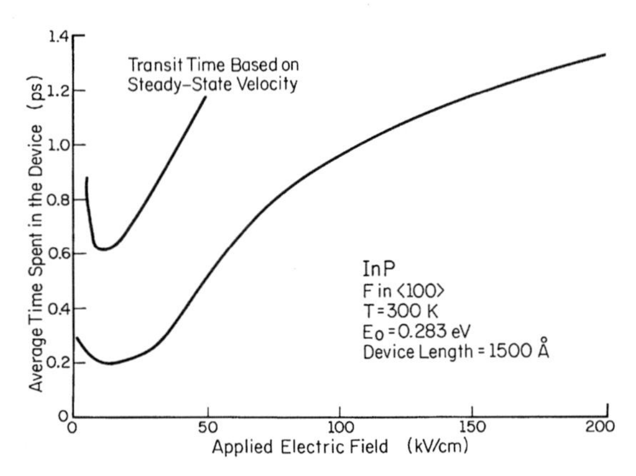

FIGURE 4.4.3 Average time spent in an InP structure 150 nm in length as a function of the applied electric field strength. The field is assumed to be constant and uniform throughout the structure and is aligned along the ⟨100⟩ direction. Initially, the electrons are injected with an energy of 0.283 eV. The transit time assuming that the electrons travel through the entire device with the steady-state velocity corresponding to the applied electric field is plotted for comparison. As can be seen, the transit time is greatly reduced for electric field strengths around 30 kV/cm, due to velocity overshoot effects.

Since the valley separation energy is greater in InP, it is expected that velocity overshoot can be attained at higher electric field strengths than in GaAs. Comparison of Figures 4.4.1 and 4.4.2 shows that velocity overshoot can indeed persist in InP at higher electric field strengths than in GaAs. It is further found that sustainable drift velocities of about 9×10^7 cm/s are achievable in InP over the first 100 nm. This is somewhat higher than that for GaAs, where the peak velocity over the first 100 nm is less than 8×10^7 cm/s. In either case, to produce sustainable velocity overshoot over the full range of the structure, the initial launching energy must be less than the valley separation energy while still above the conduction band minimum.

It is interesting to examine the carrier transit time across a 150-nm structure as a function of the applied electric field strength and launching energy. The transit time as a function of applied field for electrons injected at 0.283 eV in a 150-nm-long InP sample is shown in Figure 4.4.3. The transit time based on the corresponding steady-state velocity is shown for comparison. As can be seen from the figure, the transit time has a clear minimum for field strengths around 10 to 30 kV/cm. In addition, the transit time following high-energy injection is significantly less than the corresponding transit time based on the steady-state velocity, clearly showing the benefits of velocity overshoot. The average transit time as a function of initial launching energy at an applied field strength of 10 kV/cm in a 150-nm InP sample is shown in Figure 4.4.4. Again, a clear minimum in the transit time can be observed. In this case, the minimum transit time occurs for a launching energy of about 0.4 eV.

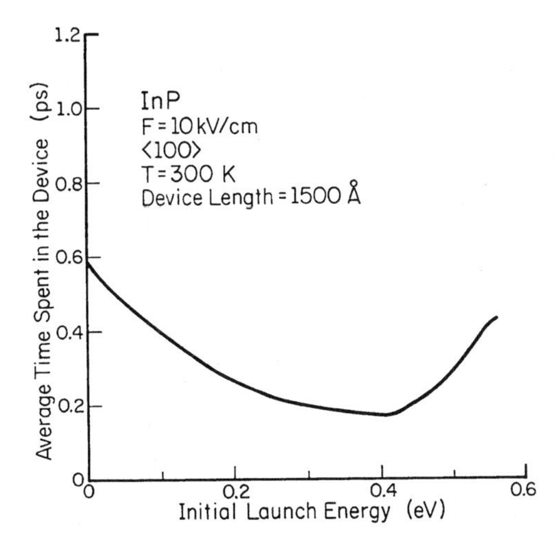

FIGURE 4.4.4 Calculated average transit time through a 150-nm InP device as a function of the initial launching energy. The field strength is assumed to be 10 kV/cm and to be applied along the ⟨100⟩ direction. As can be seen from the figure, a clear minimum appears in the transit time at a launching energy of about 0.4 eV. The increase in the transit time at higher energy arises from intervalley transfer effects; the increase at lower energy is due to the lower drift velocity at lower energy in the gamma valley.

The minimum observed in the transit time implies that the device needs to be designed carefully with respect to the applied field and launching energy to optimize nonstationary transport effects. Clearly, only select values of the applied field and launching energy result in a significant reduction in the transit time with a concomitant improvement in the frequency response of the device. In conclusion, the choice of launching energy and applied electric field greatly affects the degree of velocity overshoot in a device since these quantities strongly determine the onset of intervalley transfer.

As we have seen, the key to prolonged velocity overshoot is to delay intervalley transfer. The most obvious means by which intervalley transfer can be delayed is to use a material such as InP in which the intervalley separation energy is relatively large. However, such a choice is not always possible. Furthermore, although velocity overshoot can be reasonably maintained over a short distance, such as in the base of an HBT, it is exceedingly difficult to maintain velocity overshoot over much longer distances such as that encountered in the collector region of the device. As we discuss in Section 4.5, although the base transit time is important in improving the frequency performance of an HBT, the total emitter–collector delay time τ_{EC} is the important quantity. The emitter–collector delay time depends on parameters other than the base transit time, such as the base–collector depletion-layer transit time. Therefore, optimization of the base transit time alone is not sufficient to improve high-frequency performance. Furthermore, the reverse bias of the collector–base

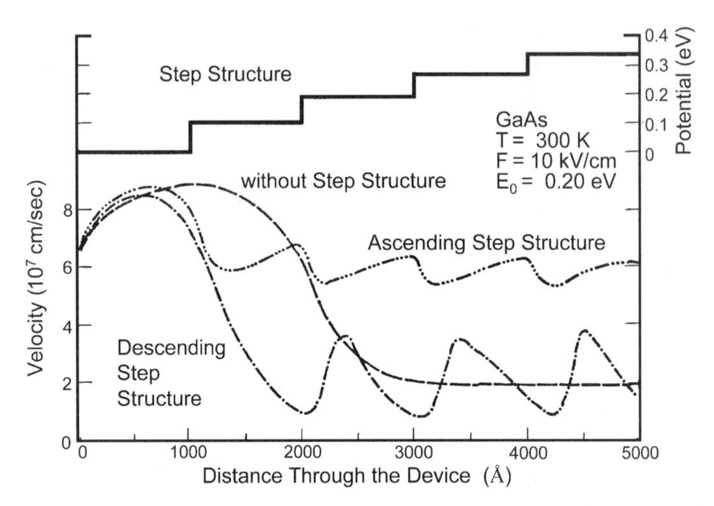

FIGURE 4.4.5 Average calculated electron drift velocity as a function of distance through the device for three cases: "climbing" up the potential stairs, down the potential stairs, and with no potential stairs present. The top part of the diagram illustrates the steplike potential structure to scale.

junction produces a high electric field that can lead to intervalley transfer, thus reducing the carrier velocity within the base–collector depletion layer. For this reason it is interesting to consider some alternative device geometries in which intervalley transfer can be delayed, thus affecting repeated velocity overshoot over significant distances.

Brennan and Hess (1983) proposed a mechanism that is suitable for achieving high speeds over large distances. This mechanism limits the electron's kinetic energy such that it remains within the Γ valley throughout the entire device. Brennan and Hess (1984) showed that impact ionization in a narrow-gap material such as InAs acts to confine electrons within the Γ valley. Since the energy gap in InAs is very small, about 0.40 eV, and the intervalley separation energy is much larger, over 1.0 eV, impact ionization occurs on average prior to intervalley transfer. The electrons gain energy from the field until they reach the threshold for impact ionization, whereupon they ionize and lose most of their kinetic energy. In this way, the electrons are confined within the Γ valley. However, this mechanism is of little practical value since it is undesirable to have carrier multiplication in most transistor applications.

A similar effect can be achieved, however, if the electrons are made to "climb" a series of potential steps under the influence of an overlaid applied electric field, as shown in Figure 4.4.5. After being launched from a high-energy barrier, the electrons are accelerated by an external electric field. They gain kinetic energy until they reach the first step. If they have sufficient energy to climb the step, they cross over into the higher-potential region, where their kinetic energy is lowered by an amount equal to the potential of the step. They continue drifting in the applied field until they reach the next step, where their kinetic energy is lowered again. The steps remove the excess kinetic energy obtained from the applied field such that the electrons remain within the Γ valley. This paradoxical effect appears to be similar in nature to the impact ionization mechanism described above for InAs.

The calculated electron drift velocity as a function of position is shown in Figure 4.4.5 for three different conditions: ascending the steps, descending the steps, and without the steps. As can be seen from Figure 4.4.5, the ascending step structure produces repeated velocity overshoot that can potentially persist over the full range of the device. In the other two cases, descending the steps and no steps present, the velocity overshoot does not persist. The key issue that determines the extent of velocity overshoot is gamma valley confinement. For the ascending step structure, gamma valley confinement is maximized, thus producing extended velocity overshoot. In the other cases, gamma valley confinement is reduced, resulting in intervalley transfer and a concomitant reduction in the drift velocity.

Maziar et al. (1986) proposed a different approach for extended velocity overshoot. In their design, the collector region has an inverted field profile. The field profiles for a uniform base and conventional collector, along with that for a uniform base and inverted field collector, are shown in Figure 4.4.6. In the diagram the base is assumed to be very heavily doped. Consequently, the depletion region is assumed to be fully within the collector. Inspection of Figure 4.4.6 shows that the slope of the field is opposite in the inverted field profile to that in the conventional device. Hence, in the inverted field collector, the electric field opposes the motion of the electrons and acts to cool the carriers instead of heat them. This works much like the potential steps discussed above. Provided that the electrons are injected into the collector region with a high kinetic energy, the inverted collector slowly cools them until they finally pass through the base–collector region. If the electrons enter the collector–base depletion region within the Γ valley, the action of the inverted field will ensure that they remain within the Γ valley for their entire flight through the device. In this way, extended velocity overshoot can be maintained. The inverted field collector can be realized by using a lightly doped base region following the heavily doped base before the heavily doped collector, as shown in Figure 4.4.7. The lightly doped base region is referred to as the *depleted base* and is doped p-type, while the collector is heavily doped n-type. The depletion region thus resides mainly within the depleted base region and the field profile resembles that shown in Figure 4.4.6*b*, where the maximum field appears at the depleted base–collector junction.

Breakdown limits the maximum voltage in a transistor. As in a diode, breakdown can occur through avalanching or tunneling. Breakdown conditions depend on the circuit configuration of the transistor. As discussed in Section 4.1, the common-base and common-emitter BJT configurations are the two most important. In the common-base configuration, the maximum collector–base

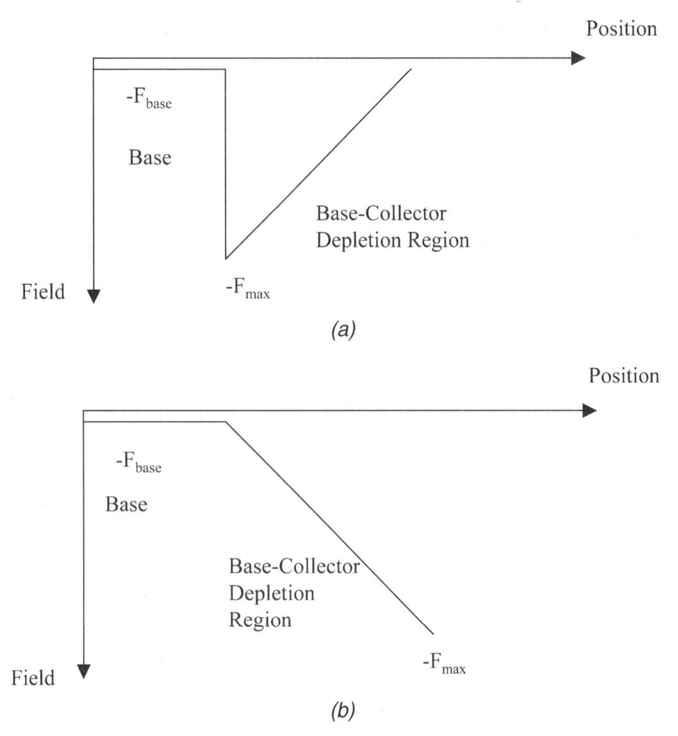

FIGURE 4.4.6 Field profile for (*a*) a conventional HBT and (*b*) an inverted field collector HBT. As can be seen from the figure, the inverted field collector acts to retard the motion of the electrons.

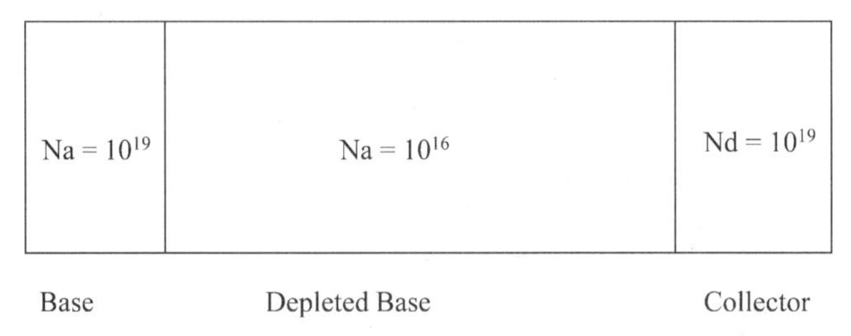

FIGURE 4.4.7 Possible inverted field collector structure showing the base, depleted base, and collector regions. Notice that the depleted base is doped lightly p-type, while the collector is heavily doped n-type. The depletion region thus resides mainly within the depleted base region and the field profile resembles that of Figure 4.4.6*b*, where the maximum field appears at the depleted base–collector junction.

voltage with the emitter open-circuited is referred to as BV_{CBO}. It is important to recognize that in the common-base configuration the behavior of the collector–base junction is essentially like that of a simple p-n junction diode.

Therefore, as the collector–base voltage increases to breakdown, the collector current increases rapidly. Provided that avalanche breakdown occurs prior to punch-through, the maximum voltage at the output of the device is just BV_{CBO}. The multiplication factor M can be expressed using an empirically based formula in terms of BV_{CBO} and the collector–base voltage V_{CB} as

$$M = \frac{1}{1 - (V_{CB}/BV_{CBO})^n} \tag{4.4.1}$$

where n is a constant and varies with the junction. The common-base current gain α_{dc} is defined from

$$I_C = \alpha_{dc} I_E + I_{CBO} \tag{4.4.2}$$

where I_{CBO} is the collector current that flows when $I_E = 0$. In the presence of multiplication, the emitter current is multiplied by M. Therefore, the common-base current gain can be rewritten as

$$\alpha'_{dc} = M\alpha_{dc} \tag{4.4.3}$$

where α'_{dc} is the common-base current gain in the presence of multiplication.

The effect of multiplication on the common-emitter configuration can be understood from the definition of β_{dc} and Eq. 4.4.3. The common-emitter current gain is given by Eq. 4.1.32 as

$$\beta_{dc} = \frac{\alpha_{dc}}{1 - \alpha_{dc}} \tag{4.4.4}$$

Substituting for α_{dc} its value under multiplication, α'_{dc}, β_{dc} becomes

$$\beta'_{dc} = \frac{M\alpha_{dc}}{1 - M\alpha_{dc}} \tag{4.4.5}$$

Notice that the common-emitter current gain goes to infinity when the product $M\alpha_{dc} = 1$, which is the breakdown condition.

The common-emitter breakdown voltage is always less than the common-base breakdown voltage (i.e., $BV_{CEO} < BV_{CBO}$). This can be understood as follows. For simplicity, let us consider a p-n-p structure. Generally, the emitter–collector voltage V_{EC} can be related to the emitter–base voltage V_{EB} and the collector–base voltage V_{CB} as

$$V_{EC} = V_{EB} - V_{CB} \tag{4.4.6}$$

Under active-mode biasing, the emitter–base junction is forward biased. Therefore, $V_{EC} \sim -V_{CB}$. One would then expect that $BV_{CEO} \sim BV_{CBO}$. However, this

EXAMPLE 4.4.1: Avalanche Breakdown vs. Punch-Through in a n-p-n BJT

For the following n-p-n BJT, determine if it breaks down from avalanche multiplication prior to punch-through by calculating the avalanche breakdown voltage and the punch-through voltage. Assume that it is biased in the common-base mode. $N_{aB} = 10^{16}$ cm^{-3}, $W_B = 1.0$ μm, critical field in silicon = 3×10^5 V/cm, and the collector doping is very heavy relative to the base.

First, determine the avalanche multiplication breakdown voltage. The maximum electric field strength in a p-n junction is given as (Brennan, 1999, Sec. 11.1)

$$F_m = -\frac{qN_{aB}x_p}{\varepsilon_s}$$

Substituting for x_p, the extent of the depletion region within the base yields

$$x_p = \sqrt{\frac{2\varepsilon_s}{q}\frac{N_{dC}}{N_{aB}(N_{aB} + N_{dC})}(V_{bi} + V_j)}$$

where V_j is assumed negative and is the voltage applied on the junction. Neglecting the built-in voltage, the critical or maximum electric field can then be written as

$$F_m = -\frac{qN_{aB}}{\varepsilon_s}\sqrt{\frac{2\varepsilon}{q}\frac{V_j N_{dC}}{N_{aB}(N_{aB} + N_{dC})}}$$

Solving for V_j yields

$$V_j = F_m^2 \frac{\varepsilon_s(N_{aB} + N_{dC})}{2qN_{aB}N_{dC}}$$

If the collector doping is heavy with respect to the base, $N_{dC} \gg N_{aB}$ and V_j can be rewritten as

$$V_j = F_m^2 \frac{\varepsilon_s}{2qN_{aB}}$$

The junction voltage at the maximum or breakdown field is just the avalanche breakdown voltage in the common-base configuration, or BV_{CBO}. Therefore, $BV_{CBO} = V_j$. Evaluating the expression for V_j, the avalanche breakdown voltage is

$$BV_{CBO} = 29.6 \text{ V}$$

(Continued)

EXAMPLE 4.4.1 (*Continued*)

The punch-through voltage is found in a manner similar to that in Example 4.3.1. Under punch-through conditions, the depletion region extends from the base–collector junction fully across the base. Thus the depletion-region length on the p-side (base side) must be equal to the base width at punch-through. Using the result developed in Example 4.3.1, the punch-through voltage is given as

$$V_{pt} = \frac{qW_B^2 N_{aB}(N_{aB} + N_{dC})}{2\varepsilon_s N_{dC}}$$

Using the fact that $N_{dC} \gg N_{aB}$, V_{pt} becomes

$$V_{pt} = \frac{qW_B^2 N_{aB}}{2\varepsilon_s}$$

Upon substituting values, $V_{pt} = 7.6$ V. Comparing the avalanche breakdown voltage to the punch-through voltage shows that punch-through occurs before avalanching in this device.

is not the case. BV_{CEO} is always less than BV_{CBO}. As V_{CB} becomes appreciable, but still less than BV_{CBO}, some of the holes that traverse the collector–base depletion region gain sufficient energy to impact-ionize. Secondary electrons and holes are produced following these impact ionization events (Brennan, 1999, Secs. 10.4 and 12.5). The secondary holes along with the primary holes continue their flight through the depletion region and are collected. The secondary electrons move in the opposite direction, however, and enter the base. In the common-emitter configuration, the base current is held constant. Therefore, the additional electrons that enter the base from the base–collector depletion region produced from impact ionization events cannot exit the base through the base contact. Instead, they must be back-injected into the emitter. This leads to an increase in the emitter hole current I_{Ep}. As discussed in Section 4.1, $\tau_p/\tau_B = \beta_{dc}$ holes pass through the base for each electron that enters the base, resulting in a large increase in the collector current. Therefore, the hole collector current increases from the transistor gain caused by feedback from the impact ionization process. As a result, the device breaks down at a lower voltage than in the common-base configuration.

A quantitative relationship between BV_{CEO} and BV_{CBO} can be obtained as follows. Equation 4.4.5 indicates that breakdown occurs when $M = 1/\alpha_{dc}$.

The multiplication M can also be expressed by Eq. 4.4.1 as

$$M = \frac{1}{1 - (V_{CB}/\text{BV}_{CBO})^n} \tag{4.4.7}$$

However, we found above that $V_{EC} \sim -V_{CB}$. Therefore, V_{CB} can be replaced by BV_{CEO} in Eq. 4.4.7 to give the multiplication when the voltage reaches BV_{CEO}. Under this condition along with $M = 1/\alpha_{\text{dc}}$, Eq. 4.4.7 becomes

$$M = \frac{1}{1 - (\text{BV}_{CEO}/\text{BV}_{CBO})^n} = \frac{1}{\alpha_{\text{dc}}} \tag{4.4.8}$$

Rearranging Eq. 4.4.8, BV_{CEO} is equal to

$$\text{BV}_{CEO} = \text{BV}_{CBO}\sqrt[n]{1 - \alpha_{\text{dc}}} \tag{4.4.9}$$

Equation 4.4.9 can be further simplified by noting that α_{dc} is generally very close to 1. Therefore,

$$\beta_{\text{dc}} = \frac{\alpha_{\text{dc}}}{1 - \alpha_{\text{dc}}} \sim \frac{1}{1 - \alpha_{\text{dc}}} \tag{4.4.10}$$

Using the approximation given by Eq. 4.4.10, Eq. 4.4.9 becomes

$$\text{BV}_{CEO} = \text{BV}_{CBO}\beta_{\text{dc}}^{-1/n} \tag{4.4.11}$$

Generally, n is larger than 1, often between 3 and 6, depending on the semiconductor material used in the junction. Consequently, BV_{CEO} is generally significantly less than BV_{CBO}.

4.5 HIGH-FREQUENCY PERFORMANCE OF HBTS

In this section we discuss the high-frequency behavior of HBTs. Generally, the intrinsic high frequency models used for HBTs are the same as those for BJTs. The only significant difference in the ac response of the two devices comes from the effects of nonstationary transport described above in Section 4.4. Since these effects cannot readily be included in analytical models of the device, for simplicity we restrict our discussion to the intrinsic model used most commonly for BJTs, the hybrid-π model.

The common-emitter configuration is the most generally employed bipolar transistor connection used in RF amplifiers. For low-frequency operation an HBT and more generally, a BJT can be represented as a simple two-port network characterized by four network parameters. As discussed in Box 4.5.1, these parameters are called hybrid or h parameters since they are all not dimensionally the same. Based on the results derived for the low-frequency behavior of a BJT in the common-emitter configuration in Box 4.5.1, the

BOX 4.5.1: Low-Frequency Small-Signal Linear Hybrid Model

The bipolar transistor can be treated as a two-port network as shown in Figure 4.5.1. We represent ac quantities as lowercase variables such as i_1 and v_1 and dc quantities as uppercase variables, I_1 and V_1. As can be seen from the figure, the inputs are i_1 and v_1 with the outputs i_2 and v_2. Generally, the independent variables are the input current i_1 and the output voltage v_2. Assuming linear conditions, the dependent variables can be related to the independent variables as

$$v_1 = h_{11}i_1 + h_{12}v_2$$
$$i_2 = h_{21}i_1 + h_{22}v_2$$

The parameters h_{ij} are called *hybrid parameters* since they do not all have the same dimensions. Let us consider in turn the physical meaning of each parameter. The first parameter, h_{11}, is the transistor input resistance with the output short-circuited and is given as

$$h_{11} = \left.\frac{v_1}{i_1}\right|_{v_2=0}$$

The second parameter, h_{12}, is the reverse voltage gain with the input open-circuited,

$$h_{12} = \left.\frac{v_1}{v_2}\right|_{i_1=0}$$

h_{21} is the negative of the forward current gain (notice that i_2 flows in the opposite direction to i_1 in Figure 4.5.1) with the output short-circuited, given as

$$h_{21} = \left.\frac{i_2}{i_1}\right|_{v_2=0}$$

Finally, h_{22} is the output conductance with the input open-circuited,

$$h_{22} = \left.\frac{i_2}{v_2}\right|_{i_1=0}$$

The h parameters for a BJT connected in the common-emitter mode are found as follows. In the common-emitter configuration shown in

(*Continued*)

BOX 4.5.1 (*Continued*)

Figure 4.1.1*b*, the input quantities are i_B and v_{BE} and the output parameters are i_C and v_{CE}. Choosing i_B and v_{CE} as the independent variables, the dependent variables i_C and v_{CE} can be written as functions of these quantities. The small-signal current i_C can be expressed as the difference between the total current $I_C(I_B + i_B, V_{CE} + v_{CE})$ and the dc current $I_C(I_B, I_{CE})$ as

$$i_C = I_C(I_B + i_B, V_{CE} + v_{CE}) - I_C(I_B, V_{CE})$$

Similarly, v_{BE} can be written as

$$v_{BE} = V_{BE}(I_B + i_B, V_{CE} + v_{CE}) - V_{BE}(I_B, V_{CE})$$

For small-signal operation the first term in the expressions for i_C and v_{BE} above can be expanded about the dc operating point, I_C and V_{BE}, using a Taylor series to first order as

$$I_C(I_B + i_B, V_{CE} + v_{CE}) = I_C(I_B, V_{CE}) + \left.\frac{\partial I_C}{\partial V_{CE}}\right|_{I_B} v_{CE} + \left.\frac{\partial I_C}{\partial I_B}\right|_{V_{CE}} i_B$$

The small-signal value of i_C is then easily found as

$$i_C = \left.\frac{\partial I_C}{\partial V_{CE}}\right|_{I_B} v_{CE} + \left.\frac{\partial I_C}{\partial I_B}\right|_{V_{CE}} i_B$$

A similar relationship is found for v_{BE} as

$$v_{BE} = \left.\frac{\partial V_{BE}}{\partial V_{CE}}\right|_{I_B} v_{CE} + \left.\frac{\partial V_{BE}}{\partial I_B}\right|_{V_{CE}} i_B$$

The dependent variables can also be written in terms of the independent parameters using the two-port network parameters as

$$v_{BE} = h_{ie} i_B + h_{re} v_{CE}$$

$$i_C = h_{fe} i_B + h_{oe} v_{CE}$$

(*Continued*)

BOX 4.5.1 (*Continued*)

The h parameters can be related to the derivatives as

$$h_{fe} = \left. \frac{\partial I_C}{\partial I_B} \right|_{V_{CE}}$$

is the small-signal current gain. The output conductance h_{oe} is defined as

$$h_{oe} = \left. \frac{\partial I_C}{\partial V_{CE}} \right|_{I_B}$$

The input resistance h_{ie} is given as

$$h_{ie} = \left. \frac{\partial V_{BE}}{\partial I_B} \right|_{V_{CE}}$$

The last parameter is the feedback voltage ratio h_{re}, and it is given as

$$h_{re} = \left. \frac{\partial V_{BE}}{\partial V_{CE}} \right|_{I_B}$$

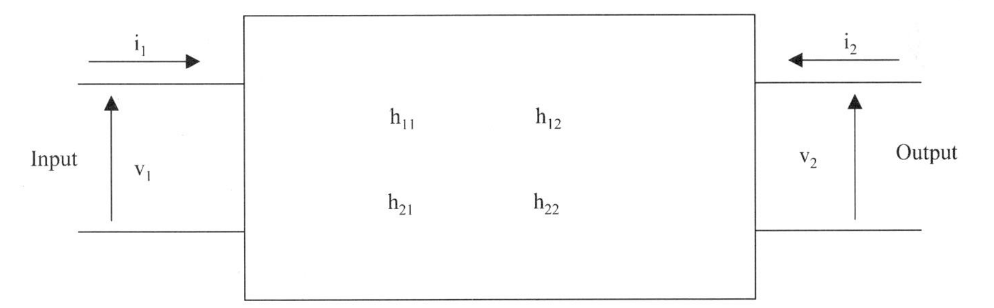

FIGURE 4.5.1 Two-port network representation of a bipolar transistor. The input parameters are v_1 and i_1, and the output parameters are v_2 and i_2. The network parameters are h_{11}, h_{12}, h_{21}, and h_{22}.

dependent variables v_{BE} and i_C can be expressed as

$$\begin{aligned} v_{BE} &= h_{ie} i_B + h_{re} v_{CE} \\ i_C &= h_{fe} i_B + h_{oe} v_{CE} \end{aligned} \quad 4.5.1$$

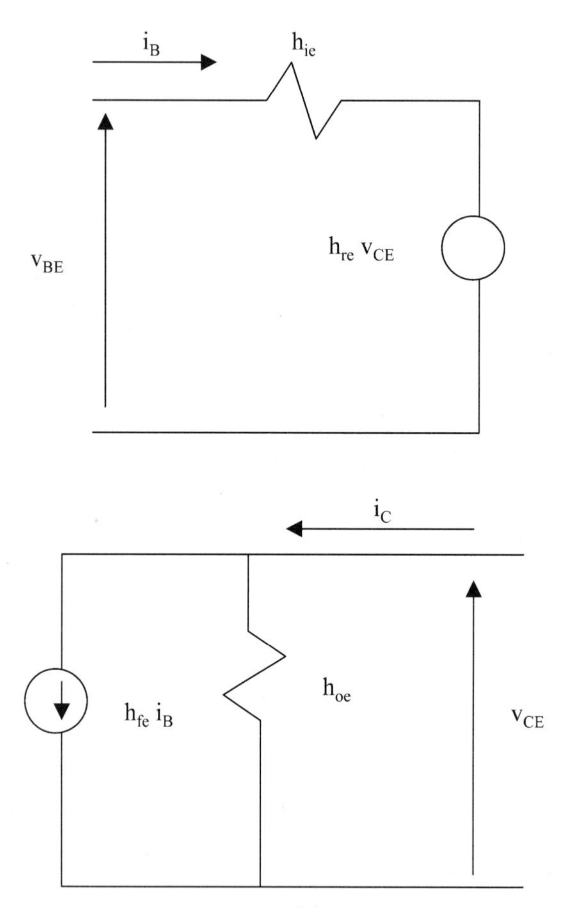

FIGURE 4.5.2 Low-frequency equivalent circuit for a BJT based on the *h* parameters developed in Box 4.5.1.

Using the loop law, the first of Eqs. 4.5.1 can be used to construct the equivalent circuit for v_{BE} shown in Figure 4.5.2. From the node law, the equivalent circuit for i_C can also be constructed as shown in Figure 4.5.2. It is important to note that at low frequencies, h_{fe} is equal to the common-emitter current gain β_{dc} defined in Section 4.1.

At higher frequencies, the simple equivalent circuit given in Figure 4.5.2 is no longer useful since it neglects junction capacitances that become more important with increasing frequency. The most commonly employed circuit model for the common-emitter configuration at higher frequencies is the hybrid-π model shown in Figure 4.5.3. Each of the circuit parameters can be determined as follows. The transconductance g_m is defined as

$$g_m \equiv \frac{dI_C}{dV_{EB}} \tag{4.5.2}$$

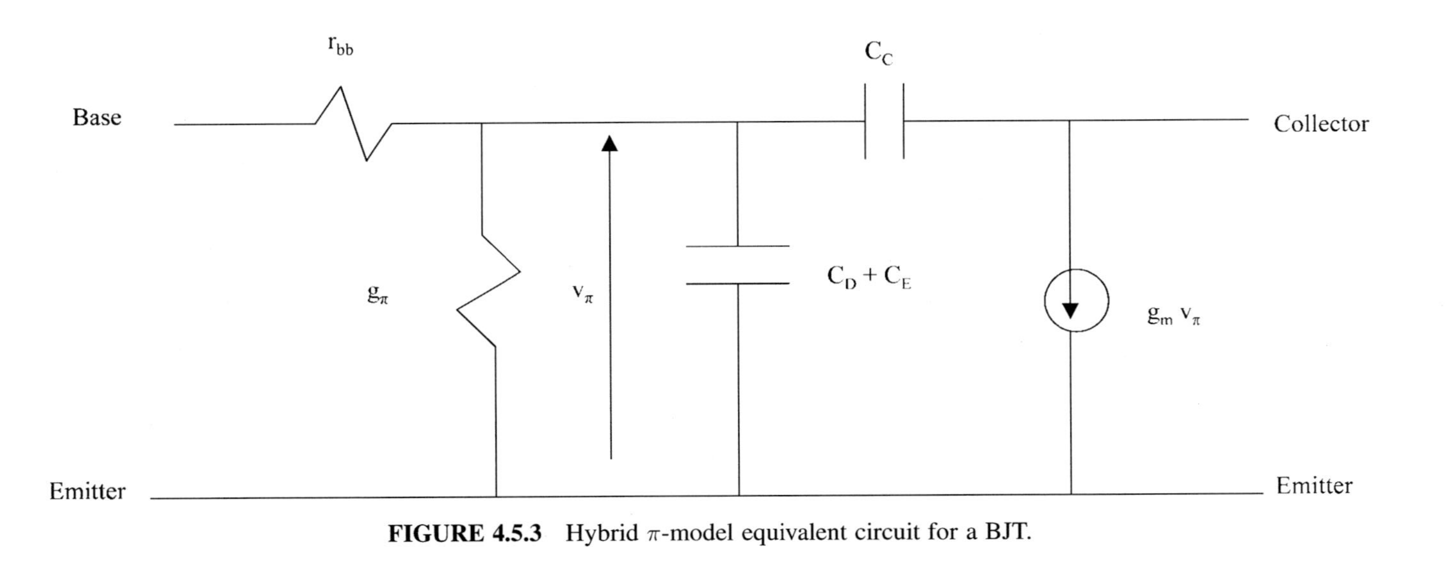

FIGURE 4.5.3 Hybrid π-model equivalent circuit for a BJT.

The collector current is given for a p-n-p transistor by Eq. 4.1.28. Under active biasing conditions where $V_{EB} \gg kT$ and V_{CB} is $\ll 0$, I_C can be approximated as

$$I_C = qA\frac{D_B}{L_B}p_B \text{csch}\frac{W_B}{L_B}e^{qV_{EB}/kT} \sim qA\frac{D_B}{W_B}p_B e^{qV_{EB}/kT} \qquad 4.5.3$$

Differentiating Eq. 4.5.3 with respect to V_{EB} yields for g_m

$$g_m = \frac{I_C}{V_T} \qquad 4.5.4$$

where V_T is the thermal voltage kT/q. The transconductance characterizes the amplifying function of the transistor; it states how the output collector current I_C varies with the input emitter–base voltage V_{EB}.

The parameter g_π is the small-signal input conductance and is defined as

$$g_\pi \equiv \frac{dI_B}{dV_{EB}} \qquad 4.5.5$$

The base current is given by Eq. 4.1.16 as

$$I_B = \frac{qAD_B}{L_B}\left[(\Delta p_E - \Delta p_c)\left(\coth\frac{W_B}{L_B} - \text{csch}\frac{W_B}{L_B}\right)\right] \qquad 4.5.6$$

with Δp_E and Δp_C given by Eqs. 4.1.2 as

$$\begin{aligned}\Delta p_E &= p_B(e^{qV_{EB}/kT} - 1)\\ \Delta p_C &= p_B(e^{qV_{CB}/kT} - 1)\end{aligned} \qquad 4.5.7$$

Again, assuming active biasing conditions, the base current can be simplified to

$$I_B = \frac{qAD_B p_B}{L_B}e^{qV_{EB}/kT}\left(\coth\frac{W_B}{L_B} - \text{csch}\frac{W_B}{L_B}\right) \qquad 4.5.8$$

With this simplification, g_π becomes

$$g_\pi = \frac{dI_B}{dV_{EB}} = \frac{I_B}{V_T} \qquad 4.5.9$$

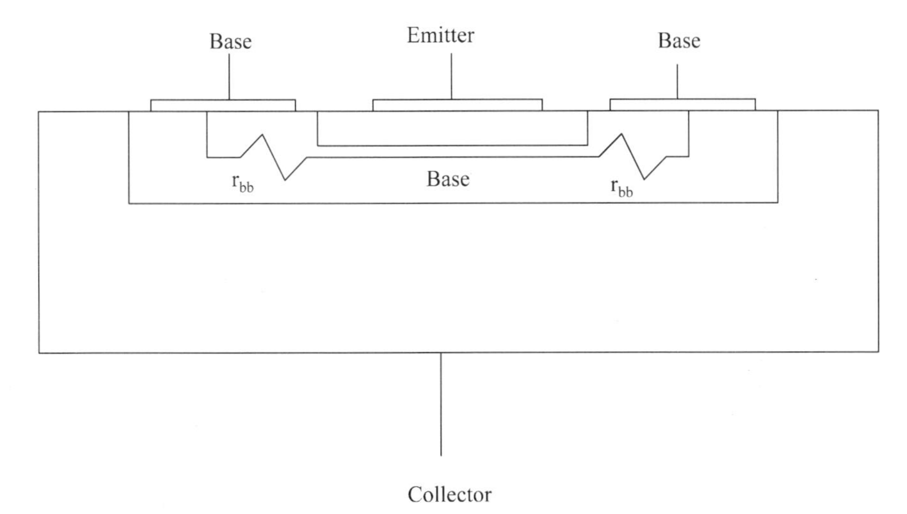

FIGURE 4.5.4 BJT showing the base resistance r_{bb} that exists between the base contact and the active region of the base.

It should be recalled that $\beta_{dc} = I_C/I_B$. It is also important to recognize that $h_{fe} = \beta_{dc}$ at zero frequency. Using these relationships, Eq. 4.5.9 becomes

$$g_\pi = \frac{I_C}{h_{fe}V_T} = \frac{g_m}{h_{fe}} \qquad 4.5.10$$

The quantity r_{bb} is the base resistance, specifies the voltage drop between the base contact and the active base region. The base resistance is shown schematically in Figure 4.5.4. As can be seen from the figure, the base resistance appears mainly from the lateral dimensions of the device. The base resistance can be reduced by either shortening the lateral dimensions or increasing the base doping concentration.

Finally, the capacitances in the model represent the collector capacitance C_C, the emitter capacitance C_E, and the diffusion capacitance C_D. The collector and emitter capacitances are the junction capacitances for the collector–base and emitter–base junctions, respectively. The diffusion capacitance describes how the base storage charge changes with emitter bias. The diffusion capacitance can be determined from

$$C_D \equiv \frac{dQ_B}{dV_{EB}} \qquad 4.5.11$$

The total charge stored in the base is given by Q_B, which can be calculated as follows. Using Eq. 4.3.4, the excess minority carrier concentration within the base for a p-n-p transistor is given as

$$\delta p_B(x = 0) = \Delta p_E \qquad \delta p_B(x = W_B) = 0 \qquad 4.5.12$$

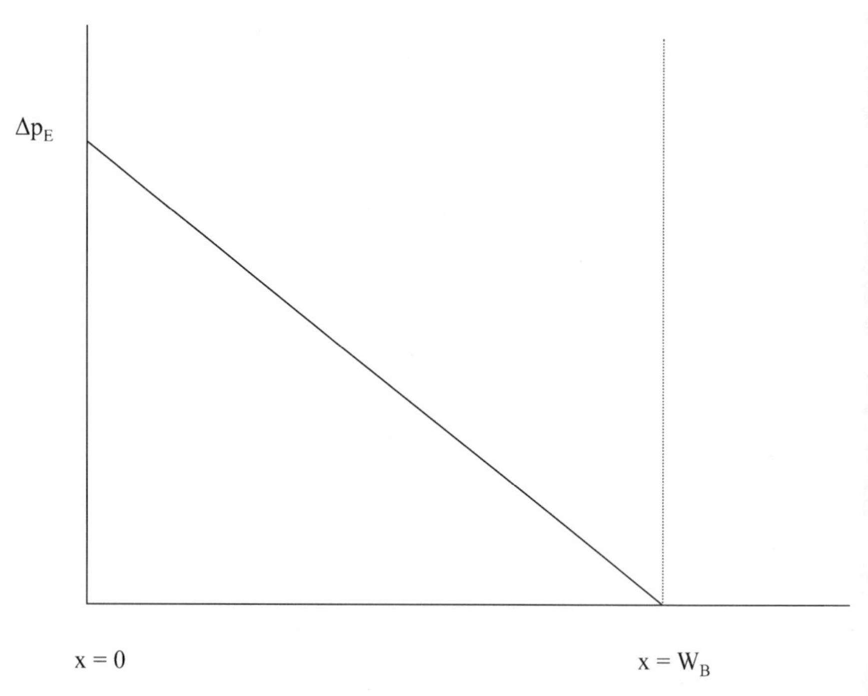

FIGURE 4.5.5 Excess hole concentration in the base region of a p-n-p BJT. Notice that the excess hole concentration is assumed to vary linearly from $x = 0$, the edge of the depletion region on the base side of the emitter–base junction to $x = W_B$, the edge of the depletion region on the base side of the collector–base junction.

where $x = 0$ is the edge of the depletion region on the base side of the emitter–base junction and $x = W_B$ is the edge of the collector–base depletion region on the base side. If a linear relationship is assumed for the minority hole concentration within the base, as shown by Figure 4.5.5, the total charge within the base is given by the area under the curve multiplied by qA (where A is the area). The resulting value for Q_B is

$$Q_B = \frac{qA\Delta p_E W_B}{2} \tag{4.5.13}$$

Using the expression for Δp_E given by Eq. 4.5.7 and neglecting the nonexponential term, the diffusion capacitance can be found as

$$C_D = qA\frac{W_B \Delta p_E}{2V_T} \tag{4.5.14}$$

But I_C can be approximated by Eq. 4.5.3 as

$$I_C \sim qA\frac{D_B}{W_B}p_B e^{qV_{EB}/kT} = qA\frac{D_B}{W_B}\Delta p_E \quad 4.5.15$$

Therefore, C_D can be reexpressed using Eq. 4.5.15 as

$$C_D = \frac{I_C W_B^2}{2D_B V_T} \quad 4.5.16$$

But $g_m = I_C/V_T$, so the diffusion capacitance finally becomes

$$C_D = g_m \frac{W_B^2}{2D_B} \quad 4.5.17$$

Given the foregoing definitions of the circuit elements, we can now examine the frequency behavior of the equivalent-circuit model. The frequency response of the device is a function of the load resistance. For this reason it is customary to eliminate the load resistance effect by examining the frequency response with the output terminals short-circuited. The resulting equivalent circuit is shown in Figure 4.5.6. Using the node law for the left-hand side of the circuit in Figure 4.5.6, the input current i_i can be expressed as

$$i_i = g_\pi v_\pi + j\omega C v_\pi \quad 4.5.18$$

where C is the total capacitance, equal to the sum of C_D, C_E, and C_C, and j is an imaginary number, used here instead of i to avoid any confusion with currents. Using the node law for the right-hand side of the circuit in Figure 4.5.6 yields an expression for the output current i_L as

$$i_L = -g_m v_\pi \quad 4.5.19$$

The ratio of $-i_L$ to i_i is equal to the frequency-dependent small-signal current gain, which we refer to as β_ω. β_ω can be found from Eqs. 4.5.18, 4.5.19, and 4.5.10 as

$$\beta_\omega = -\frac{i_L}{i_i} = \frac{h_{fe}}{1 + j\omega C/g_\pi} = \frac{h_{fe}}{1 + j\omega C h_{fe}/g_m} \quad 4.5.20$$

Recognizing that $h_{fe} = \beta_{\text{dc}}$ at zero frequency, Eq. 4.5.20 can be rewritten as

$$\beta_\omega = \frac{\beta_{\text{dc}}}{1 + j\omega/\omega_\beta} \quad 4.5.21$$

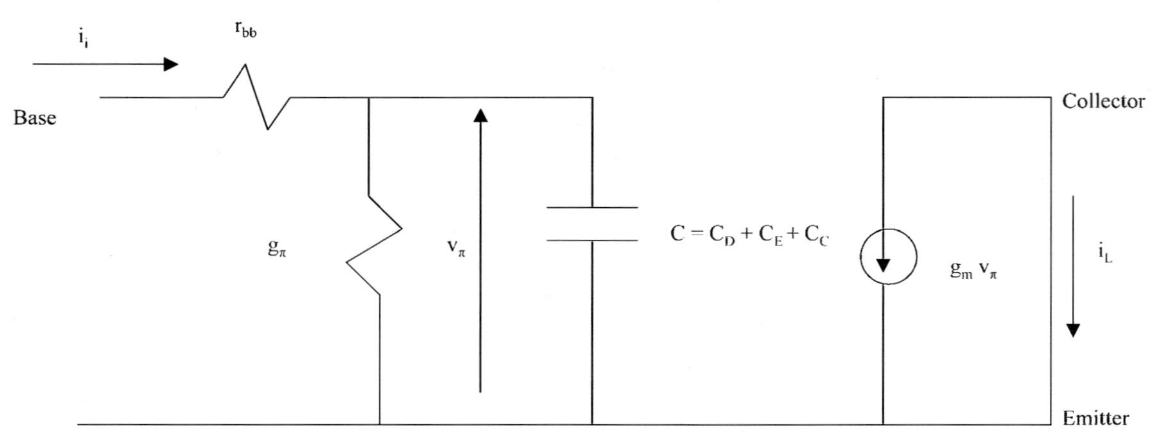

FIGURE 4.5.6 Equivalent small-signal circuit with the output terminals short-circuited. The total capacitance C is equal to the sum of the diffusion, emitter, and collector capacitances as shown in the diagram. The input and output currents are defined as i_i and i_L, respectively.

where ω_β is defined as

$$\omega_\beta = \frac{g_m}{Ch_{fe}} \quad 4.5.22$$

ω_β is called the *beta cutoff frequency*. The value of β_ω drops to $1/\sqrt{2}$ when $\omega = \omega_\beta$.

An expression for the cutoff frequency for the BJT can now be determined. As stated in Section 4.3, the cutoff frequency is defined as the frequency at which the common-emitter short-circuit current gain h_{fe} is equal to 1. Therefore, from Eq. 4.5.22 the cutoff frequency f_t can be found as

$$f_t = \frac{\omega_T}{2\pi} = \frac{1}{2\pi}\frac{g_m}{C} \quad 4.5.23$$

If we assume that the diffusion capacitance C_D is very much larger than $C_E + C_C$, the capacitance C can be approximated by C_D. With this assumption, Eq. 4.5.23 becomes

$$f_t = \frac{1}{2\pi}\frac{g_m}{C_D} \quad 4.5.24$$

The ratio g_m/C_D is given by Eq. 4.5.17 as $2D_B/W_B^2$. However, the base transit time τ_B was found in Eq. 4.3.11 to be

$$\tau_B = \frac{W_B^2}{2D_B} \quad 4.5.25$$

The cutoff frequency can now be reexpressed using Eqs. 4.5.24 and 4.5.25 as

$$f_t = \frac{1}{2\pi\tau_B} \quad 4.5.26$$

Thus we find that the cutoff frequency depends on the base transit time of the device. It is for this reason that the base transit time is often considered the most important quantity in dictating the frequency response of a BJT.

Equation 4.5.26 provides only an estimate of the cutoff frequency since it is based on the approximate hybrid-π model. A more accurate expression for the cutoff frequency is obtained using the emitter–collector delay time τ_{EC}, defined by Eq. 4.3.1 as

$$f_t = \frac{1}{2\pi\tau_{EC}} \quad 4.5.27$$

The emitter–collector delay time consists of the sum of four quantities: the emitter junction charging time τ_E, the base transit time τ_B, the collector–base junction depletion-layer transit time τ_{dC}, and the collector junction charging time τ_C.

The emitter junction charging time τ_E is defined as the time required to charge the emitter–base junction. The charging time is given as the product of the dynamic junction resistance, $r_E = 1/g_m$, of the forward-biased emitter–base junction and the emitter and collector capacitances C_E and C_C as (Shur, 1990)

$$\tau_E = r_E(C_E + C_C) \sim \frac{C_E + C_C}{g_m} = \frac{V_T}{I_C}(C_E + C_C) = \frac{kT}{qI_C}(C_E + C_C) \quad 4.5.28$$

where we assume that the common-base current gain is unity. Since the collector–base junction is reverse biased, the collector–base junction depletion layer is relatively large. Additionally, the reverse bias produces a high electric field. Therefore, it is reasonable to assume that the carriers traverse the collector–base junction depletion layer with an average velocity equal to their saturation velocity v_{sat}. With this assumption, the collector–base junction depletion-layer transit time τ_{dC} can be estimated by

$$\tau_{dC} = \frac{x_m}{2v_{\text{sat}}} \quad 4.5.29$$

where x_m is the depletion layer width. The factor of 2 in the denominator arises from averaging over the ac sinusoidal carrier current over a full period. Finally, the collector junction charging time τ_C can be estimated from the product of the collector junction capacitance C_C and the collector series resistance r_C as

$$\tau_C = r_C C_C \quad 4.5.30$$

With the definitions above, the cutoff frequency can be written as

$$f_t = \frac{1}{2\pi}\frac{1}{\tau_E + \tau_B + \tau_{dC} + \tau_C} = \frac{1}{2\pi}\frac{1}{(kT/qI_C)(C_E + C_C) + \tau_B + (x_m/2v_{\text{sat}}) + r_C C_C} \quad 4.5.31$$

where the base transit time τ_B can be expressed using either Eq. 4.3.11 or 4.3.14, depending on whether or not the base is graded.

The maximum frequency of operation of the transistor $f_{\max}$ is given by Eq. 4.2.1 as

$$f_{\max} = \sqrt{\frac{f_t}{8\pi r_{bb} C_C}} \quad 4.5.32$$

where r_{bb} is the base resistance. The complete derivation of Eq. 4.5.32 is beyond the level of this book. The interested reader is referred to the book by Liu (1999) for a detailed derivation. However, a simplified derivation of Eq. 4.5.32 using circuit analysis is presented in Box 4.5.2. It is important to recognize that whereas f_t does not depend on the base resistance, f_{max} does. Therefore, to maximize f_t, one would be tempted simply to reduce the base thickness. However, this comes at the expense of increasing the base resistance and thus lowering f_{max}. Therefore, if it is desired to maximize both f_t and f_{max}, a trade-off in the design of the base is unavoidable.

Before we end this section, it is useful to mention how the magnitudes of f_t and f_{max} compare between HBT and BJT structures. As mentioned at the beginning of this section, the most important variation in the ac behavior of an HBT compared to a BJT arises from nonstationary transport effects. Velocity overshoot and ballistic transport influence primarily the base and collector–base depletion-layer transit times. As discussed in Section 4.4, these effects can increase the carrier velocity leading to a reduction in the transit times and an increase in both f_t and f_{max}. Numerical calculations provide the best means of estimating the importance of nonstationary transport on the device operating frequency (Yokoyama et al., 1984; Maziar et al., 1986; Rockett, 1988; Nakajima et al., 1992). In addition, recent work (Vaidyanathan and Pulfrey, 1999) has shown that f_{max} in state-of-the-art HBTs requires modification of Eq. 4.5.32. A similar form for f_{max} as given by Eq. 4.5.32 is retained but with $r_{bb}C_CE$ replaced by $(RC)_{eff}$, where $(RC)_{eff}$ includes r_{bb}, C_C, and additional effects arising from parasitic emitter and collector resistances and the dynamic resistance $1/g_m$.

4.6 MATERIALS PROPERTIES AND STRUCTURE OPTIMIZATION FOR HBTS

In this section we highlight the most important material properties and epitaxial layer design issues that affect HBT performance. Table 4.6.1 shows the layer structure (thicknesses and doping concentrations) for a representative HBT device. An InP-based device is used for this example and is discussed in detail to illustrate the principal design factors.

Key to the performance of HBTs are (1) the valence band discontinuity at the emitter–base junction, and (2) the electron transport characteristics, particularly in the collector. For power applications, the breakdown voltage (associated with the critical field for breakdown) of the collector material is also important.

As with the HFET, HBT technology has advanced with refinements of epitaxial growth technology. Thus, the nearly lattice-matched AlGaAs–GaAs technology is the most mature HBT materials system. Examination of valence band discontinuities (Appendix C) shows that other materials systems offer advantages for the emitter–base junction, particularly the GaInP–GaAs, InP–InGaAs, and InP–GaAsSb heterojunctions. As described in Section 4.2, HBT gain (β) is related to either the valence band offset or the difference in

BOX 4.5.2: Circuit Analysis Derivation of f_{max}

With some approximations, the expression for f_{max} given by eq. 4.5.32 can be derived by analyzing the hybrid-π equivalent circuit shown in Figure 4.5.3. Assume that the frequency ω is sufficiently high that $\omega C_\pi \gg g_\pi$ and $\omega C_\pi \gg 1/r_{bb}$. The output resistance r_o is defined as the resistance looking into the collector due to a current i_o fed into the output and is given as

$$i_o = g_m v_\pi + v_o j\omega C_c \qquad \frac{i_o}{v_o} = \frac{g_m v_\pi}{v_o} + j\omega C_c$$

where j is the imaginary number, v_o the output voltage, and v_π the voltage as shown in Figure 4.5.3. The output conductance g_o is equal to the real part of the ratio of the output current to the output voltage as

$$g_o = \mathrm{Re}\left(\frac{i_o}{v_o}\right) = \frac{g_m v_\pi}{v_o}$$

The ratio of v_π to v_o can be determined from the voltage-divider rule for capacitors as

$$v_\pi = v_o \frac{C_c}{C_c + C_\pi}$$

Assuming that C_c can be neglected with respect to C_π in the denominator v_π is

$$v_\pi = v_o \frac{C_c}{C_\pi}$$

The output resistance $1/g_o$ becomes

$$r_o = \frac{1}{g_o} = \frac{v_o}{g_m v_\pi} = \frac{C_\pi}{g_m C_c}$$

At maximum output power the load resistance must equal the output resistance. The output power is

$$P_{\mathrm{out}} = \frac{i_c^2 r_o}{4}$$

where i_c is the collector current. The collector current equal to the output current is given by Eq. 4.5.19 as

$$i_c = i_L = -g_m v_\pi$$

(Continued)

BOX 4.5.2 (*Continued*)

The input power P_{in} is equal to

$$P_{in} = i_i^2 r_{bb}$$

where i_i is the input current defined by Eq. 4.5.18. The maximum power gain G_{max} is then equal to the ratio of the output power to the input power as

$$G_{max} = \frac{i_c^2 r_o/4}{i_i^2 r_{bb}}$$

The maximum power gain can be simplified using the relationship for i_c, which yields

$$G_{max} = \frac{g_m^2 v_\pi^2 r_o}{4 i_i^2 r_{bb}}$$

But i_i is expressed by Eq. 4.5.18. Using our earlier assumption, we can neglect the term involving g_π. Therefore, i_i becomes

$$i_i = j\omega C v_\pi$$

Substituting in for i_i into the expression for the magnitude of G_{max} above yields

$$G_{max} = \frac{g_m^2 v_\pi^2 (C_\pi / g_m C_c)}{4 v_\pi^2 \omega^2 C_\pi^2 r_{bb}}$$

Simplifying, the maximum power gain becomes

$$G_{max} = \frac{g_m}{4 r_{bb} \omega^2 C_\pi C_c}$$

The maximum frequency of operation f_{max} occurs when $G_{max} = 1$. Using the expression for f_t, the cutoff frequency given by Eq. 4.5.23, and solving for ω yields

$$\omega^2 = \frac{2\pi f_t}{4 r_{bb} C_c}$$

Recognizing that $\omega = 2\pi f_{max}$, an expression for f_{max} is finally obtained as

$$f_{max} = \sqrt{\frac{f_t}{8\pi r_{bb} C_c}}$$

TABLE 4.6.1 Representative HBT Layer Structure (InP-Based)

Layer	Material	Thickness (nm)	Doping (cm^{-3})
Cap	InGaAs	45	$N^+ = 2 \times 10^{19}$
Emitter	InP (or AlInAs)	200	$N = 5 \times 10^{17}$
Base	InGaAs	80	$P^+ = 2 \times 10^{19}$
Collector	InP (or AlInAs)	1000	$N = 1 \times 10^{16}$
Subcollector	InP (or AlInAs)	500	$N^+ = 3 \times 10^{18}$
Substrate	InP		

the bandgaps of the heterojunction constituent materials (see Eqs. 4.2.14 and 4.2.15). The former is used for an abrupt junction, and the latter applies to a graded junction.

The decision to design an HBT with either a graded or an abrupt junction will depend in part on manufacturing constraints. Abrupt junctions are relatively easily produced simply by terminating growth of the base and initiating growth of the emitter in a single step (e.g., shutter closing and opening during MBE). While the conduction band discontinuity can be used as a "launching ramp" as described in Section 4.3, a second-order effect associated with the use of an abrupt junction is the reduction in current flow associated with the conduction band barrier. Tunneling of electrons through the discontinuity will also increase the ideality factor of the emitter–base junction, as described below.

Graded junctions introduce more complexity into the growth process. In MBE, grading of $Al_xGa_{1-x}As$ compound from a lower to a higher Al composition (i.e. increasing x) can be accomplished by simply opening both the Al and Ga shutters and increasing the temperature of the Al effusion cell to increase the flux of Al atoms incident on the growth surface. A linear change in composition requires a logarithmic change in temperature due to the Arrhenius relationship between the cell flux and temperature. This approach requires excellent control of the flux with temperature, especially if the materials are not well lattice-matched.

Another approach to achieve grading is to use shuttering to modulate the alloy compositions in a "chirped" superlattice. The Al cell is shuttered open and closed, with shutter timing related to the local composition of Al. As the Al composition increases, the shutter is held open longer. The layers comprising the superlattice must be thin enough so that the electrons can tunnel through the individual layers and are not trapped in any potential well.

The turn-on voltage of the HBT is also affected by the emitter–base design and materials choice. Figure 4.6.1 shows a comparison of turn-on voltages of Si-, GaAs-, and InP-based HBTs. As the bandgap of the base material decreases, the turn-on voltage decreases. A lower turn-on voltage is advantageous for advanced low-power applications. Figure 4.6.2 shows differences in turn-on voltages for variants of the InP-based HBT family with abrupt

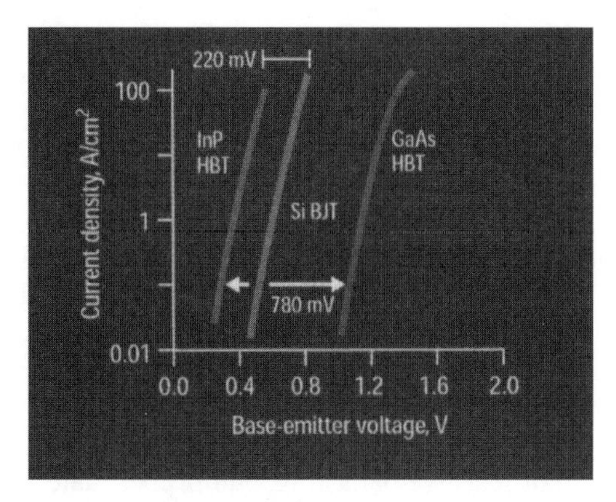

FIGURE 4.6.1 Turn-on voltage of an HBT. (Reprinted with permission from Raghavan et al., 2000.)

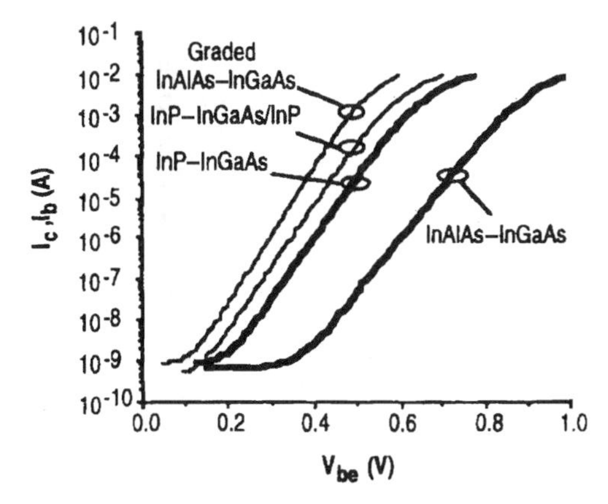

FIGURE 4.6.2 Comparision of V_{be} for InP-based HBTs. (Reprinted with permission from IEEE.)

and graded junctions. The abrupt junction with the largest conduction band discontinuity, AlInAs–InGaAs, exhibits the largest turn-on voltage, while the graded junction exhibits the smallest.

Another critical issue in the design of HBTs is the choice of dopant atom and density of dopants that can be achieved in the base. As described earlier, HBTs offer advantages over BJTs due to the ability to highly dope the base and maintain excellent emitter injection efficiency. Typical doping levels of mid-10^{19}cm^{-3} as shown in Table 4.6.1 are used for devices exhibiting high f_t and f_{max}. p-Type dopants in the GaAs- and InP-based III–V materials include Be and C. Be has been used longer than C, due to its relative ease of introduction. Be can be used as a standard solid source in MBE, while C requires advanced gas sources or compounds. However, particularly in AlGaAs–GaAs devices,

Be can exhibit significant diffusion during growth or during device operation. Approaches to limiting this effect include growth of spacer layers between the emitter and the base to allow for some diffusion without moving the p-n junction into the wider-bandgap emitter, the modification of growth conditions to minimize diffusion, or the use of higher stability dopants such as C for some material systems.

Dopant diffusion is intimately related to the long-term reliability of HBTs. Modifications of structures for enhanced long-term reliability are reduction of base dopant diffusion and improvements in device passivation. Short-term reliability is related to thermal runaway of the device during operation. The change of device gain with temperature is tied to its short-term stability. Self-heating results in a drop in I_C with increased V_{CE} and can be controlled by heat sinking, the use of resistive ballasting, and proper material design choices (e.g., the use of a high-thermal-conductivity collector).

AlGaN–GaN HBTs are limited by the ability to achieve a high hole concentration in the base. Mg is used for p-type doping in GaN but is a deep acceptor ($E_a \sim 170$ meV). This significantly limits the concentration of holes that can be achieved in GaN. Currently, hole concentrations of roughly 10^{18} cm^{-3} represent the limit, thus high gains are difficult to achieve in GaN HBTs.

As in the case of HFETs, InP-based devices offer significant advantages for HBTs, due to their favorable band offsets, transport, and thermal properties. The specific impact and relevance of the characteristics are fundamentally different than those for HFETs and provide a case study for comparisons of design elements.

InP-based devices have lower power consumption (Figure 4.6.1), better high-frequency performance, and smaller surface recombination velocity. Gummel characteristics are often used to compare the base current and collector currents of the devices as a function of V_{BE} with $V_{CB} = 0$ (see Figure 4.6.3). Gummel plots allow direct observation of gain from very low to high collector current. Ideality factors for base and collector currents can be extracted from the characteristics to aid in determining transport mechanisms and nonidealities in devices. The device in Figure 4.6.3 shows good gain (for a doping level of approximately 3×10^{19} cm^{-3}) down to low currents and a near-unity collector current ideality factor. Figure 4.6.4 shows how the ideality factors and gains are modified by emitter–base design. Figure 4.6.4*a* shows a Gummel plot for an abrupt lattice-matched AlInAs–InGaAs HBT (HBT 1), while Figure 4.6.4*b* (HBT 2) shows the same structure with a pseudomorphic 15-nm $Al_{0.7}In_{0.3}As$ emitter barrier inserted at the emitter–base junction. The change in characteristics is illustrative of important physical processes and how the information is derived from the Gummel characteristics.

First, by observing the base–emitter voltages required to achieve the same collector currents, we see that HBT 2 has a lower collector current for a given base–emitter voltage. This is consistent with the increased conduction band discontinuity at the base–emitter junction with the increased bandgap of the pseudomorphic layer.

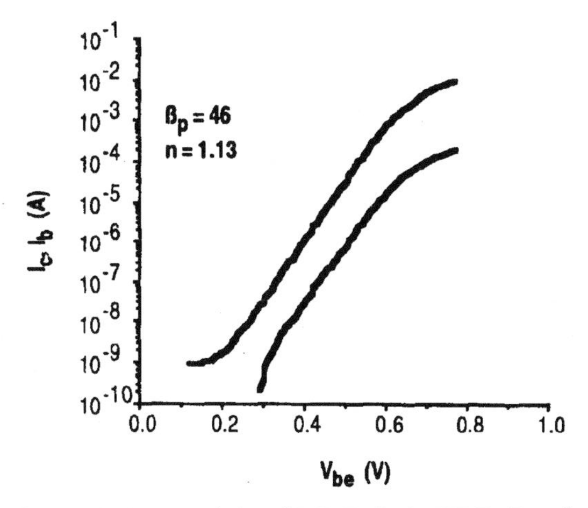

FIGURE 4.6.3 Gummel characteristic of InP–InGaAs HBT. (Reprinted with permission from IEEE.)

The base current characteristics are dominated by bulk recombination in the base and by the space-charge recombination current at the emitter–base junction. For these devices, no change is observed in the base recombination currents at moderate current levels, as inferred from the base current ideality factors. At lower current levels it is typical to observe an increase in the base ideality factor due to the predominance of recombination leakage current with $\eta = 2$. As the base current increases, this component is correspondingly smaller, and bulk recombination will dominate with a lower ideality factor of $\eta = 1$.

The collector current ideality factor is increased for HBT 2. This is due to the larger component of tunneling current ($\eta = 2$) through the pseudomorphic layer. Collector current ideality factors of unity are observed for graded structures in which thermionic emission dominates the current transport mechanisms. The current gain generally increases as the collector current increases, but flattening is typically observed in abrupt emitter–base junctions, as seen in these structures.

InP-based devices are designed with either an InP or AlInAs emitter and an InP or InGaAs collector. As shown in Figure 4.6.5, the InP collector offers advantages in breakdown voltage due to its high critical field for breakdown (Figure 2.6.3). Figure 4.6.6 shows a comparison of the thermal conductivities of InP, GaAs, and Si. Again, InP offers advantages compared with GaAs or with alloys such as InGaAs. Using the same collector and emitter materials is also advantageous in decreasing the collector–emitter offset voltage $V_{CE,\text{sat}}$ (see Figure 4.6.7) since the offset voltage arises from the asymmetry in the emitter–base and base–collector junctions.

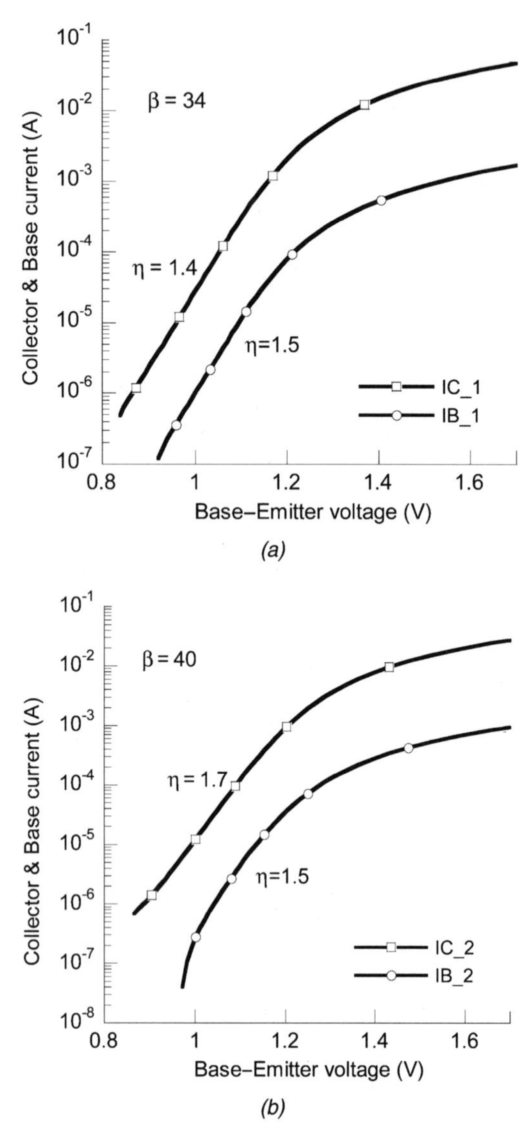

FIGURE 4.6.4 Gummel characteristics of (*a*) AlInAs–InGaAs abrupt HBT, and (*b*) $Al_{0.7}In_{0.3}As$–InGaAs abrupt HBT.

The use of InP collectors, however, introduces more complexity into the growth and design, due to the band offsets present at the base–collector InGaAs–InP interface. A barrier to electron flow from the base to the collector is present in the conduction band and can limit electron flow significantly at higher collector current densities. Thus, the base–collector design must

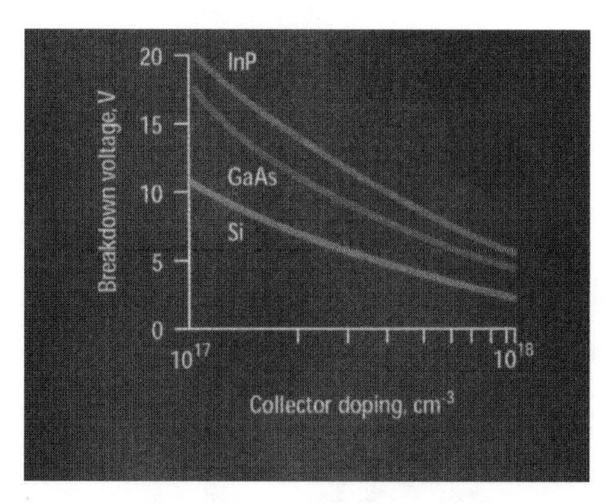

FIGURE 4.6.5 Breakdown voltage as a function of material type and collector doping in an HBT. (Reprinted with permission from Raghavan et al., 2000.)

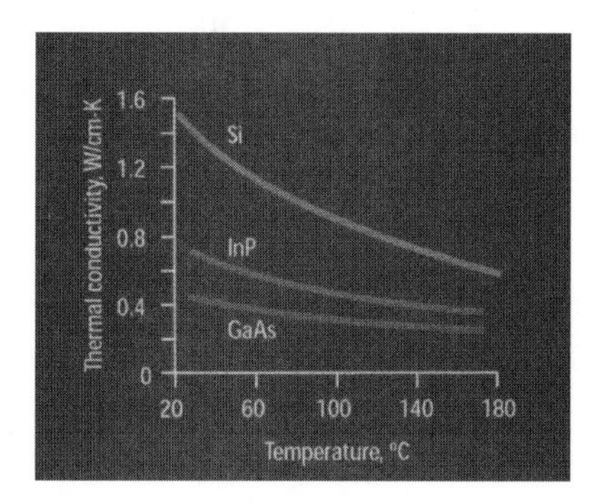

FIGURE 4.6.6 Thermal conductivity as a function of material type. (Reprinted with permission from Raghavan et al., 2000.)

be modified to reduce the barrier. Approaches include compositional grading or the insertion of doping layers to decrease the conduction band barrier.

Finally, our discussion has centered on the use of InP-based technology as a model for considering HBT epitaxial-layer design issues. Significant advances are occurring in Si-based technology due to the advent of epitaxial processes to create SiGe heterostructures. SiGe HBTs can be integrated with Si CMOS and heterostructure BiCMOS is in production. A typical SiGe HBT will have a thin, highly doped SiGe layer on an n^+ collector with a polysilicon emitter stack above the base (Paul, 1998). SiGe bases typically exploit a compositional grade (see Section 4.3) from low Ge (1 to 2%) at the emitter end to 8 to 10% at the collector end of the base. Si-based HBTs offer some advantages to III–V's, including the ease of integration into Si-based CMOS technology

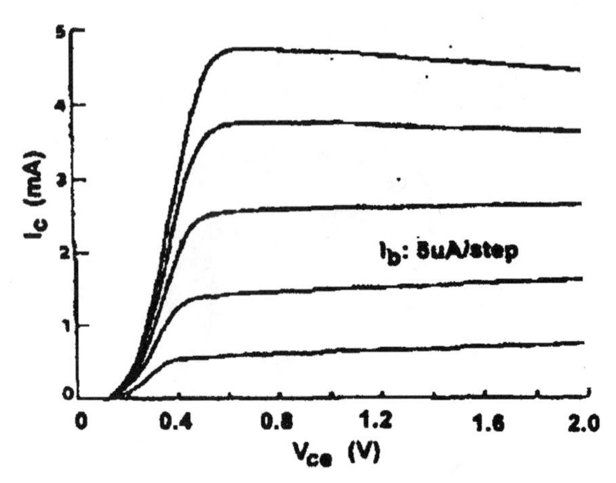

FIGURE 4.6.7 Representative I_C–V_{CE} curves for an HBT, illustrating the V_{CE} offset. (Reprinted with permission from Asbeck, 2000.)

and the resulting cost-effectiveness, and the high thermal conductivity of the material, but limitations exist due to low breakdown voltage.

PROBLEMS

4.1. Consider a p-n-p BJT operating in the common-base active region. Derive an expression for the output resistance r_{oc} in terms of I_C, W_B, L_{pB}, and dW_B/dV_{BC}. The output resistance is defined as

$$\frac{1}{r_{oc}} = \left.\frac{dI_C}{dV_{BC}}\right|_{I_E}$$

4.2. Determine an expression for the excess electron concentration as a function of position x in an n-p-n bipolar junction transistor. Apply the boundary conditions that $\delta n(x = 0)$ (the edge of the emitter–base depletion region) is equal to $\Delta n_E = n_B(e^{qV_{EB}/kT} - 1)$ and that $\delta n(x = W_B)$ is equal to $\Delta n_C = n_B(e^{qV_{CB}/kT} - 1)$. Note that n_B is the equilibrium electron concentration within the base, Δn_E the excess electron concentration at the edge of the emitter–base depletion region ($x = 0$), and Δn_C the excess electron concentration at the edge of the collector–base depletion region ($x = W_B$). Simplify your expression by assuming that the collector junction is strongly reverse biased and that the equilibrium electron concentration is negligible with respect to the electron concentration injected from the emitter.

4.3. For the p^+-n-p BJT in the circuit configuration shown in Figure P4.3, determine V_{EB}. Assume that most of the voltage drop is between the

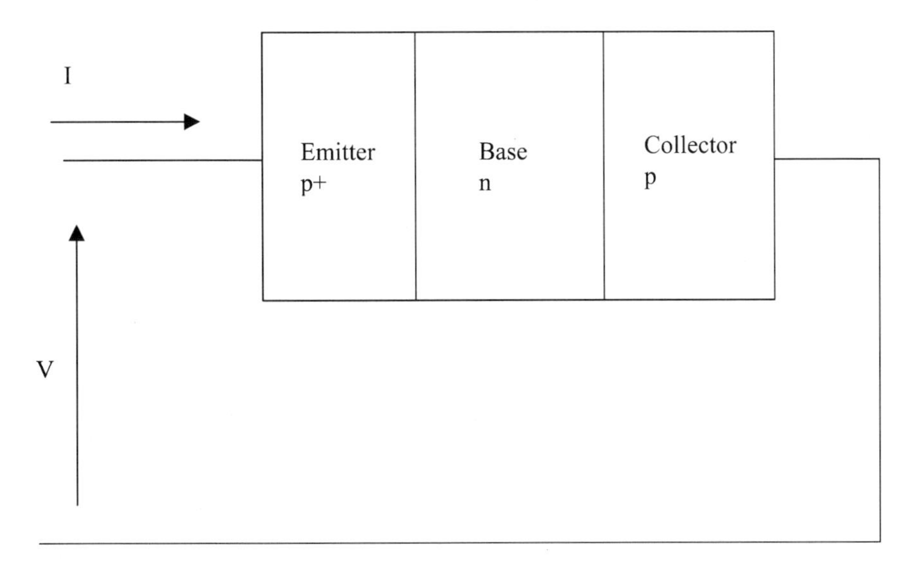

FIGURE P4.3 Circuit configuration of the BJT for Problem 4.3.

collector and base regions, implying that $V_{CB} \ll 1$. Further assume that I_C can be approximated by I_{Cp}.

4.4. In many modern BJT designs, the emitter layer is very shallow, such that the minority carrier concentration back-injected into the emitter may not recombine fully before reaching the contact. At the contact, assume that the excess concentration has decayed to zero. For a shallow p-n-p emitter structure BJT, determine how the expressions for the excess electron concentration in the emitter and I_{En} are modified. Assume that the total width of the emitter is W_E.

4.5. Determine the value of BV_{CEO} in a silicon p-n-p BJT if the collector is doped at $5.0 \times 10^{15}\text{cm}^{-3}$, the critical field for breakdown F_{cr} is 3.0×10^5 V/cm, $\beta_{\text{dc}} = 100$, and the exponent n in Eq. 4.4.11 is 4. Assume that the built-in voltage of the junction can be neglected with respect to the applied voltage and that the acceptor concentration in the collector is very much smaller than the donor concentration in the base.

4.6. Assume that a silicon p-n-p BJT is biased in the active mode such that the simplified dc currents are given by Eqs. 4.1.33. The device has $\beta_{\text{dc}} = 100$, $N_{\text{dB}} = 1.0 \times 10^{18}$ cm^{-3}, $D_B = 10$ cm^2/s, minority carrier lifetime in the base of 100 ns, area $A = 10^{-3}$ cm^2, and a base width of 100 nm. Use 1.0×10^{10} cm^{-3} for the intrinsic concentration of Si. If the emitter–base voltage is 0.7 V, determine (**a**) g_m, (**b**) g_π, and (**c**) C_D, the diffusion capacitance.

4.7. Show that the base transport factor α_T in a BJT can be approximated as

$$\alpha_T = \frac{1}{1 + \tau_B/\tau_p}$$

where τ_B is the base transit time and τ_p is the lifetime in the base. Assume that $W_B \ll L_B$.

4.8. Determine an expression for the punch-through voltage of a p-n-p BJT. *Punch-through* is defined as the condition in which the collector–base voltage is sufficiently large that the base–collector and emitter–base depletion layers touch.

4.9. Consider an AlGaAs–GaAs HBT. If the current gain of a similar BJT device made with all GaAs is 5, what must the concentration of Al within the AlGaAs be to produce a gain of 1000? Assume that the structure is n-p-n, that parameters such as the diffusivities and diffusion lengths are the same between the two devices, and that the effective density of states for GaAs and AlGaAs can be taken to be the same.

CHAPTER 5

Transferred Electron Effects, Negative Differential Resistance, and Devices

In this chapter we discuss transferred electron effects, negative differential resistance, and related devices. There are two well-known transferred electron effects that have been utilized in device configurations: k-space and real-space transfer. Both of these effects can be utilized to produce high-frequency oscillators. In addition, real-space transfer has been exploited in new transistor designs.

5.1 INTRODUCTION

Historically, the first electron transfer mechanism identified was *k-space transfer*. Ridley and Watkins (1961), and later Hilsum (1962), suggested on the basis of theoretical calculations that negative differential resistance effects could be observed in some semiconductors, such as GaAs. The mechanism producing the negative differential resistance was attributed to electron transfer from the lowest-energy central valley within the first conduction band of GaAs to a higher-energy secondary valley. As we will see below, provided that the energy separation is substantially higher than the mean thermal energy and that the effective mass of the secondary valley is much larger than that of the initial valley, negative differential resistance can be observed. Following the original proposals by Ridley, Watkins, and Hilsum, Gunn (1963) experimentally observed microwave domain formation in GaAs following the application of a dc electric field. The origin of these domains was at first not understood. Shortly afterward, Kroemer (1964) suggested that the microwave domains were a consequence of the transferred electron effect in k-space. In this chapter we illustrate how k-space transfer produces microwave oscillations.

The formation of microwave domains as a consequence of k-space transfer is often referred to as the *Gunn effect*. This effect has been utilized to produce high-frequency oscillators known as *Gunn oscillators*. These devices are essentially two terminal structures that produce a high-frequency ac signal from a dc input. We discuss below the operation of the simplest mode of a Gunn oscillator diode to illustrate the device potential of this effect.

The second transferred electron mechanism of importance to devices is called *real-space transfer*. Real-space transfer was conceived independently by Gribnikov (1972) and by Hess et al. (1979). The real-space transfer mechanism proposed by these groups is based on electron transfer between two different semiconductor materials of varying mobilities. The system is comprised of alternating heterolayers of a small-bandgap high-mobility semiconductor sandwiched between layers of a larger-bandgap lower-mobility semiconductor. An electric field is applied parallel to the heterojunction interface. The action of the applied electric field heats the electrons to sufficiently high energy such that they can be scattered out of the small-bandgap material into the wider-bandgap layer. As a consequence, the mobility changes from a high value to a low value, thus producing a negative differential resistance. The real-space transfer mechanism has subsequently been invoked in device design (Kastalsky and Luryi, 1983). Below we review the essential physics of real-space transfer and discuss some device applications of the effect.

5.2 k-SPACE TRANSFER

In this section we illustrate how negative differential resistance can occur from k-space transfer. It is important to recognize first that the k-space transfer effect arises within bulk semiconductor material. Therefore, it is an intrinsic property of the semiconductor, and as such cannot be readily engineered. As we will see below, real-space transfer is induced artificially within a semiconductor system and as such can be engineered.

Consider the one-dimensional continuity equation (Brennan, 1999, Eq. 10.2.3)

$$\frac{\partial n}{\partial t} - \frac{1}{q}\frac{\partial J}{\partial x} = 0 \qquad 5.2.1$$

where n is the electron concentration and J is the current density. Let n_0 be the equilibrium electron concentration and let us assume that there is a disturbance from equilibrium such that there is a small fluctuation in the carrier concentration. This fluctuation, $\delta n = n - n_0$, is about equilibrium. The electric field due to the fluctuation is given from the Poisson equation as

$$\frac{dF}{dx} = -\frac{q(n - n_0)}{\varepsilon} \qquad 5.2.2$$

The current density J is given in general as (Brennan, 1999, Eq. 6.3.68)

$$\mathbf{J} = q\mu_n n\mathbf{F} + qD_n \nabla_x n \qquad 5.2.3$$

Simplifying Eq. 5.2.3 to one dimension, taking the derivative of J in Eq. 5.2.3 with respect to x and dividing through by q yields

$$\frac{1}{q}\frac{dJ}{dx} = D_n \frac{d^2n}{dx^2} + \frac{1}{q}\frac{1}{\rho}\frac{dF}{dx} \qquad 5.2.4$$

where ρ is the resistivity, which is equal to $1/q\mu_n n$. Equation 5.2.4 can be modified since

$$\frac{1}{q}\frac{dF}{dx} = -\frac{n - n_0}{\varepsilon} \qquad 5.2.5$$

Substituting Eq. 5.2.5 into Eq. 5.2.4 gives

$$\frac{1}{q}\frac{dJ}{dx} = -\frac{n - n_0}{\rho\varepsilon} + D_n \frac{d^2n}{dx^2} \qquad 5.2.6$$

Substituting Eq. 5.2.6 into Eq. 5.2.1 yields

$$-\frac{dn}{dt} - \frac{n - n_0}{\rho\varepsilon} + D_n \frac{d^2n}{dx^2} = 0 \qquad 5.2.7$$

Next, let us consider the solution of Eq. 5.2.7. Notice that Eq. 5.2.7 is a partial differential equation in x and t. The equation is separable in x and t. Therefore, the general solution of Eq. 5.2.7 can be expressed as a product of two different functions, a spatial function $u(x)$ and a temporal function $T(t)$. Since n_0 is the equilibrium electron concentration, its spatial and temporal derivatives are simply zero. Thus Eq. 5.2.7 can be rewritten as

$$-\frac{d(n - n_0)}{dt} - \frac{n - n_0}{\rho\varepsilon} + D_n \frac{d^2(n - n_0)}{dx^2} = 0 \qquad 5.2.8$$

The solution for $n - n_0$ is then assumed to be $u(x)T(t)$. Substituting $u(x)T(t)$ into Eq. 5.2.8 yields

$$D_n T(t)\frac{d^2u}{dx^2} - \frac{u(x)T(t)}{\rho\varepsilon} = u(x)\frac{dT}{dt} \qquad 5.2.9$$

Dividing Eq. 5.2.9 through by $T(t)u(x)$ yields

$$\frac{D_n}{u}\frac{d^2u}{dx^2} - \frac{1}{\rho\varepsilon} = \frac{1}{T}\frac{dT}{dt} \qquad 5.2.10$$

The steady-state solution is obtained when the time derivative is set to zero. At steady state, then, Eq. 5.2.10 becomes

$$D_n \frac{d^2(n - n_0)}{dx^2} = \frac{n - n_0}{\rho\varepsilon} \tag{5.2.11}$$

The general solution to this differential equation is readily found to be

$$n - n_0 = A_1 e^{x/L_D} + A_2 e^{-x/L_D} \tag{5.2.12}$$

where L_D is equal to $\sqrt{kT\varepsilon_s/q^2 n_0}$, which is the Debye length. Applying the boundary conditions that as x goes to infinity the excess concentration vanishes implies that the coefficient A_1 is zero. Calling the excess concentration at $x = 0$, $\delta n(0)$ yields

$$n - n_0 = \delta n(0) e^{-x/L_D} \tag{5.2.13}$$

Equation 5.2.13 provides the steady-state solution for the excess electron concentration.

We next consider the temporal solution of Eq. 5.2.9. The temporal solution provides insight into how negative differential resistance can alter the behavior of the carrier concentration. The temporal dependence of Eq. 5.2.9 can be written as

$$\frac{1}{T}\frac{dT}{dt} = -\frac{1}{\rho\varepsilon} \tag{5.2.14}$$

The general solution for T is

$$T = Ae^{-t/\rho\varepsilon} \tag{5.2.15}$$

Let $\tau \equiv \rho\varepsilon$. Substituting for the resistivity, τ becomes

$$\tau = \frac{\varepsilon}{q\mu_n n_0} = \rho\varepsilon \tag{5.2.16}$$

Application of the boundary conditions provides the solution to the problem. As t approaches infinity, the electron concentration approaches the equilibrium concentration. At $t = 0$, the excess electron concentration is defined as $(n - n_0)_{t=0}$. The temporal solution for the excess concentration then becomes

$$n = n_0 + (n - n_0)_{t=0} e^{-t/\tau} \tag{5.2.17}$$

Inspection of Eq. 5.2.17 indicates that provided that τ is positive, the electron concentration will decay with increasing time ultimately recovering back to equilibrium. However, if τ is negative, the electron concentration grows, not decays! For τ to be negative, the resistivity must be negative. This occurs in a system with negative differential resistance (NDR).

EXAMPLE 5.2.1: Determination of the *RC* Time Constant

Determine the *RC* time constant of a sample that has length L and area A where the excess electron concentration can be described by Eq. 5.2.17.

The resistance of the slab of material is simply

$$R = \frac{\rho L}{A} = \frac{L}{\sigma A}$$

while the capacitance is given as

$$C = \frac{\varepsilon A}{L}$$

Therefore, the *RC* time constant is given as

$$RC = \frac{\rho L}{A}\frac{\varepsilon A}{L} = \rho\varepsilon$$

However, from Eq. 5.2.16 in the text, the lifetime τ is equal to

$$\tau = \rho\varepsilon$$

Therefore, the *RC* time constant for the sample is given as

$$RC = \tau$$

If the lifetime is negative, the *RC* time constant is then negative, as expected, since the resistance R is now negative.

There are two types of NDR in a semiconductor: voltage-controlled and current-controlled, sketched in Figure 5.2.1. Inspection of Figure 5.2.1 reveals that voltage-controlled NDR has an "N" shape whereas current-controlled NDR has an "S" shape.

In the above we have determined the conditions for NDR. The next question is whether these conditions can occur in a bulk semiconductor and how. To see how NDR can arise within a semiconductor, consider first a simple two-valley model of the first conduction band of a semiconductor. Such a system is sketched in Figure 5.2.2. The conductivity of the two-valley system, where it is assumed that the lowest energy valley is Γ and the higher-energy valley is X, can be written as

$$\sigma = q(\mu_\Gamma n_\Gamma + \mu_x n_x) \qquad 5.2.18$$

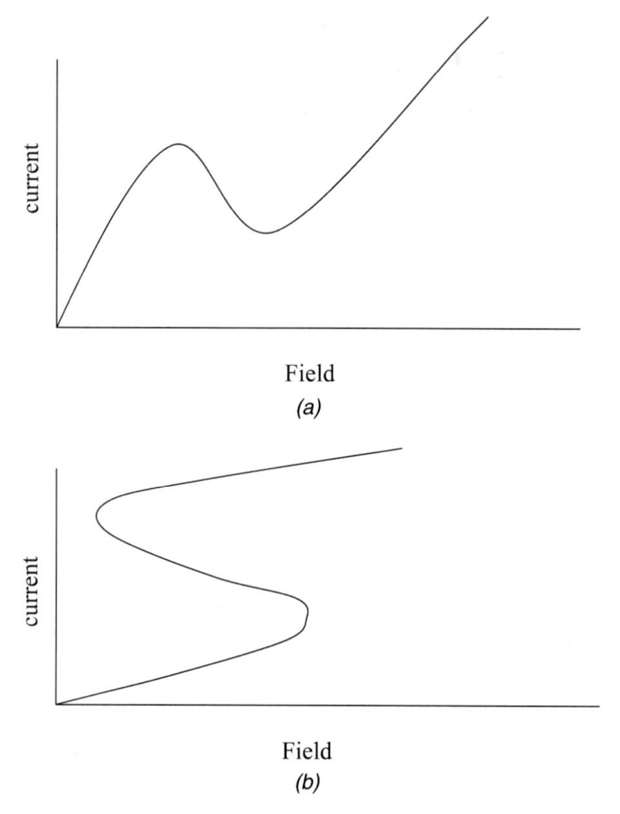

FIGURE 5.2.1 (*a*) Voltage-controlled NDR and (*b*) current-controlled NDR.

where n_Γ and n_x are the electron concentration and μ_Γ and μ_x the mobilities within the Γ and X valleys, respectively. The rate of change of the conductivity with the electric field is given as

$$\frac{d\sigma}{dF} = q\left(\mu_\Gamma \frac{dn_\Gamma}{dF} + \mu_x \frac{dn_x}{dF}\right) + q\left(n_\Gamma \frac{d\mu_\Gamma}{dF} + n_x \frac{d\mu_x}{dF}\right) \tag{5.2.19}$$

The total carrier concentration n is constant and is given as

$$n = n_\Gamma + n_x \tag{5.2.20}$$

Thus the carrier concentration remains constant and is simply distributed among the two valleys. The derivatives of n_Γ and n_x can then be related as

$$\frac{dn_x}{dF} = -\frac{dn_\Gamma}{dF} \tag{5.2.21}$$

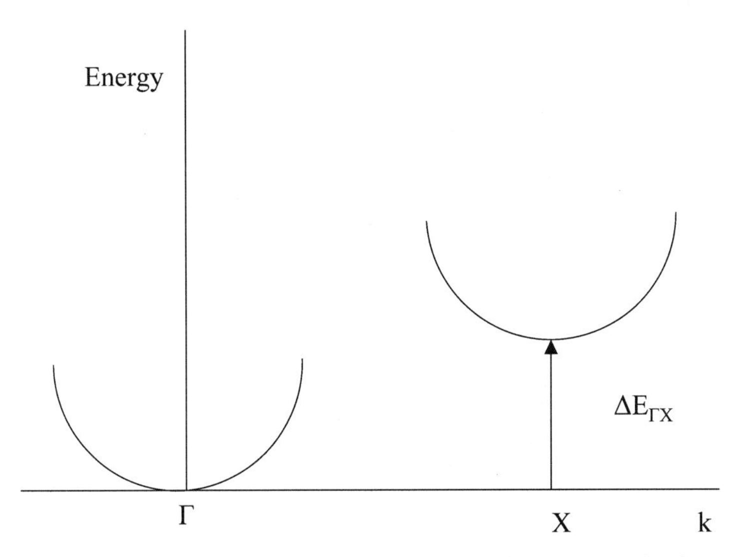

FIGURE 5.2.2 Two-valley conduction band in a semiconductor. The lowest-lying valley occurs at the Γ point (000) in k-space, while the secondary valley is assumed to lie at an energy $\Delta E_{\Gamma X}$ above the Γ valley at the X point, (100) in k-space.

For simplicity, let us assume that the field dependence of the mobility is given as

$$\mu_\Gamma \approx F^p \qquad \mu_x \approx F^p \tag{5.2.22}$$

where p is a constant. Using Eqs. 5.2.21 and 5.2.22 in Eq. 5.2.19 yields

$$\frac{d\sigma}{dF} = q(\mu_\Gamma - \mu_x)\frac{dn_\Gamma}{dF} + q(\mu_\Gamma n_\Gamma + \mu_x n_x)\frac{p}{F} \tag{5.2.23}$$

But the current density J is equal to the product of the conductivity and the field, $J = \sigma F$. Hence

$$\frac{dJ}{dF} = \sigma + F\frac{d\sigma}{dF} \tag{5.2.24}$$

The condition for NDR is that the derivative of J with respect to F is less than zero. Thus

$$\frac{dJ}{dF} < 0 \qquad \sigma + F\frac{d\sigma}{dF} < 0 \tag{5.2.25}$$

which becomes

$$-\frac{d\sigma}{dF}\bigg/\frac{\sigma}{F} > 1 \tag{5.2.26}$$

Recall that the "less than" sign changes to a "greater than" sign after multiplication by -1. Substituting into Eq. 5.2.26 the expression for $d\sigma/dF$ given by Eq. 5.2.23 yields

$$-\frac{q(\mu_\Gamma - \mu_x)(dn_\Gamma/dF) + q(\mu_\Gamma n_\Gamma + \mu_x n_x)(p/F)}{\sigma/F} > 1 \qquad 5.2.27$$

Simplifying, Eq. 5.2.27 becomes

$$\left[\frac{\mu_\Gamma - \mu_X}{\mu_\Gamma + (n_X/n_\Gamma)\mu_X}\left(-\frac{F}{n_\Gamma}\frac{dn_\Gamma}{dF}\right) - p\right] > 1 \qquad 5.2.28$$

In Eq. 5.2.28, the inequality holds provided that the mobility within the lower-energy Γ valley is greater than that within the higher-energy X valley. This can be understood from further inspection of Eq. 5.2.28. Notice that $dn_\Gamma/dF < 0$, since the rate of change of the electron population within Γ is negative with increasing electric field due to the transfer effect. Therefore, the second term in Eq. 5.2.28 (in parentheses) becomes positive. To have the product of the first two terms be positive, the mobility within the Γ valley must be greater than the mobility within the X valley. This is a general result and can be summarized as follows. The condition for negative differential resistance in a bulk semiconductor is that the carrier mobility within the lowest-energy valley of the first conduction band must be greater than the carrier mobility within the next-highest-energy valley.

The band structure of GaAs satisfies this condition for negative differential resistance. A rough sketch of the first conduction band of GaAs is shown in Figure 5.2.3. Inspection of the figure shows that the L valley lies lower in energy than the X valley. The valley separation energies are given approximately as $\Delta E_{\Gamma L} \sim 0.28$ eV and $\Delta E_{\Gamma X} \sim 0.48$ eV, and as such, carrier transfer to the L valley from the Γ valley will occur first. The electron effective masses within the Γ and L valleys are $m_\Gamma = 0.063$ and $m_L = 0.23$ times the free-space electron mass (Brennan, 1999, Sec. 8.1). Since the electron effective mass within the L valley is greater than that in the Γ valley, the electron mobility within the Γ valley is greater than that within the L valley, $\mu_\Gamma > \mu_L$. Consequently, the condition stated above for negative differential resistance (NDR) is met. Thus bulk GaAs will produce NDR through k-space transfer.

In Figure 5.2.4 the experimental and Monte Carlo calculated steady-state electron drift velocity in bulk GaAs are plotted. Notice that there is a marked decrease in the drift velocity at electric field strengths larger than ~ 3.5 kV/cm, indicating the NDR region. The field strength at which the velocity reaches its maximum, ~ 3.5 kV/cm, is called the *threshold field*. Further inspection of Figure 5.2.4 shows that the maximum drift velocity is about 2.0×10^7 cm/s.

The full criteria for NDR in bulk semiconductors are given as follows:

1. The minimum energy of the secondary valley lies several times the thermal energy above the minimum of the lowest valley. This is necessary

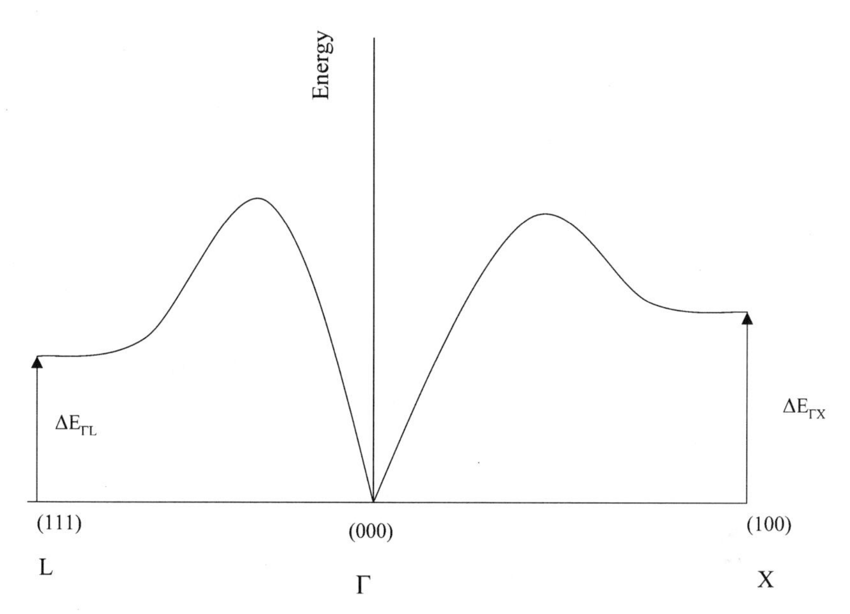

FIGURE 5.2.3 First conduction band of GaAs, showing the two satellite valleys at the L and X minima in *k*-space.

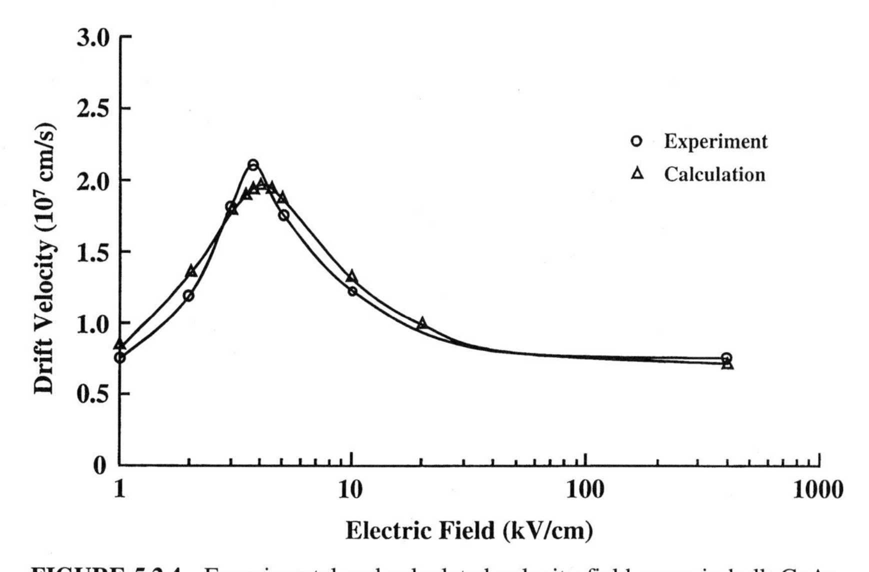

FIGURE 5.2.4 Experimental and calculated velocity field curve in bulk GaAs.

EXAMPLE 5.2.2: Conductivity Calculation for a Two-Valley Semiconductor

Determine the conductivity of a two-valley semiconductor if the mobility of the lowest-energy valley is 1200 $cm^2/V \cdot s$ and the mobility of the higher-energy valley is 300 $cm^2/V \cdot s$ given the following: applied field = 5 kV/cm, intervalley separation energy = 0.3 eV, total electron concentration = 10^{17} cm^{-3}, energy relaxation time = 5 ps, steady-state velocity = 1.0×10^7 cm/s (lower valley), $m_l = 0.067m^*$, and (upper valley) $m_u = 0.1087m^*$, and that the upper valley has a degeneracy of 3. Assume that the relaxation time approximation is valid, the system is in steady state at ambient temperature 300 K, and that the distributions can be approximated using Boltzmann statistics.

To solve this problem we will use the relaxation time approximation. It is essential to determine the average energy of the distribution. Once the average energy is known, the relative concentrations in the first and second valleys can be found. From this the conductivity can be determined using Eq. 5.2.18. Generally, one can improve on the solution by using a displaced Maxwellian distribution. For details on this approach, see the book by Lundstrom (2000, Chap. 7). Here we use a more simplified approach.

The energy relaxation rate equation is given by Brennan (1999, Prob. 6.3) as

$$\frac{dE}{dt} = qFv - \frac{E - E_0}{\tau}$$

where F is the applied field, v the steady-state velocity, τ the relaxation time, E the average energy of the nonequilibrium distribution, and E_0 the average energy of the equilibrium distribution. Solving for the average energy yields

$$E = qFv\tau + E_0$$

The equilibrium energy E_0 is given as $\frac{3}{2}$ kT at 300 K. Substituting in E becomes $E = 0.289$ eV. The relative concentrations in the first and second valleys can now be determined using Boltzmann statistics. It should be noted that, in general, since the system is in nonequilibrium, the use of an equilibrium distribution such as the Boltzmann distribution is inappropriate. However, if the field strength is relatively small, the nonequilibrium distribution can be approximated roughly by the equilibrium distribution but with a temperature equal to the carrier temperature as opposed to the equilibrium temperature. Thus we must first find the equivalent carrier temperature for the system. The average energy of the

(Continued)

EXAMPLE 5.2.2 *(Continued)*

nonequilibrium distribution is 0.289 eV. The equivalent temperature is then $\frac{3}{2}$ kT = 0.289 eV, which yields a carrier temperature of 2230 K.

The relative concentrations of carriers in the first and second valleys can be determined as

$$\frac{n_u}{n_l} = \frac{N_u}{N_l} e^{-\Delta E/kT}$$

Substituting for ΔE the intervalley separation energy, 0.3 eV, $T = 2230$ K, and for the density-of-states functions,

$$\frac{N_u}{N_l} = 3\left(\frac{m_u}{m_l}\right)^{3/2}$$

the ratio of n_u to n_l becomes

$$\frac{n_u}{n_l} = 1.3$$

Using the fact that the total concentration of electrons is 1.0×10^{17} cm^{-3}, n_u and n_l are given as

$$n_2 = 5.65 \times 10^{16} \text{ cm}^{-3}$$

$$n_1 = 4.35 \times 10^{16} \text{ cm}^{-3}$$

The conductivity of the sample is then

$$\sigma = q(\mu_1 n_1 + \mu_2 n_2)$$

Substitution yields $\sigma = 11.06\ (\Omega \cdot \text{cm})^{-1}$.

In this problem the relaxation time approximation is made. In most compound semiconductors the dominant scattering mechanism is polar optical phonon scattering at low applied field strengths. As discussed in Chapter 2, the relaxation time approximation cannot be applied to polar optical phonon scattering. Therefore, typically the approximate approach above does not hold for compound semiconductors wherein polar optical phonon scattering is the dominant scattering mechanism for carriers in the lowest-energy valley. In such cases a numerical approach must be used to predict the conductivity correctly.

since the electrons must initially reside within the lowest-energy valley prior to the application of a field. Otherwise, there will be a significant fraction of electrons populating the satellite valley in equilibrium. The valley population ratio can be determined using elementary statistical mechanics (see Brennan, 1999, Chap. 5). The ratio of the electron concentration in L to that in Γ is

$$\frac{n_L}{n_\Gamma} = \frac{N_L}{N_\Gamma} e^{-\Delta E_{\Gamma L}/kT} \qquad 5.2.29$$

where Boltzmann statistics have been assumed and N_L and N_Γ are the effective density of states within the L and Γ valleys, respectively, and T is the temperature in kelvin.

2. The energy difference between the two valleys must be less than that of the bandgap E_g. Otherwise, impact ionization can occur prior to intervalley transfer, thus mitigating the transferred electron effect. An example of this occurs in InAs (Brennan and Hess, 1984). Brennan and Mansour (1991) showed that impact ionization is very important in InAs and effectively masks any NDR. In other words, electrons undergo impact ionization prior to intervalley transfer, and as such the NDR is not observed. Although only a full transport calculation can prove this assertion, we can readily see its plausibility by inspecting the important physical parameters of InAs. These are: $E_g = 0.41$ eV ($T = 77$ K), $\Delta E_{\Gamma L} \sim 0.79$ eV, $\Delta E_{\Gamma X}$ is significantly greater than $\Delta E_{\Gamma L}$, $m_\Gamma = 0.023$, and $m_L = 0.286$. From these parameters it is clear that $\Delta E_{\Gamma L} \gg E_g$. Also, $\mu_\Gamma > \mu_L$, since the effective mass in the Γ valley is much smaller than the mass in the L valley. Notice that NDR would occur in InAs if the impact ionization process did not interfere with the carrier heating. The impact ionization process cuts off the high-energy portion of the distribution, preventing significant intervalley transfer.
3. Carrier transfer from one valley to the next must occur more quickly than the time elapsed in one period of the operation frequency.

From the above we conclude that NDR can occur in a bulk semiconductor provided that it has multiple valleys separated by an energy greater than several kT but less than the energy gap if $\mu_\Gamma > \mu_L$. Provided that the semiconductor meets the above-stated criteria, it should exhibit NDR when a field is applied that is sufficiently large so as to heat the electrons from the lowest-lying Γ valley into the secondary satellite minima.

5.3 REAL-SPACE TRANSFER

As mentioned above, there are two physical mechanisms that can produce NDR through carrier transfer. The first, k-space transfer, occurs within bulk

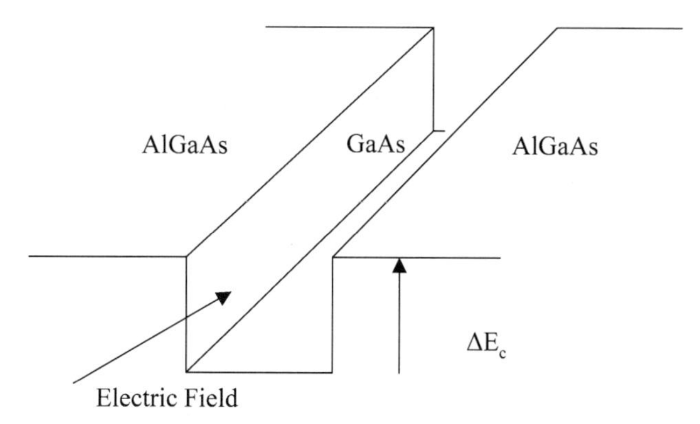

FIGURE 5.3.1 AlGaAs–GaAs double-heterostructure system. Notice that the applied electric field is parallel to the heterostructure.

material. In this section we discuss a different mechanism, called real-space transfer, that occurs in heterostructure semiconductor systems.

The basic concept of real-space transfer can be understood with the help of Figure 5.3.1. For purposes of illustration we consider the GaAs–AlGaAs material system. The structure consists of a GaAs layer sandwiched between two AlGaAs heterolayers as shown in the figure. Electrons move between the two heterolayers within the GaAs layer under the action of an applied electric field, as shown in the diagram. Under most situations, the dimensions are chosen such that spatial quantization effects do not occur. Notice that the electric field is applied parallel to the heterolayers. As the electrons within the GaAs layer drift in the field direction, they are heated by the action of the applied field. As a result, the average electron energy increases until the carriers reach energy greater or equal to the heterojunction discontinuity, ΔE_c. Once the carriers reach this energy, they can undergo a scattering that can redirect their momentum in the direction perpendicular to the heterojunction. Ultimately, the carrier can drift across the heterojunction, transferring from the GaAs layer into the AlGaAs layer. The transfer from the GaAs to the AlGaAs layer is referred to as real-space transfer (RST).

The significance of real-space transfer can be understood as follows. Recall that electron mobility depends inversely on the effective mass of the carrier. Electrons within the lowest-lying valley in GaAs, the Γ valley, have a relatively small effective mass. Therefore, electron mobility within the Γ valley of GaAs is very high. In AlGaAs, depending on the Al composition, the material is either direct or indirect. For Al compositions below about 45%, AlGaAs is direct and the Γ valley lies lowest in energy. At Al compositions greater than 45%, AlGaAs becomes indirect and the X valley lies lowest in energy. In the present discussion, let us consider only AlGaAs compositions with less than 45% Al. In these alloys, the Γ valley lies lowest in energy. The gamma valley effective mass is significantly higher within AlGaAs than in GaAs. As a result, the electron mobility within the AlGaAs is lower than within the GaAs. Therefore, conditions exist within the heterolayer system much like those necessary for bulk NDR (i.e., a two-level system exists such that the carrier mobility within the lowest energy level is substantially greater than the mobility within the higher energy level). In the present case, the GaAs layer is the lowest energy level and has a higher mobility than the higher energy level AlGaAs. Hence, NDR can occur in the heterojunction system.

Perhaps the most precise method by which one can theoretically analyze the effects of real-space transfer is to use the ensemble Monte Carlo model (Fawcett et al., 1970). Although this approach is highly accurate, its workings are not immediately transparent. Below, we present some Monte Carlo calculations, but first we present a much simpler, although more transparent picture of real-space transfer that will allow the reader to better understand its workings.

Let us make a simple calculation of the average steady-state electron drift velocity in a heterostructure system such as that of Figure 5.3.1. We make the following assumptions: (1) the system is in steady state, (2) the nonequilibrium electron distributions can be characterized using Maxwellian distributions in which the temperature is replaced by an electron temperature T_e, and (3) the energy bands are parabolic. Labeling the GaAs layer as layer 1 and the AlGaAs layer as layer 2, the average velocity within the system can then be estimated from

$$\bar{v} = \frac{v_1 n_1 + v_2 n_2}{n_1 + n_2} \qquad 5.3.1$$

where n_1 and n_2 are the electron concentrations and v_1 and v_2 are the steady-state electron drift velocities within the GaAs and AlGaAs layers, respectively. The system is similar to the two-level system discussed in Section 5.2. The ratio of the electron concentration within the AlGaAs layer to that within the GaAs layer can be determined as follows. The electron concentration is given as (Brennan, 1999, Chap. 5)

$$n = \int D(E) f(E) dE \qquad 5.3.2$$

where $D(E)$ is the density-of-states function and $f(E)$ is the distribution function. Assuming the three-dimensional density-of-states function for parabolic energy bands and a Maxwellian distribution characterized by an electron temperature T_e, the ratio of the electron concentration within layer 2 to that in layer 1 is

$$\frac{n_2}{n_1} = \left(\frac{m_2}{m_1}\right)^{3/2} e^{-\Delta E_c / kT_e} \qquad 5.3.3$$

EXAMPLE 5.3.1: Determination of the Electron Temperature in a Real-Space Transfer Device

Determine an expression for the electron temperature in a real-space transfer device. Assume that the energy relaxation time is given as τ, a uniform constant electric field is applied, and that the system is in steady state. Simplify the solution by assuming that the electron concentration in the higher-energy layer, n_2, is very much less than that in the low-energy level, n_1. Also assume that $\mu_1 > \mu_2$.

In steady state the input power must be equal to the output power. In other words, the input power from the applied electric field must be equal to the power lost through energy relaxation events. Therefore,

$$\frac{\frac{3}{2}k(T_e - T_0)}{\tau_E} = qF\bar{v}$$

Solving for T_e yields

$$T_e = \frac{2qF\bar{v}\tau_E}{3k} + T_0$$

But the average velocity is given by Eq. 5.3.4 as

$$\bar{v} = \frac{v_1 + v_2(n_2/n_1)}{1 + (n_2/n_1)}$$

which can readily be rewritten in terms of the mobility as

$$\bar{v} = \frac{\mu_1 + (n_2/n_1)\mu_2}{1 + n_2/n_1}F$$

where n_2/n_1 is given from Eq. 5.3.3 as

$$\frac{n_2}{n_1} = \left(\frac{m_2}{m_1}\right)^{3/2} e^{-\Delta E/kT_e}$$

Substituting the expression for the average velocity into that for T_e yields

$$T_e = T_0 + \frac{2q\tau_E F^2}{3k}\frac{\mu_1 + (n_2/n_1)\mu_2}{1 + n_2/n_1}$$

(Continued)

EXAMPLE 5.3.1 (*Continued*)

If we make the simplifying assumption that $n_2 \ll n_1$ and that $\mu_2 < \mu_1$, the expression for T_e becomes

$$T_e = T_0 + \frac{2q\tau_E F^2 \mu_1}{3k}\left(1 + \left(\frac{m_2}{m_1}\right)^{3/2} e^{-\Delta E/kT_e}\right)^{-1}$$

The average velocity within the system can now be estimated by substituting Eq. 5.3.3 into Eq. 5.3.1 to obtain

$$\bar{v} = \frac{v_1 + v_2 (m_2/m_1)^{3/2} e^{-\Delta E_c/kT_e}}{1 + (m_2/m_1)^{3/2} e^{-\Delta E_c/kT_e}} \tag{5.3.4}$$

Inspection of Eq. 5.3.4 shows that it satisfies the obvious limits (i.e., as the conduction band edge discontinuity increases, the average velocity is essentially that of the first layer). This is as expected since little if any real-space transfer occurs when the discontinuity is large. In the other limit, as the discontinuity approaches zero, the average velocity is given by the average velocities within each layer weighted by the density of states as reflected by the effective masses. It should be noted that Eq. 5.3.4 is only a rough approximation and cannot replace the more complete and physically accurate numerical calculations discussed below. The main limitation of Eq. 5.3.4 is that it assumes a Maxwellian distribution which is of course valid only in equilibrium. A more precise calculation can be made using a drifted Maxwellian. The reader is referred to the book by Lundstrom (2000) for details on this approach.

Calculations by Brennan and Park (1989) clearly show the effects of real-space transfer on the velocity field curve. Brennan and Park (1989) investigated real-space transfer in a strictly classical system as shown in Figure 5.3.1, where the layer widths are assumed to be sufficiently large and the doping concentration small such that two-dimensional effects do not appear. The calculations performed by Brennan and Park (1989) were made using an ensemble Monte Carlo program. The Monte Carlo method is discussed in detail by Jacoboni and Lugli (1989) and is not described here. Suffice it to say that the Monte Carlo method provides an exact solution to the Boltzmann equation limited in its accuracy only by the amount of physics that it contains.

In the specific model of Brennan and Park (1989), all principal scattering mechanisms are included: polar optical phonon, intervalley phonon, acoustic phonon, and charged impurity scattering. The details of each of these mechanisms are discussed in the book by Jacoboni and Lugli (1989). To calibrate their Monte Carlo code, Brennan and Park (1989) compared their

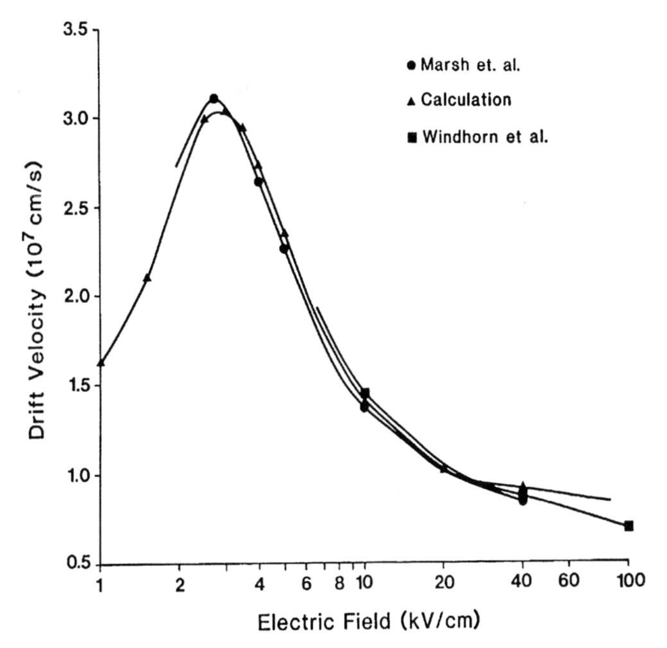

FIGURE 5.3.2 Experimental and Monte Carlo calculated electron drift velocity in bulk $In_{0.53}Ga_{0.47}As$ as a function of applied electric field.

steady-state velocity field calculations to experimental measurements made for $In_{0.53}Ga_{0.47}As$. The comparison is shown in Figure 5.3.2. As shown, the Monte Carlo calculations agree well with two sets of experimental measurements (Windhorn et al., 1982).

The effects of real-space transfer are investigated using the device structure sketched in Figure 5.3.1 but with $In_{0.15}Ga_{0.85}As$ substituted for GaAs. The alloy composition considered for the real-space transfer analysis, $In_{0.15}Ga_{0.85}As$, is different from that considered for the bulk analysis, $In_{0.53}Ga_{0.47}As$. This choice has been made for the following reasons. First, experimental measurements for the steady-state drift velocity presently exist only for $In_{0.53}Ga_{0.47}As$ (Marsh et al., 1981; Windhorn et al., 1982). Second, the most important InGaAs alloy composition for high-speed-device applications is the 15% In composition. This is the system of choice for pseudomorphic HEMTs (Rosenberg et al., 1985), one of the most useful HEMT structures. Therefore, the bulk calculations are made in order to calibrate the model, but the real-space transfer simulation is performed for the more important heterostructure system, $In_{0.15}Ga_{0.85}As$–$Al_{0.15}Ga_{0.85}As$. The narrow-gap InGaAs layer is chosen to be intrinsic, no intentional impurities are present. The AlGaAs layers are chosen to have a 15% Al composition and are assumed to be doped at about 1.0×10^{17} cm^{-3}. Therefore, impurity scattering is included within the AlGaAs layers in the Monte Carlo simulation. The structure consists of a 20-nm-wide

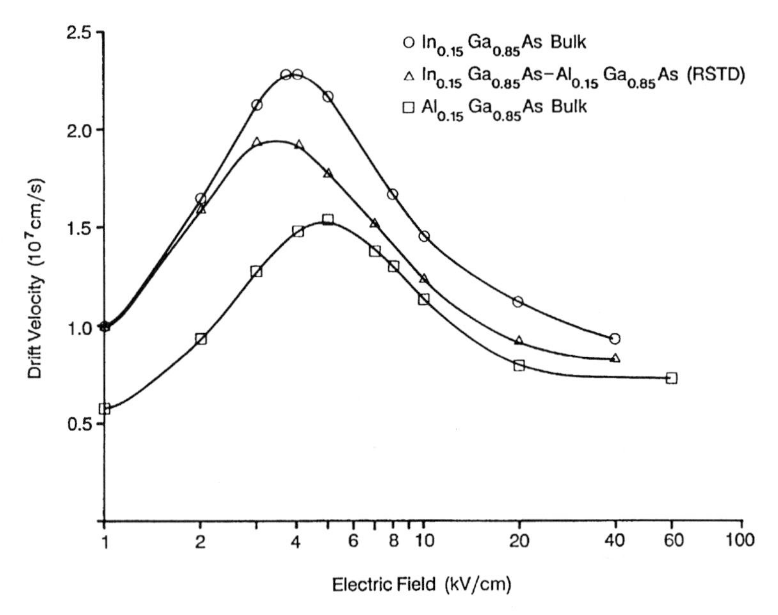

FIGURE 5.3.3 Calculated electron drift velocity in bulk $In_{0.15}Ga_{0.85}As$, $Al_{0.15}Ga_{0.85}As$, and in a real-space transfer device (RSTD) made from these materials.

$In_{0.15}Ga_{0.85}As$ layer bounded on either side by a 500-nm $Al_{0.15}Ga_{0.85}As$ layer.

To see the effects of real-space transfer on the overall electron drift velocity, the steady-state electron drift velocities within the heterostructure system and the corresponding bulk steady-state drift velocities corresponding to the constituent materials are plotted in Figure 5.3.3. At low applied electric field strengths, the carriers remain within the InGaAs layer, and the drift velocity within the heterostructure system approaches the bulk InGaAs result. As the electric field increases, the electrons begin to transfer into the lower-mobility AlGaAs layer. Inspection of Figure 5.3.3 shows that throughout the full range of applied electric field strengths considered, the electron drift velocity within the multilayered system is intermediate between that of the constitutive bulk materials, as expected and consistent with that predicted by Eq. 5.3.4. The peak electron drift velocity in the multilayered device occurs at roughly 3.25 kV/cm. At this field strength, essentially 20% of the electrons have transferred to the AlGaAs layers, as shown in Figure 5.3.4. Further inspection of Figure 5.3.4 indicates that real-space transfer between the two layers occurs prior to k-space transfer in the InGaAs layer. This is not too surprising since the conduction band edge discontinuity is assumed to be 0.27 eV, while the intervalley separation energy, $\Gamma - L$, is 0.368 V.

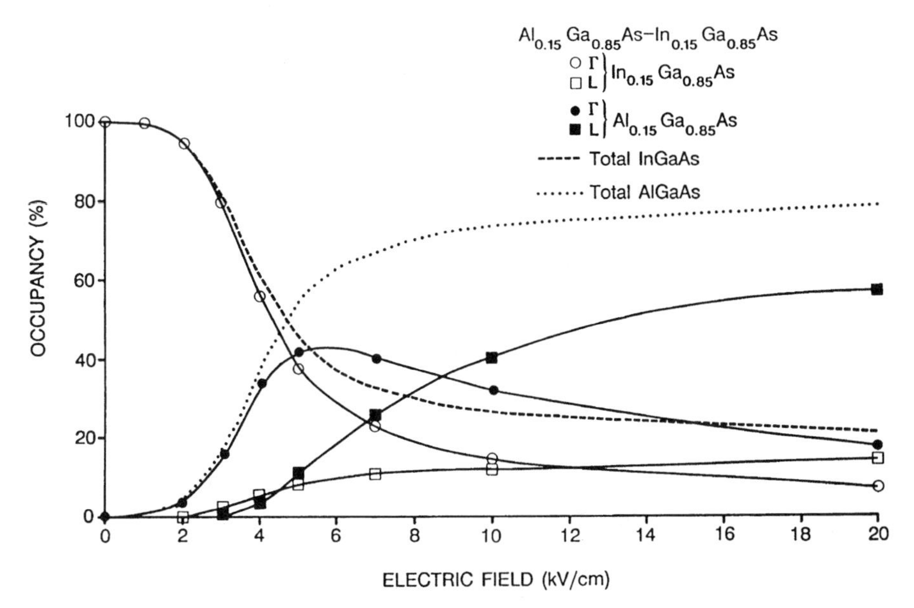

FIGURE 5.3.4 Calculated valley occupancy as a function of applied electric field for the InGaAs–AlGaAs real-space transfer device. The Γ and L valley occupancies for each layer are shown together with the total occupancy of each layer.

5.4 CONSEQUENCES OF NDR IN A SEMICONDUCTOR

What are the consequences of NDR in a semiconductor? How does it affect device performance and can it be utilized for device operation? Historically, NDR in bulk semiconductors was first hypothesized based on theoretical analysis. Gunn (1963) later observed experimentally the formation of microwave oscillations in bulk GaAs under the application of a bias. Let us now try to understand the origin of these microwave oscillations.

We start with an analysis similar to that performed in Section 5.2. Again, we begin with the expression for the current density, Eq. 5.2.3. However, let us assume that diffusion can be neglected. The current density is then

$$J = qn\mu F \qquad 5.4.1$$

Using the one-dimensional continuity equation given by Eq. 5.2.1 and the Poisson equation, Eq. 5.2.2, the time rate of change of the electron concentration, n, becomes

$$\frac{dn}{dt} = \frac{d}{dx}(\mu n F) \qquad 5.4.2$$

Expanding out the derivative in Eq. 5.4.2 yields

$$\frac{dn}{dt} = F\frac{d}{dx}(n\mu) + \mu n\frac{dF}{dx} \tag{5.4.3}$$

But

$$\frac{dF}{dx} = -\frac{q(n - n_0)}{\varepsilon} \tag{5.4.4}$$

so Eq. 5.4.3 becomes

$$\frac{dn}{dt} = F\frac{d}{dx}(n\mu) - \frac{q\mu n}{\varepsilon}(n - n_0) \tag{5.4.5}$$

Notice that the coefficient in front of the $n - n_0$ factor in the last term in Eq. 5.4.5 has dimensions of inverse time. It can thus be represented as a lifetime, τ. Substituting τ into Eq. 5.4.5 in place of the coefficient, Eq. 5.4.5 becomes

$$\frac{dn}{dt} = F\frac{d}{dx}(n\mu) - \frac{n - n_0}{\tau} \tag{5.4.6}$$

Again, we obtain a partial differential equation for the concentration n. As before, n can be written as the product of a spatial function, $u(x)$, and a temporal function, $T(t)$. The temporal function is the same as that determined in Section 5.2, and is given as

$$n = (n - n_0)e^{-t/\tau} + n_0 \tag{5.4.7}$$

If NDR is present, the mobility is negative and the concentration grows with time as

$$n = (n - n_0)e^{t/\tau} + n_0 \tag{5.4.8}$$

From the definition of the mobility, $\mu = v/F$, we can readily identify the region of NDR in the velocity field curve. For example, consider either the velocity field curve for InGaAs, shown in Figure 5.3.2, or that of GaAs, shown in Figure 5.4.1. In either material, there is a region in which the mobility is negative. Referring to Figure 5.4.1, the region marked 2 has a negative mobility and hence exhibits NDR. The region marked 1 in Figure 5.4.1 has a positive mobility. In this case τ is positive and n decays with time. In region 2, dv/dF is negative, so μ is negative. Therefore, τ is negative and n grows with time. So at low applied electric field strengths, n decays with time but at high fields, n grows with time. Therefore, at high fields a charge inhomogeneity, $n - n_0$, can grow.

The question is, though, does the charge inhomogeneity grow substantially? Under certain conditions it does. The charge inhomogeneity becomes appreciable only if the transit time is sufficiently greater than the time in which the

EXAMPLE 5.4.1: dc Power Dissipation in a Gunn Diode

Given a GaAs Gunn diode, estimate the dc power dissipation per unit volume. Assume that the length of the diode is 10 μm, the donors are all fully ionized, and the device is biased just below threshold.

From Eq. 5.4.13, the condition in GaAs for which the charge domain will grow is given as

$$N_d L \geq 10^{12} \text{ cm}^{-2}$$

Since the donors are all fully ionized, $n_0 = N_d$. Therefore, the carrier concentration at threshold must be

$$n_0 = \frac{10^{12}}{L} = \frac{10^{12}}{10^{-3}} = 10^{15} \text{ cm}^{-3}$$

In GaAs, if the device is biased just below threshold, we know that the threshold field in GaAs is about 3.5 kV/cm and that the corresponding velocity is 2.0×10^7 cm/s. The current in the diode is simply

$$I = qn_0 v_d A$$

where A is the area of the diode and v_d is the drift velocity, in this case 2.0×10^7 cm/s. The dc dissipated power is given as the product of the dc current I and the voltage V:

$$P = IV = qn_0 v_d AFL$$

where F is the field and L is the length of the diode. The power per unit volume Ω is then

$$\frac{P}{\Omega} = qn_0 v_d F$$

Substituting for each of the variables, the power dissipated per unit volume is

$$\frac{P}{\Omega} = (1.6 \times 10^{-19})(10^{15})(2 \times 10^7)(3.5 \times 10^3) = 1.12 \times 10^7 \text{ W/cm}^3$$

domain grows, τ. We call τ the dielectric relaxation time. Let t_{tr} be the transit time. The condition for growth of the charge inhomogeneity or domain is then

$$t_{\text{tr}} > \tau \qquad 5.4.9$$

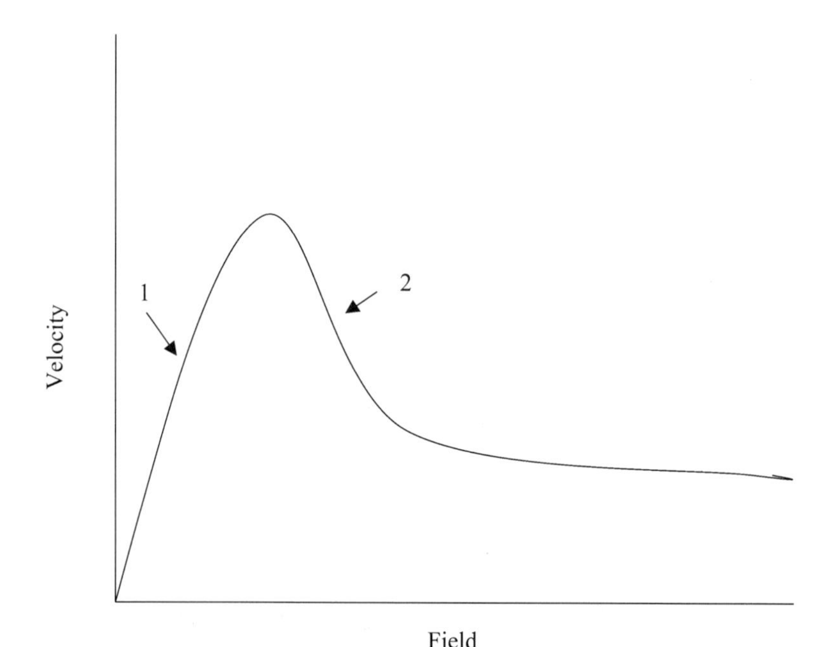

FIGURE 5.4.1 Steady-state velocity field curve in bulk GaAs. Notice that the mobility is positive in region 1 and negative in region 2. Therefore, NDR occurs in region 2.

The transit time is approximately equal to the length of the device divided by the drift velocity of the domain, v_D:

$$t_{\text{tr}} = \frac{L}{v_D} \tag{5.4.10}$$

Using as a definition of τ,

$$\tau = \frac{\varepsilon}{q\mu n} \tag{5.4.11}$$

the inequality 5.4.9 becomes

$$\frac{\varepsilon v_D}{q\mu} < Ln \tag{5.4.12}$$

which is the condition for the formation of stable domains in a bulk semiconductor as a result of NDR. Substituting values for the mobility and domain drift velocity for GaAs, one obtains the condition

$$N_d L > 10^{12}\ \text{cm}^{-2} \tag{5.4.13}$$

where N_d is the donor doping concentration and L is the length of the device.

5.5 TRANSFERRED ELECTRON-EFFECT OSCILLATORS: GUNN DIODES

The simplest mode of a transferred electron oscillator is that described above, the formation of charge domains that propagate through the device. If condition 5.4.13 is satisfied, stable charge domains form and build up as they progress through the device structure, moving from the cathode to the anode. Upon reaching the anode, the domain gives up its energy as a pulse of current in the external circuit. In this mode, typically, only one domain forms at a time.

The formation and dynamics of the domains can be understood as follows. The reader should recall that the device structure consists of a simple bar of semiconductor material with ohmic contacts at each end. Upon application of a bias, an electric field will be produced throughout the bar. The field will then be uniform unless there exist defects or doping inhomogeneities. In many Gunn diodes, an intentional nucleation site, such as a defect or doping inhomogeneity, is built into the structure. The electric field is altered near a nucleation site such that it is higher within the nucleation site than outside it. For purposes of illustration, let us assume that the electrons move from left to right in the device. If it is assumed that electrons accumulate at the nucleation site forming a dipole (as explained below), the electric field within the nucleation site is higher than that outside the nucleation site, as shown in Figure 5.5.1. The device is originally biased to be in the negative differential resistance region. The electron drift velocity is higher at lower field strengths than at higher field strengths when the device is biased within the NDR region, as can be seen in region 2 of Figure 5.4.1. As a result, the electrons within the nucleation site experience a higher electric field and thus have a lower drift velocity than those outside the nucleation site. The net flow of electrons into the nucleation site exceeds that exiting the nucleation site. Therefore, there is a "pileup" of negative charge. In front of this electron charge pileup there is a depletion of electrons, resulting in a slight amount of positive charge from uncompensated donors. As a result, a dipole forms as shown in Figure 5.5.2*a*. The electron concentration exceeds the background doping concentration at the nucleation site but is less than the background concentration to the right of the nucleation site as shown in Figure 5.5.2*b*. Thus a domain is produced. As the electron pileup increases, the electric field within the domain continues to increase, while that outside of the domain continues to decrease. As a result, the velocity of the carriers within the domain decrease while the velocity of the carriers outside of the domain increase producing an even larger space charge accumulation. Hence, the system experiences a runaway effect. Of course, there is some limit to the process; the dipole cannot increase in magnitude indefinitely. A stable situation occurs when the field within the domain increases to a value outside the NDR region and the field outside the domain decreases to a value outside the NDR region. This occurs for the points marked *A* and *B* in Figure 5.5.3. As can be seen, when the fields within and outside the domain

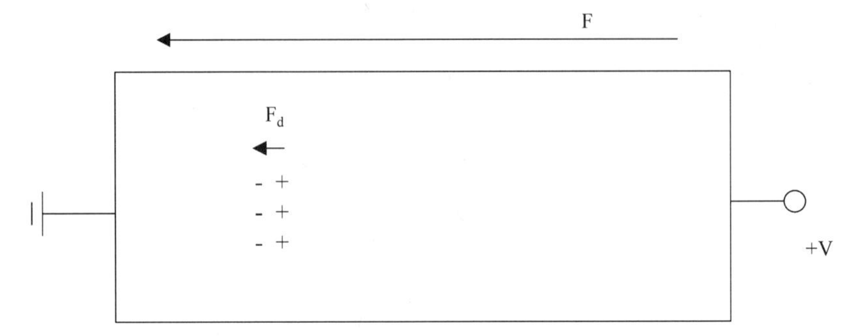

FIGURE 5.5.1 Gunn diode, showing the dipole formation. Notice that the electric field from the dipole F_d adds to the overall electric field F resulting in a higher electric field within the dipole than outside the dipole, as discussed in the text.

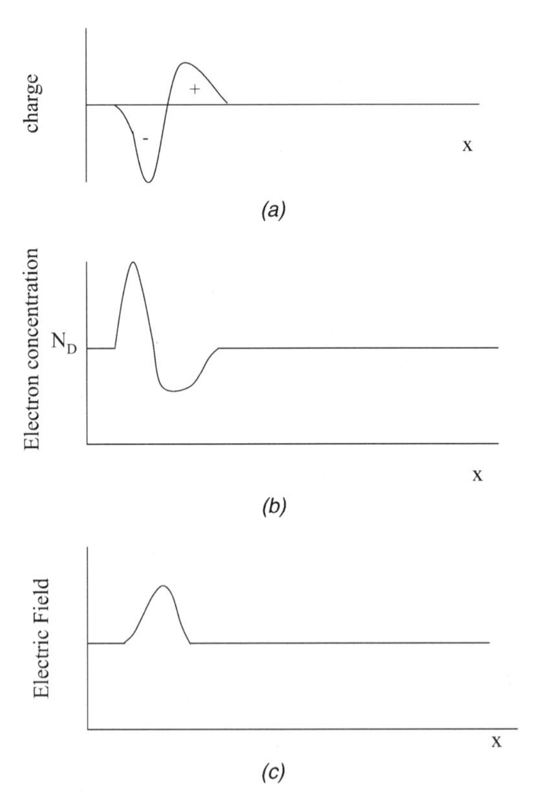

FIGURE 5.5.2 (*a*) Charge, (*b*) electron concentration, and (*c*) electric field within a Gunn diode as a function of position x along the device. Notice that once the domain forms, the field drops below the critical field everywhere except in the domain.

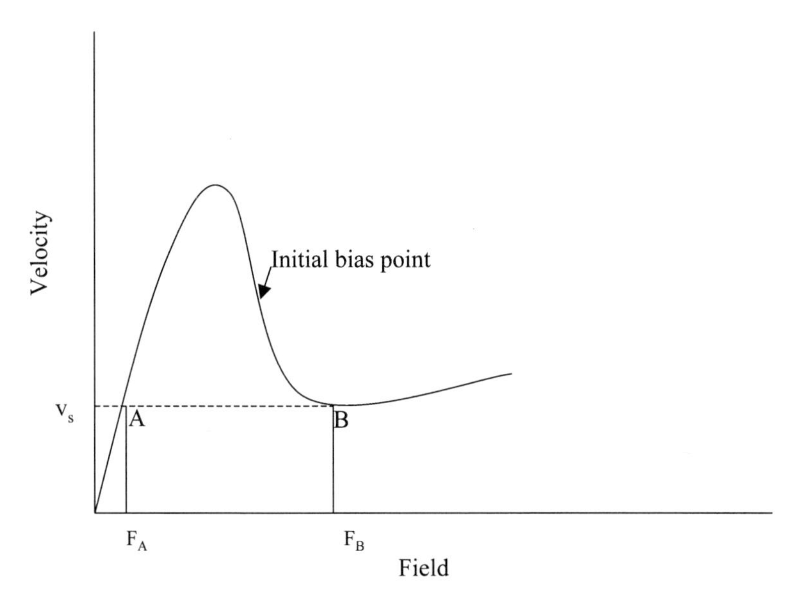

FIGURE 5.5.3 Velocity–field curve showing the initial bias point within the NDR region and the two stable bias points A and B.

equal F_B and F_A, respectively, the electrons everywhere within the device drift with velocity v_s, and no further growth occurs within the domain.

The domain will drift through the device until it reaches the anode, where it is collected. Once the domain has been collected at the anode, the field begins to rise again within the entire device until a new dipole is formed. The process then repeats itself. Domains form only one at a time under these conditions. The output current thus consists of a series of pulses corresponding to the arrivals of each domain. The frequency of the signal is given by the ratio of the saturation velocity v_s to the length of the device L as v_s/L.

It is important to understand the conditions under which the domain can be quenched. If the electric field outside the domain drops in magnitude below F_A, the velocity of the electrons outside the domain drops below that of v_s. Electrons outside the domain now have a lower velocity than those within the domain and the domain decreases in size until it is finally quenched. As a result, the domain disappears or is quenched.

There exist many different modes of operation for a Gunn diode (Streetman, 1980; Sze, 1981). Some of the different modes can be summarized as follows. For simplicity, we discuss only briefly three modes of operation different from that presented above: (1) ac transit time mode, (2) quenched domain mode, and (3) limited space-charge accumulation mode. The ac transit time mode is obtained when the transit time t is chosen to be nearly equal to the oscillation period τ and the device is operated within a resonant circuit. This mode is very similar to the mode presented above, in that a single domain forms for each cycle. The domain transits the device until it is collected at the anode, after which another domain can form. It is important to note that the frequency of oscillation of the diode is simply the inverse of the transit time, $1/t$, which is equal to v_s/L. As earlier, the transit time for this mode is simply the ratio L/v_s.

Higher-frequency operation of a Gunn diode can be obtained if the domain is quenched before it is collected at the anode. The basic workings of the quenched domain mode are as follows. Following the formation of the first domain, it is quenched before it reaches the anode, typically after traveling only about one-third of the device. The domain is quenched by swinging the applied voltage and hence the magnitude of the electric field below F_A. As the applied voltage and field increase in magnitude back above the threshold field, another domain can be nucleated; the process repeats itself. In the case where quenching occurs after the domain travels only one-third of the length of the device, three domains are created and quenched in a period equal to the transit time of the device. Hence the operating frequency is then about three times higher than that of a simple transit time mode device.

There exists another mode in which domains do not form. This mode, the *limited space-charge accumulation* (LSA) *mode*, is one of the most efficient means of utilizing a Gunn diode. In the LSA mode the frequency is chosen to be so high that the domains have insufficient time to form while the field is above threshold. The sample remains for the most part in the NDR state during a large fraction of the voltage cycle. In other modes of operation, the sample cannot remain within the NDR state for long since once a stable domain forms, the fields in the device are such that a positive resistance reoccurs. Recall that this is the condition for stability for the domain. However, in the LSA mode, stable domains never form. Much as in the quenched domain mode, the frequency depends on the resonant circuit. As a result, in the LSA mode the device can have a much higher frequency than that of the inverse transit time frequency. The requirements for operation in the LSA mode are: (1) the frequency must be high enough that stable domains do not have sufficient time to form while the signal is above threshold, and (2) any accumulation of electrons near the cathode must have time to collapse while the signal is below the threshold field. The LSA mode is suitable for generating short high-power pulses.

5.6 NEGATIVE DIFFERENTIAL RESISTANCE TRANSISTORS

New types of field-effect transistor structures have been suggested (Kastalsky and Luryi, 1983) that employ real-space transfer effects. The basic operating principle behind these devices is that the resulting carrier heating within the channel of a FET due to the action of the source–drain voltage can result in real-space transfer out of the channel by the action of the gate bias. The device structure most representative of this class of devices is called a NERFET. The NERFET is sketched in Figure 5.6.1. The device structure is essentially a HEMT device with a slight modification. Instead of forming the channel on

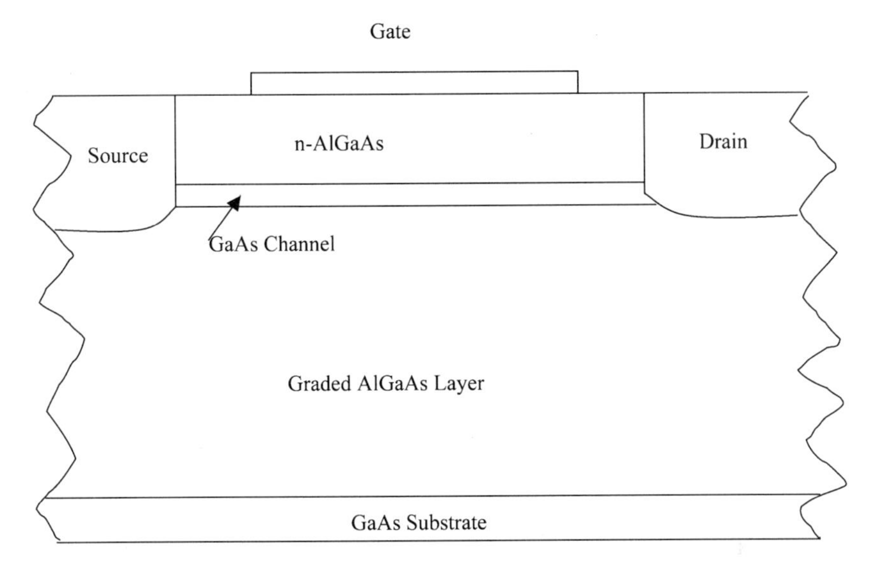

FIGURE 5.6.1 Real-space transfer field-effect transistor (NERFET).

an undoped GaAs epilayer as is typically done for a HEMT, the GaAs channel is grown on top of a graded AlGaAs layer. This graded AlGaAs layer provides a potential barrier between the channel and the substrate contact. The AlGaAs layer is graded with increasing Al concentration toward the channel. As a result, the potential barrier is largest at the channel interface and decreases downward into the device toward the substrate, as shown in Figure 5.6.2. The device can be operated in two different configurations, with the substrate contacted or with the substrate floating. When the substrate is contacted, a substrate current will flow in the device.

The operation of the NERFET can be understood as follows. The source–drain voltage acts to accelerate the electrons. If the device is biased into pinch-off, a significant voltage drop appears near the drain end of the channel, producing a very high electric field. This field strongly heats the electrons, resulting in a significant increase in the electron temperature T_e. As the electron temperature increases, the electrons can be thermionically emitted over the graded AlGaAs potential barrier. Once the electrons cross into the graded AlGaAs layer, the electric field arising from the compositional grading will cause them to drift toward the substrate. Recall that electrons "roll downhill" in energy band diagrams. Thus the channel electron concentration will decrease as charge is transferred into the substrate. This results in a significant decrease in the drain current. The resulting current–voltage (I–V) characteristic for the device is sketched in Figure 5.6.3. As can be seen from the figure, a marked negative differential resistance appears in the I–V characteristic. The drain current increases at higher drain voltages, due to thermionic emission of electrons back into the channel and hence into the drain from the substrate.

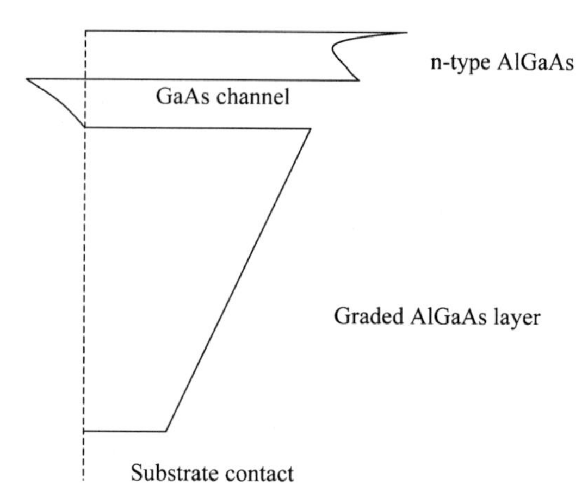

FIGURE 5.6.2 Energy band diagram for the NERFET. The drawing is oriented in the same manner as in Figure 5.6.1.

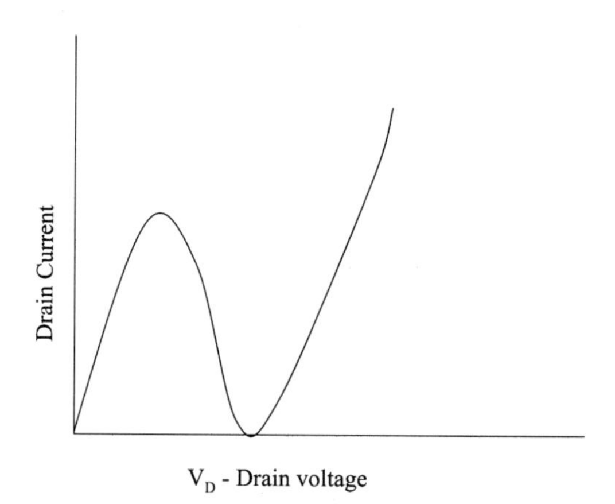

FIGURE 5.6.3 Current–voltage characteristic for a NERFET device at a single gate voltage. Notice the presence of NDR in the *I–V* characteristic. The NDR is due to a reduction in the drain current due to electron transfer out of the channel into the substrate via real-space transfer.

†5.7 IMPATT DIODES

We end this chapter with a discussion of IMPATT diodes. Although the IMPATT diode is not generally a transferred electron device, it is useful in high-frequency, high-power applications and has some similarities to Gunn diodes. Similar to the Gunn diode, the IMPATT diode is a two-terminal

†Optional material.

device and utilizes negative differential resistance effects. However, within the IMPATT NDR is induced by driving the current and voltage out of phase with one another. The IMPATT diode can be used in a microwave circuit as a high-frequency, high-power oscillator. The diode is mounted in a microwave cavity which has an inductive impedance matched to the mainly capacitive impedance of the diode so as to form a resonant system. If the diode is operated in a negative differential resistance region, it delivers power from the dc bias to the oscillation. The efficiency of the device is defined as the ac power delivered by the diode divided by the dc power dissipated.

The basic operational principle of the IMPATT diode is that microwave oscillation and amplification due to frequency-dependent negative resistance arises from the phase delay between the current and voltage waveforms in the device (Read, 1958; Haddad et al., 1970). Recall that if the voltage and current are out of phase (i.e., the ratio of V to I is negative) the device delivers power much like a battery or a solar cell. If the ratio of V to I is positive, the voltage drop is in the direction of the current flow and the device absorbs power. Therefore, the IMPATT device must be designed such that V and I are out of phase with one another by an angle between 90 and 270°. Ideally, an angle of 180° is desired to ensure maximum output power delivery. An IMPATT diode consists of two regions, an avalanche breakdown region and a transit time region, as shown in Figure 5.7.1. For illustration the device and an idealized plot of the electric field profile are sketched using a p^+-n-i-n^+ doping scheme. In this device, electrons produced within the avalanche region would then be injected into and transit across the drift region. Notice that the electric field is highest within the avalanche region and is essentially uniform but lower in magnitude within the drift region. The field within the avalanche breakdown region is sufficiently large that it provides gain through carrier multiplication by means of impact ionization. Within the drift or transit region, the field strength is sufficiently large that the carriers drift with a constant saturated velocity. The avalanche and transit time regions must be designed such that the current and voltage are driven out of phase in order to supply power to the external circuit.

Two different mechanisms are used in an IMPATT device to delay current: (1) the finite rise and decay time of the avalanche current, and (2) the finite transit time of the carriers through the drift region. In the breakdown region the field is sufficiently high that impact ionization can occur. In operation, the dc bias is sufficiently large that the peak field, arising from the sum of the dc and ac fields, F_m, is greater than the breakdown field, F_c, the critical field for avalanche breakdown, during the positive half of the voltage cycle. During the negative half of the voltage cycle, the peak field is less than the critical field F_c. The current reaches its maximum value at the point where the field within the avalanche region becomes less than the breakdown field. This occurs, of course, in the middle of the voltage cycle, one-fourth of a cycle later than the voltage maximum. Hence, the phase angle between the current

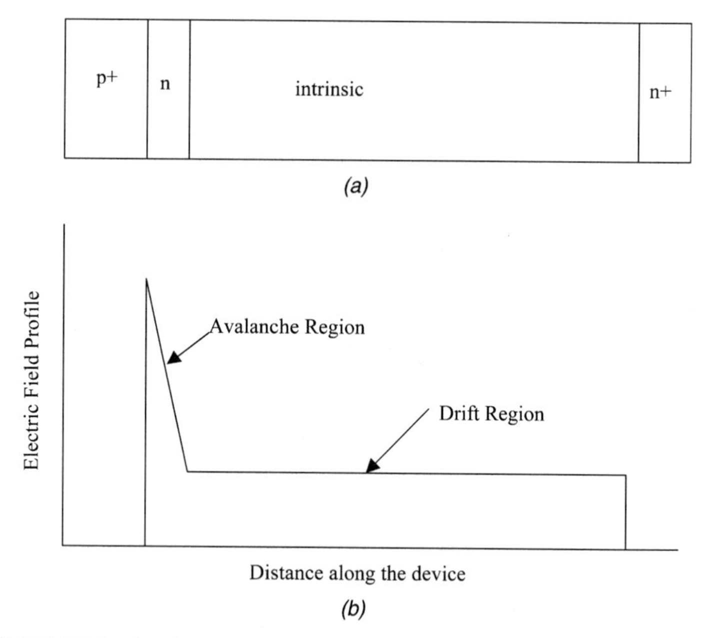

FIGURE 5.7.1 (*a*) An IMPATT diode and (*b*) its accompanying field profile.

and voltage maximums is shifted by $\pi/2$. Since the avalanche region serves as the input into the drift region, the injected current is phase-shifted by $\pi/2$ relative to the voltage upon injection. In other words, the avalanche region serves as an injecting cathode producing a pulse of current, developed during each cycle of the input ac voltage, that lags the voltage by a phase angle of $\pi/2$.

The IMPATT device must be designed such that V and I are out of phase with one another by 180°. In this way, the two reach their respective peak magnitudes at the same time but with one the negative of the other. As shown above, the first 90° is achieved within the avalanche region. The second 90° phase shift is obtained using the drift region.

Let us consider how an IMPATT diode provides output power. As mentioned above, for the IMPATT to deliver power to the external circuit, the current and voltage must be driven out of phase by 90 to 270°. The first 90° of phase shift is achieved within the avalanche region as discussed above. The remaining 90° of phase shift is obtained by adjusting the drift region width such that the finite carrier transit time across it delays the carriers by 90° with respect to the applied voltage.

First, it is important to note that for the device to exhibit a negative differential resistance, it is necessary to have an injection phase delay in addition to the transit time delay. This can be seen as follows. Assume that a current pulse is injected into the drift region from the ionization region with a phase delay angle of ϕ. Labeling the injection plane, the interface between the drift and ionization regions, as $x = 0$, the ac current pulse can be written as

$$j(x = 0) = je^{-i\phi} \tag{5.7.1}$$

The total ac current is given as the sum of the conduction and displacement currents:

$$j = j_c(x) + j_d(x) = j_c(x = 0)e^{-i\omega t} + i\omega\varepsilon_s F(x) \tag{5.7.2}$$

where $F(x)$ is the electric field as a function of position and ε_s is the dielectric constant of the semiconductor. It is further assumed that the carriers drift with velocity v_s within the drift region. Therefore, the time is given simply as $t = x/v_s$. Substituting this result and Eq. 5.7.1 into Eq. 5.7.2, the total ac current density can be rewritten as

$$j = je^{-i\phi}e^{-i\omega x/v_s} + i\omega\varepsilon_s F(x) \tag{5.7.3}$$

Solving Eq. 5.7.3 for $F(x)$ yields

$$F(x) = \frac{j(1 - e^{-i\omega x/v_s - i\phi})}{i\omega\varepsilon_s} \tag{5.7.4}$$

The ac impedance Z is defined as

$$Z = \frac{\int_0^W F(x)dx}{j} \tag{5.7.5}$$

where W is the drift region width. The total current is constant, so the ac impedance can be found as

$$Z = \frac{j\int_0^W (1 - e^{-i\omega x/v_s}e^{-i\phi})dx}{i\omega\varepsilon_s j} \tag{5.7.6}$$

Simplifying, Eq. 5.7.6 becomes

$$Z = \frac{1}{i\omega C}\left[1 - \frac{e^{-i\phi}(1 - e^{-i\theta})}{i\theta}\right] \tag{5.7.7}$$

where C is the capacitance per unit area and θ is the transit angle. C and θ are defined as

$$\theta = \frac{\omega W}{v_s} \qquad C = \frac{\varepsilon_s}{W} \tag{5.7.8}$$

In general, the ac impedance (the ratio of V to I) is complex. The real part of Z is the ac resistance R and is determined from the real part of Eq. 5.7.7 as

$$R = \frac{\cos\phi - \cos(\phi + \theta)}{\omega C \theta} \tag{5.7.9}$$

If there is no injection phase delay, the angle $\phi = 0$, and the resistance simplifies to

$$R = \frac{1 - \cos\theta}{\omega C \theta} \tag{5.7.10}$$

Notice that the expression for R given by Eq. 5.7.10 can never be less than zero. Therefore, when there is no injection phase delay, the resistance R is always greater or equal to zero and no negative differential resistance occurs. If the injection phase angle ϕ is not zero, R is negative for certain values of θ. For example, if $\phi = \pi/2$, then R has its largest negative value at $\theta = 3\pi/2$. Clearly, to have a negative differential resistance, there must be an injection phase angle different from zero.

In practice, though, how is an injection phase delay achieved? The device must be designed such that the injection of the conduction current into the drift region is delayed. Let us examine quantitatively injection phase delay due to the avalanche region. We start with the one-dimensional continuity equations for holes and electrons (Brennan, 1999):

$$\frac{\partial p}{\partial t} = -\frac{1}{q}\frac{\partial j_p}{\partial x} + G - \frac{\delta p}{\tau} \qquad \frac{\partial n}{\partial t} = \frac{1}{q}\frac{\partial j_n}{\partial x} + G - \frac{\delta n}{\tau} \tag{5.7.11}$$

where G is the generation rate, δp and δn are the excess hole and electron concentrations, respectively, and τ is the lifetime, assumed to be the same for both carriers in Eq. 5.7.11. Within the multiplication region, it is acceptable to neglect the recombination rate with respect to the generation rate due to avalanche multiplication. The generation rate due to avalanche multiplication assuming an equal ionization coefficient α for the electrons and holes is given as

$$G = \alpha(n + p)v_s \tag{5.7.12}$$

where v_s is the saturation velocity for the carriers, which is again assumed to be equal. Using Eq. 5.7.12 in Eq. 5.7.11 yields

$$\frac{\partial n}{\partial t} = \frac{1}{q}\frac{dJ_n}{dx} + \alpha v_s(n + p) \qquad \frac{\partial p}{\partial t} = -\frac{1}{q}\frac{dJ_p}{dx} + \alpha v_s(n + p) \tag{5.7.13}$$

The total current density J_T is given as

$$J_T = J_n + J_p \tag{5.7.14}$$

Adding the two relations given in Eq. 5.7.13 and integrating over the avalanche multiplication region, defined as from $x = 0$ to $x = W_A$, yields

$$\int_0^{W_A} \frac{\partial}{\partial t}(n + p)dx = \int_0^{W_A} \frac{1}{q}\frac{\partial}{\partial x}(J_n - J_p)dx + 2\int_0^{W_A} \alpha v_s(n + p)dx \quad 5.7.15$$

But the electron and hole current densities within the avalanche region are drift currents, given as

$$J_n = qnv_s \qquad J_p = qpv_s \quad 5.7.16$$

Multiplying the first term of Eq. 5.7.15 by v_s/v_s, the entire equation by q, and performing the integrations gives us

$$\frac{W_A}{v_s}\frac{\partial J_T}{\partial t} = (J_n - J_p)|_0^{W_A} + 2J_T\int_0^{W_A} \alpha\, dx \quad 5.7.17$$

The boundary conditions for the current densities are as follows. At $x = 0$, the hole current is simply equal to its saturated value prior to entering the avalanche region. Recall that the hole current is injected at the plane $x = 0$ while the electron current is injected at the plane $x = W_A$. Calling the reverse saturated electron and hole current densities within the diode J_{n0} and J_{p0}, respectively, the currents at the boundaries are given as follows:

$$\text{At} \quad x = 0: \qquad J_n - J_p = J_T - 2J_{p0} \quad 5.7.18$$

$$\text{At} \quad x = W_A: \qquad J_n - J_p = 2J_{n0} - J_T \quad 5.7.19$$

Using the boundary conditions specified in Eqs. 5.7.18 and 5.7.19 in Eq. 5.7.17, and defining the total reverse saturation current density, $J_{T0} = J_{n0} + J_{p0}$, we obtain

$$\frac{W_A}{v_s}\frac{\partial J_T}{\partial t} = 2(J_{T0} - J_T) + 2J_T\int_0^{W_A} \alpha\, dx \quad 5.7.20$$

Equation 3.7.20 can be simplified using the definitions

$$\frac{W_A}{v_s} = \tau \qquad \int_0^{W_A} \alpha\, dx = \langle\alpha\rangle W_A \quad 5.7.21$$

where $\langle\alpha\rangle$ is the spatial average of the ionization coefficient. Using Eq. 5.7.21 in Eq. 5.7.20 yields

$$\frac{\partial J_T}{\partial t} = \frac{2J_T}{\tau}(\langle\alpha\rangle W_A - 1) + \frac{2J_{T0}}{\tau} \quad 5.7.22$$

We consider the operation of the diode in the presence of an ac field with a dc offset. The field is then

$$F = F_0 + F_{\text{ac}} e^{i\omega t} \tag{5.7.23}$$

where F_0 and F_{ac} are defined as the dc and ac fields, respectively. The average impact ionization rate coefficient can be expanded in a Taylor series to first order as

$$\langle \alpha \rangle = \langle \alpha_0 \rangle + \frac{d\langle \alpha \rangle}{dF} \Delta F = \langle \alpha_0 \rangle + \langle \alpha' \rangle F_{\text{ac}} e^{j\omega t} \qquad 5.7.24$$

Substituting Eqs. 5.7.23 and 5.7.24 into Eq. 5.7.22 and taking only the ac components results in the following expression for the ac current density within the avalanche region j_A:

$$j_A = \frac{2J_T W_A \langle \alpha' \rangle}{i\omega\tau} F_{\text{ac}} \qquad 5.7.25$$

In Eq. 5.7.25 it is important to note that since there is no dc component of the displacement current, the total dc current density is given by the reverse saturation, conduction current density. The total circuit current is given as the sum of the conduction and displacement currents. The displacement current is

$$j_d = i\omega\varepsilon F_{\text{ac}} \qquad 5.7.26$$

Therefore, the total ac circuit current density j is obtained from the sum of Eqs. 5.7.25 and 5.7.26 as

$$j = j_A + j_d = \left(\frac{2J_T W_A \langle \alpha' \rangle}{i\omega\tau} + i\omega\varepsilon \right) F_{\text{ac}} \qquad 5.7.27$$

The impedance per unit area within the avalanche zone Z_A is obtained from the ratio of the ac voltage, $F_{\text{ac}} W_A$, to the ac current density j_A. Consider first only the impedance corresponding to the conduction current Z_c. Z_c is then given as

$$Z_c = \frac{i\omega\tau}{2J_T \langle \alpha' \rangle} \qquad 5.7.28$$

Notice that this is purely reactive. The conduction impedance is also inductive with an effective inductance L of

$$L = \frac{\tau}{2J_T \langle \alpha' \rangle} \qquad 5.7.29$$

This inductance is in parallel with the avalanche junction capacitance, given as

$$C = \frac{\varepsilon A}{W_A} \qquad 5.7.30$$

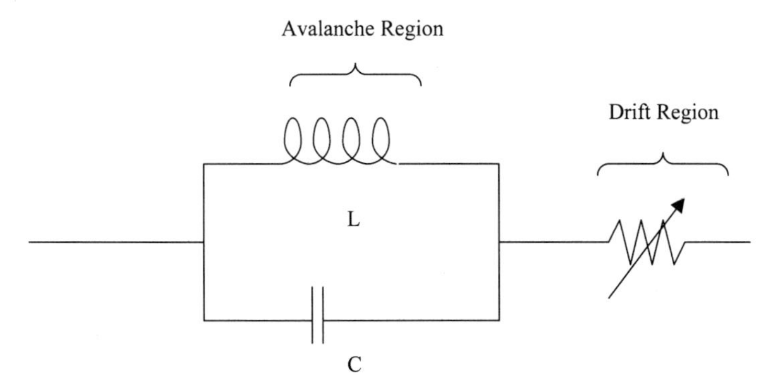

FIGURE 5.7.2 Equivalent-circuit model for the IMPATT diode.

where A is the area of the junction and ε the dielectric constant for the material. The avalanche region can then be modeled as an equivalent circuit with L and C in parallel, as shown in Figure 5.7.2. The total impedance of the avalanche region is then given as the parallel impedance of L and C. The ratio of the ac conduction current density to the total ac current density γ is determined by dividing Eq. 5.7.25 by Eq. 5.7.27 to yield

$$\gamma = \frac{j_A}{j_A + j_d} = \frac{1}{1 - \omega^2 \varepsilon \tau / 2J_T W_A \langle \alpha' \rangle} \quad 5.7.31$$

Defining the avalanche resonant frequency ω_A as

$$\omega_A = \frac{2J_T W_A \langle \alpha' \rangle}{\varepsilon \tau} \quad 5.7.32$$

γ becomes

$$\gamma = \frac{1}{1 - \omega^2/\omega_A^2} \quad 5.7.33$$

The total impedance for the avalanche region of the device, Z_A, is then given as

$$Z_A = \frac{W_A F_{\text{ac}}}{j_{\text{ac}} + j_d} = \frac{i\omega\tau / 2\langle \alpha' \rangle J_T}{1 - \omega^2/\omega_A^2} \quad 5.7.34$$

Next, we consider the behavior of the drift region of the device. Both the ac conduction current density and the displacement current density are functions of space and time. The total ac current within the drift region is equal to the sum of the displacement current density and the ac conduction current density:

$$j = j_c(x,t) + j_d(x,t) \quad 5.7.35$$

The ac conduction current density within the drift region, $j_c(x,t)$, at time t and position x can be related to the injected conduction current density from the avalanche region. It is assumed that the carriers within the drift region drift at a constant velocity v_s such that at time t they are at a new position $x' = x + v_s t$. It is further assumed that the drift velocity within the drift region is equal to the saturation velocity of the carriers assumed within the avalanche region. Thus the total ac current density within the drift region can be related to the injected ac current density from the avalanche region as

$$j_c(x,t) = j_A\left(t - \frac{x}{v_s}\right) \tag{5.7.36}$$

Clearly, the ac conduction current density within the drift region propagates as a simple wave and can be written

$$j_c(x) = j_A e^{-i\omega x/v_s} \tag{5.7.37}$$

The displacement current is again given by Eq. 5.7.26. Therefore, the total ac current density within the drift region is given as

$$j = j_d + j_c = i\omega\varepsilon F_{\text{ac}} + j_A e^{-i\omega x/v_s} \tag{5.7.38}$$

Solving Eq. 5.7.38 for F_{ac} yields

$$F_{\text{ac}} = \frac{j}{i\omega\varepsilon}(1 - \gamma e^{-i\omega x/v_s}) \tag{5.7.39}$$

which is a function of x. The ac voltage v_{ac} developed across the drift region is determined by integrating the ac electric field over x. Hence v_{ac} is given as

$$v_{\text{ac}} = \int_0^{W_D} F_{\text{ac}}dx = \frac{jW_D}{i\omega\varepsilon}\left(1 - \frac{1}{1 - \omega^2/\omega_A^2}\frac{1 - e^{-i\theta}}{i\theta}\right) \tag{5.7.40}$$

where W_D is the width of the drift region and θ is the transit angle, defined as

$$\theta = \frac{\omega W_D}{v_s} \tag{5.7.41}$$

The impedance is defined as the ratio of v_{ac} to jA where A is the area of the diode and is given as

$$Z = \frac{1}{i\omega C_D}\left(1 - \frac{1}{1 - \omega^2/\omega_A^2}\frac{\sin\theta}{\theta}\right) + \frac{1}{\omega C_D}\left(\frac{1}{1 - \omega^2/\omega_A^2}\frac{1 - \cos\theta}{\theta}\right) \tag{5.7.42}$$

where C_D is the depletion layer capacitance, defined as

$$C_D = \frac{\varepsilon A}{W_D} \tag{5.7.43}$$

The total impedance of the diode can be determined from the series combination of the avalanche and drift region impedances. Hence the total impedance is obtained by adding Eqs. 5.7.34 and 5.7.42 to yield

$$Z = \frac{i\omega\tau/2\langle\alpha'\rangle J_T}{1-\omega^2/\omega_A^2} + \frac{1}{i\omega C_D}\left(1 - \frac{1}{1-\omega^2/\omega_A^2}\frac{\sin\theta}{\theta}\right) + \frac{1}{\omega C_D}\left(\frac{1}{1-\omega^2/\omega_A^2}\frac{1-\cos\theta}{\theta}\right) \tag{5.7.44}$$

The real part of the impedance follows immediately from Eq. 5.7.44 as

$$\mathrm{Re}(Z) = \frac{1}{\omega C_D}\left(\frac{1}{1-\omega^2/\omega_A^2}\frac{1-\cos\theta}{\theta}\right) \tag{5.7.45}$$

Inspection of Eq. 5.7.45 shows that the resistance is positive for frequencies, $\omega < \omega_A$ and negative for frequencies $\omega > \omega_A$. Hence the diode exhibits a negative differential resistance at frequencies greater than the avalanche resonant frequency (i.e., the resonant frequency of the equivalent parallel *LC* circuit for the avalanche region of the device). Notice that Eq. 5.7.45 is similar to Eq. 5.7.9. As before, a phase shift within the drift region alone is insufficient to produce a negative differential resistance. Equation 5.7.45 becomes Eq. 5.7.10 if the term $1/(1-\omega^2/\omega_A^2)$ is neglected. Under these conditions it is clear that $\mathrm{Re}(Z)$ can never be less than zero and negative differential resistance cannot occur. Therefore, we conclude, as before, that the behavior of the avalanche region is crucial to developing a negative differential resistance, and hence output power, from the device.

The analysis above is based on several assumptions. These are that space-charge effects due to the carrier charges are negligible within the device and that the avalanche and drift regions of the device can be considered independently. In general, for most IMPATT diodes the latter assumption is acceptable provided that the electric field drops quickly well below the ionization threshold field within the drift region. However, space-charge effects, especially at high power levels, can be important. Generally, the solution for the frequency of operation of an IMPATT diode, including space-charge effects, is done numerically and will not be considered further here. The reader is referred to the literature, particularly the papers by Misawa (1966a,b).

Although IMPATT diodes are capable of delivering high gain at very high frequencies, their use is limited, due to noise. The most commonly employed device for high-frequency amplification is the MESFET or MODFET, since these devices have relatively low noise figures. Avalanching devices such as

IMPATTs are not as attractive in amplifiers and oscillators as FETs because of the large amount of internally generated noise arising from the randomness in the impact ionization process (Teich et al., 1986). However, use of stepped potential structures (Barnes et al., 1987) may reduce the excess noise of IMPATT diodes, leading to relatively low noise, high-frequency operation. These devices may be useful in future IMPATT applications.

PROBLEMS

5.1. Estimate the ballistic velocity of an electron in a single-valley semiconductor at $k = 0.3\ 2\pi/a$, where a is the lattice constant equal to 5.65 Å. Use a parabolic band model with an effective mass of the electron of $0.067m$. Determine the velocity of a two-valley semiconductor at the same value of k, $0.3\ 2\pi/a$, if it is assumed that the second valley has an effective mass of 0.23. In calculating the saturation velocity, assume that the occupations are proportional to the density of states of each valley.

5.2. A crude approximation can be made for the threshold field for the onset of negative differential resistance as follows. Start with the energy balance equation (Eq. 3.4.6); assume steady-state conditions, that the average energy is equal to the intervalley separation energy (0.3 eV), and that all the energy is kinetic. Given that the effective mass is $0.067m$, the low-field mobility is 8500 $cm^2/V \cdot s$, and the phonon energy is 0.035 eV, determine the critical field. Use the relationship given in Problem 3.7 to relate the energy relaxation time to the momentum relaxation time.

5.3. Determine the applied field necessary to heat the electrons in GaAs to sufficient energy such that the ratio of the concentrations of the upper (L) and lower (Γ) valley is 0.5. Assume that the valley separation energy is 0.3 eV, the low-field mobility is 8500 $cm^2/V \cdot s$, the energy relaxation time is 5 ps, and the effective masses are 0.067 and 0.1087 for the Γ and L valleys, respectively.

5.4. Determine the magnitude of the applied electric field necessary in a GaAs–AlGaAs real-space transfer system to have 10% of the total number of carriers transferred into the AlGaAs layer, given the following information: conduction band edge discontinuity = 0.25 eV, μ(GaAs) = 8500 $cm^2/V \cdot s$, m(AlGaAs) = 0.0878, m(GaAs) = $0.067m$, $\tau_E = 5$ ps, and $v = 8.5 \times 10^7$ cm/s. Neglect the equilibrium temperature of the electrons.

5.5. The critical field in bulk InP is about 10.5 kV/cm where the velocity is 2.5×10^7 cm/s. If the same condition holds for InP as for GaAs, for which a charge domain in a Gunn diode will grow, determine the dc power dissipation per unit volume if the device length is 5.0 μm. Assume that the device is biased just below threshold and that all the donors are fully ionized.

5.6. Consider a n^+-p-i-p^+ Si IMPATT diode. Determine the dc voltage required to start avalanching conditions and oscillations if the breakdown field is 350 kV/cm. The dimensions and doping concentrations are given as follows:

n^+ *region*: doping concentration $N_{d1} = 10^{20}$ cm^{-3} width $W_1 = 2\ \mu$m
p *region*: doping concentration $N_{a1} = 10^{16}$ cm^{-3} width $W_2 = 2\ \mu$m
i *region*: doping concentration $N_{a2} = 10^{14}$ cm^{-3} width $W_3 = 5\ \mu$m
p^+ *region*: doping concentration $N_{a3} = 10^{18}$ cm^{-3} width $W_4 = 100\ \mu$m

Assume that the depletion approximation can be used where appropriate and that the field within the n^+ and p^+ regions vanishes. Use 11.8 for the relative dielectric constant for silicon.

5.7. Determine the width of a dipole formed in a Gunn diode in terms of the overall applied electric field F_a, the length of the sample L, and the field strengths F_0 and F_d outside and within the dipole, respectively.

5.8. Estimate the relationship between the output power and the frequency of operation of a Gunn diode. Assume that the device is operating under transit time conditions, that the RF voltage and field produced are V_{RF} and F_{RF}, respectively, and that R is the impedance. Let L be the length of the device.

CHAPTER 6

Resonant Tunneling and Devices

The development of exacting epitaxial growth techniques such as molecular beam epitaxy (MBE) and metal organic vapor-phase deposition (MOVPE) has enabled the growth of heterostructures and multiquantum well systems. Additionally, use of in situ growth monitoring techniques such as reflection high-energy electron diffraction (RHEED) in a MBE system provides exacting control of layer thicknesses to within one monolayer. This exacting growth control enables the realization of structures with quantum-sized dimensions. Device structures can be practically realized with layer widths comparable to or smaller than the electron de Broglie wavelength. Within these structures, quantum effects arising from spatial quantization can occur. In this chapter we discuss a class of semiconductor devices called resonant tunneling devices that have dimensions typically smaller than the electron de Broglie wavelength and exploit quantum mechanical tunneling in their operation.

6.1 PHYSICS OF RESONANT TUNNELING: QUALITATIVE APPROACH

The basic structure of a resonant tunneling diode is shown in Figure 6.1.1. In its simplest implementation, the structure consists of two potential barriers sandwiching a well region. The structure is formed using two different semiconductor materials, typically a GaAs well sandwiched by AlGaAs layers. The GaAs–AlGaAs system is usually used since it forms a type I heterostructure (Brennan, 1999, Chap. 11) and is lattice matched. The conduction band discontinuity between the GaAs and AlGaAs layers produces the potential

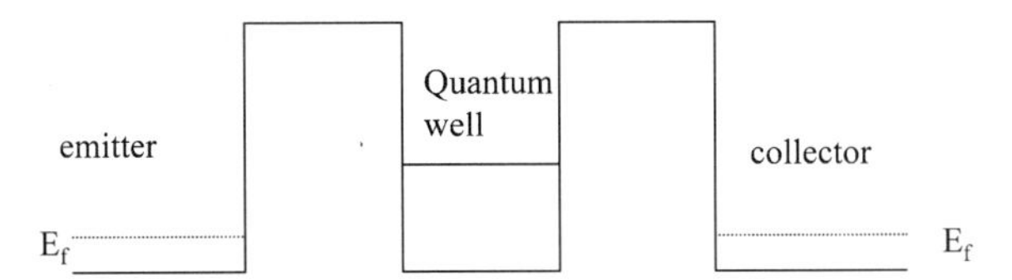

FIGURE 6.1.1 Two-barrier, single-quantum-well resonant tunneling diode under equilibrium conditions.

barrier. As the reader may recall from the discussion in Chapter 2, the magnitude of the potential barrier can be estimated as follows. The energy gap of AlGaAs as a function of Al composition x is given as

$$E_g(\text{AlGaAs}) = 1.42(\text{GaAs}) + 1.247x \qquad \Delta E_c = 0.62\Delta E_g \tag{6.1.1}$$

where 1.42 eV is the energy gap of GaAs. The potential barrier height is then given simply by multiplying 0.62 times the energy gap difference, as shown by the second relation in Eq. 6.1.1.

Under equilibrium conditions, with no externally applied potential, the device, neglecting impurities and defects, is in flat band condition, as shown in Figure 6.1.1. There are three different regions of the device: the emitter, quantum well, and collector, as shown in the diagram. Notice that the emitter and collector regions are assumed to be degenerately doped. Consequently, the Fermi levels lie above the conduction band edge within these two regions, as shown in the diagram. The device is designed such that the first quantum level lies above the Fermi levels in the emitter and collector at equilibrium. Upon the application of a bias, the energy bands bend within the barriers and well, as shown in Figure 6.1.2. Since the emitter and collector are degenerately doped, it is assumed that all of the applied bias appears across the barriers and well. If the bias is sufficiently high, the quantum level E_0 becomes aligned with the Fermi level within the emitter. As a result, electrons within the emitter can now tunnel through the first barrier into the quantum level and then into the collector.

The physics of the resonant tunneling process can be understood as follows. Tunneling occurs when the energy of an incident electron within the emitter matches that of an unoccupied state in the quantum well corresponding to the same lateral momentum. The current as a function of the applied voltage is shown in Figure 6.1.3. The system starts in equilibrium, and of course, the current is zero. As the bias is applied, the quantum well is lowered in energy until the quantum level becomes aligned with the Fermi level within the emitter. Until the quantum level aligns with the Fermi level, the current is relatively low, as shown in Figure 6.1.3 within the region marked as 1. Once the quantum level becomes aligned with the emitter Fermi level, a high current begins to flow since tunneling can now occur. As the voltage increases, the

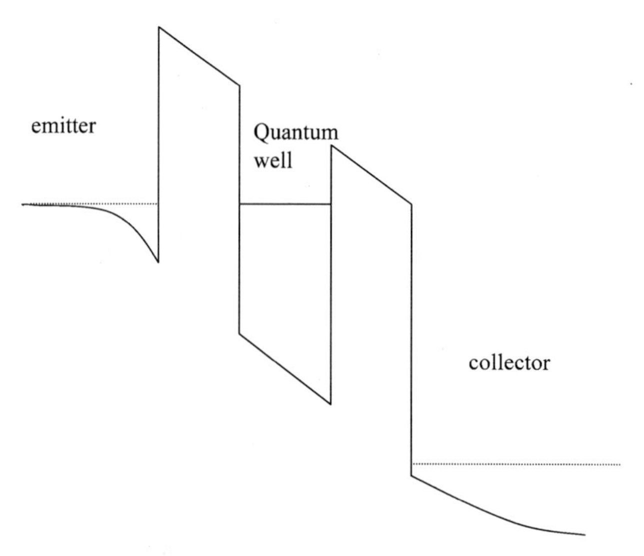

FIGURE 6.1.2 Simple single-quantum-well, double-barrier resonant tunneling diode under bias. Notice that the bias is such that the quantum level is aligned with the Fermi level in the emitter.

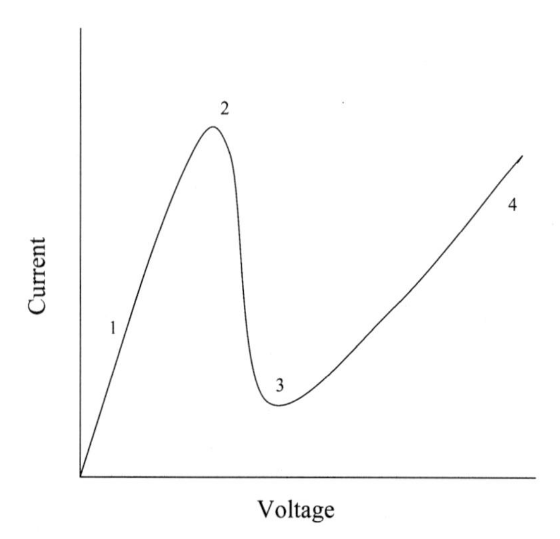

FIGURE 6.1.3 Current versus voltage characteristic for a resonant tunneling diode. Four different regions are marked in the sketch. The first region corresponds to the case of low applied voltage, where the quantum level lies above the Fermi level within the emitter. The second region corresponds to alignment of the quantum level and the emitter Fermi level. The third region corresponds to the case when the quantum level lies below the conduction band edge discontinuity. Finally, the fourth region corresponds to thermionic emission.

quantum level is continuously brought into alignment with electronic states lying below the Fermi level in the emitter and again the current is high. This is marked as region 2 in Figure 6.1.3. As the voltage is increased further, the quantum level drops below the conduction band edge in the emitter. The quantum level becomes aligned with the gap, where no allowed energy states exist. As a result, no electrons tunnel from the emitter into the quantum level and the current drops accordingly to a minimum value. This is marked as region 3 in Figure 6.1.3. As the bias is increased further, the current again becomes large, as shown by the region marked 4 in Figure 6.1.3. The high current in region 4 is due to thermionic emission current flowing over the potential barriers since the bias is now sufficiently high that the potential barrier is pulled down close to the Fermi level.

The two-barrier, single-quantum-well structure is often compared to a Fabry–Perot resonator. The two barriers act like the partially transparent mirrors through which light is coupled into and out of in a Fabry–Perot resonator. The transmissivity for electrons through the double-barrier shows resonant peaks when the perpendicular kinetic energy of the incident electron is equal to the quantum confined state energy. At these energies, the transmissivity of a double-barrier structure approaches 100% even though the transmissivity of a single barrier can be as low as 1%. The large enhancement in the transmissivity of the double-barrier structure arises physically from the fact that the amplitude of the resonant modes increases within the well due to multiple reflections of the electron wave by the potential barriers. As such, the device shows a dramatic increase in current upon resonant alignment. It is precisely for this reason that the process is referred to as resonant tunneling.

Resonant enhancement of the transmissivity of the electron waves through the double-barrier structure can occur only if the electron waves remain coherent. If there exists a high scattering rate from phonons, impurities, defects, or other electrons within the well, the phase coherence of the electron waves is disrupted. Scattering events act to randomize the phase of the electron waves and prevent buildup of the amplitude of the wavefunction in the well that would otherwise result from multiple reflections. Under relatively high scattering conditions, resonant enhancement of the electron wavefunction cannot occur and tunneling proceeds sequentially without preserving the phase coherence of the incident wave. Therefore, there are two general processes that govern electron tunneling in a double-barrier structure, resonant tunneling (coherent; Chang et al., 1974) and sequential tunneling (incoherent; Luryi, 1985). As we show, under resonant or coherent tunneling the peak transmissivity at resonance is equal to the ratio of the minimum to the maximum transmission coefficients of the two barriers, $T_{\text{min}}/T_{\text{max}}$. To achieve 100% transmissivity through the double barrier, the ratio of the transmissivities of the two barriers must be 1, implying that the transmissivities of each barrier must be equal. This is precisely the same condition as in an optical Fabry–Perot resonator. Application of an applied electric field to a symmetric double-barrier structure introduces a difference in the transmissivities of each of the barriers, thus reducing the overall peak transmissivity of the structure. Making the double-barrier structure asymmetric by making the barrier widths different can restore maximum transmissivity. However, this approach will work only to optimize the transmissivity of one level.

It is possible to ascertain whether the tunneling process in a structure proceeds sequentially or resonantly. As mentioned above, the presence of scattering can randomize the phase of the electrons, thereby rendering the electron waves incoherent. Resonant enhancement of the electron waves takes some time to establish since multiple reflections must occur. Therefore, a minimum time exists to establish phase coherence within the well. This time constant, τ_0, can be estimated from the full width, half-maximum of the transmission peak Γ_r as

$$\tau_0 \sim \frac{\hbar}{\Gamma_r} \tag{6.1.2}$$

If scatterings occur more frequently than τ_0, the electron waves become incoherent and resonant tunneling cannot proceed. The mean time between scatterings can be estimated from the total scattering rate present within the well. The total scattering rate includes both elastic and inelastic processes and will be represented as $1/\tau$. Therefore, if the scattering time τ, the reciprocal of the total scattering rate, is much shorter than τ_0, the resonant component of the tunneling process is reduced significantly. Most of the electrons tunnel only after suffering a scattering event and thus do not undergo resonant enhancement.

The total scattering rate within the well consists of both elastic and inelastic processes. The dominant inelastic scattering mechanism in GaAs, which is typically used as the well region in a resonant tunneling diode (RTD) is polar optical phonon scattering, at least for energies below the intervalley threshold energy. The polar optical phonon scattering rate in a two-dimensional GaAs system at 77 K is roughly about $6 \times 10^{12}\ \mathrm{s}^{-1}$ (Yoon et al., 1987). This rate is significantly higher than the two-dimensional acoustic phonon scattering rate, which is approximately $3 \times 10^{10}\ \mathrm{s}^{-1}$. The dominant elastic scattering mechanism is generally ionized impurity scattering depending on the purity of the sample. In most cases, the devices are grown with very high purity GaAs, and the impurity scattering can be reduced below that of the polar optical scattering rate. Therefore, for many practical situations, the total scattering rate in a GaAs–AlGaAs RTD can be approximated as that due to two-dimensional polar optical phonon scattering and is quantitatively about $6 \times 10^{12}\ \mathrm{s}^{-1}$.

The discussion above provides a reasonable picture of the tunneling mechanism and process that governs the peak current response of the RTD. The question remains, then, what process determines the high-voltage current region, marked as 4 in Figure 6.1.3. At very high applied bias, the quantum level is pulled below the conduction band edge within the emitter. As such,

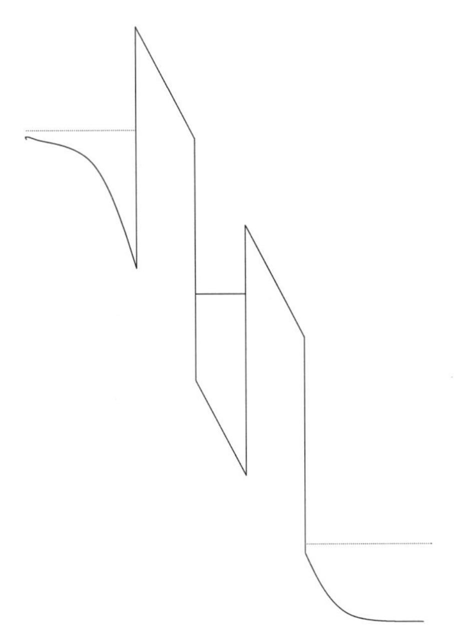

FIGURE 6.1.4 RTD under very high bias. The current is due mainly to thermionic emission under these conditions.

there are virtually no electronic states within the emitter aligned with the resonant level, and little resonant tunneling current can flow. However, the potential barrier produced by the first AlGaAs layer is also pulled down in energy relative to the Fermi level within the emitter, as shown by Figure 6.1.4. Therefore, the effective potential barrier height is reduced substantially. As a result, electrons can be thermionically emitted over the barrier, producing a current. Thus under high applied bias, thermionic emission of electrons over the barrier and injected into the collector region of the diode comprises the current.

Further inspection of the current–voltage characteristic reveals that a negative differential resistance (NDR) appears between the regions 2 and 3 in Figure 6.1.3. As in the transferred electron effect devices discussed in Chapter 2, this NDR can be exploited in an oscillator. Below we discuss the device applications of this NDR.

6.2 PHYSICS OF RESONANT TUNNELING: ENVELOPE APPROXIMATION

There exist many different approaches to modeling resonant tunneling and RTDs. The simplest picture is based on the envelope function of the electronic states. This picture, although somewhat simplified, still provides a useful description of the physics of resonant tunneling and retains some predictive power. In this section we outline some of these approaches.

The envelope function description is based on the effective mass approximation. In the effective mass approximation, the carriers are treated as if they have a different mass from that of free space (Brennan, 1999, Chap. 8). In this way, the effects of the crystalline potential can be included directly into the transport dynamics of the carrier. The envelope function model is based on the solution of the time-independent Schrödinger equation within the effective mass model. The time-independent Schrödinger equation is given as

$$\left[-\frac{\hbar^2}{2}\nabla\left(\frac{1}{m}\nabla\right)+V(r)\right]\psi = E\psi \qquad 6.2.1$$

where we have taken care to note that the mass may vary with position.

The simplest approach to calculating the transmissivity and the resonant tunneling current is to assume that the electron wavefunctions can be represented as plane waves. Implicit in this assumption is the fact that the overall bias is weak, such that the free-space electron wavefunctions remain undisturbed. As we will see below, a better approximation is to assume that the multiquantum well system is under a uniform applied bias and that the wavefunctions can be written as Airy functions. Nevertheless, it is useful first to examine what happens if plane waves are used for the electronic states.

The full details of the technique are given in Brennan (1999, Sec. 2.5). Here we only summarize the salient details. The total wavefunction is separable into the product of transverse and longitudinal components as

$$\psi = \psi_l \psi_t \qquad 6.2.2$$

The electron wavefunctions at the left- and right-hand sides of the multiquantum well system shown in Figure 6.2.1 are

$$\psi_l = Ie^{ik_1x} + re^{-ik_1x} \qquad \psi_r = te^{ik_1x} \qquad 6.2.3$$

where it is assumed that the potential is the same at either side of the system. This holds, of course, for the system sketched in Figure 6.2.1*a*, but not for that of Figure 6.2.1*b*. The coefficients r and t can be determined by solving the Schrödinger equation everywhere within the structure and applying the boundary conditions. If the overall bias is neglected, the solutions for the wavefunctions in all the well regions have the same form. Similarly, the wavefunctions for the electrons in all the barrier regions are also of the same form.

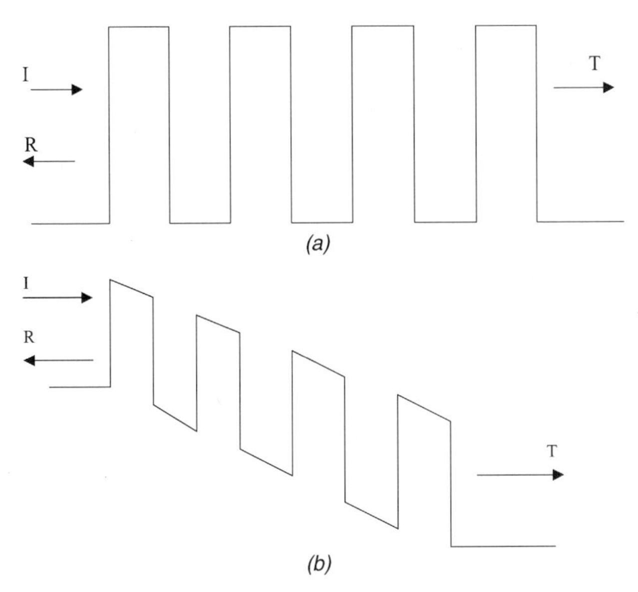

FIGURE 6.2.1 Multiquantum-well stack at (*a*) zero and (*b*) applied bias.

These are given as

$$\psi = \begin{cases} C^{+} \cos k_1 x + C^{-} \sin k_1 x & \text{well regions} \quad 6.2.4 \\ D^{+} \cosh k_2 x + D^{-} \sinh k_2 x & \text{barrier regions} \quad 6.2.5 \end{cases}$$

The solution is obtained by matching the boundary conditions at the well–barrier interfaces repeatedly throughout the structure. The boundary conditions at each interface are: (1) the wavefunction must be continuous, and (2) the probability current density must be continuous across the boundary. The second condition is equivalent to requiring the continuity of $1/m \, d\Psi/dx$ at the boundary (Brennan, 1999, Sec. 2.2). Writing the form of the wavefunctions for each region and repeated application of the boundary conditions results in a general expression relating the coefficients r and t. Using matrices the relationship between r and t can be written as

$$\begin{bmatrix} 1 \\ r \end{bmatrix} = \frac{1}{2ik} \begin{bmatrix} ik_1 & 1 \\ ik_1 & -1 \end{bmatrix} \begin{bmatrix} M_{11} & M_{12} \\ M_{21} & M_{22} \end{bmatrix} \begin{bmatrix} 1 & 1 \\ ik_1 & -ik_1 \end{bmatrix} \begin{bmatrix} t \\ 0 \end{bmatrix} \qquad 6.2.6$$

The matrix elements, M_{11}, M_{12}, M_{21}, and M_{22}, are determined from multiplying the transfer matrices connecting each of the well–barrier regions to its nearest neighbor throughout the structure. The transmission coefficient T is determined from the square of the transmission amplitude t as

$$T = \frac{1}{A}\frac{1}{A^*} \qquad 6.2.7$$

where A is given as

$$A = \frac{1}{2ik_1}(ik_1 M_{11} - k_1^2 M_{12} + M_{21} + ik_1 M_{22}) \qquad 6.2.8$$

The current density can be determined as follows. The net current density is equal to the difference between the current density entering the stack from the left-hand side and that entering the stack from the right-hand side. The general expression for the current density is $J = qnv$, where n is the electron concentration and v is the velocity. In a solid the electron velocity is given as (Brennan, 1999, Sec. 8.1)

$$v = \frac{1}{\hbar}\nabla_k E(k) \qquad 6.2.9$$

The electron density supply function is given by the product of the density-of-states function, the probability distribution function, and the transmission coefficient as

$$\int \frac{2}{(2\pi)^3} f(E) T^* T \, d^3k \qquad 6.2.10$$

Multiplying the supply function by the velocity and charge yields the current density in one direction, for example, that flowing in from the left-hand side, J_l,

$$J_l = \int \frac{2q}{(2\pi)^3} f(E) T^* T \frac{1}{\hbar}\frac{dE}{dk} d^3k \qquad 6.2.11$$

A similar expression is found for the current density flowing in from the right-hand side, J_r. The net current density is then given by the difference between J_l and J_r as

$$J = J_l - J_r = \frac{q}{4\pi^3\hbar}\int [f(E) - f(E')] T^* T \frac{dE}{dk} d^3k \qquad 6.2.12$$

where $f(E')$ is the distribution function at energy E' at the end of the heterostructure system on the right-hand side. The integral over k can be broken into integration over the transverse and longitudinal directions. To perform the integration in Eq. 6.2.12, we need to separate the energy into its longitudinal and transverse components. The energy is then given as

$$E = E_l + \frac{\hbar^2 k^2}{2m^*} \qquad 6.2.13$$

where E_l is the longitudinal energy. Equation 6.2.12 can now be written as

$$J = \frac{q}{4\pi^3\hbar}\int_0^\infty dk_l \int_0^{k_{\max}} dk_t[f(E) - f(E')]T^*T\frac{dE}{dk_l} \qquad 6.2.14$$

Next we perform the integration over k_t. It should be noted that the transmission coefficient TT^* is only a function of the longitudinal energy (that in the direction of the multiquantum well). The flux from right to left is generally much less than that from left to right. Therefore, the effect of the term $f(E')$ is neglected. The integration over the transverse direction alone can then be written as

$$\int_0^{2\pi}\int_0^{k_{\max}} \frac{k\,dk\,d\theta}{1 + e^{(E_l - E_f + \hbar^2k^2/2m)/k_BT}} \qquad 6.2.15$$

where k_B is used for Boltzmann's constant to avoid any confusion with the wavevector k. The integration over the angle θ can be done immediately. Making the assignments

$$u = \frac{\hbar^2k^2}{2mk_BT} + \frac{E_l - E_f}{k_BT} \quad \text{and} \quad du = \frac{\hbar^2 k\,dk}{mk_BT} \qquad 6.2.16$$

the integration over k_t becomes

$$\frac{2\pi mk_BT}{\hbar^2}\int \frac{du}{1 + e^u} \qquad 6.2.17$$

The integration in Eq. 6.2.17 can be evaluated using tables to yield

$$\frac{2\pi mk_BT}{\hbar^2}\left[\log\left(\frac{1}{1 + e^{-(E_l - E_f + \hbar^2k_{\max}^2/2m)/k_BT}}\right) - \log\left(\frac{1}{1 + e^{-(E_l - E_f)/k_BT}}\right)\right] \qquad 6.2.18$$

The maximum kinetic energy is equal to qV, the voltage drop across the device. Equation 6.2.18 then reduces to

$$\frac{2\pi mk_BT}{\hbar^2}\log\left(\frac{1 + e^{-(E_l - E_f)/k_BT}}{1 + e^{-(E_l - E_f + qV)/k_BT}}\right) \qquad 6.2.19$$

Substituting into Eq. 6.2.14 the expression given by Eq. 6.2.19 for the integral over k_t finally yields

$$J = \frac{qmk_BT}{2\pi^2\hbar^3}\int T^*T\log\left(\frac{1 + e^{(E_f - E_l)/k_BT}}{1 + e^{(E_f - E_l - qV)/k_BT}}\right)dE_l \qquad 6.2.20$$

It is important to recall that in the derivation of Eq. 6.2.20, the current flux originating from the right-hand side has been neglected with respect to that originating from the left-hand side. At this stage, the current density needs to be evaluated numerically by integrating the transmission coefficient times the logarithmic function in Eq. 6.2.20.

To calculate the tunneling current, an expression for the transmission coefficients T must be substituted into Eq. 6.2.20. As discussed above, the transmission coefficients can be determined within the envelope approximation by assuming plane-wave states for the form of the wavefunctions. The use of simple plane-wave states for the wavefunctions is rigorously valid only if the system has no bias across it. A better approximation is to solve the Schrödinger equation under a uniform bias to find a generally valid form for the wavefunctions and use these in the transfer matrix technique. Therefore, we next consider the solution of the Schrödinger equation for a constant applied bias.

The solution for a multiquantum well system under bias can be determined using the envelope approximation by solving the Schrödinger equation exactly in each region and then matching the boundary conditions at the interfaces. For simplicity, it is first assumed that the effective mass is constant between the well and barrier regions. Under this assumption the boundary conditions simplify to the continuity of the wavefunction and its first derivative. The multiquantum-well system under a uniform bias is sketched in Figure 6.2.1*b*. In the region to the left of the multiquantum-well system, the solution of the Schrödinger equation is simply equal to a linear combination of incident and reflected plane waves,

$$\psi_1 = e^{ik_1 x} + \mathrm{R}e^{-ik_1 x} \qquad 6.2.21$$

where

$$k_1 = \sqrt{\frac{2mE}{\hbar^2}} \qquad 6.2.22$$

and m is the effective mass of the electron. Recall that we assume that the effective mass is constant throughout the structure.

The form of the Schrödinger equation within the well–barrier region under bias can be rewritten using a change of variables as follows. To illustrate the situation consider a simple form of the Schrödinger equation for a constant, uniform electric field F given as

$$\frac{d^2\psi}{dx^2} + \frac{2m}{\hbar^2}(E + qFx)\psi = 0 \qquad 6.2.23$$

where qFx is the potential energy. Make the following change of variables:

$$x = \left(\frac{\hbar^2}{2mqF}\right)^{1/3} \xi_2 - \frac{E}{qF} \qquad 6.2.24$$

Substituting in the expression for x given by Eq. 6.2.24 into Eq. 6.2.23 yields

$$\frac{d^2\psi}{d\xi_2^2} + \xi_2\psi = 0 \qquad 6.2.25$$

The general solution of Eq. 6.2.25 is given as $\Psi(\xi) \sim \mathrm{Ai}(-\xi)$, where Ai is the Airy function.

Using the change of variables given by Eq. 6.2.24, the Schrödinger equation within the first barrier region, which we will label as region 2 (we assume that the first region is the incident region), can be written as

$$\frac{d^2\psi(\xi_2)}{d\xi_2^2} - \xi_2\psi(\xi_2) = 0 \qquad 6.2.26$$

where ξ_2 is defined as

$$\xi_2 = -\left(\frac{2mqV}{\hbar^2 L}\right)^{1/3}(x + \eta_2) \qquad \eta_2 = -\frac{L}{qV}(V_0 - E) \qquad 6.2.27$$

V/L is the applied electric field, V_0 the potential barrier height, and x the distance measured from the interface between the first barrier and the incident region. The solution of the Schrödinger equation for this potential can readily be expressed in terms of Airy and complementary Airy functions as

$$\psi_2 = C_2^+\mathrm{Ai}(\xi_2) + C_2^-\mathrm{Bi}(\xi_2) \qquad 6.2.28$$

where Ai and Bi are the Airy and complementary Airy functions, respectively.

A similar relationship is obtained for the well regions of the structure where $V_0 = 0$. As an example, we consider the first well after the first barrier. The potential energy in this region is given as

$$V(x) = -\frac{qVx}{L} \qquad 6.2.29$$

Neglecting the differences in effective mass, the Schrodinger equation becomes

$$\frac{d^2\psi(\xi_3)}{d\xi_3^2} - \xi_3\psi(\xi_3) = 0 \qquad 6.2.30$$

where we have used the subscript 3 on ξ to represent the fact that this equation holds for the first well past the first barrier region. This is then the third region of the structure. The variable ξ_3 has the value

$$\xi_3 = \left(-\frac{2mqV}{L\hbar^2}\right)^{1/3}(x + \eta_3) \qquad \eta_3 = \frac{L}{qV}E \tag{6.2.31}$$

The solution of the Schrodinger equation in the well region is then given as

$$\psi_3 = C_3^+ \mathrm{Ai}(\xi_3) + C_3^- \mathrm{Bi}(\xi_3) \tag{6.2.32}$$

The next step is to match the boundary conditions at each interface. If the effective mass is assumed to be constant in every region, the boundary conditions are the continuity of the wavefunction and its first derivative. The imposition of the boundary conditions at the interface between the region of incidence and the first barrier, $x = 0$, gives

$$\begin{aligned} 1 + R &= C_2^+ \mathrm{Ai}_2(x = 0) + C_2^- \mathrm{Bi}_2(x = 0) \\ ik(1 - R) &= C_2^+ \mathrm{Ai}_2'(x = 0) + C_2^- \mathrm{Bi}_2'(x = 0) \end{aligned} \tag{6.2.33}$$

where the prime represents the derivative which is evaluated at $x = 0$. Writing Eq. 6.2.33 in matrix form yields

$$\begin{bmatrix} 1 \\ R \end{bmatrix} = \frac{-1}{2ik}\begin{bmatrix} -ik & -1 \\ -ik & 1 \end{bmatrix}\begin{bmatrix} \mathrm{Ai}_2(x = 0) & \mathrm{Bi}_2(x = 0) \\ \mathrm{Ai}_2'(x = 0) & \mathrm{Bi}_2'(x = 0) \end{bmatrix}\begin{bmatrix} C_2^+ \\ C_2^- \end{bmatrix} \tag{6.2.34}$$

The general relationship between the coefficients for the well and barrier regions can be written as

$$\begin{bmatrix} \mathrm{Ai}(\xi_j) & \mathrm{Bi}(\xi_j) \\ \mathrm{Ai}'(\xi_j) & \mathrm{Bi}'(\xi_j) \end{bmatrix}\begin{bmatrix} C_j^+ \\ \overline{C_j^-} \end{bmatrix} = \begin{bmatrix} \mathrm{Ai}(\xi_{j+1}) & \mathrm{Bi}(\xi_{j+1}) \\ \mathrm{Ai}'(\xi_{j+1}) & \mathrm{Bi}'(\xi_{j+1}) \end{bmatrix}\begin{bmatrix} C_{j+1}^+ \\ \overline{C_{j+1}^-} \end{bmatrix} \tag{6.2.35}$$

where ξ_j and ξ_{j+1} represent the coordinate ξ evaluated for the jth and $(j + 1)$th layers, respectively, at the boundary between these layers. Equation 6.2.35 is applied successively for each well–barrier pair from the initial to the final layer. What we obtain, then, is a series of matrices, called *transfer matrices*, that couple the incident wavefunctions to the outgoing wavefunction through the multiquantum-well stack. The product of all the transfer matrices of the form given by Eq. 6.2.35 can be conveniently expressed as $S(0, L)$. $S(0, L)$ is a 2×2 matrix that results from the product of the successive matrices produced by recursively applying Eq. 6.2.35, represented as

$$\begin{bmatrix} A & B \\ C & D \end{bmatrix} \tag{6.2.36}$$

If incidence is assumed only from the left-hand side of the structure, the resulting expression for the reflection and transmission coefficients is

$$\begin{bmatrix} 1 \\ R \end{bmatrix} = \frac{-1}{2ik} \begin{bmatrix} -ik & -1 \\ -ik & 1 \end{bmatrix} S(0,L) \begin{bmatrix} 1 & 1 \\ ik' & -ik' \end{bmatrix} \begin{bmatrix} T \\ 0 \end{bmatrix} \tag{6.2.37}$$

where k is the wavevector at incidence and k' that upon transmission. The conservation rule concerning currents is given as (see Brennan, 1999, Eq. 2.2.15)

$$k(1 - |R|^2) = k'|T|^2 \tag{6.2.38}$$

The transmissivity τ is given then by the product of T and T^* and the ratio k/k' as

$$\tau = \frac{k}{k'} \frac{1}{M_{11}^2} \tag{6.2.39}$$

where the matrix M is defined as

$$M = \frac{-1}{2ik} \begin{bmatrix} -ik & -1 \\ -ik & 1 \end{bmatrix} S(0,L) \begin{bmatrix} 1 & 1 \\ ik' & -ik' \end{bmatrix} \tag{6.2.40}$$

The final result for the transmissivity is then given as

$$\tau = \frac{4(k/k')}{[A + (k'/k)D]^2 + (k'B - C/k)^2} \tag{6.2.41}$$

If the effective mass changes between regions, the correct formulation of the boundary condition is that $1/m\, d\Psi/dx$ is continuous across the boundary. The equations above for the transmissivity must be revised accordingly.

Although the foregoing approach is a useful technique for calculating the current–voltage characteristic for a resonant tunneling structure, it suffers from some important limitations. The first problem is that space-charge effects can strongly influence the behavior of the device. Space-charge effects arise from impurities and mobile electronic charge within the quantum well and emitter/barrier region. Typically, the envelope approximation can still be retained if the solution of the Schrödinger equation is made self-consistent with that of the Poisson equation.

A second issue of importance in calculating the current–voltage characteristic of an RTD is the effect of multiple bands. In the simple envelope approximation, the band structure is assumed to be parabolic, and multiband effects such as band repulsion are not included. The effect of multiple bands, such as occurs at the minimum energy within the valence band (due to light- and heavy-hole degeneracy) is not addressed within the envelope approximation.

A third issue that the formulation above does not address is the effect of dissipation. As mentioned in Section 6.1, incoherent tunneling occurs when scattering events are present. The importance of incoherent current transport

EXAMPLE 6.2.1: Energy Splitting in a Double-Well Structure

Determine the level splitting in a coupled double-quantum-well structure. The splitting energy can be estimated using degenerate perturbation theory as follows. For a two-level system the solution for the corrected energies is found from the secular equation (Brennan, 1999, Example 4.2.1) as

$$\begin{bmatrix} H'_{11} - E' & H'_{12} \\ H'_{21} & H'_{22} - E' \end{bmatrix} \begin{bmatrix} b_1 \\ b_2 \end{bmatrix} = 0$$

For a double-well device, each well is precisely the same. Therefore, the unperturbed wavefunctions must be exactly the same in each well. As the barrier width between the two wells decreases, the exponential tail of each wavefunction may extend from one well into the other. The two wells are then said to be coupled since there is now some probability that the electron can tunnel from one well to the next. Treating this overlap as a perturbation, the off-diagonal matrix elements, H'_{12} and H'_{21} become

$$H'_{12} = \langle \psi_1 | H' | \psi_2 \rangle \qquad H'_{21} = \langle \psi_2 | H' | \psi_1 \rangle$$

where H' is the confining potential of each well and Ψ_1 and Ψ_2 are the wavefunctions in each well. Since the perturbation acts only as the overlap between the two wells, the diagonal matrix elements H'_{11} and H'_{22} vanish. Therefore, the secular equation becomes

$$\begin{bmatrix} -E' & H'_{12} \\ H'_{21} & -E' \end{bmatrix} \begin{bmatrix} b_1 \\ b_2 \end{bmatrix} = 0$$

The nontrivial solution requires that the determinant of the matrix of the coefficients of b_1 and b_2 must vanish. Expanding out the determinant gives

$$E'^2 - |H'_{12}|^2 = 0$$

which yields for the corrected energies E',

$$E'_{\pm} = |H'_{12}|$$

If it is assumed further that only the ground state of the well exists and that the overlap is then simply between the ground-state wavefunctions,

(*Continued*)

EXAMPLE 6.2.1 (*Continued*)

the final energies of the states are

$$E = E_0 + E' \qquad E = E_0 - E'$$

In general, if the highest state is E_n and the overlap occurs between adjacent wells with this level, the final energies of the states are

$$E = E_n + E' \qquad E = E_n - E'$$

depends to some extent on the strength of the scattering mechanisms. Although there have been several models that incorporate inelastic scattering mechanisms within the envelope approximation, a more complete treatment of dissipation requires more sophisticated models than that of the envelope approximation. In the next section we discuss briefly how dissipation can be included in a quantum transport formulation. The reader is encouraged to consult the references for more detailed studies.

†6.3 INELASTIC PHONON SCATTERING ASSISTED TUNNELING: HOPPING CONDUCTION

Numerous techniques have emerged to treat quantum transport beyond the envelope approximation. It is beyond the scope of this book to discuss all of these techniques or even most of them. Instead, we concentrate on the fundamentals of quantum transport in treating inelastic scattering processes and direct the interested reader to the literature for details on the various methods typically employed. The primary motivation behind these more advanced approaches is that they enable inclusion of dissipation in the transport formulation. The envelope approximation techniques do not naturally extend themselves to include inelastic scattering mechanisms that strongly alter the coherence of the electron wavepacket. There are many different approaches to quantum transport that include treatments of dissipation. The basic techniques are the density matrix formulation (Kohn and Luttinger, 1957), Wigner functions (Frensley, 1987) (which can be shown to be a special case of the density matrix approach), Green's functions techniques, and Feynman path integral methods (Thornber, 1991). In this section we examine only the density matrix formulation as a means of illustrating the inclusion of inelastic scattering mechanisms in the formulation of the tunneling problem.

In Section 6.2 we discussed the envelope approximation solution for the resonant tunneling problem. One of the key assumptions made in the envelope

†Optional material.

approximation formulation is that the electron wavefunction remains coherent throughout the spatial extent of the double-barrier. In addition, it is assumed that the mean scattering time is longer than the time it takes for the electron to tunnel through the barrier. As a result, the electron coherently tunnels through the RTD. What happens, though, if the scattering rate is relatively high such that the electron does not remain coherent within the RTD? How is this handled?

One of the most important occurences of nonresonant or incoherent tunneling is in a multiquantum well system that is longer than the mean free path between collisions for an electron. In such a structure, incoherent processes can dominate carrier conduction. Conduction in a multiquantum well/superlattice can proceed via phonon-assisted hopping of electrons from localized states between adjacent wells if the potential drop over the superlattice period exceeds the miniband width. Alternatively, hopping conduction can dominate over resonant tunneling processes when the potential barriers are sufficiently high and wide such that the electronic wavefunctions are localized within each well. In either case, the conduction is due to transitions between well-defined localized spatial quantization states through the absorption or emission of phonons. Below we outline an approach for calculating the current in device structures in which the inelastic scattering rate is relatively high and the conduction is dominated by phonon-assisted hopping.

It is important to recognize that in quantum transport, one must perform two averagings. In addition to the usual quantum mechanical averaging over position to obtain the expectation value of an operator in a given state, one must also average over the set of states, much like what is done in statistical mechanics. The reader may recall from elementary quantum mechanics (Brennan, 1999, Chap. 1) that the expectation value of the operator A in the state $\Psi(x)$ is given by

$$\langle A \rangle = \int \psi^*(x) A \psi(x) dx \qquad 6.3.1$$

Equation 6.3.1 can be used if a state function can be defined for the system. There are many instances in which a state function $\Psi(x)$ cannot be defined. For example, we may be interested in the value of the observable corresponding to A for an ensemble of particles, about which we can know only statistical information. If we introduce a new set of dynamical variables q which describe the ensemble, the total state function becomes a function of both x and q, $\Psi(x,q)$. The average expectation value of the state $\Psi(x,q)$ is determined by summing over both variables x and q as

$$\langle A \rangle_{\text{ensemble}} = \iint \psi^*(x,q) A(x) \psi(x,q) dx\, dq \qquad 6.3.2$$

where $\langle A \rangle_{\text{ensemble}}$ is the ensemble average, x the variables of the subsystem, and q the variables that describe the external ensemble. The density matrix is defined as the average of the wavefunctions over the external ensemble variables:

$$\rho(x,x') \equiv \int \psi^*(q,x')\psi(q,x)dq \qquad 6.3.3$$

Using Eqs. 6.3.2 and 6.3.3, the ensemble average of the operator A is given as

$$\langle A \rangle_{\text{ensemble}} = \int \rho(x,x)A(x)dx \qquad 6.3.4$$

where we note that $x' = x$ in Eq. 6.3.4. Hence Eq. 6.3.4 can be rewritten as

$$\langle A \rangle_{\text{ensemble}} = \int dx' \int dx A(x,x')\rho(x,x')\delta(x-x') \qquad 6.3.5$$

Integrating over x' yields

$$\langle A \rangle_{\text{ensemble}} = \int dx[A(x,x)\rho(x,x)] \qquad 6.3.6$$

which is simply

$$\langle A \rangle_{\text{ensemble}} = \int dx[A\rho]_{xx} = \text{Tr}(\rho A) = \text{Tr}(A\rho) \qquad (6.3.7)$$

where Tr is the trace of the matrix $A\rho$, which is the sum over all the diagonal elements of $A\rho$. Physically, Eq. 6.3.7 means that the trace of the density matrix multiplied by the operator is simply equal to the ensemble average of the operator. The trace of the matrix is defined as the sum of all the diagonal elements of the matrix. The trace is always the same for a matrix independent of its representation. In other words, the trace is an invariant of the matrix. The density matrix for $x = x'$ is given as

$$\rho(x,x) = \int |\psi(q,x)|^2 dq \qquad 6.3.8$$

which physically represents the probability of finding the particle at position x after averaging over all the other variables of the ensemble.

The dynamics of the density matrix are obtained in a manner similar to that for a dynamical variable. The reader may recall that the general expression for the time dependence of a time-independent operator A is given as (Brennan, 1999, Sec. 3.1)

$$i\hbar\frac{dA}{dt} = [A,H] \qquad 6.3.9$$

Similarly, the equation of motion for the density matrix is given as

$$i\hbar\frac{d\rho}{dt} = [H,\rho] \qquad 6.3.10$$

The density matrix method can be used to calculate the current in the presence of inelastic scatterings as follows. For simplicity we assume that the field is applied along the z direction and that all the current flow lies along this direction. The velocity along the z direction, dz/dt, can be determined using the result of Eq. 6.3.9 as

$$i\hbar\frac{dz}{dt} = [H,z] \qquad 6.3.11$$

Therefore, the velocity v_z is given as

$$v_z = \frac{1}{i\hbar}[H,z] \qquad 6.3.12$$

The current density in the z direction can be determined from the mean value of the velocity in the z direction multiplied by the charge q and the electron concentration n as

$$j_z = -qn\langle v_z\rangle_{\text{ensemble}} = -qn\,\mathrm{Tr}(\rho v_z) \qquad 6.3.13$$

Substituting into Eq. 6.3.13 the expression for v_z given by Eq. 6.3.12 yields

$$j_z = -\frac{qn}{i\hbar}\mathrm{Tr}(\rho[H,z]) = -\frac{qn}{i\hbar}\mathrm{Tr}(\rho(Hz - zH)) \qquad 6.3.14$$

To proceed, it is useful to change to Dirac notation for the density matrix. Each element in the density matrix ρ_{mn} can be written in Dirac notation as

$$\rho_{mn} = \langle m|\rho|n\rangle \qquad 6.3.15$$

From this definition of ρ_{mn} the density matrix can be obtained as

$$\rho = \sum_{m,n}|m\rangle\rho_{mn}\langle n| \qquad 6.3.16$$

since multiplying Eq. 6.3.16 by $\langle m|$ and $|n\rangle$ on both sides yields

$$\langle m|\rho|n\rangle = \sum_{m,n}\langle m|m\rangle\rho_{mn}\langle n|n\rangle = \rho_{mn} \qquad 6.3.17$$

Note that the sums over m and n on the right-hand side of Eq. 6.3.17 simply give 1, since both $\langle m|m\rangle$ and $\langle n|n\rangle$ are equal 1 for normalized states.

The general Hamiltonian for the system we wish to describe is one that consists of an electron term H_e, a phonon term H_p, and an electron–phonon interaction term H_I. The eigenstates of the electron Hamiltonian we describe as $|v\rangle$, such that the eigenvalue equation is given as

$$H_e|v\rangle = E_v|v\rangle \qquad 6.3.18$$

where E_v are the eigenvalues of H_e. Similarly, there are eigenstates and eigenvalues associated with the phonon system as

$$H_p|\lambda\rangle = \omega_\lambda|\lambda\rangle \qquad 6.3.19$$

Finally, H_i represents the interaction between the electrons and phonons and allows for electron hopping between different localized states in the superlattice. The density matrix for the system involves two subsystems, λ and v, and can be written as

$$\rho = \sum_{\substack{vv' \\ \lambda\lambda'}} |v\lambda\rangle \rho_{vv'}^{\lambda\lambda'} \langle v'\lambda'| \qquad 6.3.20$$

where we have simply employed the result given by Eq. 6.3.16. The components of the density matrix ρ are given as

$$\rho_{vv'}^{\lambda\lambda'} = \langle v\lambda|\rho|v'\lambda'\rangle \qquad 6.3.21$$

The total Hamiltonian H_T for the system has three components, H_e, H_p, and H_I, as described above. The variable z, the position in the z direction, commutes with the phonon and interaction Hamiltonians. Thus

$$[z, H_T] = [z, H_e] \qquad 6.3.22$$

Using Eq. 6.3.22 in Eq. 6.3.14, the current density in the z direction can be written as

$$j_z = -\frac{qn}{i\hbar}\mathrm{Tr}\,\rho(zH_e - H_e z) \qquad 6.3.23$$

Expressing the density matrix using Eq. 6.3.20, Eq. 6.3.23 becomes

$$j_z = -\frac{qn}{i\hbar}\mathrm{Tr}\sum_{\substack{vv' \\ \lambda\lambda'}} |v\lambda\rangle \rho_{vv'}^{\lambda\lambda'} \langle v'\lambda'|(zH_e - H_e z) \qquad 6.3.24$$

Equation 6.3.24 is diagonal in λ since neither z nor H_e depend on λ, the coordinates describing the phonon system. Inserting a complete set of states into Eq. 6.3.24 yields (see Box 6.3.1)

$$j_z = -\frac{qn}{i\hbar}\sum_{\substack{vv' \\ \lambda}} \rho_{vv'}^{\lambda\lambda}(E_v - E_{v'})\bar{z}_{v'v} \qquad 6.3.25$$

BOX 6.3.1: Derivation of Eq. 6.3.25

To obtain Eq. 6.3.25 it is necessary to multiply the expression given by Eq. 6.3.24 by a complete set of states. A complete set of states is defined as follows:

$$\int dp|p\rangle\langle p|$$

Therefore, for the system of interest, a complete set of states involving v and λ are

$$\int dv\, d\lambda|v\lambda\rangle\langle v\lambda|$$

Multiplying Eq. 6.3.24 by the complete set of states for v and λ gives

$$\int \rho_{vv'}^{\lambda\lambda'}|v\lambda\rangle\langle v'\lambda'|(zH_e - H_e z)|v\lambda\rangle\langle v\lambda|$$

Consider the matrix element $\langle v'\lambda'|(zH_e - H_e z)|v\lambda\rangle$. It can be evaluated as follows. Since the Hamiltonian H_e and z are independent of the phonon coordinates, the matrix element can be rewritten as

$$\langle v'\lambda'|(zH_e - H_e z)|v\lambda\rangle = \langle\lambda'|\lambda\rangle\langle v'|zH_e - H_e z)|v\rangle$$

But this can be simplified by recognizing that $\langle\lambda|\lambda'\rangle$ is $\delta_{\lambda\lambda'}$. We then obtain

$$\langle v'\lambda'|(zH_e - H_e z)|v\lambda\rangle = \delta_{\lambda\lambda'}\{\langle v'|zH_e|v\rangle - \langle v'|H_e z)|v\rangle\}$$

But the states $|v\rangle$ are eigenstates of H_e and H_e is Hermitian. Using these properties, we obtain

$$\langle v'|z|v\rangle E_v - E_{v'}\langle v'|z|v\rangle$$

where E_v and $E_{v'}$ are the energy eigenstates of H_e,

$$H_e|v\rangle = E_v|v\rangle$$

The matrix element finally becomes

$$\langle v'\lambda'|(zH_e - H_e z)|v\lambda\rangle = \bar{z}_{v'v}(E_v - E_{v'})$$

(*Continued*)

BOX 6.3.1 (*Continued*)

where $z_{\nu'\nu} = \langle \nu' | z | \nu \rangle$. Substituting the expression above into Eq. 6.3.24 yields

$$-\frac{qn}{i\hbar}\sum_{\substack{vv' \\ \lambda\lambda'}} \rho_{vv'}^{\lambda\lambda'} |v\lambda\rangle\langle v\lambda| \bar{z}_{v'v}(E_v - E_{v'})\delta_{\lambda\lambda'}$$

and noticing that the sum over $\lambda\lambda'$ collapses into a sum only over λ, the expression above becomes

$$-\frac{qn}{i\hbar}\sum_{\substack{vv' \\ \lambda}} \rho_{vv'}^{\lambda\lambda} \bar{z}_{v'v}(E_v - E_{v'})$$

where we recognize that $|v\lambda\rangle\langle v\lambda|$ is 1.

Equation 6.3.25 provides a description of the current density in terms of the density matrix elements and $\bar{z}_{vv'} = \langle v' | z | v \rangle$ the matrix element of z between the electronic states v and v'.

The current density given by Eq. 6.3.25 can be reexpressed in a more useful manner as follows. It is necessary to determine the value of the matrix elements of ρ. The starting point is to consider the Liouville equation for the total Hamiltonian H_T. In steady state the Liouville equation for H_T becomes

$$[H_T, \rho] = 0 \qquad 6.3.26$$

Substituting in the expression for ρ given by Eq. 6.3.20 into Eq. 6.3.26 and using the fact that

$$H_T = H_e + H_p + H_I \qquad 6.3.27$$

yields

$$0 = \sum_{\substack{vv' \\ \lambda\lambda'}} (\langle v\lambda | H_T | v\lambda\rangle \rho_{vv'}^{\lambda\lambda'} \langle v'\lambda' | v'\lambda'\rangle - \langle v\lambda | v\lambda\rangle \rho_{vv'}^{\lambda\lambda'} \langle v'\lambda' | H_T | v'\lambda'\rangle)$$

$$6.3.28$$

Each of the matrix elements can be simplified using Eqs. 6.3.18 and 6.3.19 since

$$\langle v\lambda | H_e | v\lambda\rangle = E_v \qquad \langle v\lambda | H_p | v\lambda\rangle = \omega_\lambda \qquad 6.3.29$$

A relationship similar to Eq. 6.3.29 holds for the $v'\lambda'$ states in Eq. 6.3.28. With these substitutions, Eq. 6.3.28 becomes

$$0 = (E_v + \omega_\lambda - E_{v'} - \omega_{\lambda'})\rho_{vv'}^{\lambda\lambda'} + \sum_{\substack{vv' \\ \lambda\lambda'}} \rho_{vv'}^{\lambda\lambda'}(\langle v\lambda|H_I|v\lambda\rangle - \langle v'\lambda'|H_I|v'\lambda'\rangle) \qquad 6.3.30$$

Notice that the first collection of terms in Eq. 6.3.30 includes only off-diagonal terms in the density matrix. This is obvious since when $v = v'$ and $\lambda = \lambda'$, the first collected term vanishes, leaving only the summation term. The term under the summation contains both diagonal and off-diagonal terms. This term can be separated into its diagonal and off-diagonal terms as

$$(H_I)_{vv'}^{\lambda\lambda'}(\rho_{v'}^{\lambda'} - \rho_v^{\lambda}) + \sum_{\substack{v''\neq v' \\ \lambda''\neq\lambda'}} (H_I)_{vv''}^{\lambda\lambda''}\rho_{v''v'}^{\lambda''\lambda'} - \sum_{\substack{v''\neq v \\ \lambda''\neq\lambda}} (H_I)_{v''v'}^{\lambda''\lambda'}\rho_{vv''}^{\lambda\lambda''} \qquad 6.3.31$$

where the first term is comprised of the diagonal elements. The solution to Eq. 6.3.30 proceeds iteratively. The first iteration is such that only the diagonal terms in the summation term in Eq. 6.3.30 are included and the off-diagonal terms in the summation are neglected. Thus only the first term in relation 6.3.31 is retained in Eq. 6.3.30. The off-diagonal terms appearing in the first collective term can be expressed in terms of the diagonal terms only as

$$[E_{vv'} + \omega_{\lambda\lambda'}]\rho_{vv'}^{\lambda\lambda'} = (H_I)_{vv'}^{\lambda\lambda'}(\rho_v^{\lambda} - \rho_{v'}^{\lambda'}) \qquad 6.3.32$$

where

$$E_{vv'} = E_v - E_{v'} \qquad \omega_{\lambda\lambda'} = \omega_\lambda - \omega_{\lambda'} \qquad \rho_{vv}^{\lambda\lambda} = \rho_v^{\lambda} = \langle\nu\lambda|\rho|\nu\lambda\rangle \qquad 6.3.33$$

On the right-hand side of Eq. 6.3.32, only diagonal elements remain. Equation 6.3.32 can be simplified using a fundamental mathematical result; if $xT = A(x)$, then T is given as

$$T = A(x)[i\pi\delta(x) + P(x^{-1})] \qquad 6.3.34$$

where $\delta(x)$ is the Dirac delta function and P is the principal value function. Therefore, Eq. 6.3.32 becomes

$$\rho_{vv'}^{\lambda\lambda'} = (H_I)_{vv'}^{\lambda\lambda'}(\rho_v^{\lambda} - \rho_{v'}^{\lambda'})[i\pi\delta(E_{vv'} + \omega_{\lambda\lambda'}) + P(E_{vv'} + \omega_{\lambda\lambda'})^{-1}] \qquad 6.3.35$$

The second step of the iterative solution requires using the off-diagonal terms in Eq. 6.3.31. The details of the procedure are discussed in a paper by Calecki et al. (1984).

The relationships above can be used to determine the current density from Eq. 6.3.25. It is important to recognize that the term $\omega_{\lambda\lambda'}$ vanishes for terms diagonal in λ. The interaction Hamiltonian H_I also vanishes when the number of quanta remain fixed. Calecki et al. (1984) have derived an expression for the hopping current density by determining the form of the density matrix elements in Eq. 6.3.25 using the iterative procedure described above. The current density, after some involved manipulations, is found by Calecki et al. (1984) to be

$$j_z = -qn \sum_{\substack{v,v' \\ v \neq v'}} \sum_{\substack{\lambda,\lambda' \\ \lambda \neq \lambda'}} |(H_I)_{vv'}^{\lambda\lambda'}|^2 \frac{z_v - z_{v'}}{2} (\rho_{v'}^{\lambda'} - \rho_v^{\lambda}) \times \left[\frac{2\pi}{\hbar} \delta(E_{v'v} + \omega_{\lambda'\lambda}) + \frac{2}{i\hbar} P\left(\frac{1}{E_{v'v} + \omega_{\lambda'\lambda}} \right) \right] \quad (6.3.36)$$

The diagonal terms in v vanish in the expression for the current density since the difference term in z_v and $z_{v'}$ would be zero. The term involving H_I is the matrix element of the interaction Hamiltonian and the electronic and phonon states characterized by v and λ, respectively. The sum over λ appears since the diagonal element in λ depends upon the off-diagonal elements. The current density can be simplified by recognizing that j_z must be real. Therefore, the principal part term vanishes from real j_z. Finally, it is assumed that the phonons are in thermal equilibrium such that the density matrix components can be written as

$$\rho_v^{\lambda} = f_v N_q \quad 6.3.37$$

where N_q is the equilibrium distribution function for the phonons and f_v is the nonequilibrium distribution function for the electrons. Several terms within Eq. 6.3.36 can be grouped to reproduce Fermi's golden rule, including the equilibrium phonon distribution N_q as

$$W_{vv'} = \frac{2\pi}{\hbar} \sum_{\lambda\lambda'} N_q |(H_I)_{vv'}^{\lambda\lambda'}|^2 \delta(E_{vv'} + \omega_{\lambda\lambda'}) \quad 6.3.38$$

The final result for the current density in a nondegenerate system is given by

$$j_z = -qn \sum_{vv'} \frac{z_{v'} - z_v}{2} (f_v W_{vv'} - f_{v'} W_{v'v}) \quad 6.3.39$$

where $W_{vv'}$ and $W_{v'v}$ are the transition rates due to phonon scatterings from the state v to v' and from the state v' to v, respectively. The quantities f_v and $f_{v'}$ are the electron occupation probability functions for the states v and v'. It is further assumed that the final-state occupation probability is zero; in other words, the final states are assumed to be unoccupied initially. The factors z_v and $z_{v'}$ are defined from Box 6.3.1 as

$$z_v = \langle v|z|v\rangle \qquad z_{v'} = \langle v'|z|v'\rangle \quad 6.3.40$$

which physically represent the mean position of the electron in the states v and v', respectively.

Equation 6.3.40 can be understood physically as follows. The current density arises from electron hopping from an initial state v to a final state v', or vice versa, by the action of a scattering mechanism described by $W_{vv'}$. Hence the current flow is not via resonant tunneling but by phonon-assisted hopping between well-defined states v and v'. The formulation above can be used to calculate the steady-state current density in any system in which the current proceeds by carriers hopping between localized states. Examples of systems in which hopping transport occurs are amorphous materials, polymer chains, and superlattices.

6.4 RESONANT TUNNELING DIODES: HIGH-FREQUENCY APPLICATIONS

Resonant tunneling diodes (RTDs) can be exploited in high-frequency and digital logic applications. Let us first consider the frequency performance of RTDs. As discussed in Chapter 5, negative differential resistance (NDR) can be exploited in high-frequency oscillators. Oscillations are obtained by biasing the RTD into the NDR region while embedded within a suitable resonant circuit. The frequency response of the RTD depends on several factors: (1) the oscillation frequency of the waveguide circuit, (2) the charge storage delay in the quantum well, and (3) the transit time across the depletion region of the diode. It is reasonable then to define a maximum frequency of operation of an RTD. The maximum frequency f_{max} is given by

$$f_{max} = \frac{1}{2\pi\tau_{char}} \tag{6.4.1}$$

where τ_{char} is a characteristic time governing the resonant tunneling process. If a signal is applied to an RTD with a frequency greater than f_{max}, the carriers within the RTD cannot follow the signal and hence the device cannot respond in a manner similar to its usual dc response. At frequencies higher than f_{max}, the NDR of the RTD vanishes.

Calculation of the frequency response of an RTD has been performed using a variety of methods. The most comprehensive studies rely on numerical calculations using quantum transport schemes such as Wigner functions, direct numerical simulation of the temporal evolution of a wavepacket, and Green's functions. The details of these different approaches are too vast to be included here. We refer the reader to the references for a full description, in particular the book edited by Brennan and Ruden (2001). Instead, we present an estimate of the frequency response of an RTD based on an approximate formulation given by Liu and Sollner (1994).

Before we discuss the frequency response of an RTD it is important to make some distinction between a fully resonant and a fully sequential tunneling system. As discussed in Section 6.1, the tunneling process can proceed either sequentially or resonantly. The simplest approach to determine which of these mechanisms dominates the operation of an RTD is to assess whether the scattering time is greater or less than the carrier lifetime in the quasibound state within the RTD. In a RTD the barrier heights are of finite potential height. The electron wavefunction spreads over the entire device, penetrating the barriers. Therefore, the resonant states are not bound states but quasibound states with an associated finite lifetime. The lifetime of the quasibound state is called τ_{life}. If the scattering time τ_{scat}, defined as the mean time between collisions, is substantially larger than the carrier lifetime τ_{life}, the effects of scatterings on the frequency response of the device can be neglected. This is because the resonant tunneling process occurs more rapidly than scatterings, such that there is insufficient time for an electron to suffer a scattering during resonant tunneling. If, on the other hand, τ_{scat} is substantially smaller than the carrier lifetime, an electron would necessarily suffer many collisions before it could resonantly tunnel through the structure. Therefore, the tunneling process cannot be characterized as being resonant and is best described as a sequential tunneling process, as discussed in Section 6.1. From that discussion the lifetime can be estimated from the full-width at half-maximum of the transmission peak Γ_r as

$$\tau_{\text{life}} \sim \frac{\hbar}{\Gamma_r} \tag{6.4.2}$$

The scattering time can be estimated from the mobility of electrons within a two-dimensional system as

$$\tau_{\text{scat}} \sim \frac{\mu m}{q} \tag{6.4.3}$$

Typically, it is expected that the resonant tunneling characteristic time will decrease due to scatterings. However, in many cases the change in magnitude of the characteristic times within the resonant and sequential models is negligible. In this book we assume that the two processes, sequential and resonant tunneling, can be characterized by the same lifetimes and that the frequency response is the same in either case.

In general, there are several time scales of importance in a RTD: (1) the traversal time, the time needed to tunnel through a barrier, (2) the resonant state lifetime, and (3) the escape time. All these factors influence the overall temporal response of the device. In resonant tunneling the main contribution to the characteristic time is from the well region of the device. In resonant tunneling, the electrons become trapped in a quasibound state and persist for some time before they "leak" out of the well through the second barrier. As a result, the resonant state lifetime can be appreciably larger than the barrier traversal time and the escape time. Therefore, we estimate the characteristic time by calculating the resonant state lifetime of the RTD.

The resonant state lifetime or, equivalently, the lifetime of the quasibound state can be estimated as follows. For simplicity it is assumed that the quantization direction is along the z axis. The velocity of the electron in this direction can be estimated as

$$v_z = \sqrt{\frac{2E_n}{m}} \tag{6.4.4}$$

where E_n is the energy level of the quantized state. An attempt frequency can be defined as

$$f_{\text{attempt}} = \frac{v_z}{2L} \tag{6.4.5}$$

where L is the effective one-way distance the electron travels in the well. Notice that the attempt frequency simply represents how often the electron encounters a boundary while reflecting back and forth within the well. The effective length L is given as

$$L = L_w + \frac{1}{\kappa_1} + \frac{1}{\kappa_2} \tag{6.4.6}$$

where L_w is the width of the well and κ_1 and κ_2 are the imaginary wavevectors within the barriers. They represent the electron travel while partially penetrating the barriers. The probability per unit time of the electron escaping depends on the product of the attempt frequency (how often the electron encounters a boundary) and the transmissivity of each boundary, denoted as T_1 and T_2 (how likely it is for the electron to tunnel through the boundary). The lifetime is proportional to the inverse of the probability per unit time of the electron escaping from the quasibound level. The lifetime τ_{life} is then given as

$$\tau_{\text{life}} \sim \frac{1}{f_{\text{attempt}}(T_1 + T_2)} \tag{6.4.7}$$

If it is further assumed that the electron can escape only from the second barrier, which is usually the case when the RTD is under high bias. Then the lifetime becomes

$$\tau_{\text{life}} \sim \frac{1}{f_{\text{attempt}} T_2} \tag{6.4.8}$$

The lifetime can also be estimated from the *uncertainty principle*, which states that

$$\Delta E \Delta t \geq \frac{\hbar}{2} \tag{6.4.9}$$

Since the state is assumed to be quasibound, it has a finite lifetime. That lifetime is simply Δt. Therefore, the resonant lifetime is given as

$$\Delta t = \tau_{\text{life}} \sim \frac{\hbar}{2\Delta E} \tag{6.4.10}$$

where ΔE is the half-maximum of the transmission peak, $\Gamma_r/2$. Equating Eqs. 6.4.10 and 6.4.7 yields

$$\frac{\hbar}{2\Delta E} = \frac{1}{f_{\text{attempt}}(T_1 + T_2)} \qquad 6.4.11$$

Using Eqs. 6.4.4 to 6.4.6, f_{attempt} can be written as

$$f_{\text{attempt}} = \frac{v_z}{2(L_w + 1/\kappa_1 + 1/\kappa_2)} = \frac{\sqrt{2E_n/m}}{2(L_w + 1/\kappa_1 + 1/\kappa_2)} \qquad 6.4.12$$

Substituting Eq. 6.4.12 into Eq. 6.4.11 yields

$$\Gamma_r = 2\Delta E = \frac{\hbar\sqrt{2E_n/m}(T_1 + T_2)}{2(L_w + 1/\kappa_1 + 1/\kappa_2)} \qquad 6.4.13$$

Therefore, the resonant lifetime is simply

$$\tau_{\text{life}} \sim \frac{2(L_w + 1/\kappa_1 + 1/\kappa_2)}{\sqrt{2E_n/m}(T_1 + T_2)} \qquad 6.4.14$$

It is interesting to note that the resonant state lifetime describes both the fully sequential and fully resonant conditions to good approximation. The resonant lifetime can be determined in a somewhat different manner using a wavelike picture of the electron, as shown in Box 6.4.1.

The actual frequency dependence of an RTD is difficult to establish theoretically without employing a full quantum mechanical calculation. Nevertheless, an estimate of the upper frequency limit of performance can be made which is thought to be accurate to within a factor of 2. The simplest picture is that the maximum frequency of oscillation is given as

$$f_{\max} = \frac{1}{2\pi\tau_{\text{life}}} \qquad 6.4.15$$

The resonant state lifetime can be estimated using Eq. 6.4.14 or from the half-width at half-maximum ΔE, discussed above, as

$$\tau_{\text{life}} = \frac{\hbar}{2\Delta E} \qquad 6.4.16$$

From the lifetime, the steady-state resonant tunneling current can be estimated. Assuming that the quasibound resonant state has a relatively long lifetime, charge buildup will occur within the well. In steady state, the charge

BOX 6.4.1: Determination of the Resonant State Lifetime

Physically, a coherent electron state can be modeled much like an electromagnetic wave. Quantitative analogies can be developed between quantum mechanical electron waves in semiconductor materials and electromagnetic optical waves in dielectrics (Gaylord and Brennan, 1989). Ballistic electron waves are quantum mechanical de Broglie waves and can thus undergo refraction, reflection, diffraction, interference, and so on, much like electromagnetic waves. Therefore, a resonant tunneling diode can be modeled as a Fabry–Perot resonator. The physical description of the RTD can thus be understood using an analysis similar to that applied to an electromagnetic Fabry–Perot resonator. The transmitted field exiting one side of the Fabry–Perot resonator E_t is given as (Brennan, 1999, Sec. 13.2)

$$E_t = \frac{E_i t_1 t_2 e^{-\Gamma L}}{1 - r_1 r_2 e^{-2\Gamma L}}$$

where E_i is the incident field, t_1 and t_2 the transmission amplitudes and r_1 and r_2 are the reflection amplitudes of the first and second barriers, respectively. L is the length of the cavity and Γ is the complex propagation constant in the medium. In an optical Fabry–Perot resonator it is possible that there is gain. In this case, that of a lasing medium (Brennan, 1999, Chap. 13), the propagation constant can be written as

$$\Gamma = i\beta k_o - \alpha$$

where α is the gain coefficient. For a RTD, the electron wave does not experience any gain. The propagation constant for an electron wave can then be written as ik. The quantity corresponding to the transmitted field of an electromagnetic wave for an electron wave is the transmitted probability amplitude t_{db}. The transmitted probability amplitude can then be written as

$$t_{tb} = \frac{t_1 t_2 e^{-ikL}}{1 - r_1 r_2 e^{-2ikL}}$$

Resonance occurs when the denominator vanishes. Since the second term is complex, the denominator vanishes when both the real and imaginary parts are zero. Focusing on the imaginary part, the requirement is that the phase of $r_1 r_2 e^{-2ikL}$ must be equal to an integer multiple of π. The transmission coefficient for the system T_{2B} can be found from the square of the magnitude of t_{tb}. If we make the assumption that $t = t_1 = t_2$ and

(*Continued*)

BOX 6.4.1 (*Continued*)

$r = r_1 = r_2$, the overall transmission T_{2B} becomes

$$T_{2B} = \frac{1}{1 + (4R/T^2)\sin^2 kL}$$

where R is the square of r and T the square of t. Notice that peaks in the transmissivity appear when the product kL is equal to an integer multiple of π, as before. These peaks correspond to the resonances of the well.

The resonant lifetime of the well can be determined using the relationships above as follows. After a single round trip, the probability density of the confined electron decreases by the amount $R_1 R_2 \phi$, where ϕ is the probability density. The loss of probability density is then given as

$$(1 - R_1 R_2)\phi$$

which has occurred in the time interval $2L/v$, where v is the speed of the electron. As shown in the text, the velocity of the electron is given as

$$v_z = \sqrt{\frac{2E_n}{m}}$$

Therefore, the rate of change of the probability density within the well is given as

$$\frac{d\phi}{dt} = -\frac{1 - R_1 R_2}{2L/v}\phi$$

This simple differential equation has the solution

$$\phi = \phi(0)e^{-t/\tau}$$

where the lifetime τ is given as

$$\tau = \frac{2L/v}{1 - R_1 R_2}$$

Substituting in for v, and using the relationships between R and T, $R_1 = 1 - T_1$ and $R_2 = 1 - T_2$, along with the approximation

$$1 - R_1 R_2 = 1 - (1 - T_1)(1 - T_2) = 1 - (1 - T_1 - T_2 + T_1 T_2) \sim T_1 + T_2$$

(*Continued*)

BOX 6.4.1 (*Continued*)

the resonant state lifetime becomes

$$\tau = \frac{2L}{\sqrt{(2E_n/m)}(T_1 + T_2)}$$

Finally, the length of the cavity can be written as

$$L = L_w + \frac{1}{\kappa_1} + \frac{1}{\kappa_2}$$

where the terms $1/\kappa_1$ and $1/\kappa_2$ represent the penetration of the electron into the emitter and collector barriers, respectively. With these substitutions, the result in the text, Eq. 6.4.14, is recovered. Thus we see that we obtain the same result for the resonant lifetime assuming that the electron can be treated as a simple wave in a resonant cavity.

buildup in the quantum well σ_{QW} is related to the current density J as

$$\frac{\sigma_{QW}}{\tau_{\text{life}}} = J \qquad 6.4.17$$

where it is assumed that the lifetime is associated only with carriers exiting through the collector barrier of the RTD. Charge buildup is maximized when the transmissivity of the collector barrier is significantly less than that of the emitter barrier. In this case, charge leakage out of the well is suppressed, while charge leakage into it is high, resulting in a buildup of charge in the quantum well region.

RTDs can be used in oscillator circuits. The fact that an RTD shows a negative differential resistance (NDR) enables its use as an oscillator. To make an oscillator using a device exhibiting NDR, all that is required is that the device be connected to a tuned transmission line. The device should terminate one end of the line, leaving the other end with a large discontinuity. The output power is that part of the circulating power that leaks past the discontinuity. A simplified equivalent-circuit model for a RTD is shown in Figure 6.4.1. The conductance G shown in the diagram represents a negative differential conductance. The resistance R_s is a parasitic series resistance that includes the metal–semiconductor contact resistance, the resistance of the undepleted semiconductor between the top and bottom contacts and the active region, and any spreading resistance if the device is grown as a mesa onto a bulk substrate. If the equivalent-circuit elements are assumed to be frequency independent,

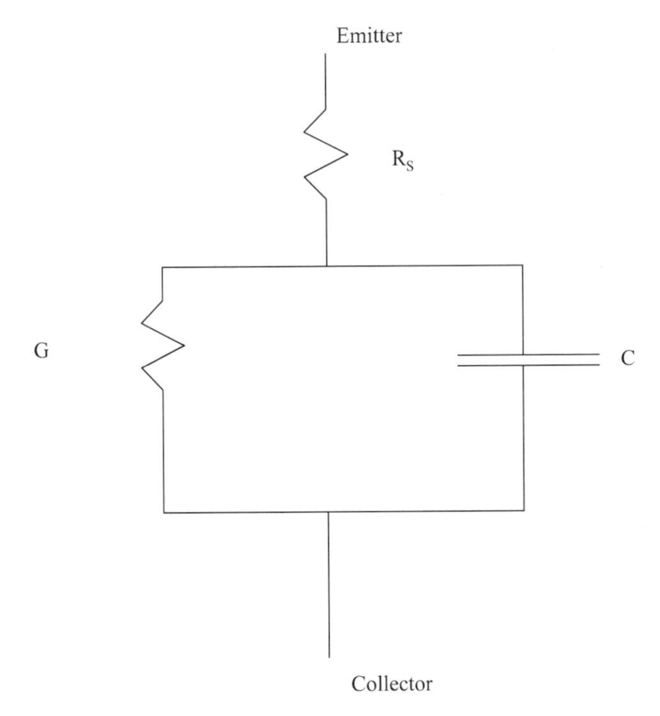

FIGURE 6.4.1 Equivalent-circuit model of an RTD.

the maximum frequency of oscillation of the circuit is given as

$$f_{\max} = \frac{1}{2\pi C}\left(-\frac{G}{R_s} - G^2\right)^{1/2} \qquad 6.4.18$$

Equation 6.4.18 places an upper limit on the oscillation frequency of an RTD circuit.

6.5 RESONANT TUNNELING DIODES: DIGITAL APPLICATIONS

Resonant tunneling diodes are of great interest in future digital logic device applications. Perhaps the most intriguing property of resonant tunneling devices for digital applications is the fact that they can be utilized to provide multivalued logic. Multiple peak resonant tunneling diodes (RTDs) provide for a new approach to digital logic design. Multiple-valued logic gates using RTDs can change logic design from binary-based arithmetic to other bases in a far more efficient manner than that possible with conventional CMOS circuitry. RTD-based circuits have much less complex interconnect requirements than comparable CMOS circuits. Other potential advantages of RTD-based

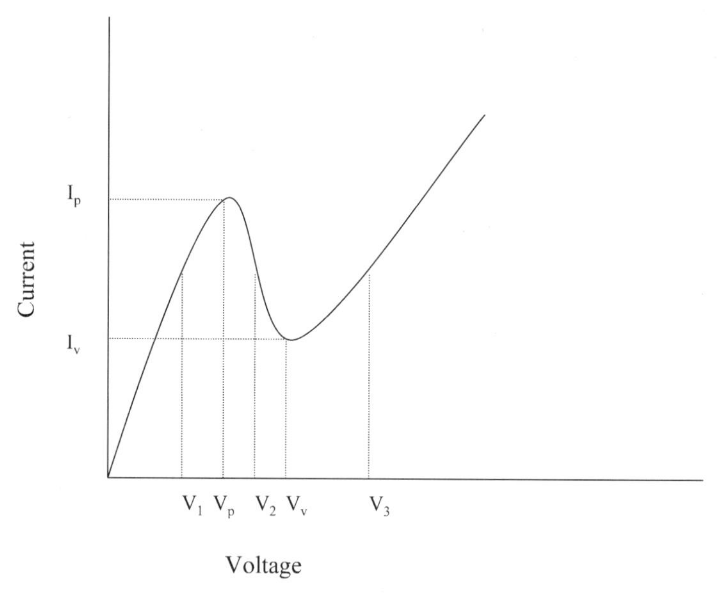

FIGURE 6.5.1 Current–voltage characteristic of an RTD showing the peak and valley currents and voltages.

logic circuits are lower power consumption, high-speed operation, and overall reduced circuit complexity. In this section we examine the potential of RTD devices in digital logic applications and some representative RTD circuit designs.

The current–voltage characteristic for a RTD is shown in Figure 6.5.1. Along both the current and voltage axes, different points are labeled. These are I_p and V_p, the peak current and voltage, respectively, and I_v and V_v, the valley current and voltage, respectively. Inspection of Figure 6.5.1 shows that there are three different voltage states, corresponding to the same current I between I_v and I_p. These voltages are V_1, V_2, and V_3, as shown in the figure. V_1 and V_3 are stable biases, while V_2 lies within the NDR region of the RTD. V_p/V_v gives the peak-to-valley voltage ratio; I_p/I_v gives the peak-to-valley current ratio. The key to operating a RTD for digital switching is that the device operates in a bistable mode; the output is latched and any change in the input is reflected in the output only when a clock or other evaluation signal is applied.

RTD logic circuits can be created in several different ways. There are essentially two general strategies. The first is to combine a RTD with conventional transistors, typically either heterostructure bipolar transistors (HBTs) or modulation-doped field-effect transistors (MODFETs). Use of a RTD in conjunction with conventional transistors provides a reduction in circuit complexity and in interconnects. The second approach is to utilize RTD devices

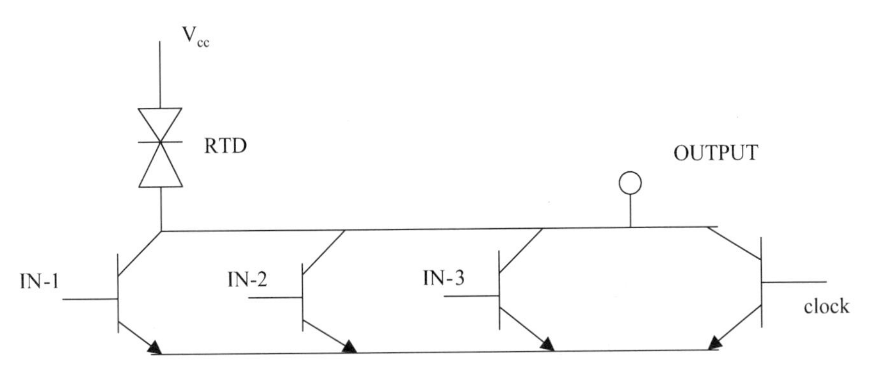

FIGURE 6.5.2 Bistable logic circuit with three inputs, labeled as IN-1, IN-2, and IN-3, respectively. The output is as shown. Depending on the current flow in the three input transistors and clock, the voltage at the output can be either high or low. If, for example, the input transistors and clock are all OFF, implying no current flow between the emitter and collector, the output is HIGH. If, however, sufficient current flows through the input and clock transistors to switch the RTD to a high-current, low-voltage state, the output is LOW.

only to construct digital circuitry. Below we discuss both general approaches and show some examples of RTD-based digital circuits.

The operation of a RTD/HBT bistable logic gate can be understood using a simple three-input circuit, as shown in Figure 6.5.2. In general, any number of input logic gates can be considered, but for simplicity we consider only three to illustrate the circuit performance. Each input heterojunction bipolar transistor has two possible states: ON, with a collector current I_c, and OFF, without any collector current. The clock transistor, shown as the rightmost transistor in Figure 6.5.2, also has two possible states. These are HIGH and LOW, representing states of high and low current I_{clkh} and I_{clkl}, respectively. Each line for the transistor characteristic corresponds to a different number of HBTs being on, as shown in the diagram. If the clock is LOW, the circuit has two stable operating points for every possible input combination. This is seen clearly in Figure 6.5.3 since the I–V characteristic for the RTD has two intersecting points with that for the HBTs. When the clock current is HIGH, the circuit has only one stable operating point when two or more inputs are high. This operation can be understood using Figure 6.5.3. When I_{clk} is HIGH and two or more inputs are high, there is only one stable bias point, as can be seen in the sketch. This stable bias point occurs at low voltage, which gives a logic zero output voltage.

The physical operation of the circuit can be understood as follows. When the clock is high, meaning that there is a high current flow through the clock transistor, from the node law the current flow through the RTD is also high. The RTD then operates near point V_1 in Figure 6.5.1. However, in the circuit

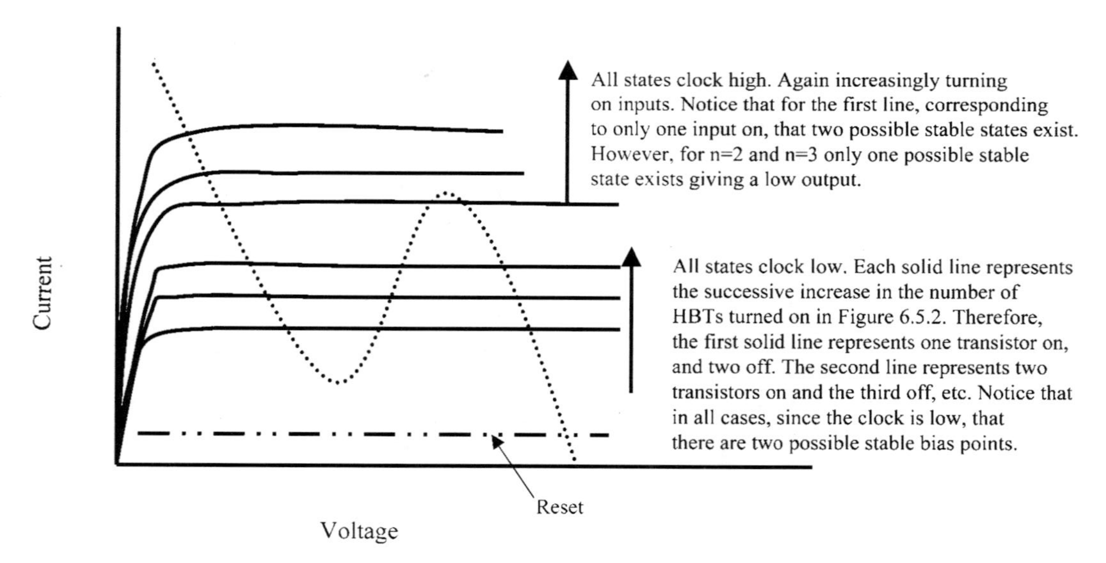

FIGURE 6.5.3 Current–voltage characteristic for the bistable logic circuit of Figure 6.5.2. Notice that for the clock-low condition, the output has always two possible stable states. When two or more inputs are high and the clock is set high, only one intersection point occurs, yielding a low output voltage.

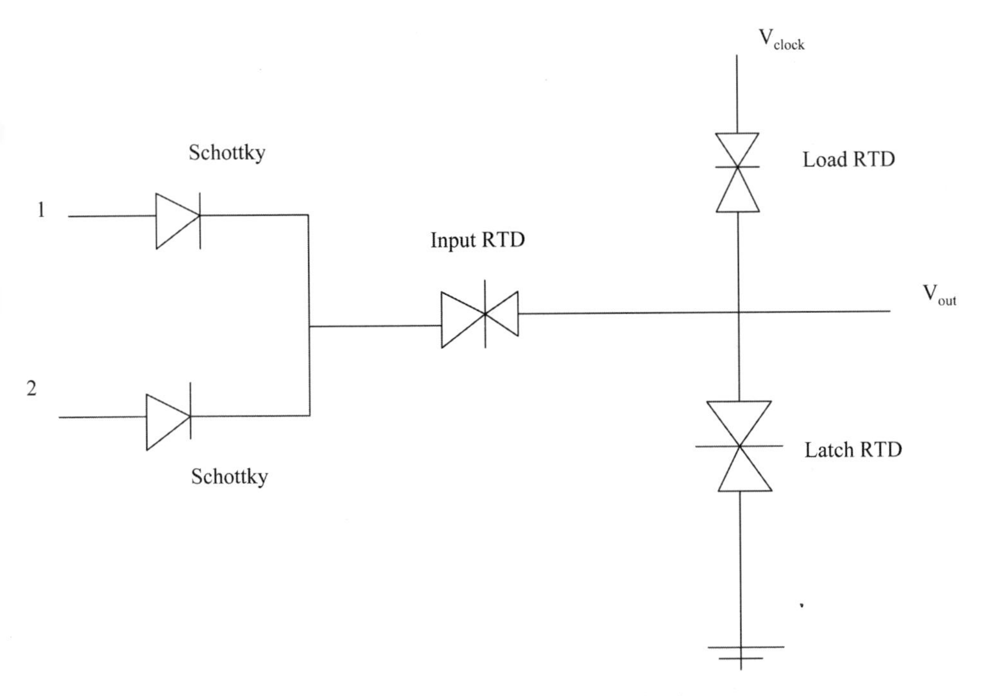

FIGURE 6.5.4 RTD-only logic circuit.

configuration of Figure 6.5.2, the voltage across the RTD is relatively HIGH, so the RTD is then operating with a high current at a high voltage, giving rise to the dashed line in Figure 6.5.3. If the input transistors are then turned on, depending on the value of the peak current, the collector currents of the input transistors, and the number of on input transistors, the RTD may switch. Notice that as the number of input transistors turned on increases, the current through the RTD increases. As the current increases in the RTD, it can exceed the peak current, the current labeled I_p in Figure 6.5.1. If the current surpasses the peak current, the circuit jumps to the far left-hand side of the I–V characteristic, where the current is larger than I_p. Under this condition, the output voltage becomes low. This is the situation for the three-input transistor scheme of Figure 6.5.2. Notice that when the clock is high and two or more input transistors are high, the circuit switches (i.e., the current through the RTD exceeds the peak current and the output is low). Notice that if only one input is high, the output remains high since no switching of the circuit occurs.

When the clock transistor is low, the previous output is maintained. In other words, if the circuit has switched so that the output voltage is low, when the clock becomes low, the output remains low. However, if the circuit was initially in a high-output state, when the clock becomes low, the output remains high.

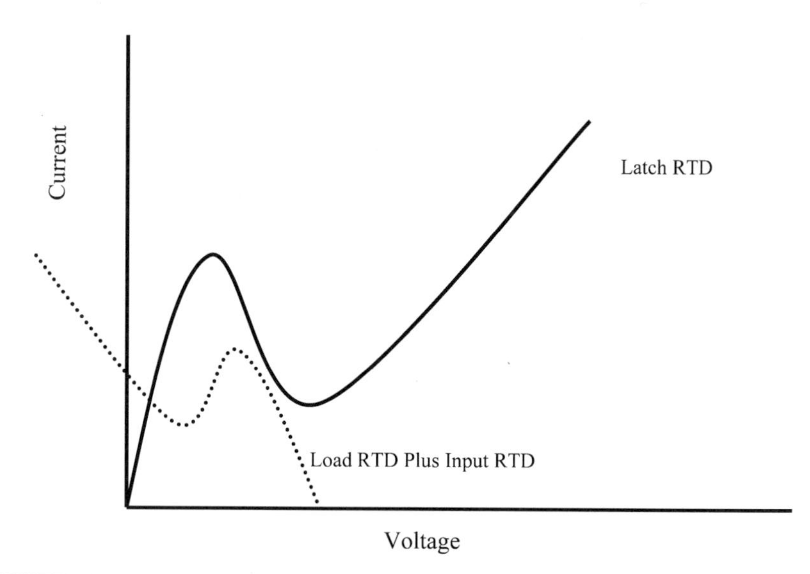

FIGURE 6.5.5 RTD current–voltage characteristics corresponding to the circuit shown in Figure 6.5.4 when the two Schottky diode input voltages are LOW. In this case, the sum of the currents from the input RTD (governed by the currents flowing from the Schottky barriers) and the clock RTD are less than the peak current of the latch RTD. Notice that the intersection and operation point occur at low voltage.

To remove the memory of the circuit, it must be biased at the reset line shown in Figure 6.5.3. Under this biasing condition all the input transistors and the clock transistor have zero collector currents. Notice that the current goes below the valley current of the RTD. The operating points are determined from the intersection of the reset line and the transistor and RTD characteristics. As can be seen from Figure 6.5.3, under the reset condition all the input transistors are switched into cutoff while the RTD operates with a high voltage and low current. Therefore, the circuit gives a high output voltage.

RTDs can be combined with MODFETs as well as HBTs. Use of MODFETs in place of HBTs has the added advantage of requiring very low power consumption and dissipation, comparable to that of CMOS circuits. The interested reader is referred to the references for details on MODFET/RTD circuits.

The second general method of utilizing RTDs for digital logic circuits is to have RTD-only circuits. A sample RTD-only circuit is sketched in Figure 6.5.4. As can be seen from the figure, only two logic inputs are considered for this example. Of course, multiple logic gates can be utilized, but for simplicity here we consider only the action of the circuit with two inputs. Let us consider two different modes of operation. The first mode is that where the combined logic input is LOW. In the circuit of Figure 6.5.4, the inputs are the currents

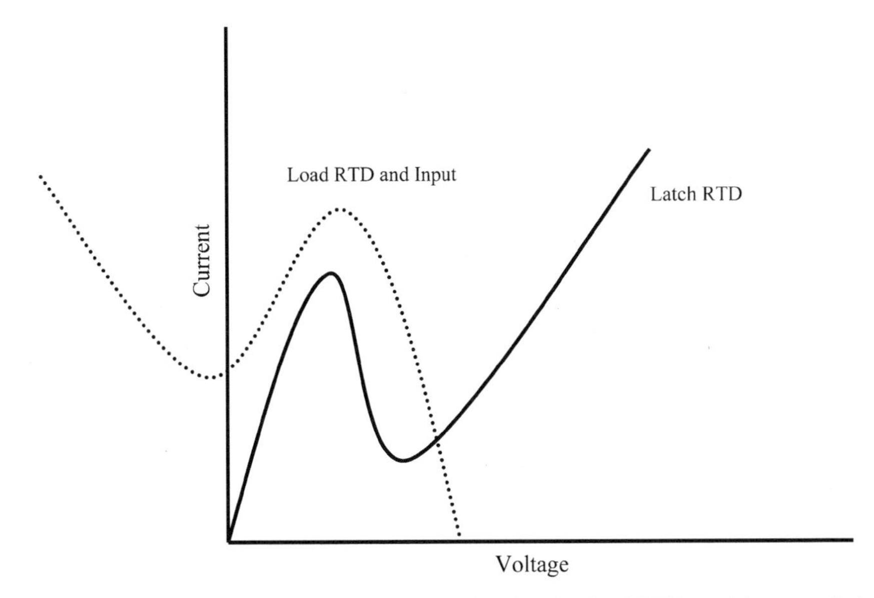

FIGURE 6.5.6 Current–voltage characteristics for the load RTD and input and the latch RTD for the RTD-only circuit shown in Figure 6.5.4. Under this condition, at least one of the two input Schottky barriers is HIGH, implying a high current flow. The combination of the HIGH input Schottky barrier diodes, input RTD, and load RTD is such that the current exceeds the peak current of the latch RTD, as shown. Therefore, the characteristics intersect at a high voltage point, and the output is HIGH.

flowing from the two Schottky diodes and the load RTD. If the Schottky inputs are both LOW, the sum of the currents from the load RTD and the input RTD is such that it is less than the peak current of the latch RTD. This is sketched in Figure 6.5.5. As shown, there is only one intersection point between the curves governing the load and input RTDs and the latch RTD. The intersection point is the operating point of the circuit. Clearly, the operating point is at low voltage and the latch RTD is then at low voltage. Since the voltage across the latch RTD is LOW, the output is also LOW.

A second case for the RTD-only circuit is that where the input logic is HIGH. In this case, either Schottky barrier diode has a high current. Under this condition, the sum of the load RTD and input currents is such that it exceeds I_p of the latch RTD. This is shown in Figure 6.5.6. Therefore, the intersection point of the two curves occurs at higher voltage, as shown. Hence the output voltage is now HIGH. The operation of the circuit under these conditions is that of an OR gate, since only if both inputs are LOW is the output LOW. Any other condition results in a HIGH output.

The above-mentioned circuit applications of RTDs mimic conventional logic circuits. The primary advantage of the RTD-based circuits is that a RTD can replace several gates in an otherwise conventional circuit. As a result, the

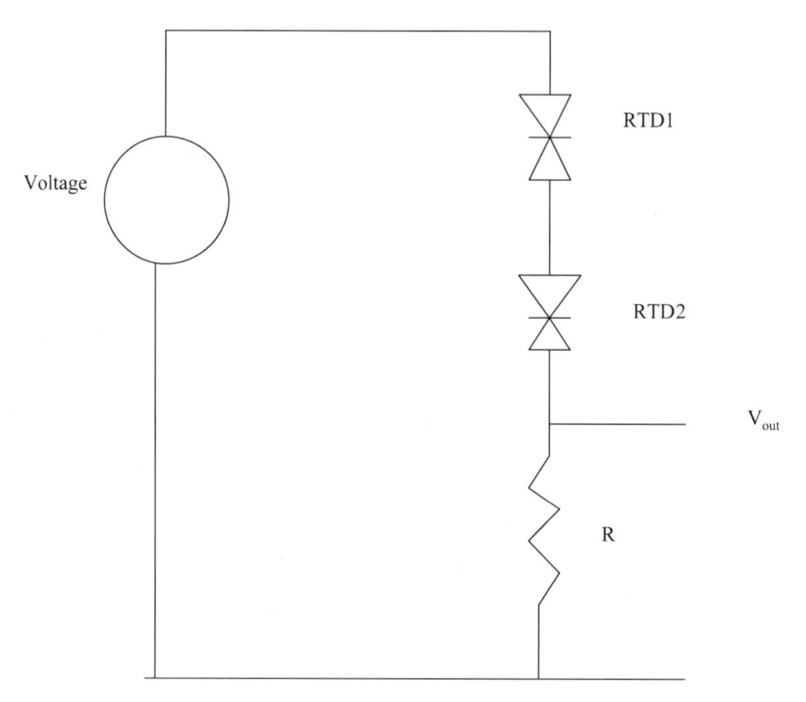

FIGURE 6.5.7 Series RTD circuit for multivalued logic applications.

circuit complexity and interconnect level can be greatly simplified. RTDs can also be used to provide multivalued logic. Multivalued logic is possible with CMOS circuitry, but it has not emerged as a feasible option to binary logic. The primary limitations of CMOS-based multivalued logic are that the circuit building blocks require a relatively large number of devices. They also operate in the threshold mode, which results in circuits with poor operating speeds and noise margins. Some multivalued CMOS circuits have been designed with reduced complexity but at the expense of requiring varying threshold voltages across the chip. RTD-based circuits overcome these limitations and are seemingly better suited for multivalued logic.

Multivalued logic operation requires that the RTDs exhibit multiple current peaks. The peaks should also be nearly equal in height, and the largest valley current must be less than the lowest peak current for proper operation. There have been several approaches to developing multivalued RTD logic circuits. The first, by Capasso et al. (1989), combined two RTDs in parallel and applied slightly different voltages across each diode to obtain two distinct NDR regions. This approach suffers from the limitation that it is difficult to achieve equal peak currents. Potter et al. (1988) have proposed an alternative scheme. In their scheme, two RTDs are combined in series. The circuit is shown in Figure 6.5.7.

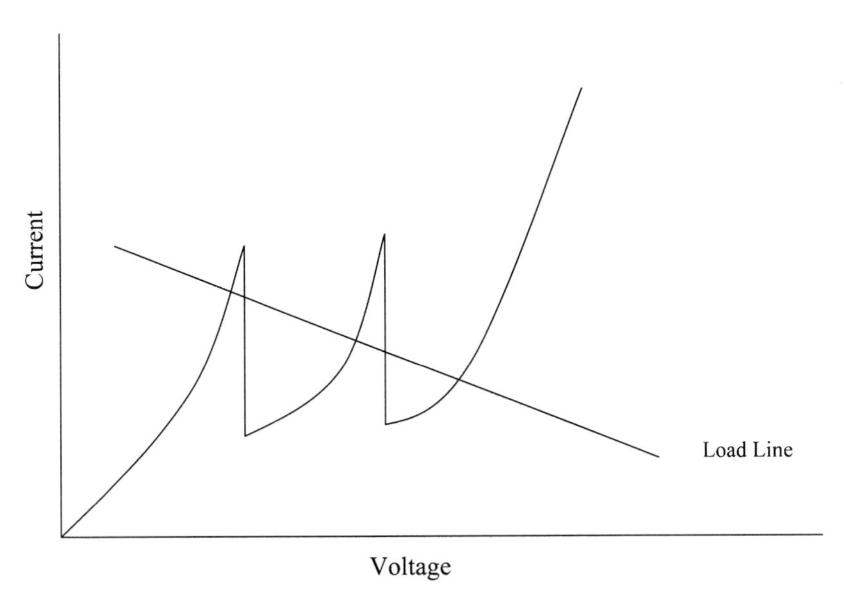

FIGURE 6.5.8 Current–voltage characteristic of the series RTD circuit and load line. Notice that two nearly equal current peaks appear.

The RTD devices used in the circuit of Figure 6.5.7 are not precisely identical. There is some difference between the two devices such that they reach the peak current at different applied biases. Since the diodes are connected in series, the same current flows through them. Hence the magnitude of the peak currents must be the same. The I–V is shown in Figure 6.5.8. If the RTDs were precisely the same, they would switch at the same voltage and only one peak would be obtained. However, since the RTDs are different, one will switch before the other, leading to two peaks in the I–V characteristic. A representative load line is also plotted in Figure 6.5.8. The load line intersects the RTD I–V curve at three stable bias points. Therefore, a three state logic gate can be created. The three-logic states correspond to the stable bias points produced from the intersection of the load line and the RTD I–V characteristic. As can be seen in Figure 6.5.8, the load line intersects the RTD I–V at five points; however, only three are stable. It is these three stable bias points that represent the different logic states. Clearly, by adding more RTDs, higher base logic can be obtained.

6.6 RESONANT TUNNELING TRANSISTORS

Although resonant tunneling diodes can be exploited in logic circuits, they compare rather poorly with three-terminal transistor devices that also utilize resonant tunneling effects. Resonant tunneling transistors provide enhanced

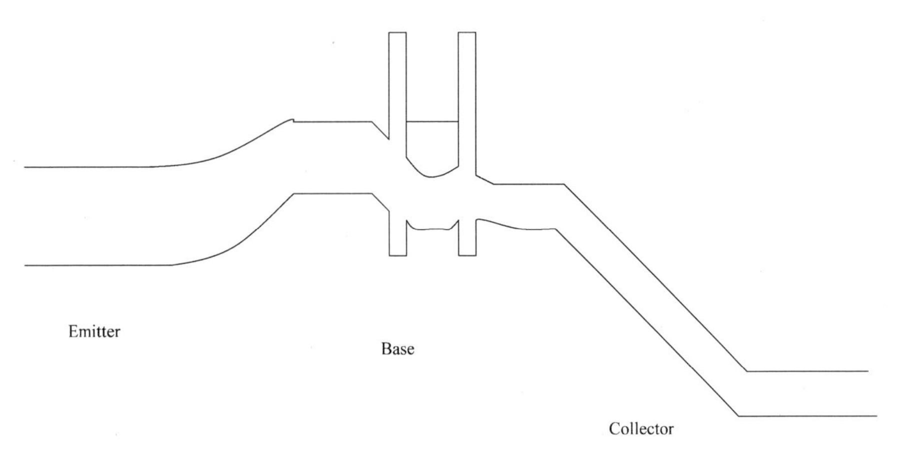

FIGURE 6.6.1 Resonant tunneling transistor with a double-barrier structure in the base region of the device. The region between the emitter and the quantum well is p-type with a bandgap larger than that of the well.

isolation between the input and the output, higher circuit gain, greater fan-out capacity, greater versatility in circuit functionality, and are better suited to large circuits than RTD-only circuits (Haddad and Mazumber, 1997). To overcome the limitations of RTD-only circuits, two different strategies have been adopted. The first is to combine RTDs with conventional three-terminal transistors to exploit the advantages of both device types. The other approach is to utilize transistor structures that incorporate resonant tunneling effects into their design. In this section we examine the latter category for possible use in digital logic applications.

Multiple types of resonant tunneling transistor structures have been reported (Capasso and Kiehl, 1985; Capasso et al., 1986b, 1989; Mori et al., 1986; Woodward et al., 1987). Perhaps the most common configuration of a resonant tunneling transistor is one in which a double-barrier resonant structure is incorporated into the base region of the device. Such a structure is shown in Figure 6.6.1. Different emitter designs have been considered (Capasso and Datta, 1990). The most common emitter designs are the tunnel injection emitter or thermionic emission injection emitter. In the device shown in Figure 6.6.1, the emitter injection occurs through thermionic emission of electrons. The basic operation of the device can be understood as follows.

The key design feature of the resonant tunneling transistor is that each resonant level of the double barrier corresponds to a different logic level. In most instances the double-barrier resonator can be designed such that several resonant levels exist in the well. Through judicious control of the emitter voltage, each resonant level within the double barrier can be brought into alignment with the conduction band edge of the emitter, resulting in a substantial current flow. At emitter voltages at which the emitter is not aligned with a resonant level, the current flow in the device is greatly reduced. The magnitude of the collector current is then controlled by resonant alignment of the double-barrier quantum levels with the conduction band edge, which is controlled in turn by the emitter voltage. Initially, the first quantum level of the double barrier may not be aligned with the emitter at low emitter bias. As a result, the current flow in the device would be relatively small. As the emitter bias changes, the first resonant level within the double-barrier structure can be brought into alignment with the conduction band edge. Consequently, a sizable current will flow in the device. Further increase in the applied emitter bias will result in a reduction in the current flow since the resonant level will now be lowered below the conduction band edge of the emitter. The current will peak again if a second resonant level is brought into alignment with the emitter conduction band edge. Again, the current will decrease once the second level is lowered below the emitter conduction band edge through increased emitter bias. The process continues until all the levels within the double-barrier are aligned. In this way, several different current peaks can appear in the I–V characteristic for the transistor. As we mentioned above, each peak represents a different logic level.

The design of the double-barrier influences the position of the resonant levels. For example, in a finite-square-well double-barrier structure, the quantum levels are not evenly spaced. The energy separation between the quantum levels decreases with increasing energy in the well. In other words, the largest energy separation is between the first and second quantum levels. If the well is made parabolic using continuous grading, the energy levels become equally spaced (Brennan, 1999, Chap. 2). Consequently, for a parabolic well, the energy levels and the corresponding current peaks are all equally spaced.

Above we discussed resonant tunneling transistors (RTTs) that contain only one double-barrier structure. Although these RTT devices can exhibit multiple current peaks, thus enabling multivalued logic, it is in practice difficult to make the multiple peaks of comparable magnitude, which is generally desirable for circuit applications. Another approach is to add two or more double-barrier structures to the device. Capasso et al. (1989) have proposed a RTT that has two double-barriers within the emitter region of the device, as shown in Figure 6.6.2. The device is in the common-emitter configuration. The RTT is not limited to only two double-barriers. Multiple double-barriers can be employed. A peak in the current–voltage characteristic occurs for each double-barrier. Therefore, if the device has two double-barriers, it will exhibit a double peak in the output I–V characteristic. If the device has n double-barriers, it will have n peaks.

The multiple peaks in the output I–V characteristic are a consequence of the fact that the electric field across the two double barriers is not uniform. The field nonuniformity is due to the action of electrostatic screening, resulting in a higher electric field in the double-barrier structure closer to the base in Figure 6.6.2. As the emitter–base voltage is altered, the resonant tunneling current peaks and decreases at two distinct applied biases. Consequently, two peaks

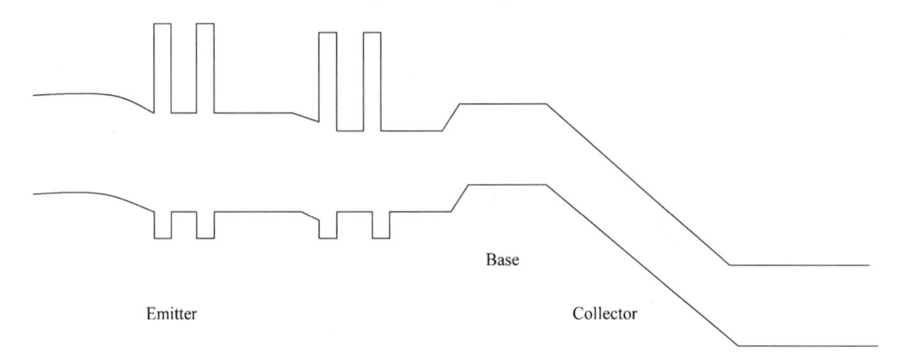

FIGURE 6.6.2 Resonant tunneling transistor with two double-barrier structures within the emitter of the device.

in the output characteristic are obtained. The double barriers are connected in series. Therefore, when resonant tunneling is suppressed at only one of the double-barriers, it would seem that the current flow would be reduced throughout the entire structure. However, inelastic tunneling current maintains continuity of the current flow within the structure.

How, though, can resonant tunneling transistors be used for multivalued logic? One of the key applications of multivalued logic is a parity generator. Parity generators are of importance in communications systems to detect transmission errors. Capasso et al. (1991) have demonstrated that a four input parity generator circuit can be constructed using a single RTT. This is in contrast to a conventional circuit that requires 24 transistors to create three exclusive OR gates. RTTs have also been employed in analog-to-digital converters and in multistate memories. The operation of a RTT as a multistate memory device can be understood as follows. As can be seen from Figure 6.5.8, a load line drawn onto the RTT I–V characteristic intersects it at several stable points. For the specific case shown in Figure 6.5.8, the load line intersects the RTT characteristic at three stable points. In general, if the RTT has n peaks, the load line intersects it at $n + 1$ stable locations. Hence the RTT can be used as a memory element in an $(n + 1)$-state logic system.

PROBLEMS

6.1. Determine the longitudinal mode spacing of a simple Fabry–Perot resonator as discussed in Box 6.4.1. (*Hint*: Consider the phase condition using the propagation constant Γ defined in the box.)

6.2. Determine a simplified expression for the current density given by Eq. 6.2.20 that holds at low temperature by modifying the logarithmic term. The final result can be left in integral form.

6.3. Using the result derived in Problem 6.2, a convenient approximation of the current density can be made. Assume that the applied voltage is very high, such that the term involving qV can be neglected. Further assume that the transmission coefficient T^*T can be approximated using a Lorentzian of the form

$$T^*T \sim T_{\text{res}} \frac{\Gamma^2/4}{(\Gamma^2/4) + (E - E_n)^2}$$

where E_n is the energy eigenstate of the well, T_{res} the on-resonance transmission, and Γ the full width, half maximum. Make the further assumption that the Lorentzian can be approximated as a delta function as

$$\delta(E_l - E_n) = \frac{1}{\pi} \lim_{\Gamma \to 0} \frac{\Gamma/2}{(\Gamma^2/4) + (E_l - E_n)^2}$$

6.4. The cutoff frequency of a RTD can be determined from the equivalent-circuit representation shown in Figure P6.4. The real part of the output impedance of the diode seen by the load must be negative to sustain oscillations. From this requirement, the cutoff frequency can be determined.

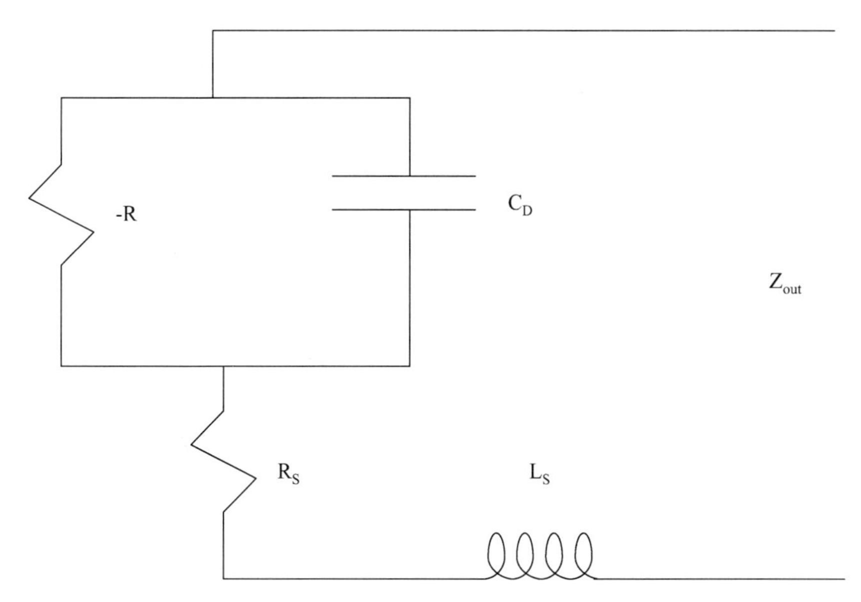

FIGURE P6.4 Equivalent circuit of a resonant tunneling diode. $-R$ is the negative differential resistance of the diode, R_S the access resistance, L_S the lead inductance, and C_D the diode capacitance.

6.5. Determine the expression for the charge density within a two-dimensional quantum well in a RTD if we assume that the probability of each level being occupied is given by the Fermi function, $f(E,E_f)$. Start with the expression

$$\rho(z) = q \sum_{E_i} |\psi(E_i,z)|^2 f(E,E_f)$$

where E_i is the energy of the ith subband, E_f the Fermi level, and z is in the direction of quantization. Evaluate $\rho(z)$ by summing over the confined states and integrating over the two-dimensional continuum.

6.6. Matrix formulation of the density matrix. Assume a normalized state χ defined as

$$\chi = \begin{pmatrix} c_1 \\ c_2 \end{pmatrix}$$

where c_1 and c_2 are amplitudes. Construct the density matrix for χ in matrix form. Given an operator A defined as

$$A = \begin{bmatrix} A_{11} & A_{12} \\ A_{21} & A_{22} \end{bmatrix}$$

determine the expectation value of A for the state χ using the density matrix formulation.

CHAPTER 7

CMOS: Devices and Future Challenges

The information revolution has been driven largely by continued progress in integrated circuit development. To date, integrated circuits have doubled functions per chip every one and a half to two years since the 1960s. This trend, formally known as Moore's law, has paced the semiconductor industry. The most ubiquitous circuitry used for digital logic applications is CMOS, the complementary metal-oxide semiconductor. The principal component of a CMOS integrated circuit is the MOSFET, the metal-oxide semiconductor field-effect transistor. The MOSFET is the fundamental switching element used to produce digital logic in integrated circuits. MOSFET switching speed and hence circuit speed increases with miniaturization. Further improvement in computer hardware (i.e., increased speed and density) requires continued progress in miniaturization. In this chapter we examine some of the boundaries that influence the continued miniaturization of MOSFET devices.

*7.1 WHY CMOS?

We begin our discussion with a basic review of the advantages of CMOS circuitry. The fundamental principles of MOSFET device performance are reviewed in the next section. Here we discuss only that which is needed to understand why CMOS circuitry has become ubiquitous in computing hardware. There are two different general types of MOSFET devices, n- and p-channel structures. These devices are sketched in Figure 7.1.1. The basic

*Section provides background material for non-ECE majors.

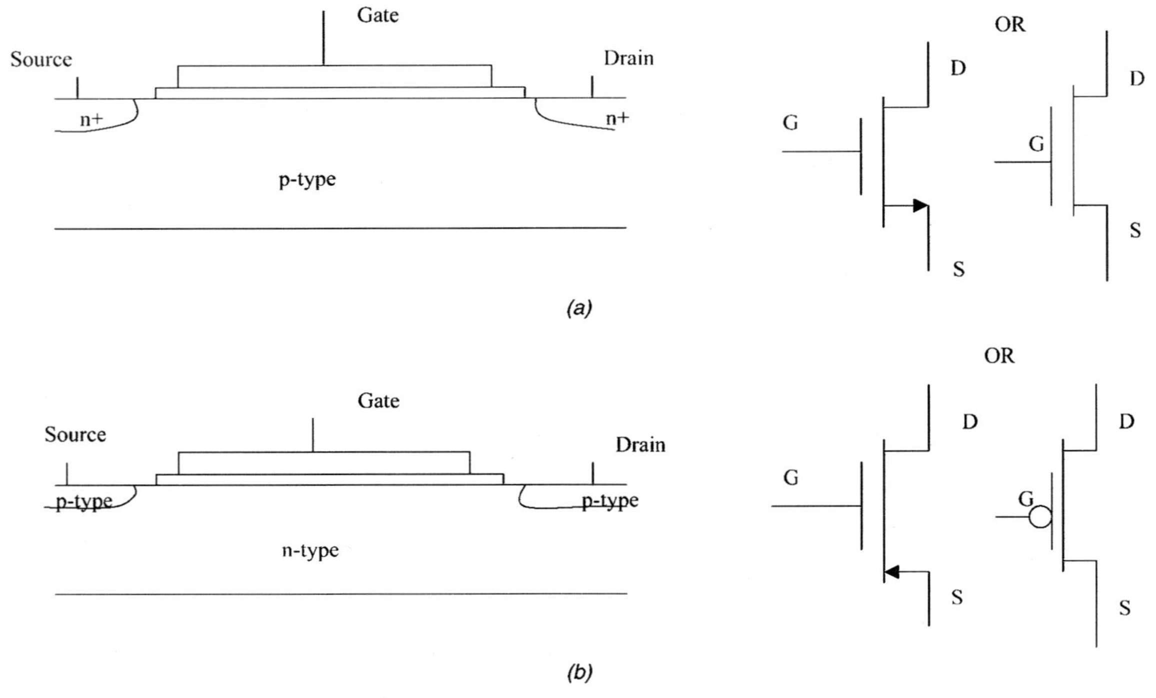

FIGURE 7.1.1 n- and p-channel devices and circuit symbols.

idea of a MOSFET device is that a gate electrode formed by a metal-oxide-semiconductor structure controls the conductivity of the underlying semiconductor region. Through judicious selection of the gate voltage, a conducting channel can be opened and closed connecting the source and drain contacts of the device. In the n-channel MOSFET, the underlying semiconductor material is p-type and the conducting channel formed between the source and drain regions is n-type. Conversely, in the p-channel MOSFET, the underlying semiconductor material is n-type while the conducting channel connecting the source and drain regions is p-type.

The typical biasing circuit for a MOSFET device and the corresponding current–voltage characteristic are shown in Figure 7.1.2. There are three general regimes of operation of the device. These are cutoff, saturation, and the triode regime. Cutoff occurs when $V_{GS} < V_T$, the threshold voltage of the device. At cutoff the source and drain are electrically isolated since no conducting channel is formed between them. Hence the drain current is low, as indicated from Figure 7.1.2. Saturation occurs when $V_{DS} > V_{GS} - V_T$. From inspection of Figure 7.1.2 it is clear that when operated in the saturation region, the MOSFET behaves as a constant-current source (i.e., the drain current is constant with increasing drain bias). The last region is the triode region. In the triode region, the drain current of the MOSFET depends on both V_{GS} and V_{DS}.

The question we seek to answer in this section is why CMOS circuitry is used so widely in digital computing hardware. The key issue that makes CMOS circuits highly attractive is that they dissipate very low dc power. The very low power use of CMOS circuitry makes them important in portable applications in which battery drain is critical to the lifetime of the system (e.g., in digital watches, portable computers, automobile clocks). Additionally, low power dissipation is very important in highly dense circuitry. As the number of devices per chip increases, very low power dissipation is crucial in maintaining room-temperature operating conditions. As the reader may recall, as the temperature of silicon increases, it ultimately becomes intrinsic, meaning that the free carrier concentration within the conduction band results mainly from interband thermal generation, negating the effect of doping, resulting in device failure. Even using extensive heat-sinking techniques, it is difficult to dissipate waste heat completely. Therefore, use of low power dissipation circuitry such as CMOS is essential for proper thermal management.

How, though, are CMOS circuits constructed such that they dissipate little dc power? Low power dissipation is accomplished by having at least one transistor in cutoff, where its drain current is extremely low, between the power source, V_{DD}, and ground for all possible logic inputs. In this way, the power source is always shielded from ground by a very high resistance path. A CMOS circuit is designed such that current flows in the circuit only during transitions or switching, with the obvious exception of leakage current.

For illustration, let us consider some representative CMOS circuits. We first examine a simple CMOS inverter gate. Let us first consider the operation of the inverter using a simple switch to represent a MOSFET. Two different states

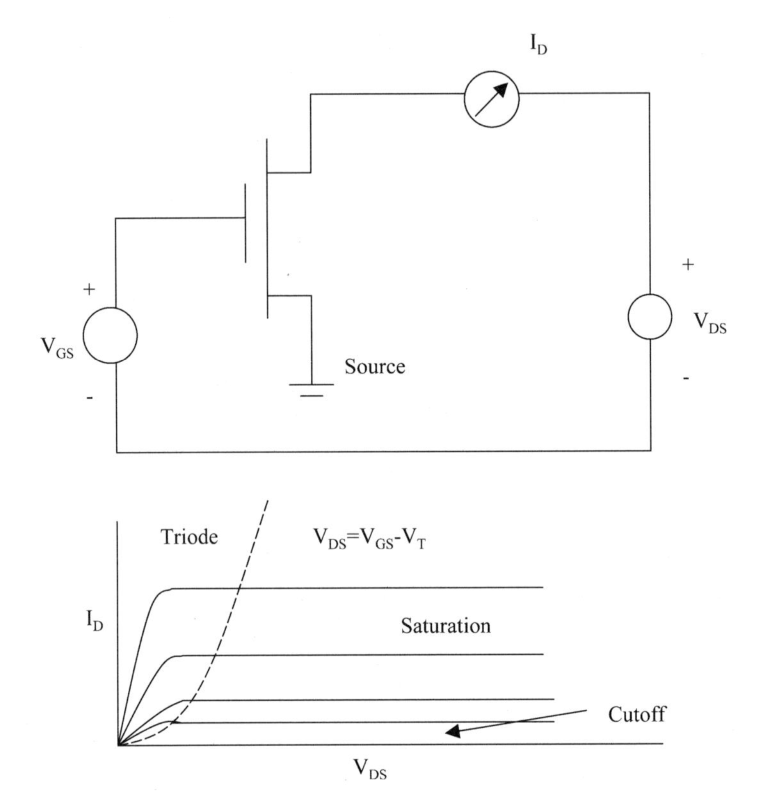

FIGURE 7.1.2 Sample biasing scheme and corresponding *I–V* characteristic.

of the circuit are shown in Figure 7.1.3: (*a*) input high and output low, and (*b*) input low and output high. From the diagrams it is difficult initially to tell how the high and low inputs are applied, but this will become clear in the discussion below. For now, we focus on the output. In the configuration shown in part (*a*) the output is connected to ground through the bottom switch, and hence a low output is obtained. In Figure 7.1.3*b* the bottom switch is open and the output is blocked from ground by the capacitor. This corresponds to a high output. So, depending on the bottom switch setting, the output can be either high or low.

The CMOS circuit implementation of this inverter is shown in Figure 7.1.4. The top transistor is a p-channel device, the bottom transistor is an n-channel device. A p-channel MOSFET has the opposite characteristics of an n-channel MOSFET. Specifically, when the gate voltage is high, the p-channel transistor is OFF (operates in the cutoff region) while the n-channel MOSFET is ON (biased in the triode region). A conducting channel in a p-channel device is formed under a low positive or negative gate voltage. Similarly, a conduction channel in an n-channel device is formed under a high positive gate voltage.

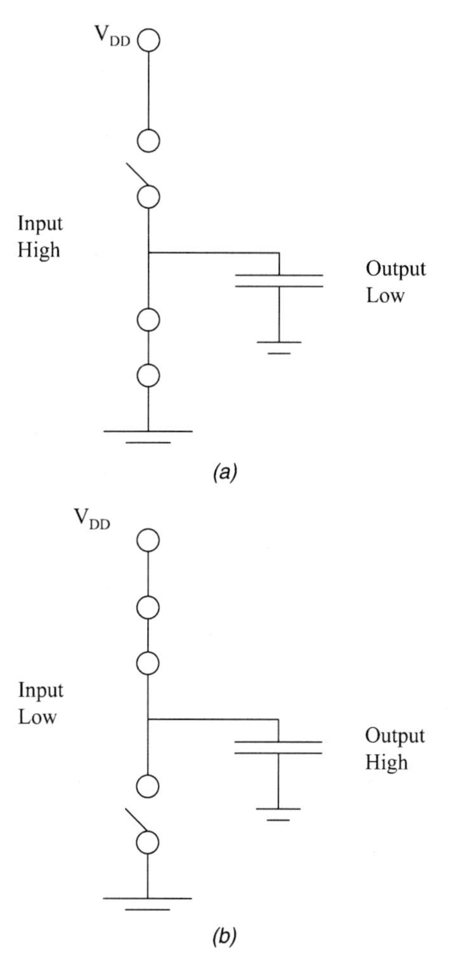

FIGURE 7.1.3 Switch representation of a CMOS inverter circuit for (*a*) output low and (*b*) output high.

Hence, when the input voltage is HIGH, the n-channel device shown in Figure 7.1.4 is ON while the p-channel device is OFF. Under these conditions, V_{OUT} is connected to ground and isolated from V_{DD}. Hence the output is LOW. Conversely, when the input voltage is LOW, the p-channel device shown in Figure 7.1.4 is ON while the n-channel device is OFF. In this case, V_{DD} appears at V_{OUT}, and the output is HIGH. Therefore, the CMOS circuit behaves as follows: input LOW-output HIGH; input HIGH-output LOW, which clearly operates as an inverter.

CMOS circuitry can be used to create various logic gates. For example, NOR and NAND gates are relatively simple to implement using CMOS circuitry. Along with the inverter, NOR and NAND functions are sufficient to

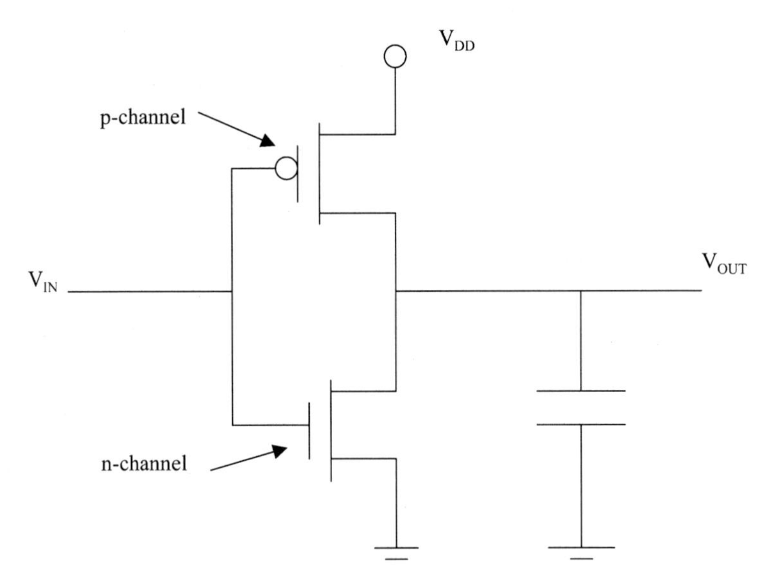

FIGURE 7.1.4 CMOS inverter circuit.

Input		Output	
A	B	OR	NOR
0	0	0	1
0	1	1	0
1	0	1	0
1	1	1	0

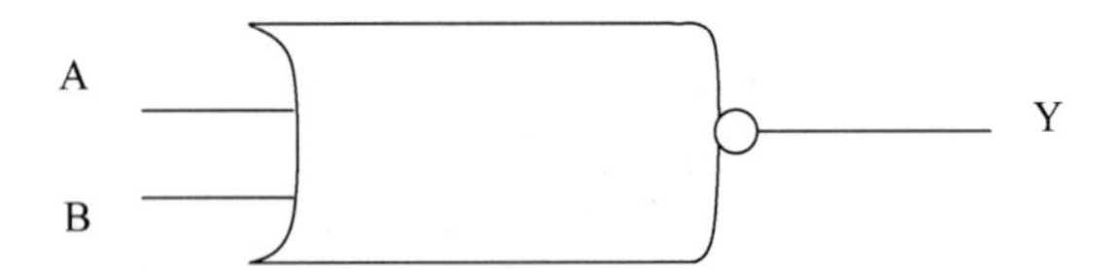

FIGURE 7.1.5 Boolean logic truth table and symbol for a NOR gate.

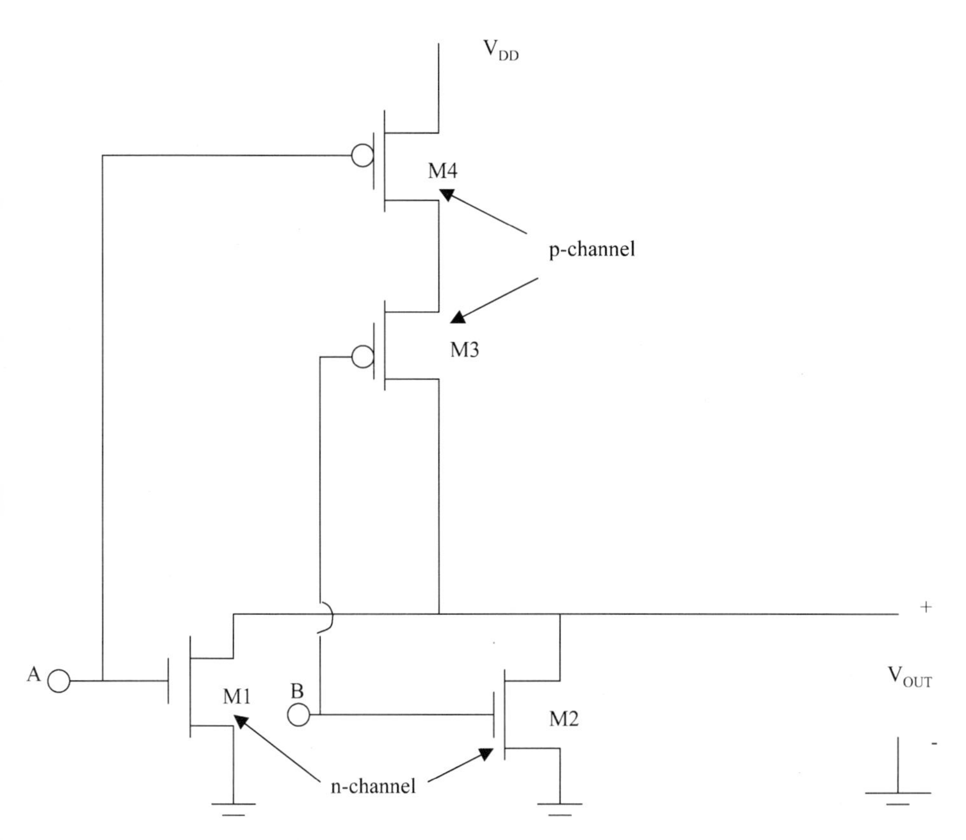

FIGURE 7.1.6 CMOS circuit implementation of a NOR gate.

reproduce all the important logic operations needed in a digital binary computer. Let us next consider NOR and NAND gates and their implementation in CMOS. The Boolean logic truth table for a NOR gate is shown in Figure 7.1.5 along with a symbolic NOR gate. The CMOS circuit implementation of a NOR gate is shown in Figure 7.1.6. For simplicity we consider only a two-input NOR gate. In this case the two inputs are marked as A and B and the output is V_{OUT}. Recall that the n-channel transistors are ON when their gate voltage is HIGH, and OFF when their gate voltage is LOW. The p-channel transistors have the opposite behavior. They are ON when the gate voltage is LOW and OFF when the gate voltage is HIGH. Consider the case when both A and B are LOW. According to the truth table in Figure 7.1.5, the output should be HIGH for a NOR gate. Let us see how the circuit of Figure 7.1.6 behaves. If inputs A and B are both LOW, then since M1 and M2 are n-channel transistors, they will both be OFF. Notice that V_{OUT} is connected to ground through the channels of either M1 or M2. Since both M1 and M2 are OFF, there is no conducting path connecting V_{OUT} and ground. Transistors M3 and

Input		Output	
A	B	AND	NAND
0	0	0	1
0	1	0	1
1	0	0	1
1	1	1	0

FIGURE 7.1.7 Boolean truth table and logic symbol for a NAND gate.

M4 are both ON since they are p-channel devices with a LOW voltage applied to their gates. V_{OUT} is connected to V_{DD} through these devices in series. Since both M3 and M4 are on, V_{OUT} is shorted to V_{DD} and hence the output V_{OUT} is HIGH. Thus the circuit yields a HIGH output for two LOW inputs. Consider what happens for any other combination, either both inputs HIGH or either input HIGH while the other input is LOW. When both inputs are HIGH, transistors M1 and M2 are ON. When *A* is HIGH and *B* is LOW, M1 is ON and M2 is OFF. Similarly, when *A* is LOW and *B* is HIGH, M1 is OFF and M2 is ON. Notice that under any of these circumstances there exists at least one path connecting V_{OUT} to ground through transistor M1 or M2 or both. Hence, when one or both inputs are HIGH, the output is LOW. Clearly, the CMOS circuit shown in Figure 7.1.6 acts like a NOR gate.

Let us next consider operation of a NAND gate. The truth table and logic symbol for a NAND gate are shown in Figure 7.1.7. Inspection of the truth table shows that only if both inputs are HIGH will the output of the NAND gate be LOW. For any other combination, the output is always HIGH. The CMOS circuit implementation of the NAND gate is shown in Figure 7.1.8. Again we consider for simplicity only a two-input NAND gate. As in the NOR gate, the circuit uses two p-channel MOSFETs and two n-channel MOSFETs. The inputs are marked as *A* and *B* while the output is V_{OUT}. Consider the case when both *A* and *B* are HIGH. Then the n-channel transistors, M1 and M2, are both ON, while the p-channel transistors, M3 and M4, are both OFF. Therefore, V_{OUT} is tied to ground through the series combination of M1 and M2 and is isolated from V_{DD} by the parallel path of M3 and M4. As a result,

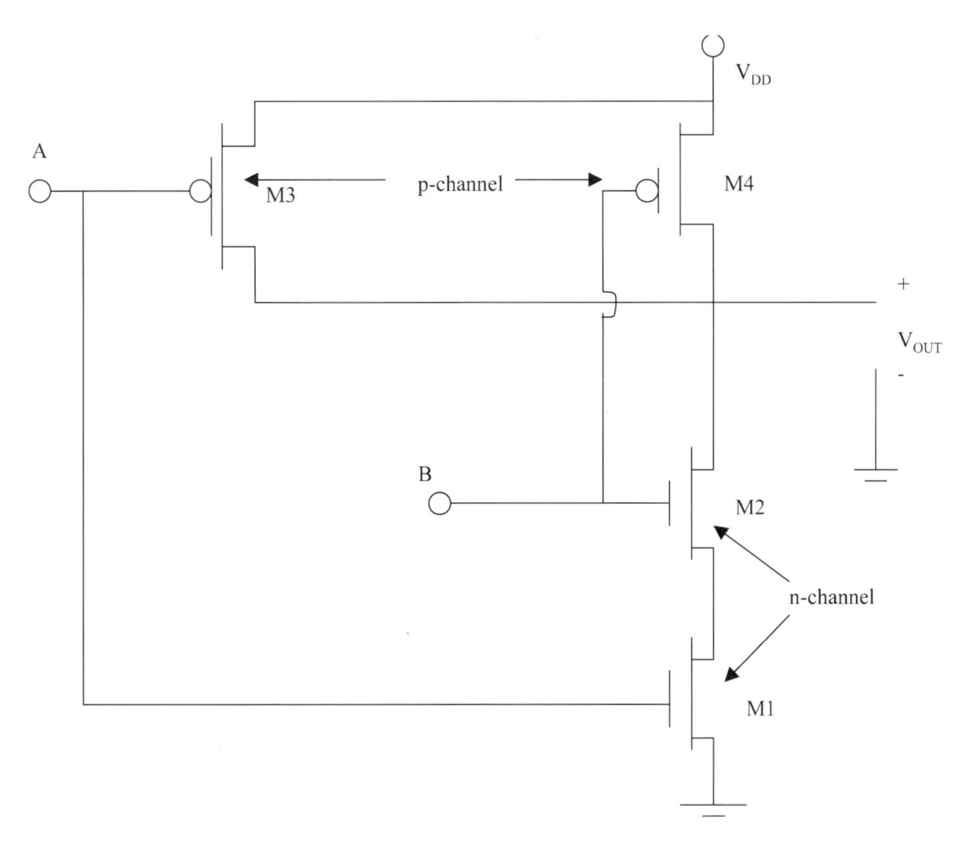

FIGURE 7.1.8 CMOS circuit implementation of a NAND gate.

V_{OUT} is LOW. For any other logical input, A or B or both LOW, then the following conditions occur. If A is LOW, M3 is ON and M1 is OFF. Thus since M3 and M4 are in parallel, independent of the state of M4, V_{OUT} is connected to V_{DD}. Additionally, since M1 and M2 are in series, if M1 is OFF, then V_{OUT} is shielded from ground, independent of the state of M2. Thus the output is HIGH. The opposite situation, B is LOW, M4 is ON and M2 is OFF, also leads to V_{OUT} connected to V_{DD} and a HIGH output. Clearly, when both A and B are LOW, the output will again be HIGH. Thus the circuit of Figure 7.1.8 behaves as a NAND gate.

Finally, one might ask why silicon CMOS is the most widely used circuitry for digital logic. Will not any complementary FET technology also be useful in digital logic? Why is it that silicon is so widely used? Although it is true that any complementary FET technology can be used for digital logic, silicon CMOS is employed almost exclusively in digital integrated circuits. The reason for this is mainly technological. Perhaps the most important features of silicon that make it so useful in commercial semiconductor products is its relative ease of fabrication and the extraordinarily high quality semiconductor and oxide material that can be reproducibly and inexpensively grown. SiO_2, the native oxide to silicon, grows readily in a nice, controllable manner. For this reason it is quite easy to reproducibly make extraordinarily high quality silicon CMOS devices that suffer from few defects. This is the essential reason why silicon CMOS technology has long dominated digital hardware. In the next section we discuss the rudiments of silicon MOSFETs. The remainder of the chapter is devoted to discussing the challenges that CMOS technology faces in the near and long-term future.

7.2 BASICS OF LONG-CHANNEL MOSFET OPERATION

In this section we present a brief discussion of the workings of long-channel MOSFET devices. More detailed discussions of the physics of MOSFETs are presented in Brennan (1999, Chaps. 11 and 14) and Taur and Ning (1998). The interested reader is referred to these sources for more information.

The MOSFET is a three-terminal device comprised of a source, gate, and drain. The basic structure of the MOSFET is shown in Figure 7.2.1. The drain current is controlled through use of the field effect. The field effect can be defined concisely as the control of the conductivity of an underlying semiconductor layer by the application of an electric field to a gate electrode on the surface. Through the application of a gate bias, the conductivity as well as carrier type of the underlying semiconductor layer can be altered. Depending on the gate bias, the semiconductor can be changed from n-type to p-type, or vice versa, a condition called *inversion* (Brennan, 1999, Chap. 11).

As discussed in Section 7.1, there are three general regions of operation of a MOSFET. In this section we give quantitative expressions for the drain current for each region of operation. Figure 7.2.2 shows a three-dimensional sketch of the MOSFET structure for purposes of defining the appropriate variables used in the expressions for the drain current. The channel length L is defined as the distance from the edge of the source region to the edge of the drain region in the device. The channel length is one of the key parameters that characterizes the device. In this section we develop expressions for the drain current that are valid for long-channel-length MOSFETs. For purposes of illustration, long-channel length will be defined as a device in which the variation of the electric field along the channel direction (y direction in Figure 7.2.2) is much less than that in the direction perpendicular to the channel (x direction in Figure 7.2.2).

The operation of a MOSFET depends on the formation of a conducting channel beneath the gate between the source and drain regions of the structure. Since the source and drain regions are of a different doping type than the substrate, in order to form a conducting channel, the semiconductor layer must be biased into inversion. There exists a minimum gate voltage, the threshold voltage V_T, for which the semiconductor becomes inverted. The threshold

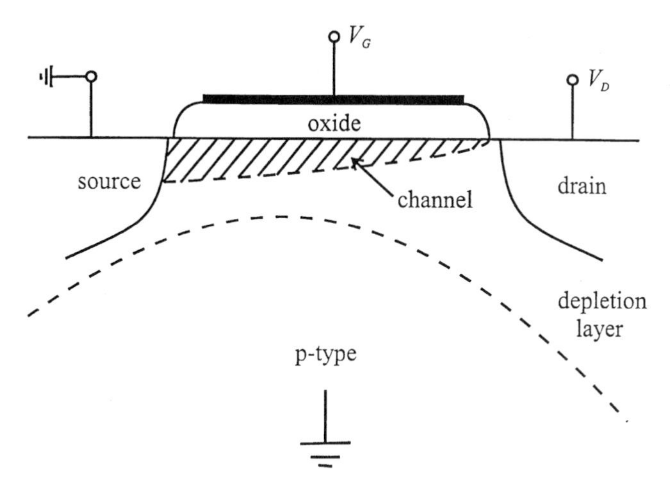

FIGURE 7.2.1 Simple n-channel MOSFET device. The conducting channel is formed at the interface between the silicon and the silicon dioxide as shown.

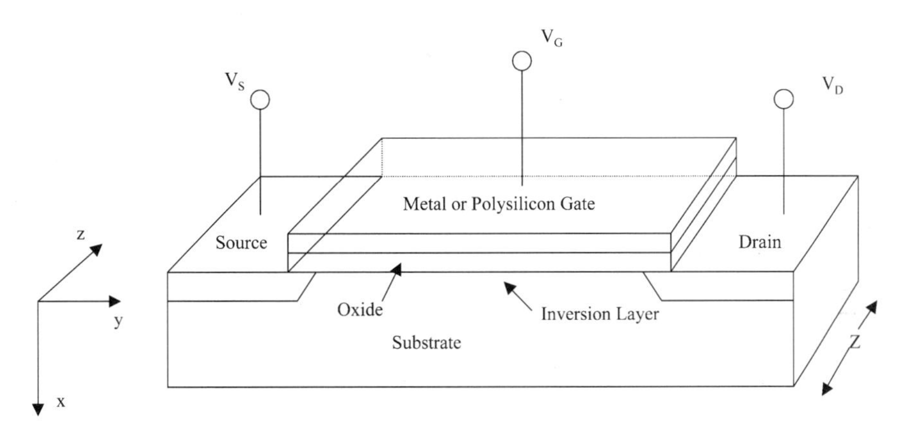

FIGURE 7.2.2 MOSFET structure showing the dimensions and important regions of the device.

voltage is equal to the sum of the flat band voltage, the voltage necessary to make the bands flat, and the amount of voltage necessary to achieve strong inversion. Typically, *strong inversion* is defined as the voltage condition such that the channel region is as n-type as the substrate region is p-type for an n-channel device, and vice versa for a p-channel device. For purposes of illustration, let us specialize our argument to an n-channel device. As shown in Box 7.2.1, the flat band voltage for a MOS structure is given as

$$V_{\mathrm{FB}} = \phi_{ms} - \frac{Q_{\mathrm{ox}}}{C_i} \tag{7.2.1}$$

BOX 7.2.1: Determination of the Flat Band Voltage

The *flat band voltage* is defined as the applied gate voltage necessary such that the energy bands of the underlying semiconductor material are flat (i.e., no band bending is present). For ideal metal–insulator–semiconductor junctions, it is assumed that the metal–semiconductor work function difference is zero. Additionally, the insulator is assumed to be perfect, with no defect charge states within the insulator or at the insulator–semiconductor interface. Under these assumptions, flat band condition occurs naturally within the semiconductor. However, the assumptions of a zero work function difference and a perfect oxide and interface are not generally applicable. Typically, the work function difference between silicon and the gate metal ϕ_{ms} is negative. This is always the case for a polysilicon gate deposited on either n- or p-type silicon. In addition, it is usually the case that the oxide layer is not perfect (i.e., in any practical situation there exist charges within the oxide and/or at the interface between the oxide and the semiconductor). Generally, there are four different types of defects in an oxide layer. These are mobile ionic charge Q_m, trapped charge Q_t, interface charge Q_{int}, and fixed oxide charge resulting from the incomplete oxidation of silicon during oxide growth. In SiO_2, the most common mobile ionic charge arises from sodium ion contamination, which can enter the oxide during growth or processing steps from impure chemicals, water, gases, or containers. The trapped charge Q_t arises from imperfections within the oxide, while the interface charge Q_{int} is due to imperfect interface formation. All of these defect charges act as positive charge within the oxide layer.

The combined effect of a negative ϕ_{ms} and positively charged defects within the oxide on the band diagram of a p-type MOS system in equilibrium (no external bias applied to the system) is shown in Figure 7.2.3. Notice that even in equilibrium there exists a band bending. The band bending arises from the combined effect of the nonzero ϕ_{ms} and the gate oxide and interface charge. For simplicity, we group all the oxide and interface charges together in one variable and refer to this as the oxide charge of Q_{ox}. Effectively, the structure behaves as if there is a net positive charge to the left of the semiconductor surface. To balance this net positive charge, an equal amount of negative charge must be induced within the semiconductor region. In this example, the semiconductor material is assumed to be p-type. Therefore, the majority carrier holes are repelled from the interface, forming a depletion layer of ionized acceptors. It is these negatively charged ionized acceptors that balance the net positive charge. From the diagram the difference between the Fermi level and the intrinsic level is greater at the interface than in the

(*Continued*)

BOX 7.2.1 (*Continued*)

bulk. The hole concentration within the semiconductor layer is given as

$$p(x) = n_i e^{[E_i(x) - E_{fs}]/kT}$$

where $p(x)$ is the hole concentration within the semiconductor layer as a function of distance from the surface, and $E_i(x)$ is the spatially dependent intrinsic level. From inspection of Figure 7.2.3, the difference between E_{fs} and $E_i(x)$ is smaller near the surface than in the bulk material. Therefore, the hole concentration is lower at the surface than in the bulk, and the majority carriers are depleted from the surface, as expected, to balance the positive charge within the oxide and gate. The bands bend down at the interface to reflect the depletion condition.

To achieve flat band, a net negative voltage must be applied to the gate electrode to compensate for the net positive charge due to the negative value of ϕ_{ms} and the positive charge within the oxide. This applied voltage, called the *flat band voltage*, has value

$$V_{\mathrm{FB}} = \phi_{ms} - \frac{Q_{\mathrm{ox}}}{C_i}$$

where Q_{ox} is the net charge due to all sources of positive charge within the oxide and interface identified above, and C_i is the insulator capacitance. Note that V_{FB} is negative for the polysilicon gate MOS system.

where Q_{ox} is the combined oxide and interface charge and C_i is the oxide capacitance. In the discussion below, we define the charge Q to be per unit area and the capacitance C to also be per unit area. In Eq. 7.2.1, the unit areas simply divide out. To achieve inversion, the bands must be bent further such that the channel region is in strong inversion, as defined above. Therefore, the applied gate voltage must be sufficiently large that it will first create the depletion region charge and additionally, attract minority carrier electrons to the interface. From charge balance, the net charge on the gate must be balanced by the net charge in the semiconductor region given by the combined depletion- and inversion-layer charge. For an n-channel device, the gate charge is positive, so the balancing charge in the semiconductor Q_s must be negative. Calling the charge on the gate Q_G, the depletion layer charge Q_d, and the inversion layer charge $Q_{\mathrm{inversion}}$, the charge balance relation can be written as

$$Q_G = -Q_s = -Q_d - Q_{\mathrm{inversion}} \qquad 7.2.2$$

The voltage across the oxide is given simply by the ratio of the charge on either side of the oxide to the oxide capacitance as

$$V_i = \frac{-Q_s}{C_i} \tag{7.2.3}$$

Therefore, there must be two additional voltage terms added to Eq. 7.2.1 to yield the threshold voltage V_T. The voltage that is necessary to form the depletion region is simply equal to the amount of depletion charge divided by the oxide capacitance, $-Q_d/C_i$. Strong inversion for an n-channel device is defined when the inverted region is as n-type as the bulk is p-type. Calling the energy separation between the intrinsic level and the Fermi level in the bulk region $q\Psi_B$, the voltage necessary to invert the surface is then simply $2\Psi_B$. Thus the threshold voltage is equal to the sum of V_{FB}, Q_d/C_i, and $2\Psi_B$. As mentioned above, V_{FB} is always the same for a polysilicon gate MOS structure. The signs for the other two terms depend on the substrate type. For an n-channel device, p-type substrate, the depletion-layer charge is due to ionized acceptors and Q_d is then negative. For a p-channel device, n-type substrate, the depletion layer charge is due to ionized donors and Q_d is then positive. Hence the sign of the term Q_d/C_i is positive for n-channel devices and negative for p-channel devices. (Note that there exists a double negative sign in the voltage term corresponding to Q_d/C_i.) The sign of the last term, $2\Psi_B$, also depends on the substrate type. For an n-channel device, p-type substrate, this term is positive, while for a p-channel device, n-type substrate, this term is negative. The threshold voltage expression for an n-channel device is then

$$V_T = V_{\text{FB}} + \frac{|Q_d|}{C_i} + 2\Psi_B \tag{7.2.4}$$

and for a p-channel device is

$$V_T = V_{\text{FB}} - \frac{|Q_d|}{C_i} - 2\Psi_B \tag{7.2.5}$$

Notice that the threshold voltage for a p-channel MOS device is always negative, since the signs of all the terms in Eq. 7.2.5 are negative. For an n-channel MOS device, V_T can be either negative or positive. When the device has a negative threshold voltage, a conducting channel exists in equilibrium and a negative bias must be applied to turn the device off (i.e., remove the conducting channel). MOSFETs that operate under this condition are called *depletion-mode transistors*. The opposite case, that of a positive threshold voltage, is what is expected for an n-channel device. Under this condition no conducting channel exists until a positive gate bias is applied to the structure. Upon the application of a positive gate bias a conducting channel is induced and the device is turned on. These devices are called *enhancement-mode transistors*.

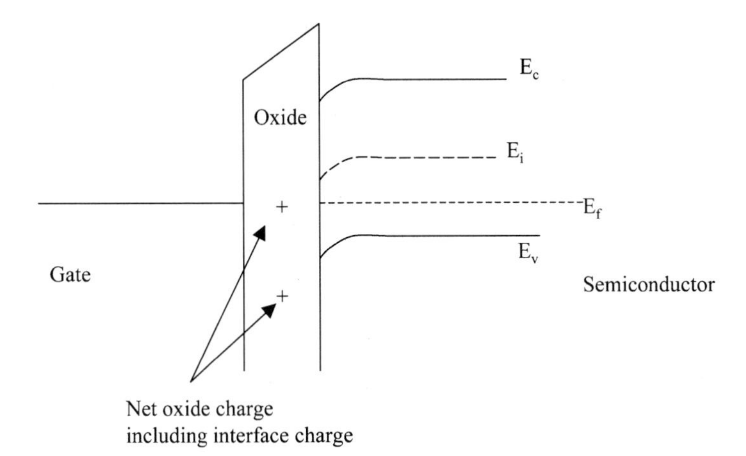

FIGURE 7.2.3 Energy band diagram for a nonideal MOS structure showing the effects of a nonzero ϕ_{ms} and gate oxide and interface charge. Note that the energy bands are bent even in equilibrium with no external bias applied.

There are several different techniques used to calculate the current–voltage characteristic for a MOSFET. Perhaps the simplest approach is to use a charge control model. This approach is discussed in detail in Brennan (1999, Chap. 14). Using the charge control model, an approximate but useful expression for the drain current I_D below saturation is given as

$$I_D = \frac{\mu_n W C_i}{L}\left(V_G - V_T - \frac{V_D}{2}\right)V_D \qquad 7.2.6$$

Equation 7.2.6 is an approximate relationship for the drain current since it neglects the variation of the depletion charge Q_d with the applied drain voltage. The depletion charge enters Eq. 7.2.6 through the threshold voltage and in general varies with position along the length of the channel.

A more accurate expression for the drain current below saturation can be found by including the variation of the depletion charge with drain bias. With the application of a drain bias, there exists a voltage drop from the source to each point along the channel, which we will call $V(y)$. To achieve strong inversion, the potential across the oxide must be equal to $2\Psi_B + V(y)$. The maximum depletion region width is then

$$W = \sqrt{\frac{2\varepsilon}{qN_a}[V(y) + 2\Psi_B]} \qquad 7.2.7$$

The depletion-layer charge per unit area Q_d is then

$$Q_d = -qN_aW = -\sqrt{2qN_a\varepsilon(V(y) + 2\Psi_B)} \qquad 7.2.8$$

The total charge per unit area within the silicon Q_s can be found as

$$Q_s = -C_i[V_G - V_{FB} - 2\Psi_B - V(y)] \tag{7.2.9}$$

The inversion charge per unit area is thus given by the difference between the total charge in the semiconductor Q_s and the depletion charge Q_d as

$$Q_i = Q_s - Q_d = -C_i[V_G - V_{FB} - 2\Psi_B - V(y)] + \sqrt{2qN_a\varepsilon[V(y) + 2\Psi_B]} \tag{7.2.10}$$

Using the results in Box 7.2.2, the drain current can be determined, including the spatial dependence of the voltage in the channel, $V(y)$, with Eq. 7.2.10 to yield

$$I_D = \frac{Z\mu_n C_i}{L}\left\{(V_G - V_{FB} - 2\Psi_B)V_D - \frac{V_D^2}{2} - \frac{2\sqrt{2qN_a\varepsilon}}{3C_i}[(V_D + 2\Psi_B)^{3/2} - (2\Psi_B)^{3/2}]\right\} \tag{7.2.11}$$

From Eq. 7.2.11, the drain current within the triode and saturation regimes can be determined. In the triode region of operation of the device, the drain current is small. Therefore, the term involving V_D^2 in Eq. 7.2.11 can be neglected. Expanding the term $(V_D + 2\Psi_B)^{3/2}$ in a Taylor series and keeping only the first-order term, Eq. 7.2.11 simplifies for small V_D to

$$I_D = \frac{\mu_n C_i Z}{L}\left(V_G - V_{FB} - 2\Psi_B - \frac{\sqrt{4\varepsilon q N_a \Psi_B}}{C_i}\right)V_D \tag{7.2.12}$$

The threshold voltage for an n-channel device is given by Eq. 7.2.4. Defining the threshold voltage as the minimum voltage on the gate such that the surface potential reaches $2\Psi_B$ and the depletion charge is that for the bulk material neglecting any channel voltage, it is given as

$$V_T = V_{FB} + 2\Psi_B + \frac{|Q_d|}{C_i} = V_{FB} + 2\Psi_B + \frac{\sqrt{4\varepsilon q N_a \Psi_B}}{C_i} \tag{7.2.13}$$

Using Eq. 7.2.13 in Eq. 7.2.12, the drain current in the triode region can be written as

$$I_D = \frac{\mu_n C_i Z}{L}(V_G - V_T)V_D \tag{7.2.14}$$

The current in the saturation regime can be determined by recognizing that under these conditions, the channel becomes pinched off near the drain. The

BOX 7.2.2: General Expression for the Drain Current in the Gradual Channel Approximation

The fundamental assumption in the derivation of the drain current typically employed is the gradual channel approximation. The gradual channel approximation simply states that the voltage drop along the direction of the channel (y direction in Figure 7.2.2) occurs much slower than that in the direction perpendicular to the channel (x direction in Figure 7.2.2). Clearly, the channel length of the device must be relatively long for the gradual channel approximation to be valid. If the channel length is very short, the expression derived below for the drain current becomes questionable. In the next section we examine the consequence of these short-channel effects. It is assumed further that the diffusion and generation–recombination currents are negligible and the drain current density is given generally as

$$J_n = -q\mu_n n \frac{dV(y)}{dy}$$

There are no sources or sinks of current, so the drain current I_D can be calculated by integrating the current density over the area in the x–z plane as

$$I_D = -Z \int_0^{x_c(y)} J_n dx$$

where Z is the dimension of the device in the z direction, and $x_c(y)$ is the channel depth, which is a function of y, the distance along the channel. The mobility is the surface mobility, since the current flow in a MOSFET occurs within the channel between the semiconductor and oxide. Generally, the surface mobility is different from the bulk mobility due to the presence of defect states at the interface and dislocations. These mechanisms act to reduce the mobility. Assuming a constant value for the surface mobility, the drain current can be written as

$$I_D = Z\mu \frac{dV}{dy} q \int_0^{x_c(y)} n(x, y) dx$$

The inversion layer charge Q_i is simply equal to

$$Q_i = -q \int_0^{x_c(y)} n(x, y) dx$$

(*Continued*)

BOX 7.2.2 (*Continued*)

With the substitution for Q_i above, the drain current becomes

$$I_D = -Z\mu \frac{dV}{dy} Q_i$$

which can be integrated over the channel length as

$$\int_0^L I_D dy = -Z\mu \int_0^{V_D} Q_i dV$$

Current continuity requires that the drain current be constant throughout. Thus the expression for the drain current above becomes

$$I_D = -\frac{Z\mu}{L} \int_0^{V_D} Q_i dV$$

saturation condition can be approximated by assuming that the drain voltage is equal to the difference in the gate and threshold voltages,

$$V_{D,\text{sat}} \approx V_G - V_T \quad 7.2.15$$

The simplest expression and the one most commonly used for the drain current in saturation, is obtained by substituting Eq. 7.2.15 into Eq. 7.2.6. With these substitutions the saturated drain current $I_{D,\text{sat}}$ is given as

$$I_{D,\text{sat}} = \frac{\mu_n C_i Z}{2L} (V_G - V_T)^2 \quad 7.2.16$$

The behavior of the device below threshold is also very important. As discussed in Section 7.1, in a CMOS circuit the power source is always shielded from ground by a MOSFET biased in the OFF state, except perhaps during switching. For this reason, CMOS circuits in principle dissipate very little dc power. However, in order that the CMOS circuit consume little dc power, it is essential that the current in the device while in the OFF state be very low. Recall that when the device is biased into the OFF state, the conducting channel no longer exists between the source and drain. This condition is called the *subthreshold condition*, since the device is operating below the threshold voltage (i.e., $V_G < V_T$). In the triode and saturation regions, the device operates above threshold, and to a good approximation, the total current can be assumed to be due to drift. However, in the subthreshold regime, the device is no longer biased into strong inversion, and the current is due primarily to diffusion. Grotjohn and Hoefflinger (1984) have derived an expression for the subthreshold current as (Brennan, 1999, Sec. 14.6)

$$I_D = \frac{\mu_n C_i Z}{L\gamma}\left(\frac{kT}{q}\right)^2 e^{q\gamma(V_G - V_T)/kT}(1 - e^{-qV_D/kT}) \qquad 7.2.17$$

where γ is defined as

$$\frac{1}{\gamma} = \frac{C_i + C_d + C_{fs}}{C_i} \qquad 7.2.18$$

In the next section we consider the complexites that arise when the channel length is decreased.

7.3 SHORT-CHANNEL EFFECTS

Many complications arise as MOSFET devices are miniaturized. These can be summarized as arising from material and processing problems or from intrinsic device performance issues. As device dimensions shrink, it is ever more difficult to perform the basic device fabrication steps. For example, as the device dimensions get smaller and the circuit gets denser and more complex, problems are encountered in lithography, interconnects, and processing. Different intrinsic device properties are affected by device miniaturization. The class of effects that alter device behavior that arise from device miniaturization are generally referred to as short-channel effects.

Many effects alter MOSFET device performance as the channel length decreases. These effects can be sorted as a function of their physical origin into three different categories. The three categories correspond to three different sources: (1) the electric-field profile becomes two-dimensional, (2) the electric-field strength in the channel becomes very high, and (3) the physical separation between the source and the drain decreases. The most important features that arise in short-channel MOSFETs are:

1. Two-dimensional potential profile
   (a) Threshold voltage reduction; the gate voltage no longer controls the total gate depletion charge but depends on the drain–source voltage
   (b) Mobility reduction by gate-induced surface fields
2. High electric fields present within the channel
   (a) Carrier-velocity saturation
   (b) Impact ionization near the drain
   (c) Gate oxide charging
   (d) Parasitic bipolar effect

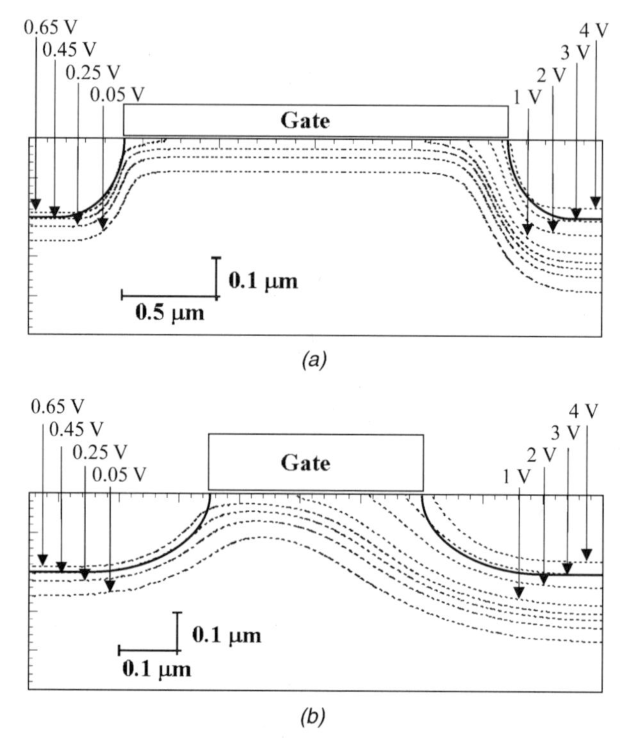

FIGURE 7.3.1 Potential profiles for (*a*) long-channel and (*b*) short-channel MOSFET device structures. Notice that the potential profile is essentially one-dimensional in the long-channel device but is much more two-dimensional in the short-channel structure.

3. Decrease in physical separation between source and drain
   (a) Punch-through
   (b) Channel-length modulation

Let us consider each of these three main categories in turn.

One of the major differences between a short- and long-channel MOSFET is the fact that the potential profile becomes two-dimensional in a short-channel device. Inspection of Figure 7.3.1 shows that the potential profile can be approximated using a one-dimensional model when the gate is long, but as the gate decreases in length, the potential profile becomes two-dimensional. Clearly, approximating the potential profile of a short-channel device using a one-dimensional model would be highly inaccurate. The physical consequences of the two-dimensionality of the potential can be understood as follows.

The two-dimensionality of the potential profile in a short-channel device is due to the fact that the region under the gate is relatively small due to the encroachment of the source and drain regions. The close proximity of the source and drain regions causes a fraction of the bulk charge density under the channel to have field lines terminated at the source and drain rather than the channel. In other words, both the gate and source–drain voltages share control of the bulk charge density below the gate. This is often referred to as the *charge-sharing model*. Consequently, as the channel length decreases, a larger fraction of the bulk charge under the channel has field lines terminated at the source and the drain junctions. The total charge below the gate controlled by the gate voltage in a short-channel device is correspondingly less than that controlled by the gate in a long-channel device. Consequently, a lower gate voltage is required to attain threshold in a short-channel device.

This effect can be considered in a different but equivalent way. Instead of picturing the change in the threshold voltage using the charge-sharing model, the behavior of the device can be examined by considering the potential barrier formed at the surface between the source and drain. When an n-channel device is "off," no channel exists since the gate voltage is below threshold; a potential barrier exists, blocking electron injection from the source contact into the bulk semiconductor. Only a small subthreshold current flows from the source to the drain regions under this condition. In a long-channel device, the potential barrier is uniform across the device, and the source and drain fields are influential only near the ends of the channel. However, as the channel length decreases, the source and drain regions encroach on the channel. The source and drain fields affect the potential underneath the gate. As a result, the potential barrier is lowered. The potential barrier is lowered further by application of a drain bias. This potential barrier lowering is called *drain-induced barrier lowering* (DIBL). The barrier lowering facilitates electron injection under subthreshold conditions. Hence the subthreshold current increases. In addition, the drain induced barrier lowering acts to lower the threshold voltage of the device.

The transverse electric fields, which become of increasing importance as the channel length decreases since the gate oxide thickness is continuously reduced, alter the carrier mobility. It has been found that the action of the gate field, normal to the channel in an MOS transistor, degrades the carrier mobility. The mobility reduction is associated, to some extent, with enhanced surface scattering at the MOS interface. The reduction in the surface mobility can be modeled as

$$\mu = \frac{\mu_0}{1 + \theta(V_{GS} - V_T)} \qquad 7.3.1$$

where μ_0 is the mobility at the threshold voltage and θ is the mobility reduction factor.

The second category of effects that appear in short-channel MOSFETs arise from the high electric field present within the channel. As the channel length decreases, if the voltage is not reduced, the channel electric field increases substantially. The first consequence of the high channel field we consider is carrier-velocity saturation. Both the electron and hole drift velocities saturate at applied electric fields in excess of about 100 kV/cm. In short-channel devices, the electric field near the drain can attain values in excess of about 400 kV/cm. The velocity–field relationship for the carriers takes the form (Selberherr, 1984)

$$v = \frac{\mu_0 F_y}{[1 + (F_y/F_c)^\alpha]^{1/\alpha}} \tag{7.3.2}$$

where F_c is the critical electric field, F_y the channel field, and α a parameter that depends on the carrier type. α has a value close to 2 for electrons and 1 for holes.

The effect of velocity saturation on the current can be illustrated by examining the $\alpha = 1$ case for holes. Although the electron case is far more important, the mathematical manipulations are much more complicated. Therefore, for simplicity we examine the consequences of velocity saturation for the $\alpha = 1$ case. It is first important to recall the simplified long-channel result for a constant mobility model for the drain current (Brennan, 1999, Eq. 14.6.10), Eq. 7.2.6,

$$I_D = \mu \frac{Z}{L} C_i \left(V_G - V_T - \frac{V_D}{2} \right) V_D \tag{7.3.3}$$

where Z is the channel width, L the channel length, V_G the gate voltage, V_T the threshold voltage, C_i the oxide capacitance, and V_D the drain voltage. In the simple model, μ, the mobility, is assumed to be constant. The field-dependent mobility is obtained from Eq. 7.3.2 as

$$\mu = \frac{\mu_0}{1 + F_y/F_c} \tag{7.3.4}$$

but the field in the y direction (in the direction parallel to the channel) and the critical field are

$$F_y = \frac{V_D}{L} \qquad F_c = \frac{v_{sat}}{\mu_0} \tag{7.3.5}$$

where v_{sat} is the saturation drift velocity. The channel charge per unit area, Q_i, can be approximated as follows. In Section 7.2 we found Q_i from the difference between the total charge in the semiconductor, Q_s, and that in the depletion region, Q_d. Neglecting $V(y)$ in the expressions for Q_d and Q_s, Eqs. 7.2.8 and 7.2.9, respectively, and using Eq. 7.2.13 for V_T, Q_i is given as (Brennan, 1999, Eq. 14.6.6) (here Q_i is defined per unit area and no multiplication by the area is required)

$$Q_i = -C_i(V_G - V_T) \tag{7.3.6}$$

However, Eq. 7.3.6 is a poor approximation to the channel charge when V_D is appreciable, as we discussed in Section 7.2. To a first approximation, the effect of the drain bias on the channel charge can be included by considering the average voltage above threshold between the gate and the channel to be $V_G - V_D/2$. Equation 7.3.6 then becomes

$$Q_i = -C_i\left(V_G - V_T - \frac{V_D}{2}\right) \quad 7.3.7$$

The drain current can be found using charge control analysis from the ratio of the channel charge Q_i to the transit time t_{tr}, multiplied by the gate area ZL, as

$$I_D = -\frac{Q_i}{t_{tr}}ZL \quad 7.3.8$$

The transit time t_{tr} is given as

$$t_{tr} = \frac{L^2}{\mu V_D} \quad 7.3.9$$

Substituting into Eq. 7.3.8 the expression for the mobility given by Eq. 7.3.4 using the results in Eq. 7.3.5, the drain current becomes

$$I_D = \frac{\mu_0 Z C_i}{L}\frac{(V_G - V_T - V_D/2)V_D}{1 + \mu_0 V_D/v_{sat}L} \quad 7.3.10$$

Equation 7.3.10 is an approximate relationship for the drain current derived to illustrate the effects of a field-dependent mobility. Notice that the drain current is less than that predicted by the constant mobility model as given by Eq. 7.3.3. Hence in a short-channel device the drain current is typically less within the saturation regime than that predicted by the constant mobility model. The drain current is smaller in the short-channel device than in the long-channel device at comparable bias conditions. This behavior can readily be seen by rewriting Eq. 7.3.10 in terms of Eq. 7.3.3 as

$$I_D = \frac{I_{D0}}{1 + \mu_0 V_D/v_{sat}L} \quad 7.3.11$$

where I_{D0} is the drain current for the constant mobility model (given by Eq. 7.3.3). Notice that the drain current is always reduced below the constant mobility model when velocity saturation effects are included.

When the drain–source voltage is high, the electric field strength within the channel can act to heat the carriers greatly as they move from the source to the drain. Carrier heating can affect the device behavior in several different ways. The three most prominent effects from channel heating are impact ionization near the drain, parasitic bipolar operation, and gate oxide charging. If the drain–source voltage is sufficiently high, impact ionization of the carriers

EXAMPLE 7.3.1: Saturation Current in a Short-Channel MOSFET

Determine an expression for the saturation current in a short-channel MOSFET assuming the simple mobility field model given by Eq. 7.3.4. In this case, the below-saturation drain current is found as in Eq. 7.3.10.

At saturation the drain current is constant. Therefore, the saturated drain voltage can be found by taking the derivative of Eq. 7.3.10 with respect to V_D and setting it equal to zero. Taking the derivative, a quadratic equation in V_D results:

$$V_D^2 + \frac{2v_{\text{sat}}L}{\mu_0}V_D - \frac{2v_{\text{sat}}L}{\mu_0}(V_G - V_T) = 0$$

Recognizing that the drain saturation voltage must be positive (n-channel device), $V_{D,\text{sat}}$ is then given as

$$V_{D,\text{sat}} = \frac{v_{\text{sat}}L}{\mu_0}\left[-1 + \sqrt{1 + \frac{2\mu_0}{v_{\text{sat}}L}(V_G - V_T)}\right]$$

The saturation current $I_{D,\text{sat}}$ can be obtained by substituting the expression above for $V_{D,\text{sat}}$ into Eq. 7.3.10. The resulting expression for $I_{D,\text{sat}}$ is

$$I_{D,\text{sat}} = C_i Z v_{\text{sat}} \frac{\begin{array}{c}[(V_G - V_T)\{\sqrt{1 + (2\mu_0/v_{\text{sat}}L)(V_G - V_T)} - 1\}] \\ -v_{\text{sat}}L/2\mu_0[\{\sqrt{1 + (2\mu_0/v_{\text{sat}}L)(V_G - V_T)} - 1\}^2]\end{array}}{\sqrt{1 + (2\mu_0/v_{\text{sat}}L)(V_G - V_T)}}$$

Notice that the square term in the numerator is proportional to L, the channel length. However, as the channel length decreases, this term decreases in magnitude. Therefore, neglecting the square term in the numerator, the saturation drain current becomes

$$I_{D,\text{sat}} = C_i Z v_{\text{sat}}(V_G - V_T)\frac{\sqrt{1 + (2\mu_0/v_{\text{sat}}L)(V_G - V_T)} - 1}{\sqrt{1 + (2\mu_0/v_{\text{sat}}L)(V_G - V_T)}}$$

near the drain can occur. The impact ionization rate has a strong field dependence (Brennan, 1999, Secs. 10.4 and 10.5; Brennan and Haralson, 2000). If the electric field strength near the drain exceeds the minimum needed to induce electron-initiated impact ionization (the actual value within a device is somewhat geometry dependent but to a great extent can be approximated by the bulk value), carrier multiplication can become appreciable. The generated

EXAMPLE 7.3.2: Velocity-Saturated Current in Short-Channel MOSFETs

Let us determine the drain saturation current in the limit as the channel length goes to zero. We start with the result derived in Example 7.3.1 for $I_{D,\text{sat}}$. Again, as the channel length goes to zero, the square term in the numerator for $I_{D,\text{sat}}$ becomes negligible, giving

$$I_{D,\text{sat}} = C_i Z v_{\text{sat}}(V_G - V_T)\frac{\sqrt{1 + (2\mu_0/v_{\text{sat}}L)(V_G - V_T)} - 1}{\sqrt{1 + (2\mu_0/v_{\text{sat}}L)(V_G - V_T)}}$$

As L approaches zero, the ratio of the two square-root terms approaches ∞/∞. Therefore, one must use L'Hôpital's rule to evaluate the current. The ratio of the derivatives of the numerator and denominator with respect to L is simply equal to 1. Thus the drain saturation current as L approaches zero is

$$I_{D,\text{sat}} = C_i Z v_{\text{sat}}(V_G - V_T)$$

Notice that the saturated drain current in a short-channel device varies linearly with $(V_G - V_T)$. In a long-channel device, the saturated drain current goes as the square of $(V_G - V_T)$, as given by Eq. 7.2.16. Therefore, the saturated drain current in short-channel MOSFETs is lower than predicted using the long-channel approximation.

It is further instructive to determine the saturation voltage $V_{D,\text{sat}}$ for a short-channel MOSFET as L approaches zero. Again using the result in Example 7.3.1, $V_{D,\text{sat}}$ is given as

$$V_{D,\text{sat}} = \frac{v_{\text{sat}}L}{\mu_0}\left[-1 + \sqrt{1 + \frac{2\mu_0}{v_{\text{sat}}L}(V_G - V_T)}\right]$$

As L approaches zero, the last term in the square root dominates over 1, so $V_{D,\text{sat}}$ reduces to

$$V_{D,\text{sat}} = \frac{v_{\text{sat}}L}{\mu_0}\left[\sqrt{\frac{2\mu_0}{v_{\text{sat}}L}(V_G - V_T)}\right]$$

which is simply equal to

$$V_{D,\text{sat}} = \left[\sqrt{\frac{2v_{\text{sat}}L}{\mu_0}(V_G - V_T)}\right]$$

Inspection of the result for $V_{D,\text{sat}}$ shows that it decreases with the channel length L.

EXAMPLE 7.3.3: Small-Signal Transconductance in Short-Channel MOSFETs

We use the simplified mobility model for a short-channel MOSFET given by Eq. 7.3.4 to determine the saturated transconductance. We start with the definition of the transconductance g_m:

$$g_m = \frac{\partial I_{D,\text{sat}}}{\partial V_G}$$

The saturated drain current for the short-channel MOSFET device was determined in Example 7.3.1 as

$$I_{D,\text{sat}} = C_i Z v_{\text{sat}}(V_G - V_T)\frac{\sqrt{1 + (2\mu_0/v_{\text{sat}}L)(V_G - V_T)} - 1}{\sqrt{1 + (2\mu_0/v_{\text{sat}}L)(V_G - V_T)}}$$

Taking the derivative with respect to V_G yields

$$g_m = C_i Z v_{\text{sat}}\frac{\sqrt{1 + 2\mu_0(V_G - V_T)/v_{\text{sat}}L} - 1}{\sqrt{1 + 2\mu_0(V_G - V_T)/v_{\text{sat}}L}}$$

$$+ C_i Z v_{\text{sat}}(V_G - V_T)\frac{\mu_0}{v_{\text{sat}}L}\frac{1}{1 + 2\mu_0(V_G - V_T)/v_{\text{sat}}L}$$

$$- C_i Z v_{\text{sat}}(V_G - V_T)\frac{\mu_0}{v_{\text{sat}}L}\frac{\sqrt{1 + 2\mu_0(V_G - V_T)/v_{\text{sat}}L} - 1}{[1 + 2\mu_0(V_G - V_T)/v_{\text{sat}}L]^{3/2}}$$

As the channel length approaches zero, we can neglect the 1 with respect to the term

$$\sqrt{1 + \frac{2\mu_0(V_G - V_T)}{v_{\text{sat}}L}}$$

in the numerators of the first and third terms. With this approximation the last two terms in the expression for g_m subtract out, leaving

$$g_m = C_i Z v_{\text{sat}}$$

which is valid as L approaches zero.

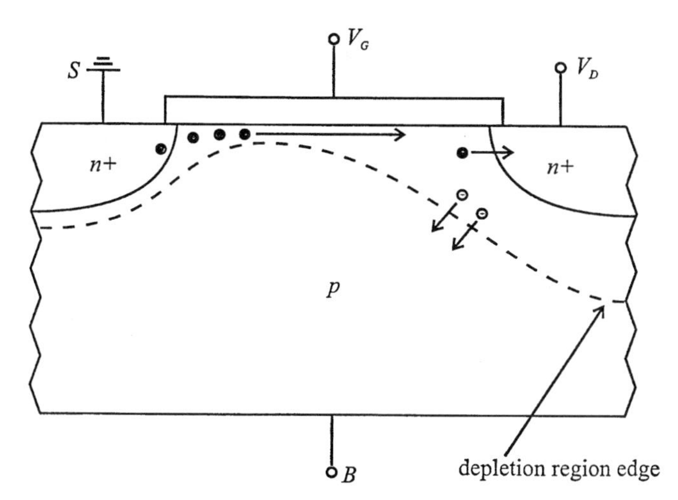

FIGURE 7.3.2 Simple MOSFET device showing the parasitic bipolar effect. The onset of carrier multiplication leads to the injection of secondary holes (open circles) into the substrate of the device, which in turn results in forward biasing of the source–substrate junction. The forward biasing of the source–substrate junction leads in turn to additional electron injection (solid circles) from the source into the channel.

electrons are swept into the drain region, increasing the drain–source current, while the generated holes are swept into the substrate as shown in Figure 7.3.2. Hence carrier multiplication at the drain end of the device results in hole injection into the substrate. The holes injected create what is called the *parasitic bipolar effect*, since the holes act to forward bias the source–substrate junction. Notice that the holes injected within the substrate increase the net positive charge within the p-type substrate. This produces a positive bias on the substrate with respect to the grounded source. Thus the substrate is biased positively with respect to the source, creating a forward-biased p-n junction. The forward bias acts to induce further electron injection from the source into the channel. The net result is a further increase in the drain current. Notice that the parasitic bipolar effect is a consequence of the fact that the substrate and source regions are of different doping type (i.e., for an n-channel MOSFET the source is n-type while the substrate is p-type).

Another important consequence of the high source–drain electric field strength is the concomitant strong carrier heating associated with this field. The very high channel electric field heats the carriers to very high kinetic energy near the drain. The carriers, electrons in an n-channel device, can be heated to sufficient kinetic energy such that they can transfer from the semiconductor channel into the gate oxide. Experimental short-time stress voltage tests show a significant change in the device threshold voltage with time. The threshold shift is attributed to gate oxide charging from hot electrons. The electrons can be injected into the oxide by the action of different mechanisms.

The first mechanism, hot electron emission over the oxide potential barrier, is essentially a classical effect. For an electron to be injected into the oxide in this way, its kinetic energy in the direction perpendicular to the interface must exceed the potential barrier height. For an Si–SiO_2 system, this potential height is larger than 3 eV, and it is likely that not many electrons survive to this energy and thus undergo hot electron emission into the oxide. Alternatively, quantum mechanical tunneling of the electrons into the oxide can occur. The probability that an electron undergoes tunneling into the oxide P_{tun} can be expressed as

$$P_{tun} = A \int_0^\infty T(E, F_{ox}) f_+(E) D(E) dE \tag{7.3.12}$$

where A is a normalization constant, $f_+(E)$ the nonequilibrium occupation probability distribution for electrons with velocity directed toward the interface, $D(E)$ the density-of-states function, and $T(E, F_{ox})$ the tunneling transmission coefficient. Notice that the transmission coefficient is a function of both the electron energy and the field in the direction of the oxide. The integral is taken from zero to infinity by assuming that the zero of energy is the energy minimum within the semiconductor and that extension of the band to infinity produces little error since the occupation probability vanishes at very high energy. Tunneling can proceed either directly through band-to-band tunneling or indirectly with the assistance of interface traps. The current density associated with tunneling processes has been found to obey an empirical relationship given by (Brews, 1990)

$$J \sim F^2 e^{-K/F} \tag{7.3.13}$$

where F is the electric field and K is an empirical value for Si–SiO_2 of 19 to 23 MV/cm. Experimentally, oxide injection is measured by comparing the gate leakage current to the channel current. The ratio of these quantities is the injection probability P.

Perhaps the most important consequence of hot carrier injection into the oxide is the deterioration of the device over time. This is often referred to as *hot-electron aging*. Although some of the electrons injected into the oxide contribute to the gate leakage current, some of the electrons injected become trapped within the oxide. As a result, the electric field beneath the oxide changes as a function of the amount of trapped injected charge. Over time with repeated hot electron stressing, the oxide charge can become appreciable, resulting in a significant change in the threshold voltage of the device. This has severe consequences for long-term device reliability. To summarize, gate oxide charging is a long-term degradation mechanism in a MOSFET.

Typically, the most successful means of combating gate oxide charging is to introduce lightly doped regions near the source and drain n^+ contact regions. Such devices are called lightly doped drain (LDD), MOSFETs, or LDMOS devices. In the usual MOS structures, the abrupt n^+-p junctions formed at the drain and source contacts result in a very high electric field in a relatively narrow region. The addition of the lightly doped n layer increases the depletion

EXAMPLE 7.3.4: Threshold Voltage Difference Between Long and Short Channel Devices

Determine an expression for the change in the threshold voltage of a short n-channel MOSFET from that for a long n-channel device.

Start with the expression for the threshold voltage for a long n-channel device, Eq. 7.2.4:

$$V_T = V_{\mathrm{FB}} + \frac{Q_d}{C_i} + 2\psi_B$$

where the sign of the depletion charge has been assumed to be negative in accordance with an n-channel device. The only difference between the long- and short-channel devices in terms of V_T is the amount of depletion charge controlled by the gate. In the long-channel approximation, it is assumed that all the depletion charge is controlled by the gate, whereas in the short-channel device some of the depletion charge is controlled by the source and drain biases. Therefore, the difference in the threshold voltage of the short-and long-channel devices, ΔV_T, is due to the difference in the amount of charge controlled by the gate. This difference can be determined as follows.

Let Q_{dL} and Q_{dS} be the long- and short-channel ionized acceptor charge per unit area in the depletion layer. The threshold voltage change is then given as

$$\Delta V_T = \frac{1}{C_i}(Q_{dS} - Q_{dL}) = \frac{Q_{dL}}{C_i}\left(\frac{Q_{dS}}{Q_{dL}} - 1\right)$$

Both Q_{dL} and Q_{dS} can be determined using Figure 7.3.3 as

$$Q_{dL} = \frac{qN_a(WZL)}{ZL} = qN_aW$$

$$Q_{dS} = qN_aW\frac{L+L'}{2L}$$

The threshold voltage change is then

$$\Delta V_T = \frac{qN_aW}{C_i}\left(\frac{L+L'}{2L} - 1\right) = -\frac{qN_aW}{C_i}\left(1 - \frac{L+L'}{2L}\right)$$

where L is the long-channel length and L' the short-channel length. Notice that the threshold voltage shift for an n-channel device is negative, implying that the threshold voltage is lower for the short-channel device than for the long-channel device. As discussed earlier, the threshold voltage is reduced with device miniaturization.

(*Continued*)

EXAMPLE 7.3.4 (*Continued*)

The expression above can be reexpressed in terms of the radius of curvature of the doped drain region r_j and the width of the depletion region W_d (see Problem 7.10). The effective channel length L' is then given as

$$L' = L - 2r_j\left(\sqrt{1+\frac{2W_d}{r_j}} - 1\right)$$

Adding L to both sides of the equation above, dividing by $2L$, and rearranging yields

$$1 - \frac{L+L'}{2L} = \frac{r_j}{L}\left(\sqrt{1+\frac{2W_d}{r_j}} - 1\right)$$

Therefore, the change in threshold voltage, ΔV_T, is finally given as

$$\Delta V_T = -\frac{qN_a W}{C_i}\frac{r_j}{L}\left(\sqrt{1+\frac{2W_d}{r_j}} - 1\right)$$

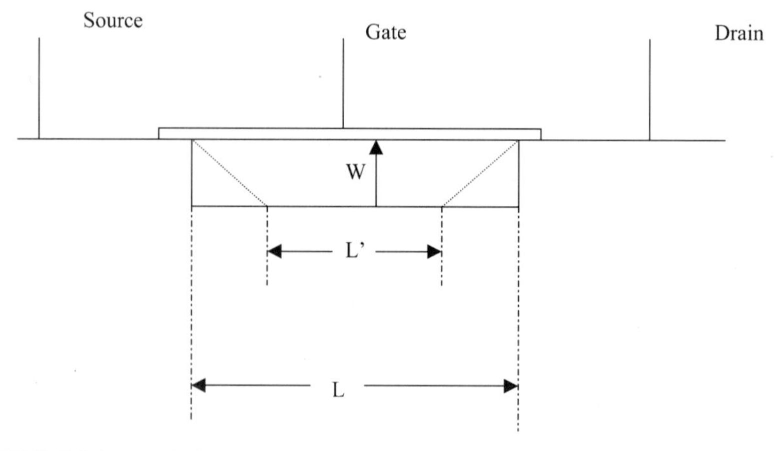

FIGURE 7.3.3 Depletion layer of a MOSFET used to determine the change in the threshold voltage between long- and short-channel devices. For simplicity, the depletion layer is chosen to be a rectangular region for the long-channel device and a trapezoidal region for the short-channel device. The dashed lines indicate the sides of the trapezoid.

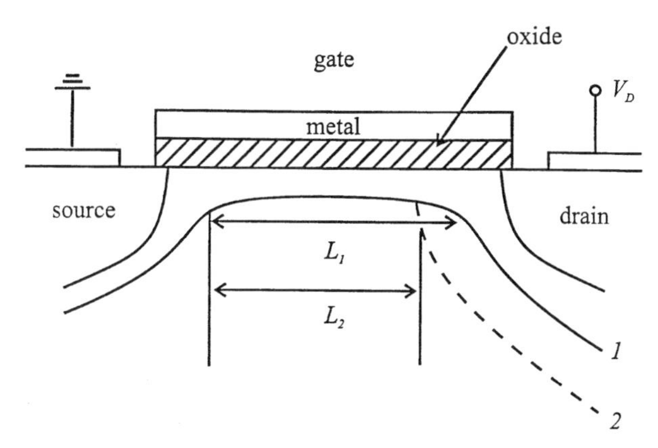

FIGURE 7.3.4 MOSFET device showing channel-length modulation. Channel-length modulation results from shortening of the effective channel length (L_2) from the original channel length (L_1) due to an increase in the drain depletion region.

layer width, thus reducing the magnitude of the electric field. Consequently, the hot carrier effects are somewhat mitigated.

Finally, we consider the last set of effects encountered in short-channel MOS devices. These arise principally from the physical decrease in the drain–source separation. The most important effects in this category are channel-length modulation and punch-through. Channel-length modulation arises from the shortening of the effective channel length of the transistor because of the increase in the drain depletion region as the drain voltage is increased. The effect is sketched in Figure 7.3.4. As can be seen from the figure, the effective channel length is reduced. The resulting channel length is simply equal to the metallurgical channel length minus the source and drain depletion region widths. To a good approximation, the depletion layer width at the source end of the device can be neglected with respect to that at the drain end. The effect of channel-length modulation on the drain current can be modeled as

$$I_D = \frac{I_0 L}{L - \Delta L} \qquad 7.3.14$$

where L is the original channel length, ΔL the difference in the channel length, and I_0 the original drain current. Notice that as ΔL increases, the drain current increases. This results in an output conductance defined as the nonzero slope of the drain current versus drain voltage for the device.

Punch-through arises when the channel length is very small. Under this condition, the source and drain depletion regions can touch, resulting in a large increase in the drain current. The origin of the punch-through condition can be understood as follows. For an n-channel device, both the source and drain regions are doped n^+, while the bulk region of the device is p-type. Therefore, the source and drain regions form n^+-p junctions. The drain must be positively biased for a drain current to flow in the device. Consequently, the drain n^+-p junction is reversed biased, resulting in an appreciable depletion region formed at the junction. Since the bulk p-region is much more lightly doped than the n^+ drain region, most of the depletion region forms in the p-region. The source contact is typically grounded, but a depletion region still exists between the n^+ source region and the p-type bulk. Clearly, the two depletion regions formed at the source and drain contacts will ultimately touch if the source-to-drain separation is made small enough, resulting in punch-through.

7.4 SCALING THEORY

In Section 7.3 we have seen that there are a number of device performance-related problems that arise upon miniaturization. Nevertheless, the industry has continuously developed integrated circuits with ever-decreasing gate lengths extending well into the submicron region. So clearly, the semiconductor industry has found a way to design and fabricate complex integrated circuits that employ MOSFETs that potentially exhibit short-channel effects. The question arises, then: How has the industry successfully developed integrated circuits that employ submicron devices? The answer is, through scaling. Scaling works as follows. One starts with a device that works at a satisfactory level. A miniaturized version is made by scaling the important features that dictate device performance (i.e., channel length, applied voltages, doping concentrations, etc.) such that similar operation is maintained. In this way, short-channel effects can be minimized in small, submicron structures.

There are several approaches to device scaling. However, the overall goal of scaling is to reduce the device dimensions and applied voltages so as to avoid device-limiting performance from short-channel effects. One approach is constant-field scaling, in which the device dimensions and device voltages are scaled by the same factor such that the electric field remains unchanged. The intention of constant field scaling is to ensure that the device will act similarly to the original structure upon miniaturization. However, some physical quantities do not scale in the same way. For example, the doping concentration must be increased by the scaling factor to ensure that Poisson's equation, and the resulting electric field and potential, remain invariant with scaling. The drift current per MOSFET width within the device is given as

$$\frac{I_{\text{drift}}}{Z} = Q_i \mu F \qquad 7.4.1$$

where Q_i is the inversion-layer charge density. The scaling is designed such that the electric field F is invariant with scaling, the mobility remains the same (since the vertical and horizontal fields both remain invariant by design), and the inversion-layer charge density remains invariant. Therefore, the right-hand side of Eq. 7.4.1 is unchanged with scaling, implying that the drift current per width is also invariant with scaling. Since W scales with miniaturization, the drift current per width of the device linearly scales down by the constant field scaling factor. However, the diffusion current does not scale in the same way. The diffusion current per width of the device is given as

$$\frac{I_{\text{diff}}}{Z} = \mu \frac{kT}{q} \frac{dQ_i}{dx} \tag{7.4.2}$$

The diffusion current per width increases by the scaling factor rather than remaining constant. This is due to the fact that the dQ_i/dx is inversely proportional to the channel length. As the device dimensions become very small, the diffusion current does not scale in a manner similar to the drift current. Consequently, at small device dimensions, the diffusion current can become important in device behavior, especially affecting the subthreshold current.

The use of constant field scaling does not solve all the problems associated with miniaturization. As shown above, the diffusion current does not scale in the same way as the drift current. Consequently, the device behavior does not remain invariant with miniaturization following constant field scaling rules. Additionally, constant field scaling requires that the power supply voltages also be scaled continuously. From a circuit design perspective, this is not highly attractive. For these reasons, a different scaling method has been typically utilized, often referred to as *generalized scaling*.

Generalized scaling (Baccarani et al., 1984) was developed to facilitate the design of MOSFETs down to 0.25-μm gate lengths. The key to generalized scaling is that the device dimensions and dopings are adjusted such that the shape of the combined vertical and lateral electric fields remains unchanged. By preserving the electric field profile, the deleterious short-channel effects are mitigated upon miniaturization. However, it should be noted that the magnitude of the electric fields is not invariant with scaling. Therefore, higher electric field strengths may exist in the scaled device from that of the original device, resulting in some trade-offs in reliability and performance, due to high field effects. In generalized scaling the magnitude of the electric field is scaled by a factor λ, while the physical dimensions are scaled by a different factor, $1/\eta$. Recall that for constant field scaling, the electric field remains invariant. Hence under constant field scaling rules the factor λ is simply 1. Since the field is scaled by λ and the dimensions are scaled by $1/\lambda h$, the potential is scaled by the ratio of λ/η. Scaling both the field and the dimensions by different factors results in the potential being scaled by the ratio λ/η. With these scaling factors, the Poisson equation becomes

$$\frac{\partial^2(\lambda V/\eta)}{\partial(x/\eta)^2} + \frac{\partial^2(\lambda V/\eta)}{\partial(y/\eta)^2} = \frac{-q(p - n + N_D - N_a')}{\varepsilon} \tag{7.4.3}$$

If an n-channel device is assumed, and the depletion approximation is used, the Poisson equation simplifies to

$$\frac{\partial^2(\lambda V/\eta)}{\partial(x/\eta)^2} + \frac{\partial^2(\lambda V/\eta)}{\partial(y/\eta)^2} = \frac{qN_a'}{\varepsilon} \tag{7.4.4}$$

where N_a' is the scaled doping concentration. For the Poisson equation to remain invariant, notice that the doping concentration must be scaled by the product of λ and η such that $N_a' = \lambda\eta N_a$, where N_a is the original doping concentration within the device.

The principal advantage of the generalized scaling rule is that the applied voltages do not have to be scaled continuously within the device to preserve performance. As mentioned above, this is highly important from a practical circuit design perspective. However, the fact that the voltage is not scaled down implies that the magnitude of the electric fields increases with miniaturization. Of course, this places different limitations on device performance, most notably in reliability and power dissipation. The higher electric field strengths present in the device can cause gate oxide charging, resulting in threshold voltage variation with time. Thus the device may experience reliability problems and accelerated aging. In addition, the power density increases in the device, resulting in greater thermal management problems from enhanced I–V heating.

Another approach to scaling is to utilize an empirical formula for the minimum channel length L_{min} (Brews, 1990):

$$L_{min} = A[r_j d(W_s + W_D)^2]^{1/3} \quad 7.4.5$$

where A is a fitting parameter, r_j the junction depth, d the oxide thickness, W_s and W_D the depletion widths of the source and drain regions, respectively, where an abrupt, one-dimensional approximation for the junction is assumed. The criterion is that the device channel length must be greater than L_{min} as given by Eq. 7.4.5. As the device dimensions shrink, principally the channel length L, in order that L be greater than L_{min} as given by Eq. 7.4.5 and thus the device avoid deleterious short-channel effects, shallower junctions, thinner oxides, and either lower voltages or heavier doping are required. Therefore, miniaturized devices must have thinner oxides and heavier dopings if short-channel effects are to be minimized. It is interesting to consider how the constraints on these quantities affect continued miniaturization. Equation 7.4.5 should not be considered a rigid rule but more of a guideline to ensure that DIBL is avoided. However, Eq. 7.4.5 is very useful since it clearly illustrates what constraints must be applied to several device parameters to effectively reduce the device dimensions, leaving device performance relatively intact.

The constraints described above apply to conventional MOSFET device structures. These constraints, to some extent, can be used as guidelines to predict the limits of conventional MOSFET device miniaturization. Conventional MOSFET scaling dictates that the oxide thickness must be decreased continuously and that the channel doping concentration be increased continuously. However, there are some obvious limits to both these trends. Clearly, the oxide thickness cannot be scaled to zero, and in fact, the oxide thickness cannot realistically be made much smaller than about 2 nm, as discussed below. Similarly, there are constraints on the magnitude of the doping concentration within the channel region since extremely high doping results in enhanced ionized impurity scattering and a concomitant reduction in carrier mobility. Even more limiting is the fact that ultimately the concentration of dopants is constrained by the solid solubility limit. The solid solubility limit is the maximum thermodynamically stable concentration of dopant atoms possible in the semiconductor. Concentrations of dopants in excess of the solid solubility limit result in the formation of clustering. Upon the onset of clustering, the free carrier concentration is no longer increased. Therefore, only a reduction in mobility occurs, without any increase in carrier concentration, a most undesirable situation.

To mitigate short-channel effects, the gate oxide thickness needs to be scaled in accordance with the channel length. The threshold voltage of a short-channel MOSFET is lower than that for an otherwise identical long-channel device. The difference in the average potential drop between a long- and short-channel device depends on the ratio of the change in the charge stored within the depletion region to the gate oxide capacitance. For a short-channel device the charge stored within the depletion region is less than for a long-channel structure. To maintain a constant threshold voltage between the two devices, the gate oxide capacitance of the short-channel device needs to be increased. The gate oxide capacitance can be approximated using the parallel-plate expression as $\varepsilon A/d$, where ε is the dielectric constant of the oxide, A the gate area, and d the oxide thickness. Notice that to increase the gate capacitance, the easiest quantity to vary is d, the gate oxide thickness. An increase in the gate capacitance can then be obtained by reducing d, the gate oxide thickness. For a MOS device with a 0.1-μm gate length, the corresponding oxide thickness required is about 3 nm, which is only about 10 monolayers.

In conventional MOSFET devices, the minimum oxide thickness is considered to be about 2 nm. Further reduction in the oxide thickness is frustrated by the gate leakage current due to electron tunneling through the oxide. The tunneling current is highly dependent on the potential barrier height and its thickness. For the Si–SiO_2 material system, the potential barrier height of the SiO_2 layer is a fixed quantity. Once the oxide thickness reaches about 2 nm, tunneling through the SiO_2 layer becomes strong, resulting in a high gate leakage current. The power consumption resulting from a high gate leakage current is intolerable in a CMOS circuit and thus must be avoided.

Since further reduction in the oxide thickness is not feasible and the potential barrier for SiO_2 on Si cannot be altered, the only feasible alternative is to find an insulator that forms a higher potential barrier on Si or, alternatively, has a much higher dielectric constant than that of SiO_2. In either case, a thicker insulator layer could then be used. However, identification of an alternative insulator that will form as controllable an interface as SiO_2 is not easy. It is well known that SiO_2, the native oxide on Si, can be grown with exacting control to form almost perfect interfaces. In fact, it can be argued that this is the primary reason that Si has emerged as the most ubiquitous semiconductor material in use today. Therefore, replacement of SiO_2 by another insulator in MOSFET devices may not be practical.

A further problem in MOSFET device miniaturization is parameter fluctuation. As device dimensions continue to shrink, random fluctuations in device parameters become more important in dictating device performance. In other words, the smaller a device becomes, the less robust its performance is to small processing fluctuations. Perhaps the most vivid example is that of random fluctuations of the dopant concentration. In many state-of-the-art MOSFET devices, the dopant atoms within the channel number only in the hundreds. Therefore, small changes in the number and location of the dopant atoms can result in significant performance fluctuations. The most sensitive parameter to dopant fluctuation is the threshold voltage. Hence, within the chip there can be a significant fluctuation in the device threshold voltage, which under most situations is intolerable in circuit design. To date, low-cost mechanisms to control the dopant concentration have not been identified.

The last effect we discuss that influences device performance upon miniaturization are the source and drain series resistances. In long-channel devices it is common to ignore the source and drain region resistances with respect to the channel resistance. However, as a device is scaled to ultrasubmicron dimensions, the source–drain series resistance can become appreciable relative to the channel resistance. As a result, the current drive of the device is degraded.

7.5 PROCESSING LIMITATIONS TO CONTINUED MINIATURIZATION

As discussed above, further miniaturization of conventional MOSFET devices is frustrated in several important ways. Aside from the physical limitations imposed by the gate oxide thickness and doping concentration magnitude, there are limitations imposed by processing and the voltage supply. Let us next consider limitations due to processing, particularly that from lithography.

One of the key drivers in reducing device dimensions has been the continued improvement of lithographic patterning. Virtually all existing CMOS circuitry is patterned using optical lithography. The most advanced production lithography equipment presently available uses KrF excimer laser sources operating at a wavelength of 248 nm. These systems can readily provide resolution near 0.25 μm. Further refinement in resolution can be achieved using an ArF source that operates at 193 nm. Feature size resolution near 0.1 μm has already been demonstrated with the ArF source and new projection techniques, specifically phase shifting. Phase shifting uses topological changes in the optical mask to alter the phase of the illuminating radiation. The resulting interference acts to sharpen the image at the wafer plane. However, the phase-shifting technique has not yet been demonstrated to be generally applicable to arbitrary device geometries that often are encountered in advanced chip design.

Using light with a wavelength shorter than 193 nm for optical lithography presents many difficulties. The most pressing problem is the identification of suitable materials for transparent refractive optical components. Presently, there is great doubt that optical techniques can be employed successfully to define feature sizes much below 0.1 μm.

Currently, there is no clear alternative choice for ultrasubmicron patterning that meets the high throughput demand required for mass production. Competing approaches are x-ray, extreme ultraviolet, electron beam lithography, and scanning tunneling microscopy (STM). X-ray is a prime candidate for high-resolution patterning. Near-contact printing enables fabrication of 30-nm feature sizes, which is sufficient to pattern CMOS gates near the perceived limits of operation. The greatest challenge to x-ray lithography lies in mask fabrication. Current x-ray masks are thin membranes patterned with an x-ray-absorbing material. Precise control of mechanical stress in the membrane must be maintained to preserve accuracy. X-ray proximity printing gives a 1 : 1 replication. Therefore, very stringent control of defects is required to ensure precise pattern definition.

A second approach being considered for future device patterning is extreme ultraviolet lithography (EUV). Reflective optics is used at a 13-nm wavelength with a fourfold reduction scheme. The major technological hurdles to the implementation of this approach are the radiation source and mask fabrication.

Electron-beam (e-beam) lithography provides a third choice for future lithography. E-beam lithography has been the most widely used lithography tool for ultrasubmicron device fabrication within the laboratory. Excellent patterning has been achieved down to about 10 nm, limited primarily by e-beam–resist–substrate interactions using current resist systems. The major limitation of e-beam as a commercial lithography tool is throughput. Rapid, fine patterning of chips with a high level of complexity cannot presently be accomplished using e-beam lithography. Projection systems proposed for e-beam lithography can potentially increase throughput. These systems utilize projection systems with fourfold reduction optics with a mask patterned with electron-absorbing material on an otherwise electron transparent substrate. It is expected that these systems can resolve features to within 50 nm.

Scanning tunneling microscopy (STM) is a variation on e-beam lithography wherein a low-energy beam of electrons is used. The particular advantage of using low-energy electrons for high-resolution imaging is that the detrimental effects of electron scattering that occur in high-energy beams are mitigated. STM provides a convenient means of generating a low-energy beam of electrons that can be focused into a beam with a 10-nm diameter. STM has been demonstrated to yield pattern features down to 20 nm. Again, the primary limitation of STM in commercial processing is throughput. However, it is possible that by assembling a network of independent probes that operate in parallel, a reasonable throughput can be obtained.

It is difficult to project what capability mass production lithography will have in the future. No clear lithographic technology has emerged that will supplant optical lithography for mass-production CMOS. Projections indicate that use of known optical lithography techniques will enable patterning down to 100 nm. To keep pace with Moore's law, such reduction will need to be accomplished by 2003–2005. At around that time, a significant departure from Moore's law may occur. This is because the lithographic tools that would be needed have not even been clearly identified yet but would almost certainly need to be in place soon to maintain the trend represented in Moore's law.

In addition to miniaturizing component devices, fabrication of dense circuits requires shrinking the cross section of interconnecting wires. It is obvious that as device dimensions shrink, the interconnect lines between them would also be reduced in length. Both the wire and insulator thicknesses are scaled down as well. This is because fringe capacitances and crosstalk between wires would increase disproportionally unless the wire thicknesses are scaled accordingly. However, the interconnecting wires must be capable of supporting very high current densities, almost 200 times larger than that allowed in household wiring. Such high current densities can adversely affect the wire through electromigration of the constituent atoms within the wire. As the diameter of the wire is reduced, significant electromigration can result in voids within the wire, which can ultimately lead to breaks. To date, the most ubiquitous metal used for interconnects is aluminum. This is because aluminum is highly ductile and has a very low electrical resistance. However, aluminum is highly sensitive to electromigration, which can subsequently lead to wire degradation.

To circumvent electromigration-induced degradation, several strategies can be adopted. The most obvious approach is to limit the current flow in the wires to a value that is safely below that necessary for electromigration. This approach is not attractive since limiting the current magnitude places stringent limits on the ability to charge stray capacitances and switch other transistors. Alternatively, use of other interconnect metals is possible. One such choice is copper. Copper has the two favorable qualities of lower resistance and greater insensitivity to electromigration. However, mass volume manufacture of circuits using copper interconnects is just beginning. Much progress has been made in overcoming the fabrication difficulties encountered in etching copper, but work remains. Resolution of these limitations requires considerable effort before copper supplants aluminum in high-volume CMOS production.

Increased speed performance of CMOS circuitry depends critically on reducing interconnect delays between devices. There is a significant signal delay associated with the charging capacitance of the connecting wires, which is specified by the resistance–capacitance (RC) product. The RC product for local wires, defined as interconnecting wires placed between devices, is invariant with scaling. Although the RC delay of local wires does not scale, the RC limit for aluminum-based technology is well below the intrinsic delay of 0.1-μm-gate-length CMOS. Therefore, scaling of local wires will not limit the overall speed of a CMOS circuit.

In addition to local wires, there exist global wires. Global wires are defined as wire interconnects with lengths on the order of the chip size. These wires do not scale down with increasing circuit complexity since chip sizes typically remain the same or slightly increase. The RC delay of global wires increases by the square of the scaling factor. Consequently, global wire delay can become a serious limitation to further CMOS circuitry reduction.

Several solutions have been suggested to overcome the speed limitations imposed by global wires. The most immediate approach is simply to reduce the number of global interconnects by redesigning circuits with this in mind. Another approach is to avoid scaling the cross-sectional area of the global wires altogether. In fact, the global wires can be scaled up in cross-sectional area while the local wires are scaled down in cross-sectional area. The ultimate limits of performance of global wires are then reached when they are scaled up such that they approach the transmission line limit.

PROBLEMS

7.1. Consider an n-channel MOSFET made using an Al gate with an SiO_2 oxide. Given: $N_A = 10^{17}\text{cm}^{-3}$, $L = 1.0\ \mu\text{m}$, $Z = 10\ \mu\text{m}$, oxide thickness = 20 nm, $\mu_n = 400\ \text{cm}^2/\text{V}\cdot\text{s}$, $Q_i = 5\times 10^{11} q\ \text{C/cm}^2$, $kT = 0.0259$ eV, $q = 1.6\times 10^{-19}$ C, $\phi_m = 4.1$ V, $\chi_s = 4.15$ V, $E_g = 1.12$ eV, $n_i = 10^{10}\ \text{cm}^{-3}$, free-space dielectric constant = 8.85×10^{-14}F/cm, relative dielectric constant in oxide = 3.9, and relative dielectric constant in the semiconductor = 11.8. Determine (*a*) the threshold voltage V_T, and (*b*) I_D for $V_G = 2$ V and $V_D = 6$ V.

7.2. A MOSFET is made using Al–SiO_2–Si. Given: $N_A = 10^{17}\ \text{cm}^{-3}$, $L = 1\ \mu\text{m}$, $Z = 10\ \mu\text{m}$, oxide thickness = 20.0 nm, electron mobility of $400\text{cm}^2/\text{V}\cdot\text{s}$, $Q_i = 5\times 10^{11} q\ \text{C/cm}^2$, $\phi_m = 4.1$ V, $\chi_s = 4.15$ V, $E_g(\text{Si}) = 1.12$ eV, $kT = 0.0259$ eV, $n_i = 10^{10}\ \text{cm}^{-3}$, $q = 1.6\times 10^{-19}$ C, free-space dielectric constant = 8.85×10^{-14} F/cm, relative dielectric constant in oxide = 3.9, and relative dielectric constant in the semiconductor = 11.8. Determine (*a*) V_T, (*b*) I_D for $V_G = 2$ V and $V_D = 6$ V, and (*c*) I_D for $V_G = 2$ V and $V_D = 1$ V.

7.3. A p-channel silicon MOSFET has an Al gate with $N_D = 10^{15}\ \text{cm}^{-3}$, oxide thickness = 8 nm, $Z/L = 10$, hole mobility = $450\ \text{cm}^2/\text{V}\cdot\text{s}$, and $Q_i = 5\times 10^{11} q\ \text{C/cm}^2$. Determine (*a*) thc turn-on voltage, (*b*) the value of V_{sat} at $V_G = -5$ V, (*c*) the value of V_{sat} at $V_G = -7$ V, and (*d*) I_{sat} for $V_G = -5$ V.

7.4. An NMOS has $Z = 10\ \mu\text{m}$, $L = 2\ \mu\text{m}$, and $C_{\text{ox}} = 10^{-7}\ \text{F/cm}^2$. At $V_D = 0.1$ V, the drain current is given as at $V_G = 1.6$ V, $I_D = 40\ \mu\text{A}$, and at $V_G = 2.6$ V, $I_D = 90\ \mu\text{A}$. Calculate (*a*) the electron mobility, and (*b*) the threshold voltage. Assume saturation and that the device has a positive threshold voltage.

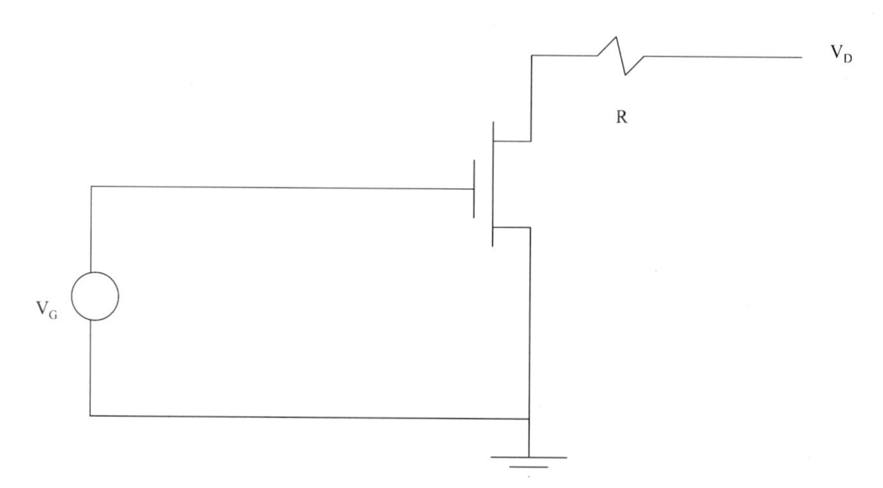

FIGURE P7.5 Figure for Problem 7.5. V_G is the gate voltage, V_D the drain voltage, and R the resistance.

7.5. Consider the circuit diagram shown in Figure P7.5. Determine the value of the voltage V_D if the transistor is biased just at saturation. Assume that the device is an n-channel MOSFET with the following parameters: $V_g = 2$ V, $R = 1$ kΩ, $V_T = 1$ V, and $Z\mu_n C_i/2L = 1.72 \times 10^{-5} A/V^2$.

7.6. Given an n-channel MOSFET with the following parameters: $N_a = 10^{17}$ cm^{-3}, $\mu = 500$ cm^2/V · s, $\phi_{ms} = -0.95$ V, $Q_i = 10^{11}q$ C/cm^2, $Z = 50$ μm, $L = 2$ μm, $d_{ox} = 20$ nm, $n_i = 10^{10}$ cm^{-3}, $E_g = 1.12$ eV, $\varepsilon_{ox} = 3.9$, $\varepsilon_{si} = 11.9$, ε (free space) $= 8.85 \times 10^{-14}$ F/cm, $kT = 0.0259$ eV, $q = 1.6 \times 10^{-19}$ C. Determine the applied gate voltage if the device is in saturation with a current of 5 mA.

7.7. Often in short-channel MOS devices it is useful to employ a high–low doping profile in the channel region. Determine the depletion region width in terms of the surface potential if the doping profile consists of N_s doping from $x = 0$ (Si–SiO$_2$ interface) to $x = x_s$ and N_a doping from $x = x_s$ to $x = W_d$, the edge of the depletion region. The surface potential can be determined using

$$\psi_s = \frac{q}{\varepsilon_{\text{Si}}} \int_0^{W_d} xN(x)dx$$

The advantage of the high–low doping profile is that the depletion charge can be increased while the depletion layer width is reduced.

7.8. Consider an n-channel MOSFET with the following parameters: $Z = 10$ μm, $L = 0.5$ μm, $\mu = 750$ cm^2/V · s, $C_i = 1.5 \times 10^{-7}$ F/cm^2, $v_{sat} = 9 \times 10^6$ cm/s, and $V_T = 1$ V (long-channel result). Compare the saturated drain current and transconductance under both long- and short-channel conditions. For the short-channel case, assume the simplified form for $I_{D,\text{sat}}$ given in Example 7.3.2. Determine the threshold voltage shift if $N_a = 1.0 \times 10^{16}$ cm^{-3} and $L' = 0.3$ μm. Use $V_G = 5$ V. Assume that the depletion layer width is given by its maximum value under strong inversion. Let $n_i = 10^{10}$ cm^{-3} and $\varepsilon = 11.8$.

7.9. Determine how the short-channel saturated drain current, cutoff frequency, dc power, and oxide capacitance per unit area scale if constant field scaling is employed. Note that in constant field scaling the voltages and device dimensions are all scaled by the same factor, κ.

7.10. Using geometrical considerations, show that the effective length, L', used in the expression for the threshold voltage change in Example 7.3.4, can be written as

$$L' = L - 2r_j\left(\sqrt{1 + \frac{2W_d}{r_j}} - 1\right)$$

where L is the long-channel length, r_j the radius of curvature of the n^+ drain region, and W_d the depletion region width surrounding the drain. Assume that the widths of the depletion region underneath the gate and surrounding the drain are the same.

7.11. Derive Eq. 7.2.11. Start with the result derived in Box 7.2.2 and the expression for the inversion layer charge, Q_i, given by Eq. 7.2.10. Using Eq. 7.2.11, derive Eq. 7.2.12 for the triode region of the MOSFET. As stated in the text, use a Taylor series expansion and keep only the first-order term. Neglect the term involving $V_D^2/2$.

CHAPTER 8

Beyond CMOS: Future Approaches to Computing Hardware

From the discussion in Chapter 7 it is clear that current state-of-the-art computing hardware utilizes CMOS technology. To meet an ever-increasing demand for higher computational efficiency and increased dynamic memory storage, the packing density of CMOS circuitry has been increasing exponentially. Consequently, the device dimensions, particularly the channel length, has continuously been scaled down in size. As we discussed in Chapter 7, the digital integrated-circuit industry is fast approaching serious challenges to continued device miniaturization. As a result, both evolutionary and revolutionary approaches are being considered for replacing conventional CMOS technology. By evolutionary changes we mean that the basic approach of CMOS circuits will be retained but with nontrivial modifications to the device structures. The details of some of these approaches are discussed in Section 8.1. Alternatively, it can be argued that an entirely new approach to computing hardware is required, necessitating a revolutionary approach to device engineering. In the remaining sections of this chapter, we examine some of the most attractive revolutionary approaches to future computing hardware that have been identified to date. It is important to note that much of the material in this chapter is highly speculative. Our intention is to provide an introductory discussion to these concepts to alert the reader to future possibilities. Some, all, or perhaps none of these approaches will be utilized in the future.

8.1 ALTERNATIVE MOS DEVICE STRUCTURES: SOI, DUAL-GATE FETS, AND SIGE

As discussed above, there are several constraints to continued scaling of conventional CMOS FETs, mainly the minimum gate oxide thickness needed to control the gate leakage current, the solid solubility limit for the channel doping concentration, and a minimum voltage supply to manage the noise margin of the device. There exist alternative technologies that may succeed conventional CMOS devices in future ultrashort-channel devices. Perhaps the most attractive alternative to conventional CMOS devices is a variant on Si-based MOSFETs. In this section we discuss one such variant of Si-based MOSFET structures: silicon-on-insulator (SOI) FETs.

The features that make SOI CMOS devices highly attractive as a potential substitute for conventional CMOS are that SOI uses the same basic technology as conventional CMOS. Therefore, SOI devices can leverage the great technological strides that have been realized in Si-based integrated circuits. SOI utilizes the same substrate material, oxide, and fabrication techniques as conventional CMOS but offers substantial improvement in device-level performance since it has the added advantage of nearly complete electrical isolation. In addition, SOI devices are less susceptible to radiation-induced failure, such as soft errors, than conventional CMOS devices. A *soft error* is defined as a change in the logic state of a device due to the accumulation of charge resulting from carrier generation following the transit of a high-energy particle or ionizing radiation through the device. The enhanced radiation hardness of SOI devices makes them suitable for space flight applications. However, radiation tolerance is also important in terrestrial applications to avoid α-particle-induced errors. Consequently, SOI devices are possibly better suited to miniaturization than are conventional devices.

In an SOI MOSFET device, the active structure is grown onto an insulating layer, SiO_2, which is grown on top of the substrate material. A sketch of a typical SOI structure is shown in Figure 8.1.1. The active portion of the device, the thin Si layer grown on top of the buried oxide layer, is relatively thin. There are two different manners in which the device is operated. These are partially and fully depleted. Depending on the thickness of the Si layer and its doping concentration, the Si layer can be either completely depleted of free carriers or only partially depleted. Partially depleted SOI devices typically are made with thicker Si active layers. In the fully depleted SOI devices, the active Si layer is sufficiently thin and the doping concentration sufficiently low that it is depleted of free carriers. In either case, partially or fully depleted SOI, the primary benefits of SOI devices over conventional CMOS devices are that the source and drain regions are grown onto an insulating layer, resulting in a substantially lower parasitic capacitance, better device isolation, and excellent immunity to radiation. The lower junction capacitance of SOI devices provides an important switching speed advantage of SOI devices over comparable conventional CMOS devices. Partially depleted SOI devices typically have a floating body, as discussed below. The floating body can couple to the gate potential, resulting in a higher ON/OFF current ratio than that of conventional CMOS devices. As a result, they are more suitable for lower drain voltage operation.

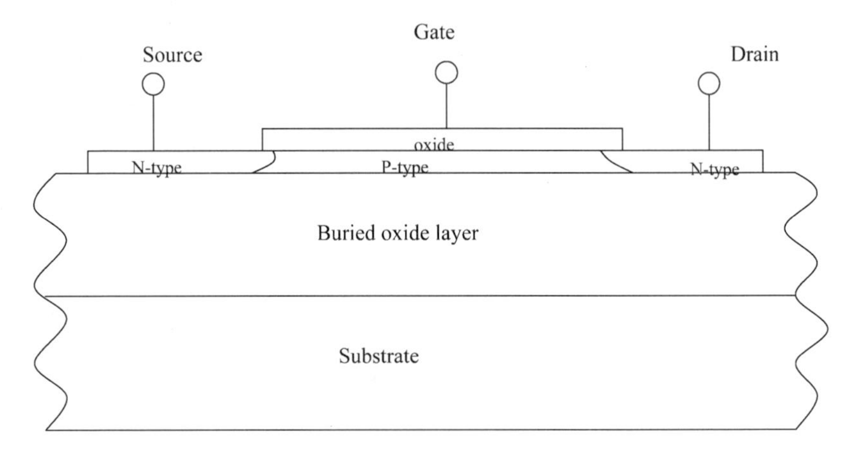

FIGURE 8.1.1 SOI MOSFET structure. Notice that the active region of the device is relatively thin and is grown onto the buried oxide layer, which results in excellent device isolation.

Although SOI devices have been shown to offer superior performance to conventional CMOS, they have some important limitations. In the partially depleted SOI device the substrate is floating electrically. Since the region of the device beneath the conducting channel, usually referred to as the *body*, is not depleted or contacted, it can charge up with charge generated through impact ionization events near the drain of the device. Charging of the floating body can result in premature breakdown as well as an increase in the subthreshold slope at high drain bias. Since the degree to which the floating body of the device is charged is variable, different devices may exhibit different behaviors, thus complicating circuit performance. One particular problem that a floating-body SOI device experiences occurs during switching. The body potential rises with the gate potential, which acts to reduce the threshold voltage, resulting in an increase in the transient current. This effect is called *drain current overshoot*, and its magnitude is a function of the earlier states of the device. Thus the device performance is history dependent: its performance depends upon how recently and how often the device has been switched through conditions under which impact ionization can occur. Generally, the floating-body effects present in partially depleted SOI are undesirable since they lead to nonreproducible behavior.

Alternatively, if the active Si layer is sufficiently thin, an SOI MOSFET device is fully depleted of free carriers. In the fully depleted device there is no floating-body effect. Unfortunately, in a fully depleted SOI device, short-channel effects can become worse unless the silicon and oxide layer thicknesses are carefully designed. In fully depleted SOI devices the silicon and oxide layer thicknesses need to be scaled appropriately to ensure acceptable short-channel behavior. Quantitatively, the thickness of the silicon layer needs

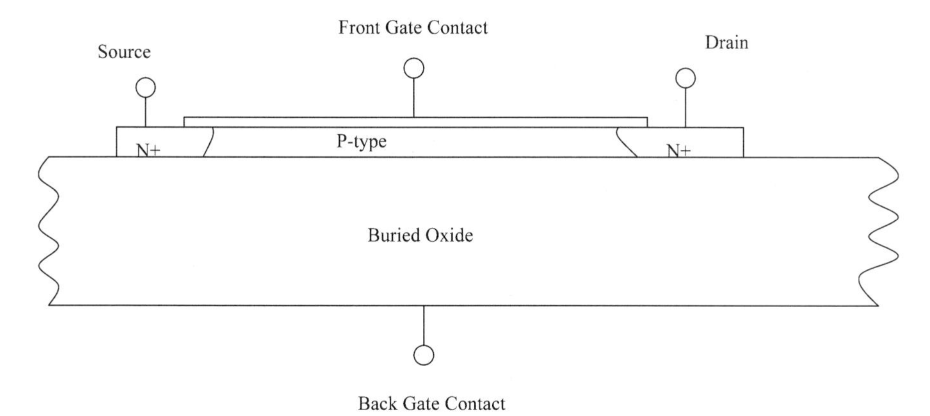

FIGURE 8.1.2 Back-contacted SOI MOSFET device.

to be smaller than the junction depth that would occur in a comparable conventional CMOS device. The buried oxide layer should be thinned to a thickness of about the drain-to-channel depletion width.

In many MOSFET devices, in addition to the usual gate formed on the top of the device, there can be another gate on the bottom of the device. The gate formed on top of the device is usually referred to as the *front gate*, the one on the bottom, the *back gate*. Both gates control the channel potential. A typical back-gated structure is shown in Figure 8.1.2. Dual-gate devices have also been made. A typical dual-gate device is shown in Figure 8.1.3. As can be seen from Figure 8.1.3, both the front and back gates control the channel. The two gates act to shield the source region of the channel from the action of the drain, providing a high degree of isolation of the source and drain. As a result, short-channel effects are reduced in this structure. The dual-gate device has a relative scaling advantage of about a factor of 2 over conventional CMOS devices. Additionally, dual-gate devices in theory do not require channel doping to operate and can thus be highly resistant to random dopant-induced parameter fluctuations. There are multiple types of dual-gate structures. An example of such a structure is sketched diagrammatically in Figure 8.1.3. Alternative dual-gate structures can be formed using a vertical geometry.

Another possible extension of conventional CMOS technology is to replace the Si channel with a SiGe alloy. Si_xGe_{1-x} is not lattice matched to silicon. Therefore, there is an inherent lattice-mismatch in Si–SiGe heterostructures. There are in general two different heterostructures that can be formed using the Si–SiGe system. A type I heterostructure can be formed by growing strained SiGe onto a relaxed Si layer. The type I heterostructure formed within the Si–SiGe system is typically exploited in heterostructure bipolar devices. This is because the band alignment of a type I heterstructure is best suited to improving bipolar device behavior since back injection of the holes from the

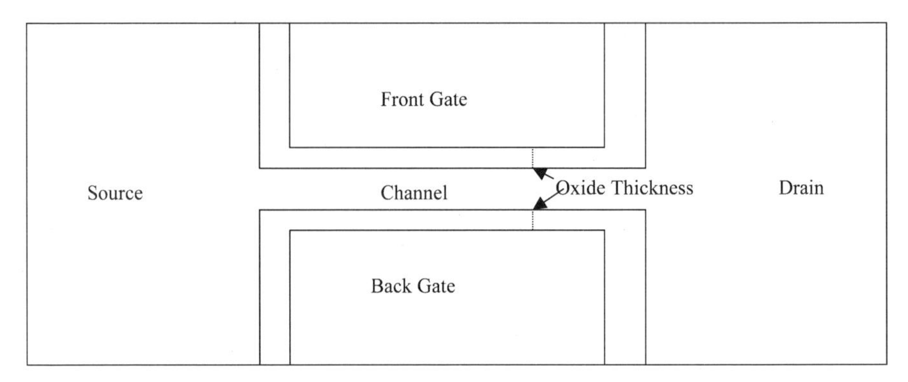

FIGURE 8.1.3 Cross section of a dual-gate MOSFET device. Notice that the channel is controlled by both the front and back gates.

base is reduced while high-energy injection of the electrons from the emitter and concomitant base transit time reduction can be realized. Alternatively, a type II heterostructure can be formed by growing strained Si on a relaxed SiGe layer. The type II heterostructure results in a very high electron mobility within the strained Si. Typically, the type II heterostructure formed in Si–SiGe is used to improve the performance of field-effect transistors.

The electron mobility within the strained Si layer is significantly higher (Ismail et al., 1991) than that in bulk intrinsic silicon by about a factor of 2, than n-type Si-doped 10^{12} cm^{-2} by about an order of magnitude, and by about a factor of 5 over that in Si inversion layers. The tensile strain field within the Si layer modifies the conduction band structure. The increased mobility arises from strain-induced splitting of the degeneracy of the lowest-lying valleys within the first conduction band in Si. In unstrained Si there are six equivalent valleys that lie lowest in energy in the $\langle 100 \rangle$ directions. Two valleys, called the *normal valleys*, have their longitudinal axis normal to the growth plane. The other four valleys, called the *parallel valleys*, lie within the growth plane. When the Si layer is under tensile strain, the normal valleys are shifted down in energy while the parallel valleys are shifted up in energy. The degree to which the degeneracy is broken (i.e., the shifting energy ΔE_s) is a function of the Ge mole fraction present in the SiGe layer on which the Si is grown. Since the degeneracy of the valleys is lifted by the action of the strain field, the lower-lying normal valleys are more readily occupied than the higher-lying parallel valleys. Depending on the Ge concentration, the energy separation between normal and parallel valleys can be made significantly larger than the mean thermal energy, thereby ensuring excellent confinement of the electrons within the normal valleys over a wide range of applied field strengths. Transport parallel to the interface is governed by the smaller transverse effective mass within the normal valleys. The electron mobility is then greater for the strained Si layer than for bulk Si. Therefore, in the strained Si layer, at relatively low applied electric fields, the electrons move within the normal valleys and experience a high level of mobility.

The enhanced electron mobility within strained Si can be used to increase the speed of FET devices. The effect of the increased mobility can best be understood from the transconductance. The transconductance in a FET is defined as

$$g_m = \left.\frac{\partial I_D}{\partial V_g}\right|_{V_D} \qquad 8.1.1$$

Using the result given by Eq. 7.1.3, the transconductance can be found as

$$g_m = \mu_n \frac{W}{L} C_i V_D \qquad 8.1.2$$

Inspection of Eq. 8.1.2 shows that a higher mobility yields a higher transconductance for the device. Therefore, strained Si–SiGe FETs have an intrinsically higher transconductance and thus higher frequency of operation than conventional MOSFETs. Additionally, the enhanced mobility translates into a reduction in the switching resistance, resulting in faster device switching speeds. The increased mobility is a major factor in reducing the source resistance.

8.2 QUANTUM-DOT DEVICES AND CELLULAR AUTOMATA

As discussed above, one of the principal limitations to increased computing capability is the difficulty encountered in interconnecting devices in a high-density array. In the standard computing algorithm, each device must be contacted to enable independent switching of its state. As the density of devices increases, an attendant increase in the interconnect complexity occurs. It is expected that the interconnect complexity ultimately will circumvent further reduction in device size and packing density. It is likely, then, that an ultimate limit in packing density will occur in standard CMOS circuitry.

One approach by which this packing density limit can potentially be overcome is through the exploitation of a different computing paradigm. If computing can be accomplished without the necessity of contacting each device, it is conceivable that much higher device packing densities can be achieved. In this section we discuss two such options. These are quantum-dot cellular automata and spin-polarized systems. In either approach, contacts need only be made to the perimeter devices of the structure; devices lying inside the array need not be separately contacted. First, let us consider the workings of quantum-dot devices and arrays of these devices.

The most common architecture for a quantum-dot-computing machine is an interconnected two-dimensional array (Lent and Tougaw, 1997; Porod, 1997). The fundamental principle behind the operation of a quantum-dot array is that information is contained in the arrangement of charges and not in the flow of

EXAMPLE 8.2.1: Energy States in a Quantum Dot

Let us determine the allowed energy states in a quantum dot. One of the most common means of fabricating quantum dots is to use a split-gate approach. In these structures, the potential due to the space-charge region under and around the gates varies quadratically with position as

$$V(x, y) = c_1 x^2 + c_2 y^2$$

The Schrödinger equation for a system having the potential above is similar to that for a two-dimensional harmonic oscillator. The solution of the one-dimensional quantum mechanical harmonic oscillator is given in Brennan (1999, Sec. 2.6). Additionally, using the result of Example 2.6.1 from Brennan (1999), the constants c_1 and c_2 can be written in terms of the frequencies ω_x and ω_y as

$$c_1 = \tfrac{1}{2} m \omega_x^2 \qquad c_2 = \tfrac{1}{2} m \omega_y^2$$

With these subsitutions, the Schrödinger equation becomes

$$\left(-\frac{\hbar^2}{2m} \nabla^2 + \frac{1}{2} m \omega_x^2 x^2 + \frac{1}{2} m \omega_y^2 y^2 \right) \psi = E \psi$$

The eigenenergies are then given simply as (Brennan, 1999, Example 2.6.1)

$$E = (n_x + \tfrac{1}{2}) \hbar \omega_x + (n_y + \tfrac{1}{2}) \hbar \omega_y$$

where n_x and n_y are quantum numbers that have integer values $0, 1, 2, 3, \ldots$. If the potential can be characterized by the constants c_1 and c_2, the energy can be rewritten in an equivalent form as

$$E = \left(n_x + \frac{1}{2} \right) \hbar \sqrt{\frac{2c_1}{m}} + \left(n_y + \frac{1}{2} \right) \hbar \sqrt{\frac{2c_2}{m}}$$

charges (i.e., current). This means that the devices interact by direct Coulomb coupling and not by currents flowing through wires. In this approach the actual dynamics of the computing operation proceeds by direct Coulomb coupling between neighboring cells without introducing wired interconnections between each cell. As such, a distinct advantage is achieved since complex interconnections can be avoided.

EXAMPLE 8.2.2: Angular Momentum States in a Quantum Dot

As discussed in Example 8.2.1, for a quantum dot formed using a split-gate device, the potential under and around the gates varies quadratically. The Hamiltonian for the dot is determined using creation and annihilation operators, a and $a^\dagger$, respectively, in the usual way as for a harmonic oscillator. As shown by Brennan (1999, Sec. 2.6) for a one-dimensional harmonic oscillator with motion in the x direction, the Hamiltonian is given as

$$H = \hbar\omega_x(a_x^\dagger a_x + \tfrac{1}{2})$$

Generalizing the Hamiltonian for two dimensions yields

$$H = \hbar\omega_x(a_x^\dagger a_x + \tfrac{1}{2}) + \hbar\omega_y(a_y^\dagger a_y + \tfrac{1}{2})$$

The energy states in the dot arise from the quadratic potential underneath the split gate, producing quantization in the plane with concomitant harmonic oscillator–like energy eigenstates and confinement in the vertical direction arising from a heterostructure. Therefore, the energy is quantized in all three directions, with the energy eigenvalues corresponding to motion in the plane given by a harmonic oscillator–like solution while those in the vertical direction are given by bound states in the z direction.

The angular motion of electrons associated with the orbital angular momentum allows us to introduce rotating creation and annihilation operators (Ferry and Goodnick, 1997) as

$$a = \frac{1}{\sqrt{2}}(a_x - ia_y) \qquad a^\dagger = \frac{1}{\sqrt{2}}(a_x^\dagger + ia_y^\dagger)$$

$$b = \frac{1}{\sqrt{2}}(a_x + ia_y) \qquad b^\dagger = \frac{1}{\sqrt{2}}(a_x^\dagger - ia_y^\dagger)$$

In Problem 6.3 the following commutation relations are proven:

$$[a, a^\dagger] = [b, b^\dagger] = 1 \qquad \text{and} \qquad [a, b] = [a, b^\dagger] = 0$$

One can generalize these commutators for other mixtures of a, b, $a^\dagger$, and $b^\dagger$. Using the rotating creation and annihilation operators defined above, we can rewrite H as follows. First we construct $a^\dagger a$ and $b^\dagger b$ as

$$a^\dagger a = \tfrac{1}{2}(a_x^\dagger a_x + a_y^\dagger a_y + ia_y^\dagger a_x - ia_x^\dagger a_y)$$

$$b^\dagger b = \tfrac{1}{2}(a_x^\dagger a_x + a_y^\dagger a_y - ia_y^\dagger a_x + ia_x^\dagger a_y)$$

(*Continued*)

EXAMPLE 8.2.2 (*Continued*)

Adding $a^\dagger a$ and $b^\dagger b$ yields

$$a^\dagger a + b^\dagger b = a_x^\dagger a_x + a_y^\dagger a_y$$

Therefore, if it is assumed that $\omega_x = \omega_y = \omega$, the Hamiltonian H can be written as

$$H = (a^\dagger a + b^\dagger b + 1)\hbar\omega$$

The corresponding energy eigenvalues of H are then

$$(n_a + n_b + 1)\hbar\omega$$

where n_a and n_b are integers that describe the number of quanta in the rotational states a and b defined above. The quantum numbers n_a and n_b can be combined into a new quantum number, n, that is simply the sum of the two, to give for the energy eigenvalues,

$$(n + 1)\hbar\omega$$

The quantum number n has any integer value starting at zero.

A similar expression can be determined for L_z using the rotating creation and annihilation operators. L_z is given in terms of the creation and annihilation operators as (Brennan, 1999, Sec. 2.6)

$$L_z = i\hbar(a_y^\dagger a_x - a_x^\dagger a_y)$$

Subtracting $b^\dagger b$ from $a^\dagger a$ yields

$$a^\dagger a - b^\dagger b = i(a_y^\dagger a_x - a_x^\dagger a_y)$$

Therefore, L_z becomes

$$L_z = (a^\dagger a - b^\dagger b)\hbar$$

But from the expression for H, the quantum number for $a^\dagger a$ is n_a and that for $b^\dagger b$ is n_b. Thus L_z becomes

$$L_z = \hbar(n_a - n_b)$$

(*Continued*)

EXAMPLE 8.2.2 (*Continued*)

The difference between n_a and n_b is, of course, another integer, which we call m. Therefore, the allowed values of L_z are

$$L_z = m\hbar$$

Hence we see that the z component of the angular momentum is quantized in a quantum dot provided that the potential is assumed to be quadratic. The allowed values of m are the subject of Problem 8.4.

Before we discuss the arrangement of quantum dots into an array, it is necessary first to describe the operation and fabrication of quantum dots. Exacting materials growth capabilities have enabled the realization of atomically thin layers of material. As is well known, when the device dimensions are comparable to the electron de Broglie wavelength, spatial quantization effects occur. In a system comprised of an atomically thin layer of a narrow-bandgap semiconductor sandwiched between two layers of a wider-bandgap semiconductor, spatial quantization levels appear in the narrow-gap semiconductor material. Such a system is said to be a *two-dimensional quantum well system*. The electron motion is constrained only in one direction, that parallel to the well dimension (see Chapter 6). The electron motion in the other two directions is not quantized since the electron motion is not constrained in either of those two directions. Etching the resulting layers to form either wires or dots as shown in Figure 8.2.1 can further constrict electron motion. As can be seen in Figure 8.2.1, quantum wires are structures in which the electron motion is quantized along two directions leaving one direction free. A quantum wire is thus a one-dimensional system. Further constriction can be obtained by etching the wires to form dots, as shown in Figure 8.2.1. In a quantum dot the electron is quantized in all three dimensions. Such a system is said to be *zero-dimensional*.

The simplest implementation of the quantum-dot cellular automata scheme is to arrange the dots into a four-site cell with each dot at the corner of a square. Each cell contains two additional mobile electrons that can tunnel between the four different sites in the cell. It is assumed that the spatial separation between each cell is sufficiently large that tunneling of the mobile electrons between different cells is prohibited. Therefore, the electrons must be localized within each cell. The minimum energy configuration of the cell is such that the electrons occupy antipodal sites (diagonally apart sites), as shown in Figure 8.2.2. As can be seen from Figure 8.2.2, if the two electrons are placed into adjacent dots, the Coulomb repulsion is relatively high, leading to an unstable arrangement. The lowest-energy configuration occurs when the two electrons

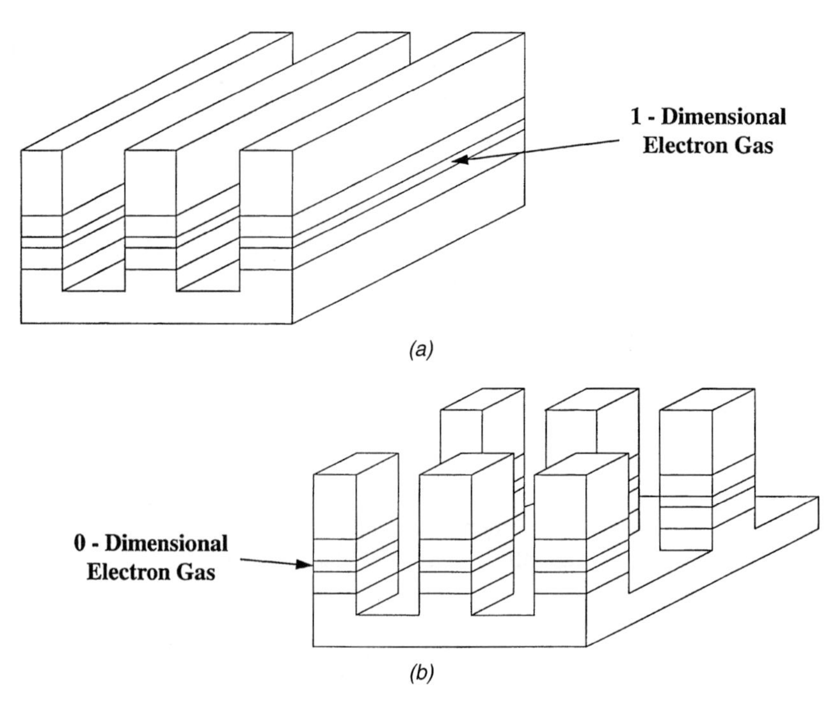

FIGURE 8.2.1 (*a*) Quantum wire and (*b*) quantum dot. The quantum wire is a one-dimensional system and the quantum dot is a zero-dimensional system. (Reprinted from Porod, 1997, with permission from Elsevier Science and W. Porod.)

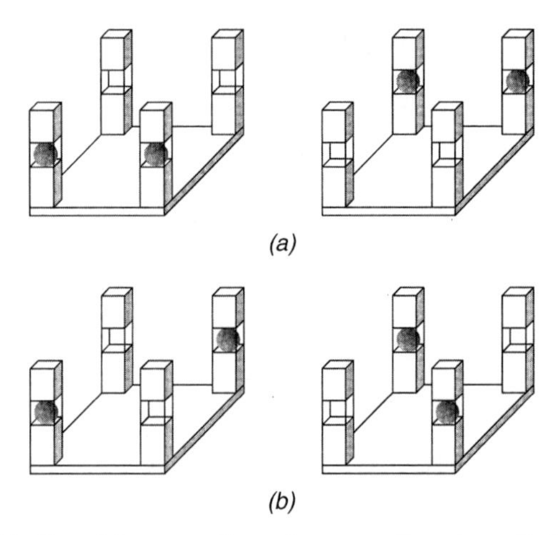

FIGURE 8.2.2 (*a*) Unstable states for a four-quantum-dot cell. Note that the close proximity of the two electrons leads to a high-repulsion, high-energy state and is thus unstable. (*b*) Stable states for a four-quantum-dot cell. Note that there are two stable states, each with electrons at two of the four corners. The larger spatial separation of the two electrons reduces the Coulomb repulsion, thus leading to a lower-energy state.

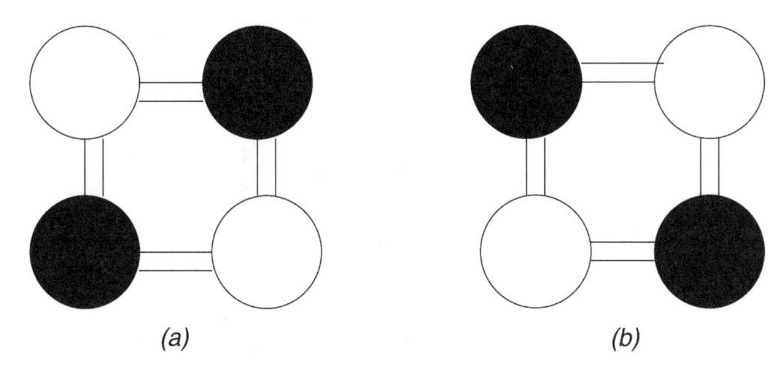

FIGURE 8.2.3 Two different polarization states in a four-site cellular automata: (*a*) $P = +1$; (*b*) $P = -1$. Both sites have the same energy of configuration when independent.

occupy opposite corners of the cell, as shown in Figure 8.2.2*b*. Two such arrangements are possible. These are shown in Figure 8.2.3, where the shaded dots represent dots that hold an additional electron. As shown in Figure 8.2.3, the different arrangements can be referred to as $P = +1$ or $P = -1$. Therefore, two different states of the cells exist which correspond to two different logic states, $+1$ or -1.

Cells that are adjacent to one another tend to align. Cells that are arranged diagonally to one another tend to antialign. From these simple rules, the cell polarization can be used to encode binary information. Let us consider how information propagates from one cell to another. Consider the system shown in Figure 8.2.4*a*. Initially, both cells have electrons at the same opposite corners as shown in the diagram. If cell 1, called the *driver cell*, is reset by an external agent as shown in Figure 8.2.4*b*, a Coulomb repulsion is produced between cells 1 and 2. This repulsion forces cell 2 to reset as shown in Figure 8.2.4*c*. Therefore, the information fed into the driver cell, cell 1, propagates to cell 2. The propagation would then continue from cell 2 to its next nearest-neighbor cell, and so on, along a linear array of cells. Several simple cell arrays are shown in Figure 8.2.5. In each case, the cell farthest to the left in the diagram is the driver cell. The state of the driver cell is fixed and forms the input to the array. The driver cell determines the state of the entire array in the manner shown in Figure 8.2.4.

The simplest array is that shown in Figure 8.2.5*a*, that of a simple cellular wire. Notice that the state of the driver sets the state of each of the following cells within the wire. An input state of "1" is transferred to the output as a "1," as shown in the diagram. It is important to recognize that no charge is transferred between cells, only information. Hence there is no need for interconnect circuitry in this scheme. The propagation of information occurs by the arrangement of the charge configuration in each cell. The driver cell acts to set all the succeeding cells by direct Coulomb interaction. A simple inverter is shown in Figure 8.2.5*d*. In this case, the input state "1" is switched to a "0" at the output. Notice that because the cells are arranged diagonally,

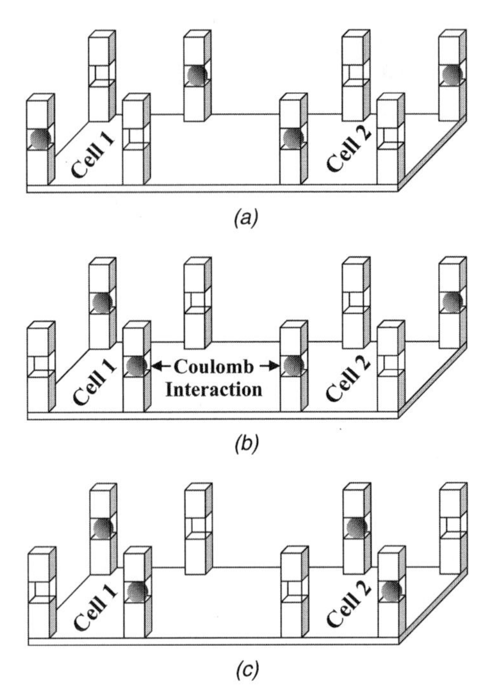

FIGURE 8.2.4 Two-quantum-dot cellular automata. (*a*) Initial state of two cells in the quantum-dot cellular automata machine. (*b*) Cell 1 is set externally into another state, causing a Coulomb interaction between the electrons in the two dots as shown. (*c*) The Coulomb interaction between the two dots causes cell 2 to reset as shown.

as can be seen in Figure 8.2.6, the charges antialign since this leads to the lowest-energy configuration of the cells, in accordance with the second rule stated above.

Conventional AND and OR logic gates can be obtained using the quantum-dot logic functions. Therefore, all the important logical operations can be made using quantum-dot cellular automata (QCA). The question is, though, how can computations be performed using this scheme? The basis for computation in the cellular automata scheme lies with two key concepts: (1) computing with the ground state and (2) edge-driven computation. The basic structure of the quantum-dot computation scheme is outlined in Figure 8.2.7. As can be seen from the figure, the QCA is connected to the external world through input and output cells. It is important to notice that no interconnects need be made to any cells within the QCA array. The only interconnects are made to the input and output cells, thus greatly reducing the interconnect complexity of the system. Further inspection of Figure 8.2.7 reveals that the input to the computation is addressed by setting the edge cells to the QCA array. The solution of the computational logic problem is obtained physically from the collective ground state of the QCA array. The basic workings of the QCA

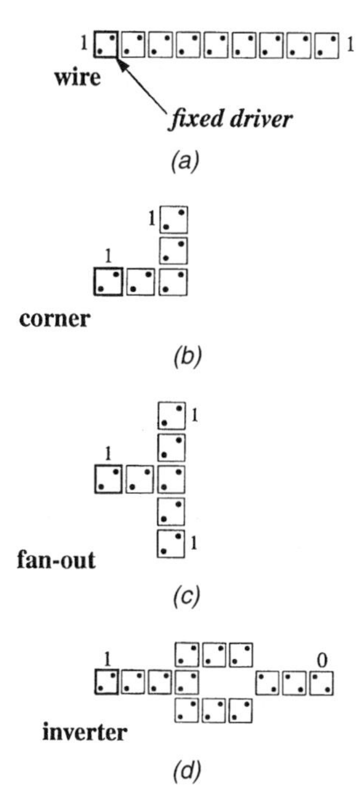

FIGURE 8.2.5 Arrangements of quantum dots showing different functionality: (*a*) wire, (*b*) corner, (*c*) fan-out, and (*d*) inverter. In all situations the leftmost cell provides the input and is called the *fixed driver*. Notice that in the wire, a "1" input results in a "1" at the output. For the corner, a "1" input leads to a "1" at the output. In the fan-out configuration, a "1" at the input leads to "1"s at both outputs. Finally, for the inverter, an input "1" is inverted to a "0" at the output. (Reprinted from Porod, 1997, with permission from Elsevier Science and W. Porod.)

computer can be summarized as follows. First, the polarization state of the input cells is fixed in accordance with the input logic. This is what is meant by edge-driven computation. Second, since the internal cells within the QCA array are not independently contacted, they cannot be supported indefinitely in a high-energy nonequilibrium state. Instead, the array must eventually collapse into a ground state that is uniquely specified by the input cells. In other words, the QCA array relaxes to some stable ground state consistent with the condition of the input cells. It is important to note that many ground states are, in general, available for the array, but for each initial state condition on the input cells, the resulting ground state of the array is unique. Therefore, the array will be found in only one unique ground state for any given initial state of the input cells. This resulting ground state is "read" by the output cells by sensing the polarization of the periphery cells of the QCA array.

The crucial aspect of the QCA computing engine is that the computational result resides in the ground state of the array obtained following its excitation

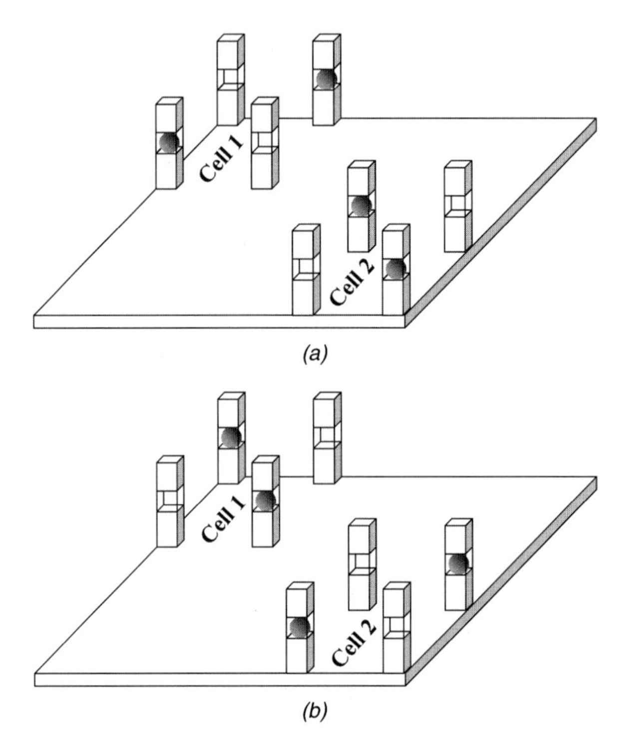

FIGURE 8.2.6 (*a*) Two cells arranged diagonally. Since the spatial separation of the electrons is maximized, the two cells are at their lowest energy. Note that the two cells are antialigned. (*b*) Alternative arrangement of the electrons in the two diagonal cells. Note that again the electrons are farthest apart, leading to the lowest-energy configuration of the system, which is degenerate with that shown in part (*a*). Again the two cells are antialigned.

by the input cells. The particular ground state that the array relaxes into depends uniquely on the input cell states and not on the dynamics by which the system relaxes. The actual mechanism by which the array relaxes plays little role in determining the result. Therefore, no external control of the array is needed, hence no interconnects are required.

It is important to recognize that ground-state computing is a fundamental requirement of a computing system that has an interconnectless architecture. Since the internal devices within the array are not connected to an external power source, there is no mechanism to retain them perpetually in excited states. As a result, each device will ultimately decay into its ground state. The ground-state computing algorithm simply utilizes this concept to construct a computing machine.

An additional key issue in understanding the operation of a QCA array as a computing machine is that the information propagates from the input

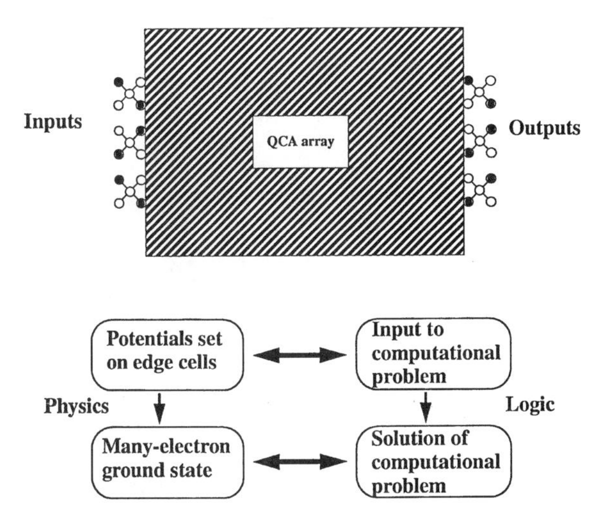

FIGURE 8.2.7 Format of a quantum-dot cellular automata.The only interconnections to the external environment required are to the input and output cells that are connected to the array. (Reprinted from Porod, 1997, with permission from Elsevier Science and W. Porod.)

cells to the output cells not by current flow, as in conventional CMOS circuitry, but by an electrostatic interaction between cells. The electrostatic interaction occurs between electrons in a spontaneously polarized system of quantum dots. Through this interaction, if the polarization condition on the input dot is initially set, that input will propagate through the array flipping dipoles in the following cells, much like a soliton wave. Such a propagation would result in a rapid transfer of information from the input cells into the array.

Two different strategies have been proposed for switching in a QCA array. These are by tunneling, wherein electron transfer occurs between dots, and by a long-range electrostatic interaction, wherein the information transfer occurs from dipole interactions. Tunnel transfer of electron charge between dots can occur if the dots are made sufficiently "leaky" such that a nonzero probability exists that an electron can tunnel from one dot to another. As discussed by Porod (1997), the electrons within the QCA array are allowed to tunnel between dots within a single cell but not between cells. Since the tunneling probability is a function of dot separation, this is accomplished by placing the dots in relatively close proximity within each cell and spacing the cells somewhat farther apart. Therefore, the long-range Coulomb interaction affects the polarization state between cells. To minimize the energy of the cell, if necessary, electrons will tunnel from one dot into another to produce the cell with the lowest-energy polarization state, as induced by neighboring cells. Notice that within this model, electron tunneling occurs between dots. A second strategy that avoids tunnel transfer was first proposed by Bakshi et al.

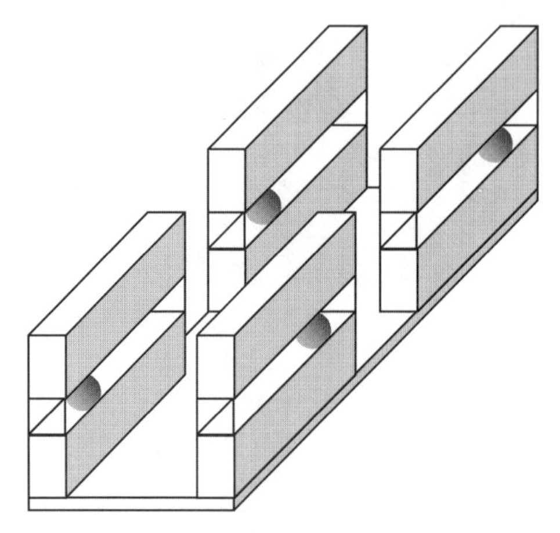

FIGURE 8.2.8 Single cell of a quantum dash structure. Notice that in the two dashes in the foreground, the charges are repelled, forcing them spatially apart. As a result, the charge in the lower-right cell then repels the charge in its nearest-neighbor dash lying above it. The resulting configuration of the charges is then given as shown.

(1991). In their approach they employ *quantum dashes*, quantum structures in which each dot is elongated along one dimension, as shown in Figure 8.2.8. The Coulomb force between electrons within adjacent dashes is greater than the confining force along the dash direction of elongation. Therefore, the interdash Coulomb force can overcome the confining force shaping the electron configuration within the dash. The resulting displacement of electron charge in each dash forms an electric dipole that is replicated between dashes along the array. The array then collapses spontaneously into an antiferroelectric polarization arrangement.

The most severe drawback to ground-state computing systems is the possibility that the system will get "stuck" in metastable states, ones different from the ground state that do not decay quickly. The presence of metastable states can severely hamper the computing performance of the array since it may not reliably collapse into the ground state in a timely manner. As a result, the output cells may sense an improper metastable state response to the input condition, thus rendering an incorrect computational result. Of course, with time, the array would ultimately relax into the proper ground state, but that might require an inordinate duration of time, not particularly attractive for high-speed computing applications.

In addition to the metastable state problem, there exist other limitations to quantum-dot computing schemes. One of the key problems encountered in the coupled quantum-dot approach is that one electron must be sequentially transferred from one dot to another. Trapping of the transferred electron in a dot other than the desired final dot would result in a serious failure. The devices must be switched slowly to avoid trapping, which necessarily slows the transfer of information within the computer. There are indications that the net transfer rate between devices is orders of magnitude slower than in present-day CMOS devices. A second key limitation to quantum-dot computing architectures is reproducibility. To affect charge transfer by tunneling between different dots under similar conditions, the dots must be made uniformly. In most embodiments of the quantum-dot architecture, the barrier widths are defined by lithography, which cannot generally provide monolayer resolution necessary to precisely control the magnitude of the potential. Therefore, it is questionable that such devices can be utilized in a high-density integrated circuit where billions of devices must be fabricated with nominally identical characteristics. Use of quantum dashes instead of dots with the concomitant avoidance of tunnel transfer eliminates this second limitation. However, the quantum dash polarization array is highly sensitive to temperature. It must be operated at liquid He temperatures of about 4.2 K, to avoid thermal smearing of the polarization state. In addition, the quantum dash array is still sensitive to the device geometry (i.e., the spacing and shape of each dash greatly determines the coupling).

Due to the above-mentioned limitations, alternative strategies need be considered. Some researchers have proposed using spin polarization in place of charge polarization. Spin polarization arrays operate in some ways like the quantum dash array of Bakshi et al. (1991). In spin polarization devices nearest-neighbor sites have opposite spin states. If a single cell can be set with a specified spin state, it is possible to cause a spontaneous transition of an array of cells through antiferromagnetic ordering. The spin polarization array works as follows (Bandyopadhyay et al., 1994).

A weak external magnetic field is applied to the system. This ensures that the electron spin in each cell will collapse into one of two possible states, either aligned with the field or antialigned. It is important to recognize that the electron can be in only one of two possible states. Therefore, the spin system naturally provides a bistable system for performing binary logic. Switching occurs through changing the spin state of the electron. For example, the device can be switched by changing the spin state from up to down with respect to the external magnetic field direction, or vice versa. As in the quantum dot or dash arrays, interconnections between devices are avoided. Information is transferred from one cell to another by inducing spin flips between cells. A cell in this case is defined as a single element containing one electron with spin either up or down. The premise of the approach lies in the fact that it is energetically favorable for the two nearest neighbors to have opposite spin states. Thus when the spin state of one electron changes, this affects the energy of its nearest-neighbor electrons. If the change is such as to produce a higher-energy configuration (i.e., both electrons have the same spin state) the system will try to relax by a spin flip of the neighbor. The neighbor electron can change its spin state by emitting a phonon or magnon. The neighbor then induces a spin flip in its nearest neighbor and so on throughout the array. Hence the spin state propagates through the array in a dominolike fashion until

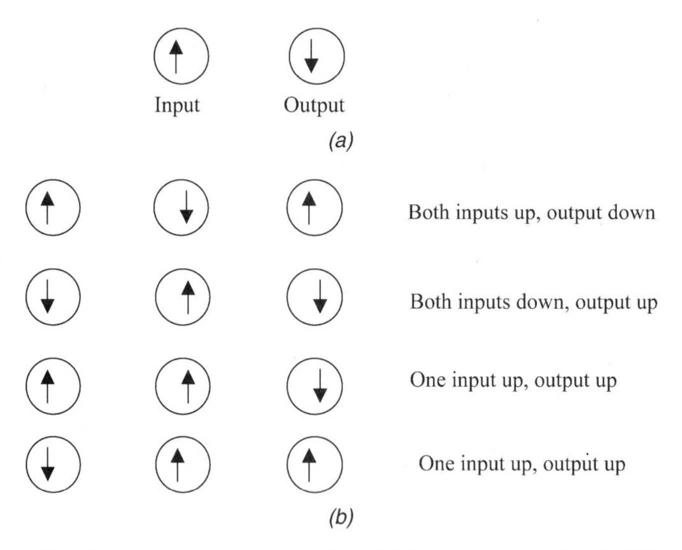

FIGURE 8.2.9 Simple spin systems used for (*a*) inversion and (*b*) NAND logic functions. In part (*b*) the first and third cells in each row correspond to the inputs, while the center cell is the output. Note that when one input is up while the other is down for the NAND gate, the output is always up. This is because the overall dc field is assumed to point such as to favor the up state.

the full system has reached a new ground-state spin configuration. As in the case of the quantum dot arrays, the collective spin state of the array represents the response to the system from the input. Information is thus transmitted across the entire array by perturbations that behave as *spin waves*.

In spin polarization devices, it is the spin state of the outermost electron within each cell that interacts with its neighbor. As is the case for quantum wells and atoms, only electrons in the highest-energy states interact, since the overlap of the wavefunctions is appreciable only between these electrons. Therefore, the system is designed using cells such that only a single-electron exists within the highest energy level, all other lower energy levels being filled. As in the quantum dot arrays, the spin polarization devices can be configured to produce all of the logic functions essential for computing.

Figure 8.2.9*a* and *b* illustrate the simplest implementation of spin systems that perform the inversion and NAND logic operations. An inverter gate can be made using only two cells. Recall that the minimum energy of the two-cell system occurs when the spin states are opposite. Therefore, if the first cell has spin up, the second cell will have spin down, and vice versa. Calling the first cell the input cell and the second cell the output cell, the system clearly produces an inversion operation. A simple NAND gate using three spin cells is shown in Figure 8.2.9*b*. In this case, the inputs are the first and third cells shown in the diagram. The output is the center cell. If the two inputs are both up, the output is down, as shown. Similarly, if the two inputs are down, the output is up. Now if the inputs are mixed, one up and the other down, it at first appears that the output can be either up or down, since the result is energetically degenerate. However, the overall dc magnetic field removes this degeneracy. If it is further assumed that the dc field is pointed such that the up state is favored, the output is up for either arrangement, as shown in Figure 8.2.9. Therefore, the three-cell arrangement produces a NAND gate.

Although the approaches to computing discussed above offer many advantages over conventional CMOS, they have some important limitations. The quantum-dot arrays (QCAs) typically require electron tunneling between dots within each cell, although no tunneling occurs between cells. As a result, these structures may have problems with reproducibility. The tunneling requirement makes these arrays unsuitable for room-temperature operation. It is expected that QCA arrays will have to operate at very low temperatures to perform properly. Additionally, the QCA arrays may be relatively slow. The switching speed of these arrays is limited by the lack of tolerance to electron trapping, which could result in very poor response times.

The spin polarization arrays seemingly offer some important advantages over QCAs. There are a number of advantages to the use of spin polarization arrays:

1. The devices can operate at room temperature.
2. The devices have a fabrication tolerance. The size and shape of each cell is not critical provided that it can host a single electron.
3. The extrinsic switching speed is much greater than that of a QCA. The improved switching speed stems from the fact that no charge transfer occurs in the array (i.e., there is no tunneling from one element to another).
4. The logic variable is spin, which is a robust quantity. The system can then operate with a high noise margin and reliability.
5. As for QCAs, a very high density and compaction can be achieved without interconnects.
6. The array utilizes ground-state computing, as do QCAs. Therefore, there is no need to refresh the array as is needed in conventional memory cells.

Finally, it is important to note that the cellular automata schemes all require ground-state computing. This is necessary since there are no interconnects to each individual cell within the array. Therefore, no external power can be delivered to a cell to maintain it indefinitely within an excited state. Thus all the cells need to reside within a stable state determined only by the input conditions on the periphery cells. This is the basis for ground-state computing. The most important limitation to ground-state computing is, however, the fact that the system can get "stuck" in an intermediate, meta-stable state different from the proper ground state. As such, the output cells will read an improper result, leading to a computational failure.

8.3 MOLECULAR COMPUTING

Molecules have been suggested as potential candidates for nanometer-sized hardware. The major qualities of molecules that make their potential use in future applications attractive are that they provide for simplified fabrication, low cost, and relatively unlimited availability, and that devices made from molecules are very much smaller and more compact than existing CMOS structures. For example, a CMOS OR gate requires several orders of magnitude more surface area than that of a comparable molecular logic gate. Molecular electronics thus offers the potential for extremely compact computing capability (i.e., a microprocessor on a pinhead). Coupled with their expected lower power consumption, molecular devices offer a revolutionary trend in electronics: highly miniaturized, low-power devices that are relatively inexpensive to fabricate and mass produce. The fact that molecular materials can readily be engineered and occur in many different forms with a variety of properties enables potential realization of a vast number of possible structures.

The principal advantages of molecular materials include:

1. *Size*. The dimensions of individual molecules are in the nanometer range. For example, they are about 4 nm on average. These device dimensions are about two orders of magnitude smaller than that which can be obtained using silicon CMOS technology. Quantum dots and spin systems are also typically larger than these dimensions.
2. *Three-dimensional structures*. Molecular materials are inherently three-dimensional. In contrast, in silicon-based technology, much fabrication effort and cost is required to produce three-dimensional geometries.
3. *High packing density*. The combined features of small size and three-dimensionality make very high packing densities possible. Some estimates are that increased packing density of six to nine orders of magnitude can be achieved for molecular materials over CMOS.
4. *Bistability and nonlinearity*. Bistability and nonlinearity can be utilized to perform switching functions. Both of these properties are commonly available in molecules.
5. *Anisotropy*. It is important to recognize that the electronic and optical properties of a molecule are inherent in the molecular structure instead of being fabricated by the processing technologies as in CMOS.
6. *Upward construction*. Organic synthesis enables growth of microstructures from the small upward. In standard CMOS, device and circuit functionality is sculpted from a relatively large piece of material.

7. *Self-organization.* Molecules present some new, important functions not found in other materials: self-organization, self-synthesis, and redundancy factors well known in organic and biological molecules that could potentially be applied to molecular electronic devices.
8. *Low power dissipation.* Biological molecules require much lower amounts of energy to switch than do CMOS circuits. Estimates are that the total power requirements for molecular switches will be about five orders of magnitude lower than for CMOS switches.
9. *Molecular engineering.* It is possible that molecules can be tailored or engineered to perform specific tasks or have specific properties. In other words, molecules can be grown selectively and made to possess, inherently, desired qualities to perform a task.

There are generally two different approaches to molecular computing. The first approach is to exploit molecular systems to replicate a functionality similar to that given by existing CMOS digital circuitry. In this approach, molecular systems would be utilized to create smaller, less expensive, higher-density memory and processing chips for digital computing, following design functionality similar to that used in contemporary CMOS circuits. In other words, a conventional computing architecture would be implemented using molecules in place of CMOS devices. This has the obvious advantage of a vast reduction of device sizes, resulting in high-density circuitry. We refer to this approach as molecular electronics or moletronics.

The second approach is perhaps more revolutionary. It envisions using molecular systems to duplicate the powerful information-processing capabilities of biological systems. Examples of these are pattern and object recognition, self-organization and replication, and parallelism. These are functions that are not easily realized using today's computers and circuitry. We refer to this approach as tactile computing.

The primary difference between conventional and tactile-based molecular computing platforms can be summarized as follows. Conventional electronic digital computers can be thought of as structurally programmable machines. This means that the program controls the behavior of the machine. A compiler translates the input code into machine language that is expressed in terms of the states of simple switching devices and their connections. The machine computes symbolically, and the result depends to some extent on the human input.

Tactile computing machines are quite different. These machines are not structurally programmable. In biomolecular computing, pattern processing is physical and dynamic as opposed to the symbolic and passive processing in a conventional machine. Programming depends on evolution by variation and selection. Such a molecular computer, called a *tactile processor*, can be thought of as a computer driven by enzymes; the inputs are converted into molecular shapes that the enzymes can recognize. An enzyme thus performs sophisticated pattern recognition since it scans the molecular objects within its environment, interacting only with molecules that have a complementary shape. Recognition is thus a tactile procedure.

It is important to recognize that biomolecular or tactile computing proceeds along an entirely different line than conventional computing. Essentially, tactile computing provides a high degree of pattern recognition that can work relatively quickly; an enzyme encounters and interacts with a substrate molecule providing molecular switching action. The "program" is in the molecule itself; the computation occurs by the recognition of one molecule by an enzyme and their subsequent chemical reaction. Conventional computing requires massive amounts of code to instruct a machine to recognize a single molecule. Therefore, tactile computing provides a different, yet complementary approach to computing than that used by conventional machines.

It is interesting to pause in our discussion and recognize that tactile computing has been going on around us for quite some time, mainly in biochemistry. In a sense, DNA molecules act as biomolecular memory cells. All the information necessary to create a living creature is stored within DNA molecules, and as such they serve clearly as a biomolecular memory. Interestingly, these biomolecular memories persist over very long times, have error correction schemes, and are self-replicating. The information stored within DNA molecules can be accessed and processed, a function repeatedly performed in cell function and division. Additionally, tactile computing, as in the recognition of a foreign molecule, occurs in the immune system. Antibodies identify and remove foreign molecules from living systems by tactilely identifying that an antigen is a foreign substance that may be potentially dangerous.

Although tactile computing using biomolecular materials offers exciting future possibilities, it is perhaps much harder to see how it can be implemented. For this reason, we omit further discussion of molecular processing or tactile computing and concentrate on moletronics. In the remainder of this section we examine some of the salient features of moletronics.

As mentioned above, *moletronics* is defined here as the exploitation of molecular systems to effectively replicate conventional computing approaches. It may be best first to define what we mean by *conventional computing*. By this we mean the broad picture of a machine that requires a CPU (central processing unit), storage (memory devices), and bandwidth (communications links). Further, we imply that these functions are all programmable in the sense that the machine follows a series of steps to complete a task. To accomplish a task it is necessary to have molecular devices. It is the task of these devices to provide logic functionality such as NOR gates, flip-flops, and memory cells. It is quite possible that for calculation the machine will rely primarily on lookup tables in a vast memory storage system rather than recalculate everything from the beginning each time. Nevertheless, the machine is still conventional in the sense that it will follow a series of instructions to complete a task. To replicate all the important logic functions, three fundamental gates are needed: AND, OR, and XOR gates. These gates can, in turn, be made using two different devices, resonant tunneling and rectifying diodes. Below we sketch out how

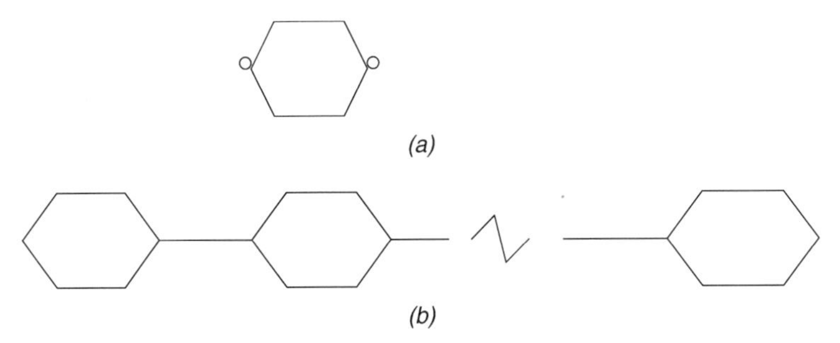

FIGURE 8.3.1 (*a*) Phenylene group and (*b*) polyphenylene. The two free binding sites on the phenylene group are represented by the open circles on the ends of the ring. Each ring represents a C_6H_4 molecule. Notice that with two free binding sites, each phenylene group can be bound to two others.

molecular-based resonant tunneling diodes (RTDs) and rectifying diodes can be made.

We begin our discussion of moletronics by examining the types of molecules used and their properties. Generally, both conducting and insulating molecules are needed to form devices. Presently, the two most promising conducting molecular species are polyphenylenes and carbon nanotubes. The polyphenylenes are essentially chainlike molecules formed by linking a basic molecular unit, phenylene (C_6H_4), which is a derivative of the benzene ring but with two free binding sites (see Figure 8.3.1). Given that each phenylene has two free binding sites, it can be linked to two other phenylenes to form a chain, as shown in Figure 8.3.1*b*. These chains can then be used to form molecular wires. Polyphenylene wires are fairly conductive. Conduction in these molecules proceeds by electrons moving through extended molecular orbitals that span or nearly span the length of the entire molecule. The extended molecular orbitals are called π-type and lie above and below the molecular plane when the polyphenylene is in a planar conformation. Basically, the extended π-orbital states occur in much the same way that extended states occur in a superlattice [i.e., when the atoms (wells in a superlattice) become close enough spatially, the wavefunctions overlap, leading to extended states] (Brennan, 1999, Sec. 7.3). The resulting extended states are no longer degenerate in energy. As in a superlattice, if many atoms contribute π-orbitals, corresponding to a relatively long polyphenylene chain, a range of extended states is formed, producing a band. If the chain is relatively small, containing few atoms contributing π-orbitals, only a relatively small set of extended states is produced. In any event, these extended states provide a conducting path through the molecular chain. Hence the polyphenylene chains are conducting.

Alternatively, chains formed with aliphatic organic molecules can produce an insulator. Aliphatic molecules are singly bonded molecules that do not contain π-bonds but have what are called *σ-bonds* instead. The σ-bonds lie along the axes of the molecules and are not easily extended between atoms. This is because each σ-bond terminates on a positively charged nucleus on either end, and thus their spatial extent is "interrupted" by the nucleus. Chaining together methylene (CH_2) molecules can form an aliphatic molecule. Inserting aliphatic molecules into a polyphenylene chain can also form insulating molecules. The aliphatic molecules interrupt the extended states formed by the π-bonds, thus breaking the conducting pathway through the polyphenylene chain.

Using both polyphenylene and aliphatic molecules, molecular RTDs and rectifying diodes can be made. As we discussed in Section 6.5, RTDs can be used to form logic gates. Therefore, the development of molecular RTDs enables a pathway toward developing molecular logic gates. In addition, logic gates can be formed using rectifying diodes. By combining both RTDs and rectifying diodes, a full range of logic gates can be developed.

Let us first consider a molecular resonant tunneling diode. In Chapter 6 we showed that sandwiching a quantum well between two potential barriers forms an RTD. Typically, the quantum well is formed with GaAs and the barriers with AlGaAs. By making the GaAs layer sufficiently thin, spatial quantization effects occur. A peak in the current–voltage characteristic results when a voltage is applied to the device as the confined state becomes resonant with the Fermi level within the emitter contact. At biases away from resonance, the current decreases, leading to a nonlinear current–voltage characteristic. The basic structure of the molecular RTD is shown in Figure 8.3.2*a*. For comparison a solid-state RTD is shown in Figure 8.3.2*b*. It is assumed that the polyphenylene chains on either side are terminated at a gold contact to produce ohmic behavior. Notice that each polyphenylene chain is connected in the middle of the device to a CH_2 aliphatic molecule. As discussed above, the CH_2 molecules are insulating. These molecules play the role of the potential barriers formed in the solid-state RTD. Sandwiched between the CH_2 molecules is another phenylene group. The center phenylene group plays the role of the GaAs well in the solid-state RTD. Given the small size of the center phenylene group, spatial quantization occurs just as in the solid-state RTD.

The basic operation of the molecular RTD is very similar to that of the solid-state RTD. In either device, peak current flows when the spatial quantization level is resonantly aligned with the occupied levels within the emitter. For biases either below or above the resonant alignment condition, the current decreases. Current flow in the molecular RTD can be understood as follows. Within the polyphenylene chains extended molecular states exist. As in a solid, most of the lower-energy states are occupied while most of the higher-energy states are unoccupied. The highest occupied molecular orbitals are called *HOMO states*; the lowest unoccupied molecular orbitals are called *LUMO states*. In either polyphenylene chain, both HOMO and LUMO states exist. The small spatial size of the center phenylene group results in spatial quantization effects (i.e., only certain discrete energy levels exist). When the HOMO levels within the emitter (the left-hand side of the device) become

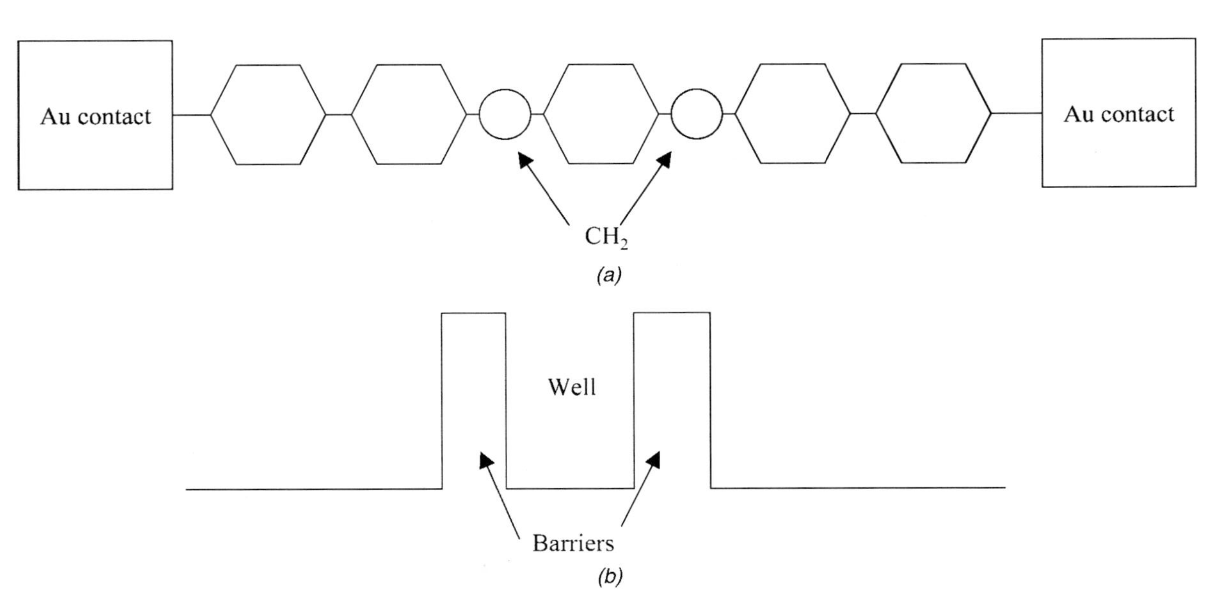

FIGURE 8.3.2 (*a*) Molecular RTD formed using polyphenylene chains, two CH_2 molecules (barriers), and a center phenylene group (well). (*b*) Solid-state RTD structure. Notice that the CH_2 molecules correspond to the barriers in the solid-state RTD and the center phenylene group corresponds to the well.

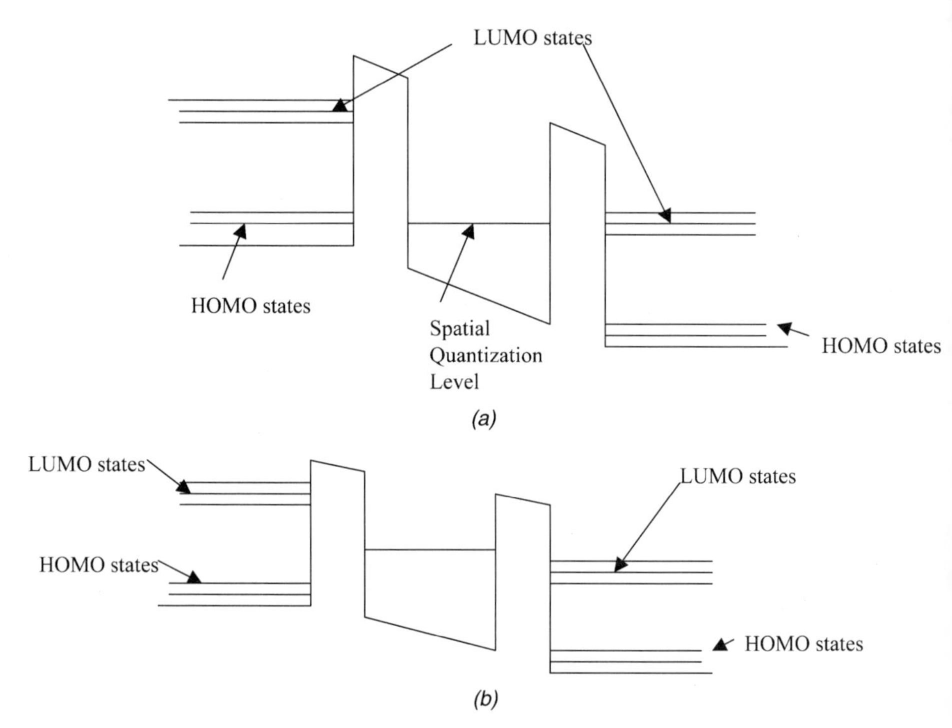

FIGURE 8.3.3 (*a*) Molecular RTD wherein the CH_2 insulators are represented as potential barriers, and the two polyphenylene chains are the two regions to the left and right of the barriers. The situation shown corresponds to resonant alignment and a peak current flows in the device. (*b*) Molecular RTD not resonantly aligned. Notice that the spatial quantization level is not aligned with the HOMO states. Under these conditions the current flow is much lower than in part (*a*).

coincident with the spatial quantization levels produced in the center phenylene group, a resonant current can flow as shown in Figure 8.3.3*a*. This is identical to the situation for a solid-state RTD. In a solid-state RTD, resonant current flows when the spatial quantization levels within the well are resonantly aligned with the highest-lying energy levels within the emitter (i.e., the Fermi level). If the bias is such that the spatial quantization levels are pulled below the HOMO levels or above the HOMO levels as shown in Figure 8.3.3*b*, resonant tunneling cannot occur within the structure, resulting in a strong decrease in the current. Clearly, resonant alignment of the spatial quantization levels and the HOMO levels within the emitter produces a maximum current flow. Increasing or decreasing the voltage about this bias point results in a decrease in the current. Therefore, as in the solid state RTD, a nonlinear current–voltage characteristic results.

The second molecular device type we consider is a molecular rectifying diode. As in a p-n junction diode (Brennan, 1999, Chap. 11), two different doped regions are required, a p-type region and an n-type region. As in semiconductors, the electron concentration can be altered by the introduction of foreign agents. In semiconductors, one adds impurity atoms called *dopants*. In molecular systems, one adds molecular groups that attach themselves at specific places within the chains, altering the electron concentration. Groups that add electrons to the system are called *electron donating groups*, and groups that remove electrons are called *electron withdrawing groups*. Electron donating and withdrawing groups are analogous to donors and acceptors in semiconductors. As in a semiconductor, a p-n junction can be formed by placing a chain with withdrawing groups together with a chain with donating groups. A semi-insulating group is inserted to form a potential barrier between the donating and withdrawing group chains. The potential barrier formed by the semi-insulating group serves to maintain the charge imbalance between the two sides of the junction. In other words, the differing electron densities between the two sides of the junction would equilibrate if the potential barrier did not exist, thus removing any diode action. In a p-n junction diode, the built-in potential plays the role of the semi-insulating group. The basic scheme is shown in Figure 8.3.4*a* along with a solid-state p-n junction for comparison (Figure 8.3.4*b*).

Before we discuss the current flow in the molecular diode it is important to understand the energy-level diagram under equilibrium. In equilibrium the Fermi levels are of course aligned (Brennan, 1999, Chap. 11), as shown in Figure 8.3.5. The LUMO levels in the donating chain are at a higher energy than those in the withdrawing chain, as shown in the diagram. This is because the electron density is increased within the donating chain, resulting in increased electron–electron repulsion and hence a higher total electron energy. Similarly, the lower electron density within the withdrawing chain results in a lower total electron energy. Consequently, there exists an energy difference between the two sides of the molecular diode (i.e., the levels within the donating chain lie at a higher energy than the levels within the withdrawing chain).

Current flow within the molecular diode can be understood as follows. A molecular diode has two bias conditions, forward and reverse, similar to the case of a solid-state diode rectifier. As in a solid-state diode, the current flow is asymmetric with respect to the applied bias. Under forward bias, the current flow is high; under reverse bias the current flow is much lower. The key to understanding the asymmetry in the current flow is the fact that the difference in energy between the Fermi level and the LUMO levels on the side with the withdrawing group is less than the energy difference between the Fermi level and the LUMO levels on the side with the donating group. The bias required to align the Fermi level with the LUMO levels on the withdrawing side is less than that required to align the Fermi level with the LUMO levels on the donating side. Therefore, for the same magnitude of bias, the current flow within the device will be highly nonlinear and asymmetric.

Let us consider each bias in turn. The insulator groups, I, and the semi-insulating group, SI, act as potential barriers in the structure. Therefore, in the

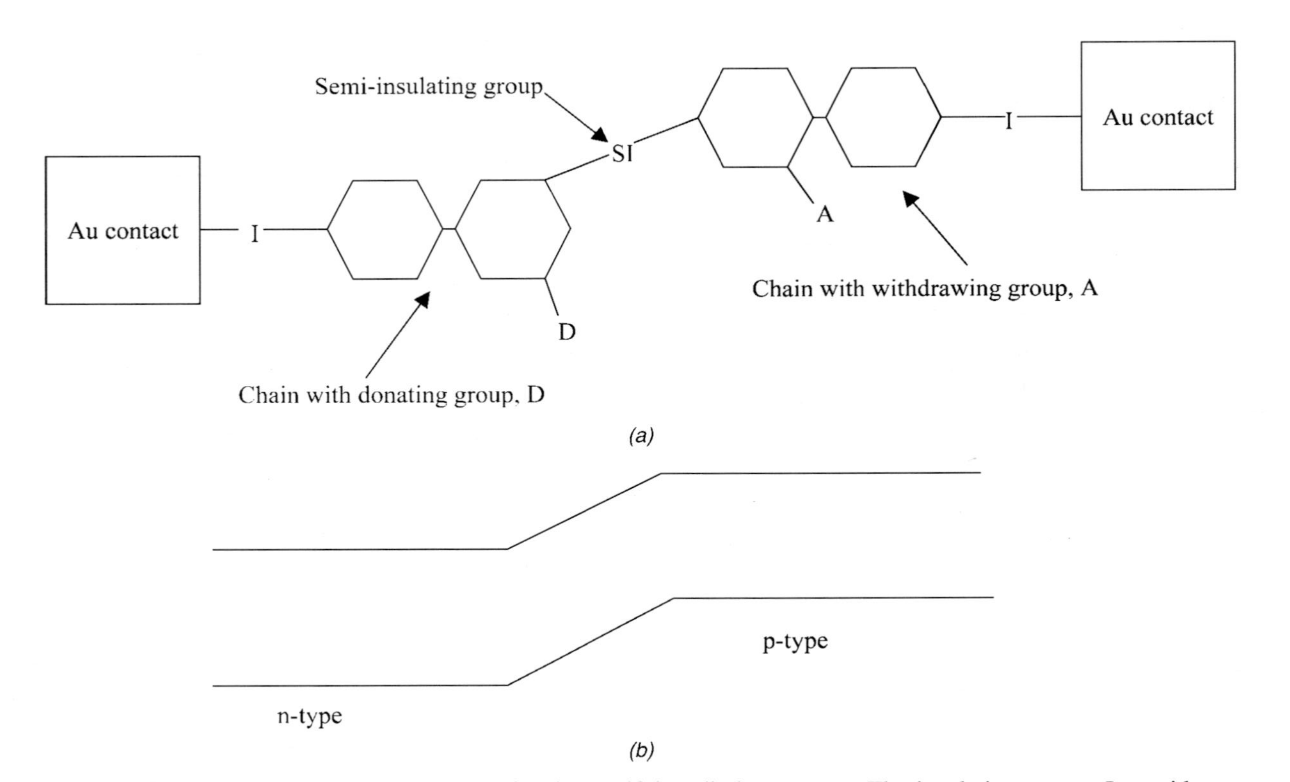

FIGURE 8.3.4 (*a*) Molecular arrangement for the rectifying diode structure. The insulating groups I provide potential barriers for tunneling into and out of the diode. The semi-insulating group (SI) in the center of the device plays the role of the built-in potential of the diode maintaining the electron density imbalance within the junction. (*b*) Solid-state p-n junction diode. The p-type layer is on the right-hand side and the n-type layer on the left.

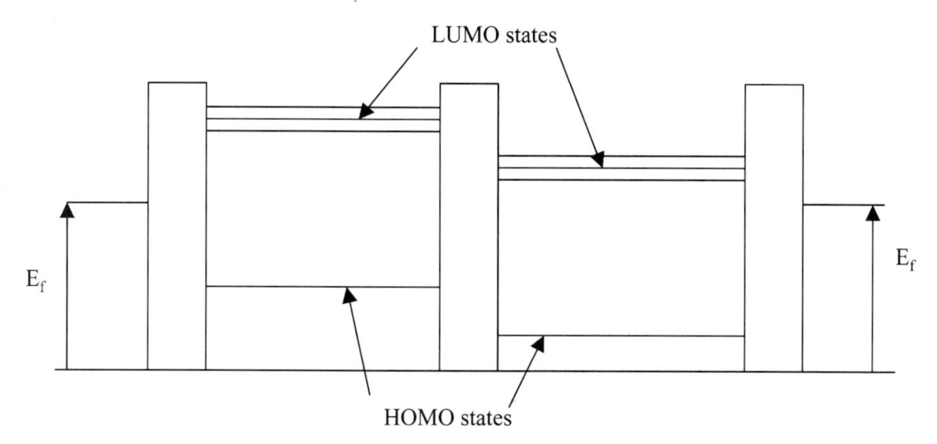

FIGURE 8.3.5 Molecular diode energy levels under equilibrium. The center potential barrier, marked SI, corresponds to the semi-insulating layer. The two end potential barriers, marked I, are the insulating layers between the two chains and the contacts. In equilibrium notice that the Fermi levels are aligned. The LUMO levels on the donating chain lie at a higher energy than the LUMO levels on the withdrawing chain.

band diagram for the device, they are shown as barriers. The donating chain (left-hand side of the diode) and withdrawing chain (right-hand side of the diode) are conducting with HOMO and LUMO states. Forward bias occurs when a high potential is applied to the left-hand contact with respect to the right-hand contact. A positive potential acts to lower the electron energies. Hence the occupied energy levels within the left-hand contact are lower than those within the right-hand contact, as shown in Figure 8.3.6*a*. Since the contacts are metals, the uppermost occupied levels in the contacts are the Fermi levels. When the Fermi level in the right-hand contact is sufficiently raised so as to align with the LUMO levels in the withdrawing chain, electrons can tunnel from the contact into the LUMO levels (see Figure 8.3.6*a*). These electrons can then tunnel through the SI potential barrier into the donating-chain LUMO levels. Finally, the electrons tunnel through the last potential barrier into the positively biased left-hand contact. Notice that under what is called forward bias, only a relatively low bias is necessary for a current to flow. As shown by Figure 8.3.5, the Fermi level lies closer to the LUMO levels in the withdrawing chain than in the donating chain. Hence, only a relatively low bias is necessary to align LUMO levels in the withdrawing chain with the Fermi level in the contact.

The reverse-bias conditions can be understood as follows. Under reverse bias, a high voltage is applied to the right contact with respect to the left-hand-side contact as shown in Figure 8.3.6*b*. The Fermi level is lowered on the right contact and raised on the left contact as shown in the diagram. For the same magnitude of bias as in the forward case, the Fermi level on the left contact is *not* aligned with the LUMO levels. As such, no resonant alignment occurs

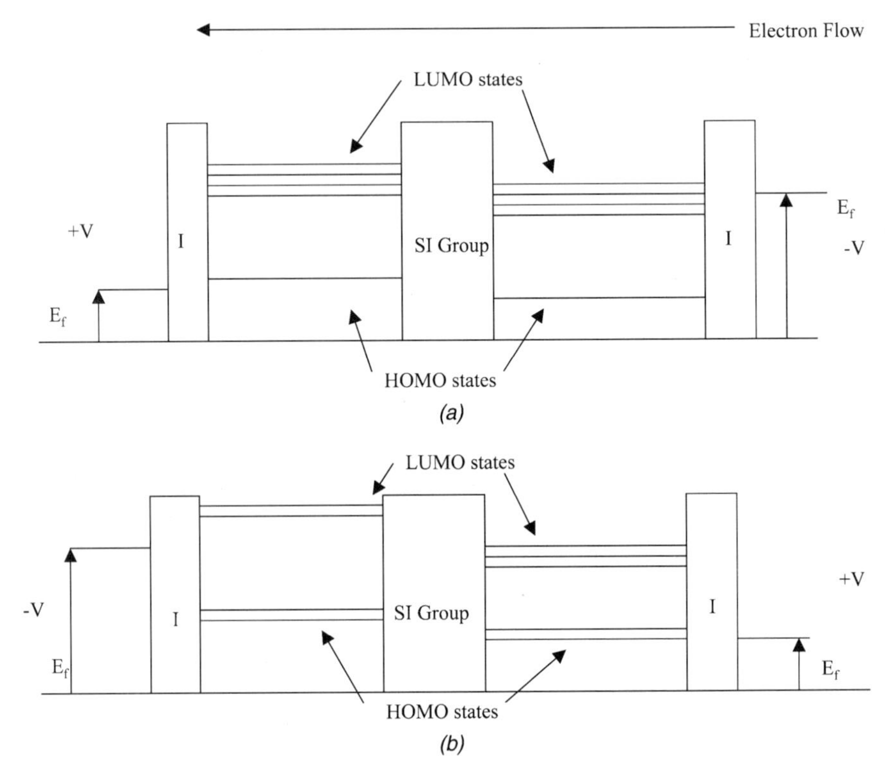

FIGURE 8.3.6 Electronic structure of the molecular diode. (*a*) Forward-bias conditions for a molecular diode. Electron flow is from the right to the left. (*b*) Reverse-bias conditions for a molecular diode. Intended electron flow is from left to right. In both cases the same magnitude of bias is applied. Notice that resonant alignment does not occur at the same magnitude of bias as in part (*a*).

and little current flows in the device. The principal current flow mechanism under reverse bias is nonresonant. Of the nonresonant mechanisms, arguably the most important is that due to hopping conduction, which was discussed in Section 6.3.

Given that molecular RTDs and diode rectifiers can be made, the next question is whether logic gates can be constructed. Conceivably, some of the same logic circuits discussed for solid-state RTDs in Section 6.5 can be realized using molecular RTDs. Alternatively, it is well known that simple AND and OR gates can be constructed using rectifying diodes. A simple example of a two-input AND gate using solid-state diodes is discussed in Box 8.3.1. An OR gate can be made in a similar manner. Finally, a simple inverter gate can be made using RTDs and diode rectifiers.

A different approach for utilizing molecular systems to perform logic functions has recently been suggested by Collier et al. (1999). Their approach can

BOX 8.3.1: Diode–Diode Logic

As an example of how rectifying diodes can be used to construct logic gates, let us consider a two-input AND gate made using rectifying diodes. The truth table for an AND gate was presented in Figure 7.1.7. The basic circuit diagram for the two-input AND gate is shown in Figure 8.3.7. Operation of the gate can be understood using the ideal diode model. A positive voltage is applied across the resistance R as shown in the diagram. If either V_A or V_B is low, or both are low, diodes 1 and 2 are forward biased. As a result, there is little voltage drop across the diodes, and in the ideal approximation there is no voltage drop across the diodes. Therefore, the point V_C is also low. Thus for either input low or both inputs low, the output is low. Alternatively, if inputs V_A and V_B, are both high, diodes 1 and 2 are both reverse biased. Under the ideal approximation, the two diodes are open circuits. Therefore, V_C is high. Hence, if both inputs are high, the output is high. This constitutes AND gate performance.

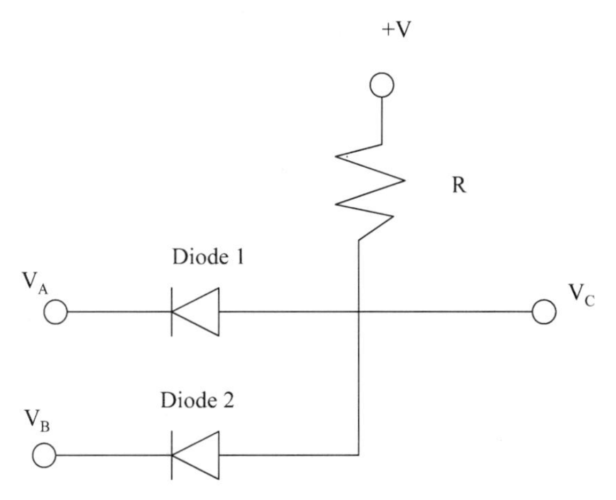

FIGURE 8.3.7 Two-input diode AND logic gate. The inputs are V_A and V_B. The output is V_C.

be summarized as follows. Collier et al. (1999) recognized that the key advantages of molecular systems are their density, adaptability, and the fact that interconnects can be avoided in molecular circuitry. These three ingredients underlie their approach. In their design, the positioning of wires and switches and their operation is not precisely controlled as it is in conventional CMOS circuitry. The fracture and breakage of a single wire can result in the failure

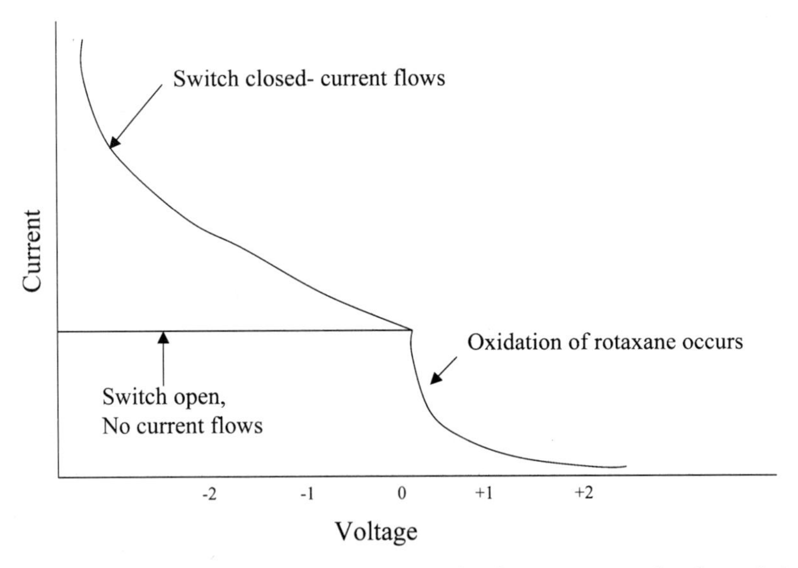

FIGURE 8.3.8 Current–voltage characteristic of a rotaxane molecular switch.

of an entire chip. As the density of devices and the interconnect complexity increases, the probability of wire breakage increases, resulting in lower yield and higher cost. Therefore, avoidance of the need for precisely controlled wiring provides a major advantage.

The basic device architecture of the device consists of a pair of Al electrodes. One electrode is coated with Al_2O_3, the other is coated with a thin layer of titanium, Ti. Sandwiched between the two coated electrodes is a layer only one molecule thick of organic molecules called *rotaxanes*. This organic molecular layer was deposited as a Langmuir–Blodgett film. Langmuir–Blodgett film deposition is a technique for transfering monolayers of a specific molecular type one after another from a water surface onto solid substrates to form monolayer-thick molecular assemblies. The rotaxane molecules act to isolate or connect the two electrodes electrically. Depending on the charge conductivity state of the rotaxane molecules, an electrical current can pass from one of the coated electrodes to the other or the electrodes become electrically isolated. The default condition is that the rotaxane molecules provide a conducting path between the two electrodes. The application of a relatively large positive voltage across the device oxidizes the rotaxane molecules, permanently removing electrons and altering their electrical conductivity. Once oxidized, the rotaxane molecules can no longer conduct a current, and the two electrodes become electrically isolated.

The current–voltage characteristic for the device is sketched in Figure 8.3.8. As can be seen from Figure 8.3.8, a negative voltage applied to the electrode with the Al_2O_3 coating initially behaves as a closed switch (i.e., it conducts

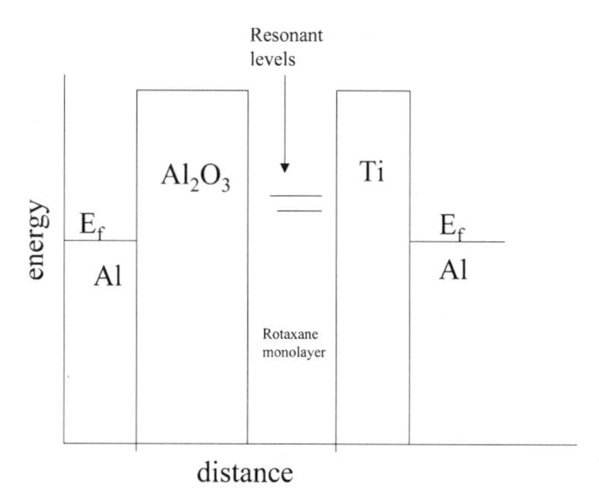

FIGURE 8.3.9 Energy versus distance, showing the rotaxane molecular switch.

an electrical current). If the current is increased to about +1 V, the rotaxane molecules become oxidized. Once oxidized, the molecules no longer conduct an electrical current and the switch is effectively open. Now when the bias is reversed back to a negative value, the switch remains open—no current flows. The switching process is irreversible. Hence the basic functionality of a transistor is achieved: a third element—in CMOS circuits a gate, here the magnitude of the positive applied voltage—alters the electrical conductivity between two electrodes—in CMOS, the source and drain contacts, here the two different-coated electrodes.

The electrical conduction mechanism in the rotaxane molecules can be understood as follows. An electron can resonantly tunnel from one electrode to the other through a molecular state of the rotaxane molecule. Now we can understand the importance of coating the two electrodes. Each electrode is coated so as to produce a tunneling potential barrier on the surface of each electrode, as shown in Figure 8.3.9. Also shown in the figure are the Fermi levels within the Al electrodes. The Al_2O_3 layer is significantly wider than the barrier formed by the Ti–rotaxane interface. The resonant levels are sketched within the rotaxane monolayer as shown in the figure. It is through these states that electrical conduction from one electrode to the other proceeds. Under a small bias, these molecular levels resonantly align with the Fermi levels in the first electrode, resulting in a sharp increase in the current. Once the rotaxane molecules are oxidized, these resonant tunneling states are no longer accessible since they become filled.

Collier et al. (1999) have also constructed AND and OR logic devices using these molecular systems. They found that the devices had a very high noise tolerance since the current levels between the ON and OFF states was substantial. It is projected that many highly dense logic gates can be made using the foregoing procedure provided that the electrode dimensions can be effectively reduced. One suggestion is that carbon nanotubes, thus ensuring nanoscale dimensions, replace the electrodes. It should be noted that the deposition process is very inexpensive and highly reliable. Therefore, mass quantities of devices can be made at very low cost. Unfortunately, the present devices can switch in only one direction and are not reversible. Therefore, these devices are not currently an acceptable replacement to CMOS logic. Nevertheless, some recent developments have resulted in the realization of these types of molecular computing schemes using molecules that offer reversibility (Williams, 2000). It is expected that by using reversible molecular switches, molecular computing machines using fat tree architecture (Section 8.4) may become a reality, providing high computational power.

8.4 FIELD-PROGRAMMABLE GATE ARRAYS AND DEFECT-TOLERANT COMPUTING

There exist two somewhat opposite approaches to computer hardware design. The first is to utilize integrated-circuit chips that are highly versatile such that they can perform many different tasks well. Most microprocessors used in general-purpose computing platforms are necessarily highly versatile. Microcode instructions can enable a general-purpose microprocessor to execute virtually any operation that a programmer can conceive. However, this extensive versatility comes at the expense of computational efficiency. General-purpose microprocessors are relatively slow since they must be configured suboptimally in speed to enable high flexibility.

In contrast, custom hardware circuits, called application-specific integrated circuits (ASICs) are designed to perform a limited set of tasks as quickly as possible. ASICs are thus optimized for speed rather than flexibility. An ASIC is carefully tuned for a specific task. It can thus accomplish that task faster than a general-purpose processor and consume far less power. The primary drawback to using ASICs in many applications is that if the problem becomes modified even slightly, an ASIC typically cannot perform the modified task after its construction. Therefore, modified tasks require new ASIC chips, which can drive up the cost of the system.

Recently, an intermediate solution between the hardwired, fast execution, but inflexible ASIC chips and the flexible but relatively slow execution general-purpose processors has been advanced (Villasenor and Mangione-Smith, 1997). This new option is called a *field-programmable gate array* (FPGA). A FPGA is a highly tuned hardware circuit that can be modified during use in a variety of ways. FPGAs consist of arrays of configurable logic blocks that implement the basic logic gate functions (i.e., switches with multiple inputs and a single output). The key difference in FPGAs and ASICs is that the logic blocks in FPGAs can be rewired and reprogrammed continuously during the lifetime of the chip. ASICs, on the other hand, are designed for a specific task and permanently perform only a limited set of actions. There is no manner in which an ASIC can be reconfigured after it has been assembled. As such, if the tasks are modified, an existing ASIC cannot be utilized and a new one must be inserted. If the system uses FPGAs instead, the FPGA can be reconfigured through software applications to a relatively short period, on the order of milliseconds to 100 μs.

The basic structure of a FPGA consists of a large number of configurable logic blocks and a programmable array of interconnections that can link the blocks in any pattern desired. Essentially, the FPGAs are connected by a large number of wires and switches arranged in a crossbar network that ensures that the blocks can be connected in any pattern desired. Such an architecture, called fat tree architecture, is discussed in detail below. The designer can thus reprogram the configuration at any time using software instructions; no physical changes must be made to the hardware or the interconnects.

There are various ways in which a FPGA can be used. One approach is to have the hardware reconfigure itself real time to refine its performance as a function of the task and its execution. In this way, the efficiency of the system can be greatly enhanced and the machine can effectively "learn" a procedure to perform the task as efficiently as possible. The easiest implementation of FPGAs is to enable switching between functions on command. Essentially, the computer would execute a particular program, complete and exit that program, reconfigure for a different task, and repeat the cycle. If the reconfiguration time is relatively fast, the FPGA can be used in a time-share mode, swapping tasks rapidly as is typically done in a multitask computer. Configurable computing has the obvious advantage over ASICs of reducing hardware requirements. For example, a single FPGA can be configured to perform several different tasks that would each require an individual ASIC. As a result, tremendous cost savings can potentially be realized, owing to the reduction in the number of chips needed in the system.

As we discuss below, FPGAs have the additional advantage that this technique enables defect-tolerant computing. General-purpose chips must operate nearly to perfection. Although some error correction can be used, the most versatile general-purpose chips need to operate nearly perfectly to ensure maximum flexibility. As the computational demands of the chip increase, the cost to manufacture large volumes of chips reliably increases proportionally. Computer designers often quote what is commonly referred to as *Moore's second law*, the fact that the cost of the fabrication facilities used to build chips is increasing faster than the growth in demand for chips. Although substantial, facilities cost historically have been a small fraction of the total cost of the manufacture and distribution of silicon integrated circuits. Presently, the cost of new fabrication facilities has escalated into the billions of dollars. It is estimated that by 2010 the cost of a single fabrication facility will be in the range $30 billion to $50 billion, and few manufacturers will have the financial means or the courage to make this kind of investment. Even a consortium of manufacturers may balk at such a high investment for only about a factor of 2 improvement in performance from earlier generations.

Substantial cost savings in manufacture can be realized if the requirement of nearly perfect operation of the chip is alleviated. Consequently, computer designs that utilize chips that contain some defects may prove to be far less expensive than those that require nearly perfect chip performance. The ability to configure a chip to operate around defects and imperfections would greatly improve the manufacturing yield and consequently, the cost of the chips. Configurable computing can provide a means of operating a chip around its defects. Perhaps the most important advantage of configurable computing will then be the potential improvement in computing power at reduced cost. This could be of immense importance to the future of the computing industry.

As discussed above, the manufacturing cost of integrated-circuit chips is a strong function of their degree of perfection. If nearly perfect performance is required, as the complexity of the chip increases, the yield decreases and the resulting cost escalates. However, if a defective chip can still be used, the yield would then increase greatly and the cost of the chips would decrease substantially. In addition, it is likely that the manufacturing costs associated with fabrication plant construction and operation would also be mitigated. The question is, then: Can a computing machine be made that utilizes defective chips? Recently, a group of researchers at Hewlett-Packard and UCLA have announced a defect-tolerant computer design (Heath et al., 1998). These researchers have developed a machine that uses conventional silicon integrated-circuit technology. The machine, called *teramac* (tera multiple architecture computer), is designed to operate at 10^{12} operations per second. The teramac was constructed using components that were defective but inexpensive. Many of the chips used in the machine were "throwaways," chips that did not meet specifications and thus were disposed of rather than sold. Due to the manner in which the chips were interconnected and the utilization of configurable computing algorithms, the defects within the chips were mapped out in advance and suitable computational pathways were identified. The teramac configured itself to avoid defective devices or zones in each chip, thereby providing reliable computing pathways. In this way, many defective chips could be connected to produce a powerful, yet very inexpensive parallel computer.

One of the key issues in the design of the teramac is the manner in which the chips are connected. The basic problem encountered in configurable computing is how to maximize the number of interconnections such that the machine can have the greatest number of configurations, and hence have the greatest flexibility. The manner in which the logical units of the teramac were linked, called the *fat tree architecture*, can be understood best by comparing it to a family tree architecture. In a family tree architecture, each child branches from one parent. Easy communication between parent and child then results. However, communication between children of the same parent is not direct. Instead, a message must be passed to the parent first and then to the other child. Communication to the grandparents must also proceed through the parent. Not only does this slow communication, but it makes it extremely vulnerable to disruptions. For example, if the pathway between the parent and grandparent becomes disabled, communication between a whole branch of the tree and the rest of the family is destroyed. Such a scenario is not likely to unfold in a computer in which nearly perfect chips are employed. However, if one utilizes defective chips, it is highly likely that important pathways between parts of the structure will be blocked. Consequently, a different architecture from that of the family tree must be employed in a machine like the teramac, which employs defective chips.

Fat tree architecture avoids the vulnerability of family tree architecture by replacing the single nodes of the parents by several nodes. The structure has a high degree of connectivity between nodes. If one of the pathways is disabled in a fat tree architecture, this would have little effect, since many other routes and connections exist. Therefore, fat tree architecture is relatively immune to pathway interruptions and hence is capable of working even when many defects are present. For these reasons, the teramac utilizes a fat tree architecture.

An important part of the operation of the teramac machine is the location and identification of the defects in the chips. In teramac, all "repairs" are made using software. A software routine locates and maps the defects. Once all the defects are identified, the software also maps configurations that avoid the defects, providing computational pathways.

Teramac has demonstrated that a computing machine can operate reliably even in the presence of many defects. There are several useful lessons that the successful operation of teramac illustrates.

1. It is possible to build a powerful computer that contains a high level of defects provided that there is sufficient bandwidth communication between nodes.
2. A high degree of connectivity is more important than regularity in a computing machine. It is not essential that each component be connected in an orderly manner as long as each component can be located and characterized. If the testing algorithm is sufficiently robust, it can determine after the fact how the machine is connected and which parts can contribute to its execution. Therefore, the architecture of the machine need not be highly organized.
3. The most essential components for a nanoscale computer are the switching and interconnect mechanisms.
4. The teramac computer has introduced a new algorithm in computer manufacture. It is: build the computer; find and map the defects; configure the resources with software; compile and execute the program. In this way, imperfect hardware can be utilized, thus greatly reducing the cost of the machine.

Configurable computing has the added advantage of being naturally extendable to molecular systems. Molecular systems are particularly attractive for future computing applications since they provide for simplified fabrication, offer low cost, have relatively unlimited availability, offer massive densities of computing elements, can be self-reproducing, and are self-organizing. Using configurable computing, computational pathways can be mapped out in an otherwise highly complex system after its construction. In other words, the machine itself will not have to be fully designed to begin with; only enough interconnections between computing elements need be established. The machine will then be configured using software algorithms that search and map out computational pathways avoiding defective paths. In some ways such a machine will work like a human brain; the neural pathways are first created and numerous such pathways exist. Over time, different pathways are activated and used, while many others are never employed through a learning process. The key to highly complex processing in such a system is to enable many pathways and interconnects from each node. A molecular system is, in many ways, the system best suited for such processing.

8.5 COULOMB BLOCKADE AND SINGLE-ELECTRON TRANSISTORS

In nano-sized structures the current flow can be blocked by the electrostatic charging energy of a single electron. This current blockage due to the charging energy associated with a single electron is called the *Coulomb blockade*. In a tunnel junction, the current flowing between the two leads can be blocked due to the action of the Coulomb blockade. Perhaps the simplest demonstration of the effects of the Coulomb blockade can be seen from tunneling of electrons from one lead to another through a metal island as shown in Figure 8.5.1. The action of the Coulomb blockade can be understood as follows. The small metal island isolated from the two leads by an electrically insulating material is sufficiently small such that the addition of a single electron to the island will change its energy state significantly. The energy state of the metal island changes upon the addition of a single charge by (see Box 8.5.1)

$$E = \frac{q^2}{2C} \qquad 8.5.1$$

where C is the total capacitance of the island. In the case where the metal island is small and its capacitance is small, the energy of addition of a single electron to the island can be substantial. An electron can be added to the island only if it has energy above the Fermi level within the contact lead of $q^2/2C$. If an electron does not have this energy, tunneling from the leads into the island cannot occur and no current can flow. Consequently, the current is blocked. Physically, the current blockage can be understood to arise due to an energy gap appearing in the energy spectrum, as shown in Figure 8.5.2. As can be

BOX 8.5.1: Calculation of the Charging Energy

It is very easy to see, using a strictly classical argument, that the energy needed when a single electron charge is transferred to the metal island is $q^2/2C$. We assume that the system can be modeled as a two-electrode capacitor. The amount of work required to transfer a single charge q' is given as

$$dW = V\,dq'$$

The potential difference V is given simply as q'/C. Therefore, the differential work done can be written as

$$dW = \frac{q'}{C}dq'$$

The total work done, equal to the energy, is found simply by integrating dW. The result is

$$W = \int_0^q dW = \int_0^q \frac{q'}{C}dq' = \frac{q^2}{2C}$$

Of course, this is equal to the well-known value for the energy stored in a capacitor,

$$W = U = \tfrac{1}{2}CV^2$$

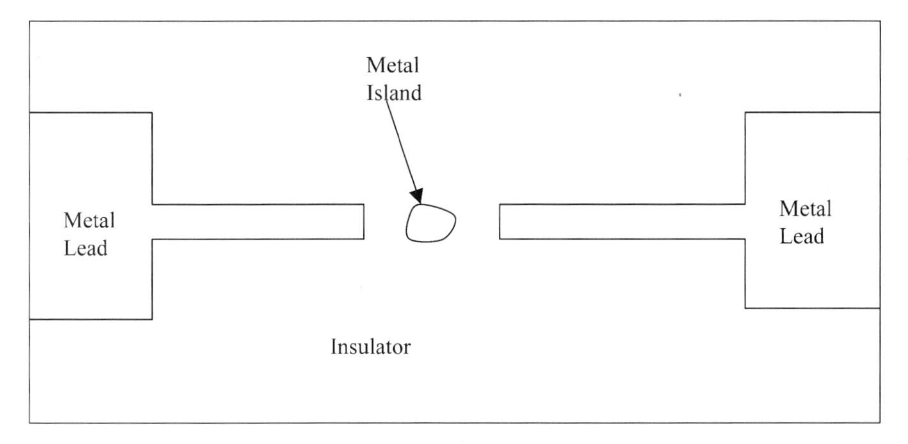

FIGURE 8.5.1 Tunnel junction showing a metal island separated by two metal contact leads. The metal island is sufficiently small that its charging capacitance for a single electron requires a significant amount of energy. As such, single electron charging of the metal island can block the current flow between the two leads. This is called the *Coulomb blockade*.

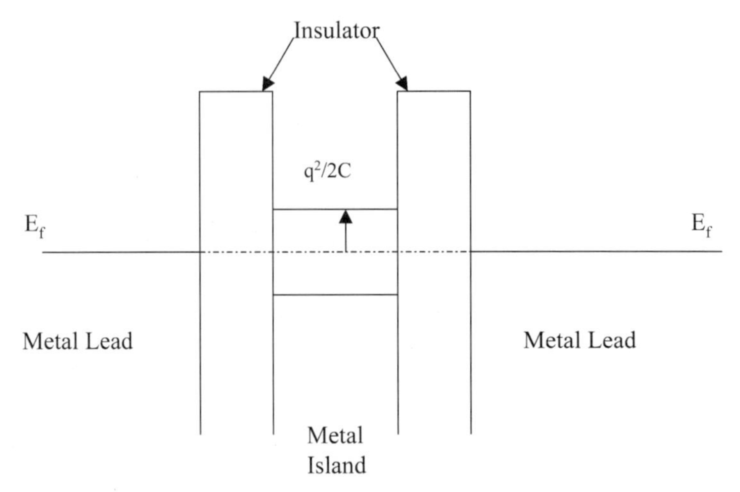

FIGURE 8.5.2 Energy band structure for the tunnel junction shown in Figure 8.5.1. Notice that an energy gap appears in the energy spectrum within the island. An electron can only tunnel onto the island if it has an energy of $q^2/2C$ greater than the Fermi level in the leads. A hole can tunnel only if it has an energy $-q^2/2C$ below the Fermi level.

seen from the figure, an energy gap of $q^2/2C$ opens above the Fermi levels for electron transfer and an energy gap of $-q^2/2C$ occurs for hole transfer. The energy gap has a net value of q^2/C. Since the tunneling electron must have a discrete energy of $q^2/2C$, to enter the island, a transition can occur only if this energy is available. Otherwise, an electron cannot tunnel onto the island, resulting in a blockade of the current.

Although the device structure shown in Figure 8.5.1 is useful for illustrating the Coulomb blockade effect, a more interesting structure is that of a gated MOSFET with a narrow inversion layer. The actual structure is a dual-gated device arranged such that a quantum wire (a one-dimensional electron gas system similar to that shown in Figure 8.5.1) is formed connecting the source and drain regions. A figurative sketch of the device is shown in Figure 8.5.3*a*, and a cross-sectional sketch is shown in Figure 8.5.3*b*. In Figure 8.5.3*a* the inversion layer is illustrated as a block sandwiched between the source and drain regions and controlled by the gate. From inspection of Figure 8.5.3*b*, operation of the device can be understood. Application of a positive bias on the top gate while the bottom gate is grounded or negatively biased results in inverting the surface only in the thin region between the two portions of the lower gate [marked as the quantum wire in part (*b*) of the figure]. Therefore, electrons can flow from the source to the drain along this thin wire. An interesting feature of the gated MOSFET device is that an additional control exists (i.e., the gate voltage). Application of a gate voltage alters the number of electrons within the inversion layer between the source and drain contacts.

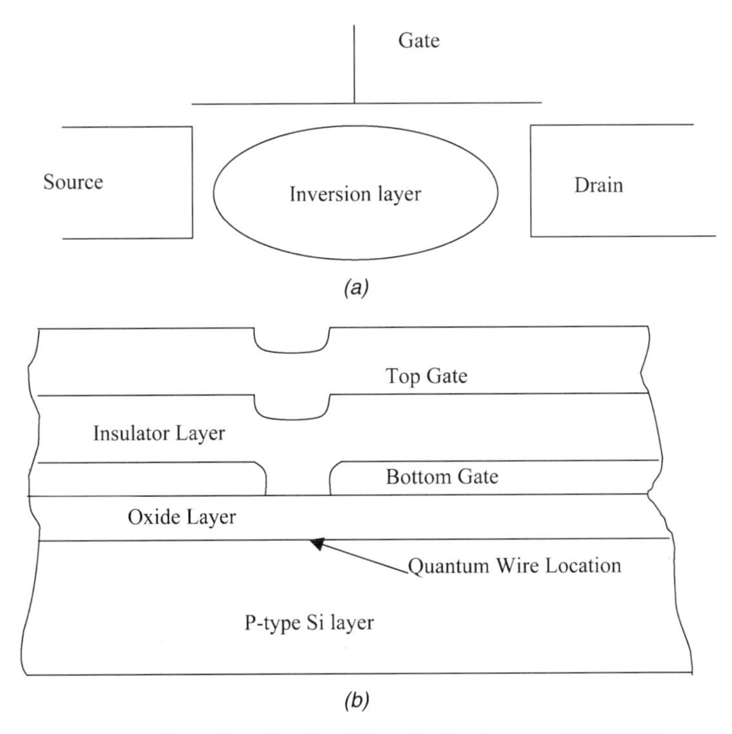

FIGURE 8.5.3 (*a*) Figurative sketch of the single-electron MOSFET structure. (*b*) Cross-sectional sketch of the quantum wire MOSFET device based on the work by Kastner (1992).

We define the gate voltage V_g as the potential difference between the gate and the electron system within the inversion layer. The electrostatic energy of a charge q within the inversion layer of the transistor is then given as

$$E = -qV_g + \frac{q^2}{2C} \qquad 8.5.2$$

since the top gate must be positively biased. The second term in Eq. 8.5.2 is the charging energy of the inversion layer formed between the source and drain. Notice that the first term in Eq. 8.5.2 is negative, implying an attractive interaction between the charge and the gate voltage, whereas the second term is positive, implying a repulsive interaction between charges within the inversion region.

The conductance of the quantum wire MOSFET device exhibits periodic peaks as a function of the applied bias. It is highly interesting that the conductance exhibits periodic peaks. Aperiodic peaks can be attributed to noise, but periodic peaks are the usual signature of quantized phenomena. In this case, the quantized quantity is the charge itself. The origin of the oscillations

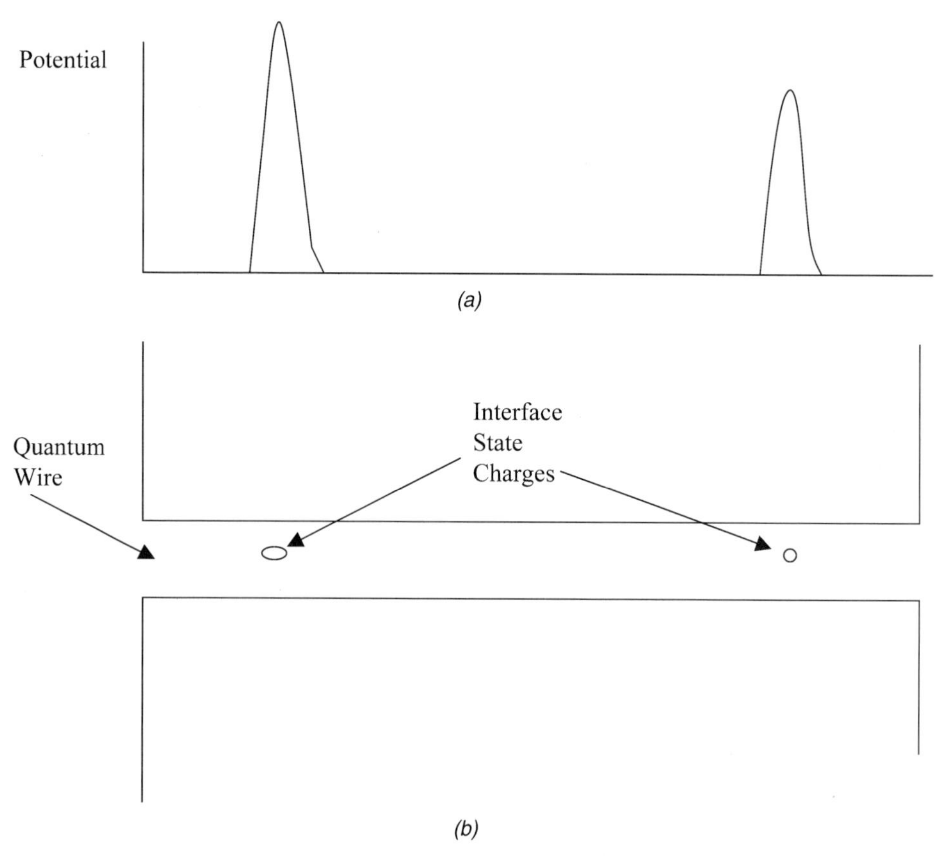

FIGURE 8.5.4 (*a*) Potential and (*b*) spatial location of the interface state charges within the quantum wire channel of the MOSFET device. Notice that the two charges create potential barriers restricting electron transport in the channel.

can be understood as follows. The conductance fluctuations were determined to be temperature dependent (Kastner, 1992) and as such were attributed to the action of a trapping mechanism. The observed behavior arises from the temperature dependence of the impurity charge distribution at the Si–SiO_2 interface. These interface charges create potential barriers within the inversion layer, forming tunnel barriers against electron flow through the wire. Therefore, to add an additional electron into the inversion layer channel, it must tunnel through the potential barrier produced by the interface impurity state. In the structures examined, typically two impurity states exist within the channel. Therefore, on average, the inversion layer channel contains two potential barriers. For an electron to flow from the source to the drain contacts, it would have to tunnel through both potential barriers, as shown schematically in Figure 8.5.4. The potential barriers formed by the interface charge states play the

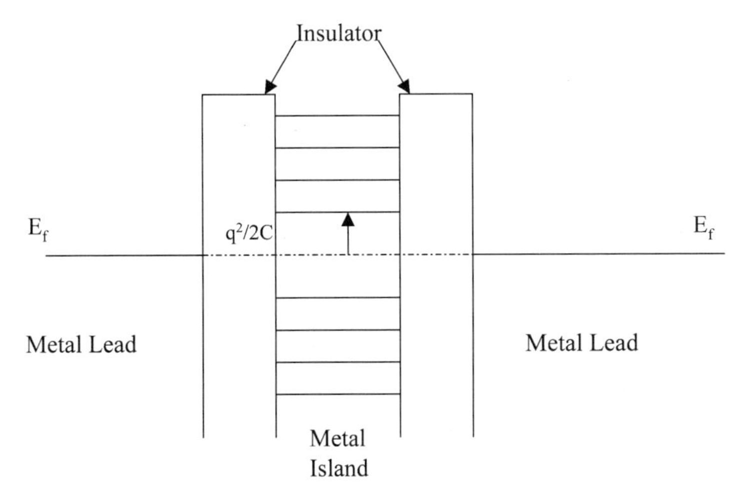

FIGURE 8.5.5 Energy band structure for the tunnel junction shown in Figure 8.5.1, including the effects of spatial quantization. Notice that only discrete energy states are allowed above and below the Coulomb blockade gap.

same role as the insulator layers in the metal island structure shown in Figure 8.5.1. In the structure of Figure 8.5.1, the metal island is isolated from the two metal leads by the insulator potential barrier. In the quantum wire MOSFET structure, the inversion layer channel is isolated from the source and drain regions by the potential barriers formed by the interface state charge. A similar physical situation results: To add charge to either isolated region, the metal island or inversion-layer channel, an electron must tunnel through the potential barrier. More significantly, the Coulomb blockade will resist electron tunneling unless the electron energy matches that of the energy gap level shown in Figure 8.5.2. As in the metal island tunnel barrier device, to add a charge to the inversion-layer channel within the MOSFET requires an energy of $q^2/2C$.

In the discussion above, the explanation for the behavior of the MOSFET is mostly classical. With the exception of the obvious quantization of charge, quantum effects have not yet been addressed. Owing to the dimensions of the structure, however, it is reasonable to expect the appearance of spatial quantization effects in the device. It was mentioned in the above that the conductance undergoes oscillations in the MOSFET structure. The period of these oscillations is related to the charging of the inversion layer; a single period corresponds to the addition of a single electron to the channel. Additionally, the magnitude of the conductance peaks changes with time. The variation of the magnitude of the conductance peaks is associated with quantum phenomena.

The small spatial size of the isolated island results in spatial quantization effects. Consequently, the allowed energy levels above and below the Coulomb blockade gap are quantized, only discrete energy values are allowed, as shown in Figure 8.5.5. For an electron to tunnel into the Coulomb island, the energy

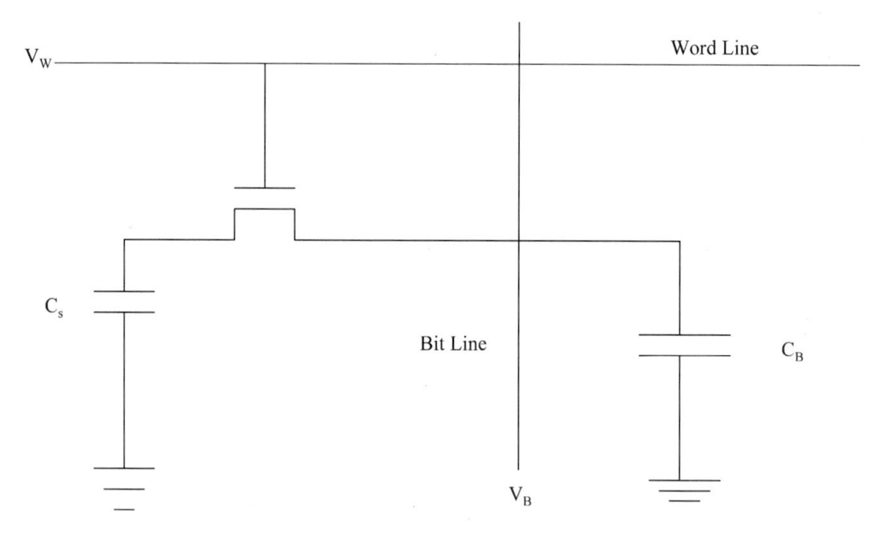

FIGURE 8.5.6 Simple single-transistor DRAM cell.

of the quantized levels must match that of the Fermi energy. The conductance of the device depends on the tunneling matrix element coupling the emitter Fermi level to a discrete energy state within the island. However, potential fluctuations due to defects or impurities cause the tunneling probability to vary randomly between allowed energy states. As a result, the magnitude of the conductance varies between quantum levels, which is observed as a variation in the height of the conductance peaks.

Before we end this section it is useful to consider applications of devices that utilize the Coulomb blockade. One of the potentially most important applications of single-electron transistors is in memory cells. The mere fact that single-electron transistors are extremely small and can be operated by the control of single charges implies that they can potentially be used to construct highly dense memory circuits. Potentially, such devices could be used to build gigascale semiconductor memory chips. One of the most attractive features of single-electron transistors for memory cell applications is that these devices use a very small number of electrons to represent one bit of information. Consequently, a single-electron transistor dissipates a very small amount of power, can potentially operate at very high speed (since the charging and discharging of the device can occur rapidly), and is compatible with existing CMOS device architectures. Since single-electron transistors can be compactly made and operate at ultralow power, three-dimensional packing of these devices is potentially possible. In addition, the operating threshold voltages for these devices depend mainly on the Coulomb charging energy, which is controlled by the island dimensions. In contrast, in most tunnel junction devices, such as resonant tunneling diodes and transistors, the performance depends critically on controlling the tunnel barrier widths. For high levels of integration this

BOX 8.5.2: Simple DRAM

Perhaps the simplest implementation of a memory cell is that given by a single-gate dynamic random access memory (DRAM). The advantage of a DRAM over a static random access memory (SRAM) is that a DRAM cell can be implemented using a single transistor. Therefore, a very high memory packing density can be attained with a concomitant reduction in cost and improvement in memory storage capability. The readout from a DRAM cell destroys the information content of the cell. In addition, the information must be refreshed periodically even when the system power is on. This is quite different from a SRAM. In a SRAM, the information can be held indefinitely as long as power is supplied to the system. However, SRAMs are more complex than DRAMs and have only about one-fourth as large a packing density. Therefore, in most commercial applications, DRAMs are preferred since they offer a substantially greater amount of memory per chip area. A representative DRAM cell is shown in Figure 8.5.6. The Word line shown in the diagram controls operation of the memory cell. The read and write functions are performed using the Word line. The actual data that are read or written into the memory are held on the Bit line. Notice that the Word line is attached to the gate of the transistor and as such, controls the state of the device. The information content is stored in the capacitor, C_s. The write operation can be understood as follows. V_W is high, called V_{DD}, which turns the transistor ON (the gate voltage is positive for an *n*-channel device). When the transistor is ON, the capacitor C_s, is tied to the bit line and V_B. A "1" is written if V_B is high, equal to V_{DD}, whereas a "0" is written if V_B is low, tied to ground. The read operation requires that the Bit line be precharged to an intermediate voltage, typically $V_{DD}/2$. The Word line is then raised high, to V_{DD}, turning the transistor ON. Thus capacitor C_s is tied to the Bit line and the charge on capacitor C_s is shared with that on the Bit line, C_B. If C_s initially holds a "1," that is, it is charged up, the voltage on the Bit line increases. Alternatively, if C_s initially holds a "0," that is, it is uncharged, the voltage on the Bit line decreases. The change in voltage is sensed and a "1" or "0" is read. Notice that the information content of the DRAM is destroyed upon reading. Therefore, each cell must have its information content restored, which is typically accomplished using a sense amplifier.

becomes highly difficult. Therefore, single-electron transistors are potentially more robust than RTD devices to processing fluctuations.

The workings of a simple single-transistor DRAM are discussed in Box 8.5.2. Here we summarize how a single-electron transistor can be used to

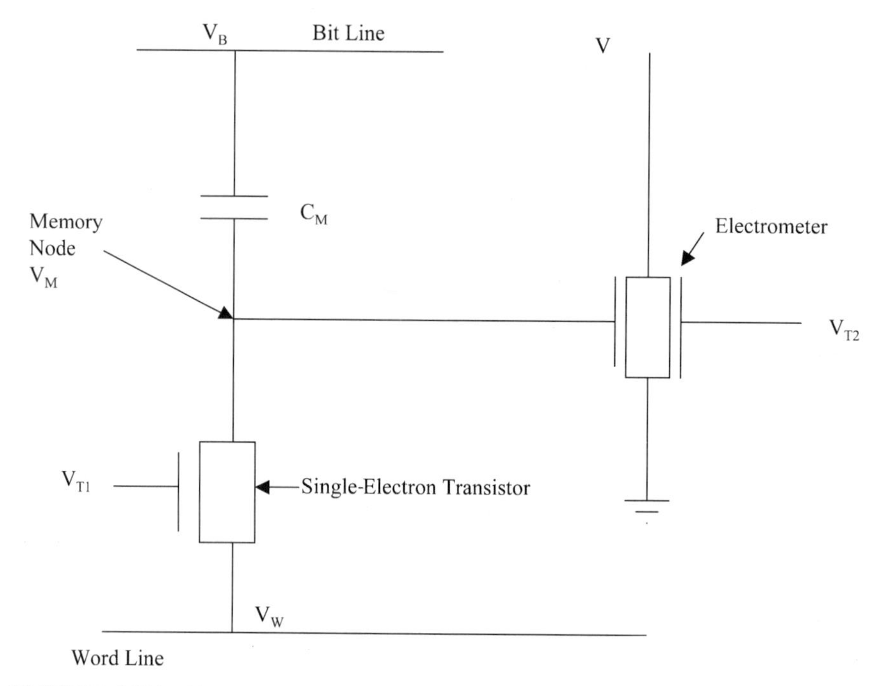

FIGURE 8.5.7 Single-electron memory cell. The charge is stored on the memory node and can only be added or subtracted by action of the single-electron transistor. The electrometer reads the charge state of the memory node.

make a simple DRAM memory cell. There are multiple approaches to utilizing single-electron transistors for memory cells. Nakazato et al. (1993) originally proposed a single-electron transistor memory cell structure which has been modified by Stone et al. (1999). The basic workings of their design can be understood as follows. The memory cell architecture of Stone et al. (1999) is similar to that of a conventional single-transistor DRAM cell, as shown by Figure 8.5.7. In the single-electron transistor memory cell, charge is stored in a memory node. Notice that a single-electron transistor controls the charging and discharging of the memory node and is connected to the Word line. This ensures that the charge on the memory node cannot escape, due to the blocking action of the Coulomb blockade. Charge flows onto the memory node when the voltage of the node, V_M, is greater than the Coulomb blocking voltage. This is accomplished by supplying to the Bit line, V_B, a positive voltage that is larger than the sum of V_W and the Coulomb gap voltage. Negative charge is then pulled onto the memory node. This constitutes the WRITE function of the memory. As a result, V_M becomes negative. Alternatively, supplying a negative bias to the Bit line can erase the memory since the negative charge is pushed off the memory node through the single-electron transistor. In this case, V_M becomes positive. Readout is accomplished using an additional single-electron transistor, that marked as an electrometer in Figure 8.5.7, to determine whether or not the memory node contains excess electrons. The electrometer is gated by the memory node. The second gate is biased to establish the operating regime of the electrometer, which is biased such that a large drain current flows. A change in the charge state of the memory node changes the gate voltage on the electrometer. This change is reflected as a change in the drain current flowing through the electrometer.

The memory cell described above operates at very low temperature, liquid helium (4.2 K) and below. The principal reason why the device of Nakazato et al. (1993) is limited to very low temperature operation is that the input voltage swing of the electrometer depends on the total capacitance related to the storage node. Even if a very small node is used with a concomitant small capacitance, parasitic capacitances related to the node are often too high to allow high-temperature operation since the change in voltage that can be detected is small with respect to the thermal voltage except at low temperature. As such, the operating temperature of this memory cell is less than about 10 K. Such low-temperature operation is not highly attractive for most computing applications. Certainly, room-temperature operation is much more attractive. Yano et al. (1994) first proposed a room-temperature single-electron transistor memory. The key issue in developing a room-temperature single-electron device is the size of the device. Room-temperature operation can be achieved only if the dimensions of the structure are less than 10 nm.

The scheme of Yano et al. (1994) utilizes a floating dot. The device geometry and resulting band structure are shown in Figure 8.5.8. A quantum dot is made within the oxide layer of the gate region of a MOSFET structure. Application of a high source-to-drain voltage results in substantial carrier heating. If an electron within the channel is heated sufficiently, it can attain a kinetic energy greater than the barrier height and be transferred into the dot through the mechanism of real space transfer. Alternatively, the electron can be transferred into the dot after heating through quantum mechanical tunneling. Once the dot is charged, the potential of the dot is lowered, blocking further transfer of electrons from the channel. This results physically from the Coulomb blockade effect. Charging of the dot changes the threshold voltage of the device. Sensing the current difference between the charged and uncharged conditions enables reading of the memory. This device scheme functions much like a flash memory cell.

Tiwari et al. (1996) proposed a similar structure that utilizes silicon nanocrystals within the oxide layer to form the quantum dots. A silicon nanocrystal formed within the oxide layer of the gate of a MOSFET is charged by electron tunnel injection from the inversion layer of the device. This is similar to the floating gate that appears in a CMOS flash memory. Depending on the gate bias, the nanocrystal can be charged by a single electron, corresponding to the write memory function, or discharged by a single elec-

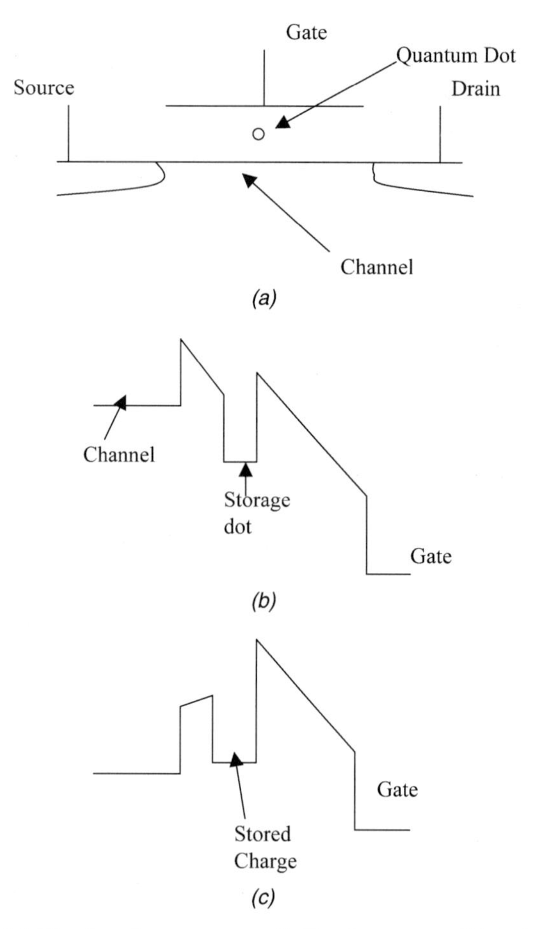

FIGURE 8.5.8 (*a*) Floating dot memory cell. Band structure (*b*) before (positive gate voltage) and (*c*) after writing into the dot.

tron, corresponding to the erase memory function. By changing the charge state of the nanocrystals, the threshold voltage of the device is shifted by a measurable amount, which is easily sensed. Due to the small size of the nanocrystal, there is a considerable Coulomb charging energy needed to insert an electron. This ensures that a significant threshold voltage is necessary to write the cell that can readily be made greater than the thermal energy. By making the charging energy larger than the thermal energy, an electron cannot enter the memory cell randomly (i.e., without there being an applied bias). This ensures that the memory cell is robust to thermal fluctuations. In the structures of Yano et al. (1994) and Tiwari et al (1996), room-temperature operation can be obtained provided that the dimensions of the dots are small.

8.6 QUANTUM COMPUTING

We conclude this chapter with a discussion of quantum computing. The essential idea behind what is commonly called quantum computing is that information can be stored, manipulated, and read out using quantum states as opposed to classical states. Such a computing machine is inherently quantum mechanical (i.e., not only are the devices quantum mechanical but the information is encoded quantum mechanically in the quantum states themselves). The fundamental principle that underlies use of quantum states as a computational resource is *quantum entanglement*, which can be understood as follows. Two quanta that are spatially separated and are arranged such that no communication between them is possible can nevertheless be strongly correlated in an inseparable state (see Box 8.6.1). Thus when a measurement is made on one of the quanta, thus collapsing its wavefunction into a definite state, instantaneously this affects the measurement of the other quanta in the pair. Such a pair of quanta are said to be in an entangled state, as discussed in Box 8.6.1.

A further question now arises: Given that two quanta can exist in an entangled state such that the measured state of one instantaneously affects the state of the other, how come such a thing is not observed in the macroscopic world? After all, quantum mechanics forms the fundamental theory of physics governing all phenomena, so how is it that the macroscopic world does not exhibit such strange behavior? In the macroscopic world, strong correlations between two events do not occur unless there is a common cause dictating this behavior. In the example of Box 8.6.1, one would be tempted to conclude that the occurrence of a strong negative correlation between the spins of the two quanta arises from some common cause, either direct communication between the two quanta or some preset condition. The fact that for numerous instances this strong correlation is always observed leads one to expect such a common cause is responsible. However, according to quantum theory, the state of the quanta is completely unknown until a measurement is made on it. It is important to recognize that the nature of this uncertainty is more fundamental than just the ignorance of the observer. The quanta itself, effectively does not "know" what state it is in (Brennan, 1999, Chap. 1). Recall that the quanta is in a state that is the linear combination of the spin-up and spin-down states, and only after a measurement does it collapse into a definite state. Therefore, according to quantum theory, there is no preset condition dictating the state of the quanta. The first possibility, that of direct communication, can be ruled out by the design of the experiment. The spatial separation of the two quanta is made sufficiently large, and the settings of the measuring instruments are changed rapidly such that the decision to measure the spin component along a certain direction with one detector is not made in time for it to influence the other quantum.

The second explanation, that of some hidden causes, often referred to as *hidden variables*, can also be ruled out. This requires a far more complicated

BOX 8.6.1: Einstein–Podolsky–Rosen Paradox and Quantum Entanglement

The simplest implementation of what has become known as the *Einstein–Podolsky–Rosen* (EPR) *paradox* is as follows. We consider the following thought experiment. Consider the decay of a single quantum that initially is known to have zero spin into two identical quanta each of which has spin angular momentum. The system initially has no net spin angular momentum. From the conservation of angular momentum law, the two emitted quanta must have net zero spin angular momentum. Let us say that the emitted quanta have spin $\frac{1}{2}$ such that measurement of the spin of either quanta would result in either spin up or down. For the net angular momentum of the two quanta to be zero, if one quantum has spin up, the other must then have spin down. The paradoxical situation arises as follows. Recall that unless a measurement is made on a system, its quantum state is undetermined (i.e., it exists in a superposition of the possible states available to it) (Brennan, 1999, Chap. 1). Therefore, in the present situation, each of the emitted quanta exist, prior to a measurement, in a state that is the superposition of both spin up and spin down. Now, let us arrange an experiment in which the two emitted quanta are spatially separated from each other by a significant distance, greater than can be traversed by light in the time it takes to perform a measurement. According to special relativity theory (Brennan, 1999, Chap. 3), nothing can travel faster than light, and no information can be transferred faster than the speed of light. Therefore, the two quanta are separated such that a measurement made on one cannot influence the measurement made on the other. However, as mentioned above, the net angular momentum of the system was zero to begin with and must, by conservation of angular momentum, continue to be zero. So if one measures the spin of one of the quanta and it is found to be spin up, we know immediately, without performing any measurement on the other quanta, that it must have spin down. This is, however, apparently in conflict with our assumption that each quantum exists in a quantum mechanical state that is the superposition of both spin up and spin down. Until a measurement is made collapsing the wavefunction of the quanta, its spin state is indefinite. Yet, after measuring only one of the quanta, the spin state of the other is known without a measurement. This is the basis of the EPR paradox.

The two emitted quanta are said to be an *entangled pair* or, equivalently, an *entangled state* since the state of one depends on the state of the other. Several experiments have been made to resolve the EPR paradox. In these experiments it has been found that the pair is always entangled such that measurement of the state of one sets the state of the other. In

(*Continued*)

BOX 8.6.1 *(Continued)*

other words, the state of one quantum is always correlated to the state of the other quantum, independent of the spatial separation of the particles and even in the case where the nature of the measurement is not decided until after they are out of interaction range. This nonlocal property of quantum mechanics has been found to prevail consistently. Basically what is thought to occur is that the two quanta cannot be considered as separate; rather, they are an entangled or strongly correlated state. We can think of these quanta as entities of a single physical system that becomes progressively spatially extended until a measurement is made on the system. Upon the application of a measurement, the system collapses and the collective state is no longer coherent. *Decoherence* is said to have occurred.

explanation. The reader is referred to the references, particularly the articles by d'Espagnat (1979) and Shimony (1985). However, the salient features of the argument are as follows. Basically, what is found is that the rules of quantum mechanics predict different results from that of what are often called *local realistic theories* for EPR experiments. Local realistic theories incorporate a common-cause model and are thus similar to hidden variables theories. The local realistic theories include *separability* (i.e., no influence of any kind can propagate faster than the speed of light) and hence once sufficiently separated, two particles can have no communication whatsoever. According to local realistic theories, the strong correlation of the quanta, which has been observed repeatedly and consistently, must be due to a preset condition of some hidden variable. Otherwise, there can be no explanation of why the two quanta always show a strong correlation. For certain EPR experiments, local realistic theories and quantum mechanics predict different outcomes. These experiments are designed to compare the predictions of local realistic theories and quantum mechanics to what is known as *Bell's inequality* (see Box 8.6.2). The rules of quantum mechanics predict that Bell's inequality does not always hold for EPR experiments in which two spin directions are measured, one for each electron in an entangled pair. In contrast, the local realistic theories predict that Bell's inequality always holds for these types of measurements. It has been found that the experiments agree primarily with the predictions of quantum mechanics, in direct disagreement with local realistic theories. Therefore, we are led to conclude that two quanta spatially separated such that they cannot communicate with one another nevertheless are entangled; the state of one is correlated with the state of the other even though no common cause or hidden variable exists.

Returning to our original question, why doesn't the macroscopic world show this strange quantum mechanical behavior? The simplest answer to this

BOX 8.6.2: Bell's Inequality

Consider spin measurements made along one of three axes, X, Y, and Z. The outcome of an individual measurement can then be either up or down, represented here as + or −. Thus if the spin is measured along the X axis and has spin up, it is represented as $X+$, with spin down as $X-$, and so on. Although it is impossible from the rules of quantum mechanics to measure two components of the spin simultaneously (see Brennan, 1999, Sec. 3.1), for ease in our argument, we assume for the moment that this can be done. Therefore, we will assume that the spin components along, for example, the X and Y axes are measured and found to be $X+$ and $Y-$. The spin component along the Z axis, although not measured directly, can have only two possibilities, + or −. Hence the measured particle must either have spin components, $X+,Y-,Z+$ or $X+,Y-,Z-$. The total number of electrons with spin components, $X+,Y-$, represented as $N(X+,Y-)$, can be written as the sum

$$N(X+,Y-) = N(X+,Y-,Z+) + N(X+,Y-,Z-)$$

where $N(X+,Y-,Z+)$ and $N(X+,Y-,Z-)$ are the number of electrons with spin components, $X+,Y-,Z+$ and $X+,Y-,Z-$, respectively. Clearly, the relation for $N(X+,Y-)$ simply states the fact that the number of elements in the set is equal to the sum of all the elements in its subsets. A similar relationship can be found for each different pair, $X+,Z-$, for example, as

$$N(X+,Z-) = N(X+,Y+,Z-) + N(X+,Y-,Z-)$$

or $Y-,Z+$,

$$N(Y-,Z+) = N(X+,Y-,Z+) + N(X-,Y-,Z+)$$

From the relationships above, $N(X+,Z-)$ must be greater than or equal to $N(X+,Y-,Z-)$, and $N(Y-,Z+)$ must also be greater than or equal to $N(X+,Y-,Z+)$. Substituting in for $N(X+,Y-,Z-)$ and $N(X+,Y-,Z+)$, the inequalities for $N(X+,Z-)$ and $N(Y-,Z+)$, $N(X+,Y-)$ can be rewritten as

$$N(X+,Y-) \leq N(X+,Z-) + N(Y-,Z+)$$

The result above for $N(X+,Y-)$ cannot be measured directly. However, one can perform an experiment on correlated or entangled pairs of electrons and not just single electrons. Using an entangled pair, the spin of

(*Continued*)

BOX 8.6.2 (*Continued*)

one electron along the X axis is first measured. For example, let us say that the spin along the X axis for one of the electrons in the entangled pair is measured to be +, giving $X+$. Since the two electrons are entangled, there is no need to measure the spin along the X axis for the second electron. It must be $X-$. If the spin along the Y axis of the second electron is measured instead to be −, yielding $Y-$, the spin of the first electron along the Y direction must be $Y+$. It can be shown that the number of pairs with say $X+, Y+$ spin components, denoted here as $n(X+,Y+)$, obeys the same type of inequality derived above for single electrons. Thus $n(X+,Y+)$ obeys

$$n(X+,Y+) \leq n(X+,Z+) + n(Y+,Z+)$$

This is called *Bell's inequality.*

Local realistic theories predict that Bell's inequality will always be obeyed, while quantum mechanics predicts that Bell's inequality will be violated for a specific choice of axes. Quantum mechanics predicts in these circumstances that there are more pairs of electrons with spin $X+,Y+$ than there are of $X+,Z+$ and $Y+,Z+$ combined. Interestingly, most experimental measurements performed over the past 20 to 30 years indicate that Bell's inequality is violated in precisely the manner predicted by quantum mechanics. Therefore, if indeed this is correct, it seems evident that two particles that have become entangled will show correlated behavior when spatially separated that does not arise from hidden variables.

question is that at the macroscopic level the quantum mechanical states are no longer coherent. In other words, the interactions between particles result in decoherence, and strong correlations no longer occur. Decoherence of quantum states arises when the state interacts with its external environment. Even the slightest interaction can cause the quantum state to collapse into decoherence. Therefore, in the macroscopic world, it is highly unusual for a system to persist in a coherent quantum mechanical state.

With the understanding described above we are now ready to examine quantum computing, at least in an introductory way. Quantum computing exploits the entanglement of coherent quantum states as a computing resource. Quantum computing proceeds using the different quantum states of atoms, particle spins, or photon polarizations as the basic computational currency. In conventional computing, a "1" is represented by a positive output voltage and a "0" by a low output voltage. Similarly, in quantum computing, a "1" could be represented by spin up and a "0" by spin down, for example. However, there is a key difference between quantum and classical computing. In classical computing the state of the device is always either a "1" (high voltage) or a "0" (low voltage). In quantum computing, the state of the system is a linear superposition of both possible states. In other words, the spin state is a coherent superposition of the spin-up and spin-down states. Until it is measured, the spin state is inherently unknown, collapsing only after a measurement into either spin up or spin down. We call such a state a *quantum bit*, or *qubit*, since it has only two possibilities, up or down, "1" or "0," much like a bit in a classical machine. However, the major difference between a qubit and a bit is that the qubit exists in a superposition of the spin-up and spin-down states, whereas for a classical bit, it is either "1" or "0." Qubits can be formed with any two-state quantum mechanical system. For example, if there are two spins in the system, the two qubits can be in a superposition of four possible states, 00, 01, 10, or 11, where 1 represents spin up and 0 spin down. Thus the two qubits represent four possible numbers. If there are three spins in the system, the three qubits can be in a superposition of eight possible numbers: 000, 001, 010, 100, 011, 101, 110, and 111. For the general case of N spins in the system, there are 2^N possible states of the wavefunction, which enables the representation of 2^N numbers simultaneously. The quantum computer exists in all these states simultaneously until a measurement is performed. Therefore, we see the interesting result that in quantum computing only a relatively few qubits are required to represent very large numbers. Additionally, the qubits can represent all possible values of the input variables to the problem we are interested in solving. Therefore, it is possible that the solution to the problem can be found in one step (i.e., upon measuring the system, it collapses into the desired final outcome). Thus a complex problem can be resolved in one step. This massive parallelism at the bit level provides an immense improvement in the computational speed of quantum computing.

How, though, is a single answer arrived at? The basic premise of quantum computing is that by using a carefully crafted algorithm, the final measurement of the system results in the outcome desired. This can be accomplished as follows. Initially, the collective state of the system exists in a simultaneous superposition of all possible states. Each of the states superposed can interfere destructively or constructively with the other states. Processing the information such that all undesired solutions interfere destructively, leaving only the desired solution, induces the desired final state upon measurement. To achieve this in general is extremely challenging and presents a major primary hurdle to quantum computing. In addition, it is important that the system remain in an entangled, coherent collective state until measurement is complete.

The basic mechanics of how a qubit can be set can be explained as follows. Let us consider a very simple two-level spin system, consisting of spin up and spin down, which for simplicity we represent as 1 and 0. A simple two-level system can be made based on the spin of the nucleus of an atom, called *nuclear spins*. The nucleus can assume either a spin-up (1) or spin-down (0) state. In some systems it is possible to contact individual spins. Using the techniques of nuclear magnetic resonance (NMR) a commonplace technology used in imaging and chemical analysis, it is possible to affect the spin state of individual nuclei. A time-dependent perturbing magnetic field with magnitude equal to the energy separation of the two states is applied to the system. This time-dependent perturbation can induce a transition between states (Brennan, 1999, Chap. 4). If the qubit is initially in 0 it can be induced into 1, and vice versa, by the action of the perturbation. In a dc field, the spin will either align or antialign with the field, giving either a 1 or 0, respectively. A particle initially in one state can be placed into the superposition state by the application of an RF pulse of appropriate frequency and duration. In this way, a single qubit can be set. A two-qubit operation can also be performed. The RF pulse is used to alter the state of one of the nuclei in a two-nuclei pair. Spin–spin coupling provides internuclei coupling, which can be used to mediate logic operations. Therefore, the state of the first nucleus will be influenced by the second nucleus via spin–spin coupling.

Let us further consider a two-qubit system. If both qubits are placed into superposition through the action of an RF pulse, neither qubit is in a definite state. If one of the qubits is measured, its state collapses into a definite state. Since the qubits are initially entangled, the state of the second qubit is also definite. For example, consider a system comprised of two nuclear spins that form the two qubits. Initially, if the two nuclei collectively have no net spin, then when one of the nuclei is measured to have spin up, the other nuclei must collapse into a spin-down state. Thus the ambiguity of the second nucleus is removed by a single measurement made on the other nucleus in the entangled pair. In a strange sense, quantum entanglement acts to “wire together” different qubits in the system.

The interaction between two nuclei can be used to create a controlled NOT gate. Calling the first nucleus the input nucleus and the second nucleus the reference nucleus, a series of RF pulses can be used to change the state of the input nucleus, depending on the state of the reference nuclei. Gershenfeld and Chuang (1998) used an NMR technique with a liquid containing chloroform molecules ($CHCl_3$) to create a two qubit system. The nuclei of the hydrogen and carbon atoms in the chloroform molecule are used as the qubits. Carbon 13 is used such that the carbon atom has a net spin. Initially, the system is set with the spin of the carbon atom pointing up and that of the hydrogen atom either up or down with respect to an external dc magnetic field. In this example, the carbon atom is the input nucleus and the hydrogen is the reference nucleus. An RF pulse is then applied perpendicular to the dc field. Depending on the duration of the RF pulse, the spin of the carbon atom will be rotated downward into the horizontal plane, as shown in Figure 8.6.1. The carbon atom then precesses, much as a top does, about the vertical dc field. The carbon atom will rotate horizontally and point in one direction or its opposite, depending on the spin orientation of the hydrogen atom. At this instant, another RF pulse is applied to the system which will rotate the carbon nucleus an additional 90° in the vertical. The rotation in the vertical is in the

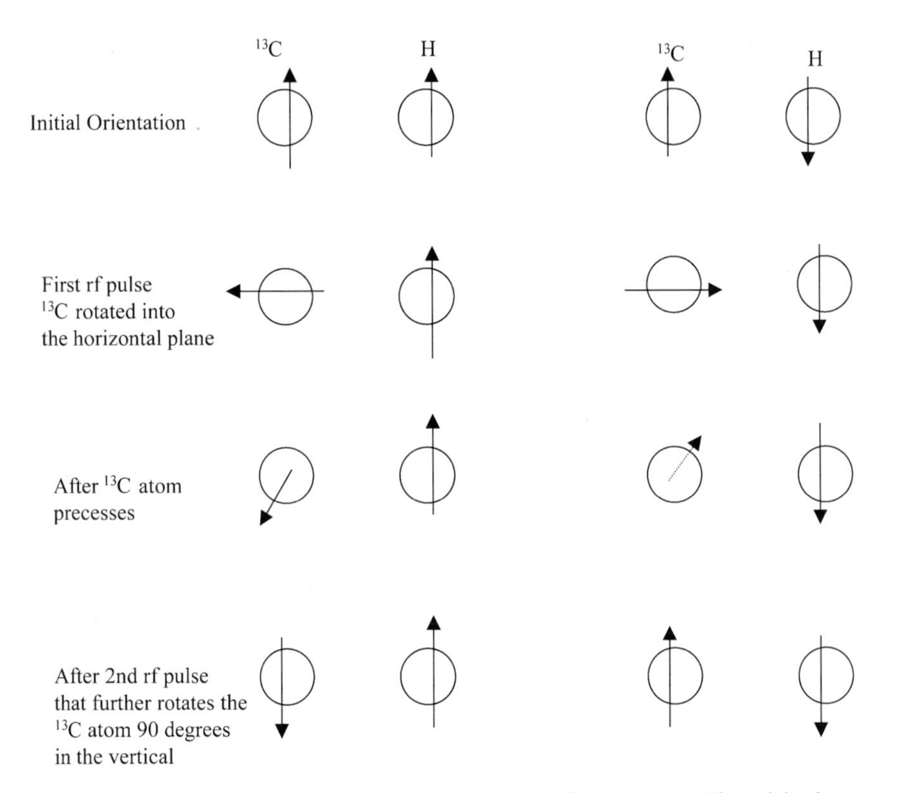

FIGURE 8.6.1 Spin states, denoted by the arrow, of the carbon 13 and hydrogen atoms as a function of the applied RF magnetic field.

same sense (i.e., clockwise or counterclockwise) independent of the initial orientation of the carbon nucleus. Thus, depending on whether the carbon atom had rotated in the plane in a counterclockwise or clockwise direction following the first RF pulse, after the second RF pulse it will end up either up or down. The sequence of events is shown in Figure 8.6.1 for two cases, hydrogen atom with spin up or down. Therefore, depending on the spin state of the hydrogen atom, the carbon nucleus will be flipped into the opposite state. Thus, if the hydrogen atom has spin up, the carbon will be flipped into the spin-down state, and vice versa. These operations constitute a controlled NOT gate since the state of one input controls whether the other input is inverted at the output. Similarly, an exclusive OR gate (XOR) can be created using a coupled two-spin system. Various logic gates can thus be constructed.

There exists a very important fundamental difference between quantum computing and classical computing. In a classical machine, every computation follows a well-defined pathway in time from the initial condition to the final result in a methodical manner. In a quantum computer, the calculation can be split into multiple pathways, owing to the superimposed, entangled quantum states that proceed in parallel with time.

Finally, we are left with the question of what added value quantum computing presents? It is well known that "classical" computers, those that function through the processing of classical bits of information, can handle vast amounts of data and perform many complex calculations. (Classical computers are often termed *Turing machines*.) Given that no quantum computer presently exists and that it is expected to be very difficult and costly to construct one, if it is possible at all, the question arises: Is there a need for a quantum computer? Additionally, what new functionality does a quantum computer add that is not available using a classical computer? To answer this question, we first need to know whether classical computers can simulate all possible natural behaviors. Classical computers have been used successfully for some time to describe the behavior of quantum systems, albeit under considerable simplifying assumptions. Nevertheless, it is questionable that quantum computers can perform tasks that are not possible on classical computers. Without going into the full details of what is computable [the interested reader is referred to the references, particularly the one by Steane (1998)], suffice it to say that it is unlikely that any class of scientific problems exist that cannot, in principle, be solved on a classical computer. However, this is not to say that all scientific problems can *in practice* be solved. There are many problems that are computationally very difficult to do and in practice may appear to be impossible. Therefore, although quantum computing may not increase the set of computational problems that can be examined, it may provide a more computationally efficient means of solving very difficult problems. In other words, quantum computers may provide a much faster means of calculating difficult problems than would otherwise be practically impossible on classical machines.

The question is, then, are there problems that a quantum computer can do that would be impractical on a classical machine? Interestingly, there are. The most famous of these problems is the factorization problem. The factorization of a large number into its primes is an exceedingly difficult task on a classical computer. It is well known that the degree of difficulty in factoring a number into its primes grows exponentially in the number of computational steps on a classical computer. For example, a number with 130 digits can be factored in a matter of days using a classical machine. However, if the number of digits is doubled, the problem is intractable on a classical computer, since the computational expense on present state-of-the-art workstations rises to about 1 million years! Therefore, encoding messages using factoring of large numbers into primes provides an excellent cryptographic strategy. In fact, this is the basis for most Internet security systems.

It was found by Shor (1999) that using a quantum computer, an N-digit number could be factored into its primes in a time of order N^3. In contrast, using a classical computer, the time requirement grows faster than any power with an increasing number of digits. The important implication of Shor's theory is that a quantum computer can greatly jeopardize the security of most existing encryption schemes that rely on the factoring of large numbers into their primes.

Aside from the factorization of large numbers, quantum computers may also be of importance in searching algorithms, an example of which is the searching of an unsorted database. Grover (1997) has shown that a quantum computer can search an unsorted database far faster than can a classical computer. An example of an unsorted database is a telephone directory, wherein one has the phone number but not the name. Searching for a name using only a phone number is a computationally expensive problem on a classical computer, requiring a time of order N for N names. Using a quantum computer, Grover (1997) has shown that searching such a directory of N names, would require a time on order of the square root of N, which implies a significant improvement as the number of names, N, increases.

Clearly, there exist important problems to which quantum computing can make significant contributions. The next question is: Can a quantum computer be built, and will it work? Presently, the answers to these questions are not clear. There are numerous obstacles to the construction of a working quantum computer that contains sufficient power to be useful (Preskill, 1998). Although modest machines using only a handful of qubits have already been built, scaling these machines to the range of computational power necessary to supplement classical computers remains questionable. The principal limitation to quantum computing is the fact that highly elaborate entangled quantum states must be prepared and maintained within the machine. A practical quantum computer would need to prepare and maintain states that involve the superposition of a large number of qubits in a coherent, entangled state that could persist for sufficient time to perform the calculation. Such a state is subject to random and spontaneous relaxation into decoherence. A single relaxation event, triggered by virtually any external environmental influence, would ruin the coherence of the state and thus spoil the computation. In essence, the machine would have to be completely isolated from the external environment and *any* influence that could lead to decoherence.

It is possible that error-correcting algorithms can be employed to protect highly entangled encoded states from decoherence. Of course, error correction necessarily increases the complexity of the machine and potentially slows the calculation. Nevertheless, one can argue that these limitations can be overcome and that such a machine could be possible. The debate over the potential of quantum computing is sure to continue. Only time will tell whether or not quantum computing will emerge as a practical technological supplement to classical computers. For a more comprehensive discussion of the pros and cons of quantum computing, the reader is referred to articles by Lloyd (1993), Haroche and Raimond (1996), and Preskill (1998).

PROBLEMS

8.1. Show that in a quantum dot, the operator corresponding to the z component of the angular momentum L_z can be written as

$$L_z = ih(a_y^\dagger a_x - a_x^\dagger a_y)$$

8.2. For a quantum dot, determine the commutation relation between the Hamiltonian H and L_z. Use the Hamiltonian determined in Example 6.2.1 and L_z determined in Problem 6.1.

8.3. Determine the following commutators for the creation and annihilation operators a and b that are used to describe the angular motion of an electron in the parabolic potential associated with a quantum dot dual-gate transistor. Operators a and b are given as

$$a = \frac{1}{\sqrt{2}}(a_x - ia_y) \qquad b = \frac{1}{\sqrt{2}}(a_x + ia_y)$$

Evaluate the following: $[a, a^\dagger]$, $[a, b]$, $[a, b^\dagger]$.

8.4. Determine the values allowed for the quantum number m that describes the states for L_z in a quantum dot. Assume that the dot has a quadratic potential.

8.5. For two electrons in a four-quantum-dot cell, determine the ratio in the potential energy between the diagonal and adjacent configurations by treating the electrons as point particles. Assume that the separation between adjacent dots is R and that the four-dot cell is square.

8.6. For a four-qubit quantum computer, write the wavefunction that describes the system. Use constants c_i for the coefficients for each term. Use the following notation to represent a single state of the machine, $|0000\rangle$, for a state with all bits zero, and so on. How many states comprise the wavefunction? What does the square of the coefficient c_i mean physically?

8.7. We can give a crude estimate of the decoherence time due to phonon scattering of a quantum computer as follows. Let T be the relaxation time of a single qubit and t the operation time of a single gate. Let R be the product of the number of qubits and the number of gate operations that the machine must make to perform a task. Let us assume that for a practical task, R would have to be about 10^{10}.

(a) If we assume that the operation time of a single gate, t, is 1 ms (which is roughly the time for an ion trap typically used in quantum computing), determine the coherence time.

(b) If instead of an ion trap, a semiconductor system is employed such that the operation time is on the order of 10 ps, determine the coherence time.

(c) In a semiconductor system, assume that decoherence occurs from single-phonon scattering. If at 300 K the mean time between scatterings is 10^{-13} s, determine the mean time between scatterings at 4 K if the phonon energy is 35 meV, and if all the temperature dependence is absorbed in the Bose–Einstein factor. Assume that only absorption processes occur. Neglect the chemical potential. (As can be seen from this problem, phonon scattering is not the most important decoherence effect. In most proposed quantum computers, decoherence arises primarily from spontaneous emission processes rather than from phonon scattering events.)

CHAPTER 9

Magnetic Field Effects in Semiconductors

In this chapter we discuss the effects of magnetic fields in semiconductors. We start with a discussion of the effect of magnetic fields on quantum states and introduce the concept of Landau levels. Next we discuss the Hall effect, both classical and quantum mechanical. The Hall effect is the most basic magnetic field effect in a semiconductor.

9.1 LANDAU LEVELS

In this section we examine the consequence of the application of an external magnetic field on the energy levels in a semiconductor. It is found that application of a magnetic field alters the allowed energy levels by introducing additional states equally spaced in energy. These additional states are called *Landau levels*. As we will see below, the action of the applied magnetic field results in a harmonic oscillator equation. The resulting energy levels for the system, as for any harmonic oscillator, are equally spaced.

It was found in Brennan (1999, Chap. 1) that the transition from classical mechanics to quantum mechanics can be made by first determining the classical expression for a physical observable and then rewriting it using the corresponding quantum mechanical operators. As a further requirement, the operator must be Hermitian. In classical mechanics the generalized momentum can be written as

$$m\mathbf{v} = \mathbf{p} - \frac{q\mathbf{A}}{c} \qquad 9.1.1$$

where A is the vector potential, defined as

$$\mathbf{B} = \nabla x \mathbf{A} \qquad 9.1.2$$

and c is the speed of light (cgs units are used). The corresponding quantum mechanical form of the momentum can be written as

$$mv_{\text{op}} = p_{\text{op}} - \frac{qA}{c} \qquad 9.1.3$$

Therefore, the Hamiltonian of a free electron in a uniform magnetic field can be constructed as

$$H = T + V = \frac{1}{2m}\left(p - \frac{q}{c}A\right)^2 \qquad 9.1.4$$

where the potential V is zero since the electron is assumed to be free. If the magnetic field is assumed to be uniform and point along the z axis, from Eq. 9.1.2 the vector potential is given as

$$\mathbf{A} = -yB\hat{\mathbf{i}} + 0\hat{\mathbf{j}} + 0\hat{\mathbf{k}} \qquad 9.1.5$$

Substituting Eq. 9.1.5 for the vector potential in Eq. 9.1.4, the Hamiltonian becomes

$$H = \frac{1}{2m}\left[\left(p_x + \frac{qyB}{c}\right)^2 + (p_y^2 + p_z^2)\right] \qquad 9.1.6$$

The corresponding Schrödinger equation becomes

$$\frac{1}{2m}\left(p_x^2 + p_y^2 + p_z^2 + 2\frac{qyBp_x}{c} + \frac{q^2B^2y^2}{c^2}\right)\psi = E\psi \qquad 9.1.7$$

Completing the square in the Hamiltonian operator of Eq. 9.1.7 yields

$$\left[\frac{p^2}{2m} + \frac{q^2B^2}{2mc^2}\left(y + \frac{cp_x}{qB}\right)^2 - \frac{p_x^2}{2m}\right]\psi = E\psi \qquad 9.1.8$$

Noticing that the operators p_x and p_z commute with the Hamiltonian, the eigenfunctions ψ are simultaneous eigenstates of H and p_x and p_z. Therefore, the eigenfunctions can be written as

$$\psi = e^{i(k_x x + k_z z)} f(y) \qquad 9.1.9$$

Substituting Eq. 9.1.9 for ψ into Eq. 9.1.8 yields

$$\left[\frac{p_y^2}{2m} + \frac{q^2B^2}{2mc^2}\left(y + \frac{cp_x}{qB}\right)^2\right] f = \left(E - \frac{\hbar^2 k_z^2}{2m}\right) f \tag{9.1.10}$$

If we make the following substitutions, realizing that the magnetic field B is uniform,

$$K \equiv \frac{q^2B^2}{mc^2}$$
$$y_0 \equiv -\frac{cp_x}{qB} \tag{9.1.11}$$

the bracketed term on the left-hand side of Eq. 9.1.10 becomes

$$\frac{p_y^2}{2m} + \frac{1}{2}K(y - y_0)^2 \tag{9.1.12}$$

which is the Hamiltonian for a harmonic oscillator. Therefore, the energy eigenvalues are found as discussed in Brennan (1999, Chap. 2) by making the association that the frequency Ω is given by

$$\Omega = \sqrt{\frac{K}{m}} \tag{9.1.13}$$

The energy eigenvalues are then given as

$$E_n - \frac{\hbar^2 k_z^2}{2m} = \hbar\Omega\left(n + \frac{1}{2}\right) \tag{9.1.14}$$

which can be rewritten as

$$E_n = \frac{\hbar^2 k_z^2}{2m} + \hbar\Omega\left(n + \frac{1}{2}\right) \tag{9.1.15}$$

Notice that Eq. 9.1.15 contains two terms. The first term is the energy corresponding to free motion along the z axis. Recall that if the magnetic field is applied along the z axis, no external force due to the magnetic field acts on the electron in this direction, neglecting the relatively weak interaction of the spin magnetic moment and the field. Therefore, the electron motion along the z axis is free and the energy corresponding to this motion is simply given by the kinetic energy of motion in the z direction. The second term corresponds to the energy of a simple harmonic oscillator. The applied magnetic field acts on particles moving in the x–y plane. From the Lorentz force law for a charged particle in a uniform magnetic field, the force is perpendicular to the velocity of the particle, so the acceleration is also perpendicular to the velocity. Consequently, the magnetic field causes the particles to move in circular orbits.

The wavefunctions allowed for the electron are in general given by Eq. 9.1.9. The form of the function $f(y)$ is determined from the solution of Eq. 9.1.10. Using the substitutions given by Eqs. 9.1.11 and 9.1.12, the resulting equation for $f(y)$ is

$$\left[\frac{p_y^2}{2m} + \frac{1}{2}K(y - y_0)^2\right] f(y) = \left(E - \frac{\hbar^2 k_z^2}{2m}\right) f(y) \qquad 9.1.16$$

As mentioned above, the solution of this differential equation is that of a harmonic oscillator. The corresponding wavefunctions are given as

$$\psi \sim e^{ik_x x} e^{ik_z z} H_n(\xi) e^{-\xi^2/2} \qquad 9.1.17$$

where ξ is defined as

$$\xi = \sqrt{\frac{m\omega}{\hbar}}(y - y_0) \qquad 9.1.18$$

Equation 9.1.15 can be further generalized to account for the energy of the electron spin magnetic moment in the presence of the external magnetic field. As discussed by Brennan (1999, Chap. 3), the energy of an electron due to interaction of its spin magnetic moment with an external magnetic field is, $g\mu_B B_z$, where μ_B is the spin magnetic moment, g the Landé g factor, and B_z the component of the magnetic field along the z direction.

It is instructive to analyze the motion classically to confirm that the motion is circular. With no external electric field, the Lorentz force law is

$$\mathbf{F} = \frac{q}{c}\mathbf{v} \times \mathbf{B} \qquad 9.1.19$$

Again it is assumed that the magnetic field is oriented only along the z direction. If the velocity is assumed to be within the x–y plane, the force law gives

$$|\mathbf{F}| = \frac{mv^2}{r} = \frac{qvB}{c} \qquad 9.1.20$$

where we have recognized that a centripetal acceleration results. Solving for r yields

$$r = \frac{mvc}{qB} \qquad 9.1.21$$

The angular velocity is obtained from the ratio of v to r as

$$\Omega = \frac{v}{r} = \frac{qB}{mc} \qquad 9.1.22$$

The result for Ω above is readily seen to be the same as that given by Eq. 9.1.13. From the definition of K, Eq. 9.1.13 becomes

$$\Omega = \sqrt{\frac{K}{m}}$$
$$K = \frac{q^2B^2}{mc^2} \qquad 9.1.23$$
$$\Omega = \sqrt{\frac{q^2B^2}{m^2c^2}} = \frac{qB}{mc}$$

which is precisely the same as Eq. 9.1.22. Therefore, the motion of a charged particle in a uniform magnetic field is circular with an angular frequency of qB/mc. Quantum mechanically, the energy is quantized with values given by the solution of the harmonic oscillator. The trajectory of the particle is helical. The particle spirals along the z axis.

It is interesting to examine the effect of the applied magnetic field on the density of states. We first examine the behavior of a three-dimensional system. The density of states for a three-dimensional system is given by (Brennan, 1999, Sec. 5.1)

$$\frac{dN}{dE} = \frac{2V}{(2\pi)^3}\frac{d^3k}{dE} \qquad 9.1.24$$

where the factor of 2 in the numerator is due to the spin. If the interaction of the spin magnetic moment and the external field is neglected, the motion of the electron along the z direction is unaffected by the magnetic field. It is assumed that the crystal is bounded with dimensions L_x, L_y, and L_z. The number of states for a given value of k_z is

$$N_z dk_z = 2\frac{L_z}{2\pi}dk_z \qquad 9.1.25$$

where the factor of 2 comes from summing both positive and negative k_z. The number of states in the plane perpendicular to the magnetic field N_{xy} between k_r and $k_r + dk_r$ can be determined as

$$N_{xy} = 2\pi k_r dk_r \frac{L_x L_y}{(2\pi)^2} \qquad 9.1.26$$

where k_r is the magnitude of the k-vector in the plane perpendicular to the magnetic field. Recall that the energy separation between adjacent Landau levels is equal to $\hbar\Omega$. Therefore, we must have

$$\frac{d(\hbar^2k_r^2/2m)}{dk_r}dk_r = \hbar\Omega \qquad 9.1.27$$

Taking the derivative in Eq. 9.1.27 and using the last of Eqs. 9.1.23 yields

$$\frac{\hbar^2 k_r}{m} dk_r = \frac{\hbar q B}{mc} \tag{9.1.28}$$

Using Eq. 9.1.28 to solve for $k_r dk_r$ and then substituting into Eq. 9.1.26, N_{xy} becomes

$$N_{xy} = 2\pi \frac{qB}{\hbar c} \frac{L_x L_y}{(2\pi)^2} = \frac{qB}{\hbar c} \frac{L_x L_y}{2\pi} \tag{9.1.29}$$

The density of states is then given by the product of the factors given by Eqs. 9.1.25 and 9.1.29 after converting to energy as

$$D(E_z) dE_z = \sum_n \frac{VqB\sqrt{2m}}{(2\pi\hbar)^2 c\sqrt{E_z}} dE_z \tag{9.1.30}$$

where the sum is over the quantum states that arise from the Landau levels present in the expression for E_z, and V is the volume given by the product $L_x L_y L_z$. For mks units, the factor $1/c$ is not present. The energy is given by Eq. 9.1.15. Using Eq. 9.1.15, E_z can be written as

$$\frac{\hbar^2 k_z^2}{2m} = E - \left(n + \frac{1}{2}\right)\hbar\Omega \tag{9.1.31}$$

Substituting into Eq. 9.1.30 the expression for energy given in Eq. 9.1.15, the density of states for a three-dimensional system in the presence of a magnetic field finally becomes

$$D(E) dE = \sum_n \frac{VqB\sqrt{2m}}{(2\pi\hbar)^2 c\sqrt{E - \hbar\Omega(n + \frac{1}{2})}} dE \tag{9.1.32}$$

Again, the factor of $1/c$ in Eq. 9.1.30 is not present when mks units are used. Inspection of Eq. 9.1.32 shows that singularities appear in the density of states. When the energy is equal to that of a Landau level, a sharp increase in the density of states occurs due to the singularity in the denominator as shown by Figure 9.1.1. In realistic systems, energy broadening removes the divergence, however, the periodic behavior of the density of states is retained.

It is interesting to compare the allowed states in the presence of the external magnetic field and without the external field present. For a free three-dimensional electron gas, the Fermi surface is simply a sphere. The action of the applied external magnetic field changes the allowed energy states of the system. Since the magnetic field does not affect motion along the z direction, the motion in this direction corresponds to that of a free particle as given by the term with k_z in Eq. 9.1.15. For motion in the x–y plane, the analysis above

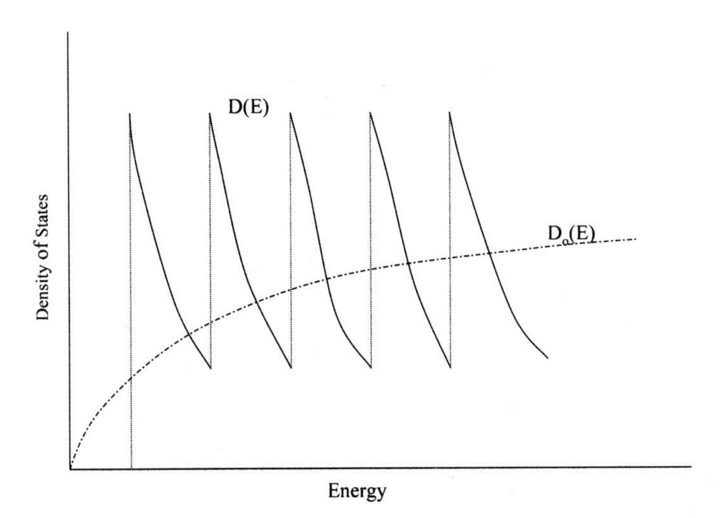

FIGURE 9.1.1 Density of states for a three-dimensional system without an applied external magnetic field (dashed dotted curve), $D_0(E)$, and in the presence of an external magnetic field $D(E)$ (solid curve).

shows that the energy is quantized as $h\Omega(n + \frac{1}{2})$. The allowed states in k-space along the k_x and k_y directions are located on circles with radii given as

$$k_x^2 + k_y^2 = \frac{2m}{\hbar^2}\hbar\Omega\left(n + \frac{1}{2}\right) \qquad 9.1.33$$

Therefore, the allowed states lie on concentric cylinders about the k_z axis as shown in Figure 9.1.2.

The density of states for the system in the presence of an external magnetic field (Figure 9.1.1) can now be further understood. As can be seen from Figure 9.1.2, no allowed states exist inside the cylinder corresponding to $n = 0$. The density of states must then be zero for energies less than $1/2\hbar\Omega$, as shown in Figure 9.1.1. The density of states rises abruptly once the first cylinder is reached, which gives rise to the sharp increase in the density of states shown in Figure 9.1.1. Further sharp increases in the density of states follow as new cylinders are reached. In contrast, the density of states for a free three-dimensional system increases with the square root of E, as shown by the function labeled $D_0(E)$ in Figure 9.1.1. The total number of states under the two conditions, with and without the magnetic field, remains the same.

It is also important to determine the degeneracy of a Landau level. We recall that no two electrons can occupy the same identical quantum state simultaneously, due to the Pauli principle (Brennan, 1999, Chap. 3). Therefore, the number of states available limits the number of electrons that can occupy a collection of states. From the density-of-states plot shown in Figure 9.1.2, it is

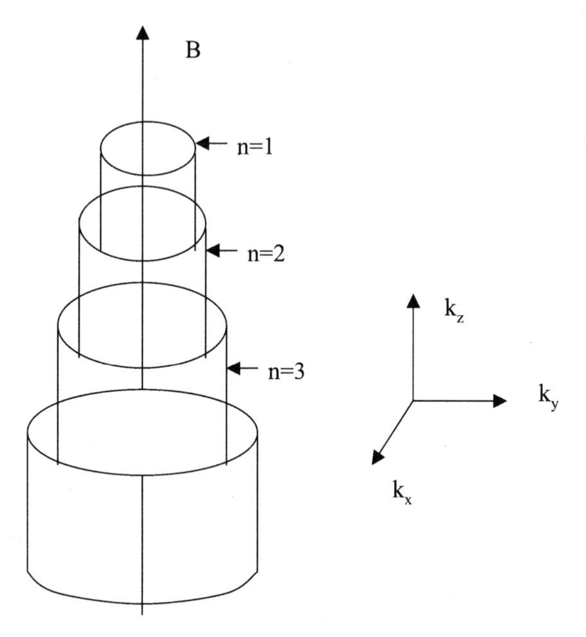

FIGURE 9.1.2 Effects of an applied magnetic field along the z axis. Notice that the allowed energy surfaces become a set of concentric cylinders with axes parallel to the magnetic field.

clear that there exist energy gaps between each concentric cylinder. Therefore, given that there are a finite number of states in each cylinder, upon filling an entire cylinder there is a jump in energy before another available state exists. The total number of states available in each Landau level can be calculated as follows. We assume periodic boundary conditions for the system. If the system is assumed to be cubic with walls of length L, the periodic boundary conditions imply that

$$\psi(x, y, z) = \psi(x + L, y + L, z + L) \qquad 9.1.34$$

Consider motion in the x direction where the particle motion is free. Substitution of the free particle wavefunction in the x direction into Eq. 9.1.34 yields

$$k_x = \frac{2\pi n_x}{L} \qquad 9.1.35$$

where n_x is an integer. The parameter y_0 must lie between 0 and L, since it is associated with the center of the circular classical motion. Therefore, the maximum value of y_0 is L. Using the expression for y_0 given by Eq. 9.1.11,

EXAMPLE 9.1.1: Quantum Dot in a Magnetic Field

Determine the eigenenergies of an electron in a circular quantum dot with a symmetric parabolic potential in the presence of a uniform static magnetic field B. The vector potential A is assumed to be symmetric and is given as

$$\mathbf{A} = \tfrac{1}{2}(-By\hat{\mathbf{i}} + Bx\hat{\mathbf{j}})$$

The kinetic energy operator can then be written as (see Problem 9.1)

$$\frac{1}{2m}\left(\mathbf{p} - \frac{q}{c}\mathbf{A}\right)^2 = \left[-\frac{\hbar^2\nabla^2}{2m} + \frac{iq\hbar B}{2mc}\left(x\frac{\partial}{\partial y} - y\frac{\partial}{\partial x}\right) + \frac{1}{4}\frac{q^2B^2}{2mc^2}(x^2 + y^2)\right]$$

The potential energy has two terms, one due to the confining parabolic potential of the quantum dot and the other due to the interaction of the electron spin magnetic moment with the external field. The potential energy is then

$$V = \frac{1}{2}m\omega^2(x^2 + y^2) + \frac{g\mu_B}{\hbar}\mathbf{B}\cdot\mathbf{S}$$

where the last term is the electron energy due to the interaction of the spin magnetic moment and the external field (Brennan, 1999, Sec. 3.6). The total Hamiltonian for the quantum dot is then

$$\left[-\frac{\hbar^2\nabla^2}{2m} + \frac{iq\hbar B}{2mc}\left(x\frac{\partial}{\partial y} - y\frac{\partial}{\partial x}\right) + \frac{1}{4}\frac{q^2B^2}{2mc^2}(x^2 + y^2)\right] + \frac{1}{2}m\omega^2(x^2 + y^2) + \frac{g\mu_B}{\hbar}\mathbf{B}\cdot\mathbf{S}$$

The terms involving $x^2 + y^2$ can be combined as follows. Factoring out the $x^2 + y^2$ term given as

$$\left(\frac{1}{4}\frac{q^2B^2}{2mc^2} + \frac{1}{2}m\omega^2\right)(x^2 + y^2)$$

We then set the first term in the equation above equal to $\frac{1}{2}m\Omega^2$:

$$\frac{1}{2}m\Omega^2 = \frac{q^2B^2}{8mc^2} + \frac{1}{2}m\omega^2 = \frac{1}{2}m\left(\frac{q^2B^2}{4m^2c^2} + \omega^2\right)$$

(Continued)

EXAMPLE 9.1.1 (*Continued*)

Defining ω_c as the cyclotron frequency,

$$\omega_c \equiv \frac{q^2B^2}{m^2c^2}$$

then Ω is

$$\Omega \equiv \sqrt{\omega^2 + \frac{\omega_c^2}{4}}$$

With these assignments, the Hamiltonian becomes

$$H = -\frac{\hbar^2\nabla^2}{2m} + \frac{iqB\hbar}{2mc}\left(x\frac{\partial}{\partial y} - y\frac{\partial}{\partial x}\right) + \frac{1}{2}m\Omega^2(x^2 + y^2) + \frac{g\mu_B}{\hbar}\mathbf{B}\cdot\mathbf{S}$$

But the term involving the derivatives of y and x can be rewritten in terms of L_z, the z component of the angular momentum operator, as (Brennan, 1999, Sec. 3.1)

$$\frac{iL_z}{\hbar} = \left(x\frac{\partial}{\partial y} - y\frac{\partial}{\partial x}\right)$$

The Hamiltonian can now be written using L_z as

$$H = -\frac{\hbar^2\nabla^2}{2m} + \frac{1}{2}m\Omega^2(x^2 + y^2) - \frac{1}{2}\omega_c L_z + \frac{g\mu_B}{\hbar}\mathbf{B}\cdot\mathbf{S}$$

Recognizing that the vector product of B and S is equal to

$$\mathbf{B}\cdot\mathbf{S} = \pm\frac{\hbar}{2}B$$

and that if the charge is negative the L_z term changes sign, H becomes

$$H = -\frac{\hbar^2\nabla^2}{2m} + \frac{1}{2}m\Omega^2(x^2 + y^2) + \frac{1}{2}\omega_c L_z \pm \frac{g\mu_B B}{2}$$

Finally, the energy eigenvalues for this Hamiltonian are

$$E = \left(n + \frac{1}{2}\right)\hbar\Omega + \frac{1}{2}\hbar\omega_c m' \pm \frac{g\mu_B B}{2}$$

where m' is the quantum number associated with the z component of the angular momentum and is primed here to differentiate it from the mass. The quantum number n is, of course, associated with the parabolic potential, and the last term gives the energy due to the spin of the electron.

EXAMPLE 9.1.2: Quantum Wire in the Presence of a Magnetic Field

Consider a system like that in the text, in which a magnetic field is applied along the z direction. However, in this case, the system is a heterostructure wherein a two-dimensional electron gas is formed at the interface. Therefore, motion in the z direction is quantized so that in place of a continuum of energy states, only discrete energy states exist. Further, let us assume that the only direction that is not constrained is along the x axis. Etching sidewalls into the structure restricts the motion along the y axis. Thus the system of interest forms what is called a quantum wire; that is, motion is constrained in two directions, z and y, with free motion only along x.

If the system is two-dimensional, the energy eigenvalues are given by Eq. 9.1.15 except that motion along the z axis is quantized, giving

$$E_n = E_i + \hbar\Omega(n + \tfrac{1}{2})$$

where E_i are the quantum subband energies along the z direction. We start with Eq. 9.1.16,

$$\left[\frac{p_y^2}{2m} + \frac{1}{2}K(y - y_0)^2\right] f(y) = E_n f(y)$$

where $K = m\Omega^2$. Let us assume that the potential along the y direction, $V(y)$, due to confinement along the y direction is $\frac{1}{2}m\Omega_0^2 y^2$. We recall that y_0 is given by Eq. 9.1.11 as

$$y_0 = -\frac{cp_x}{qB}$$

With these definitions, the Schrödinger equation becomes

$$\left[\frac{p_y^2}{2m} + \frac{m}{2}(\Omega^2 + \Omega_0^2)y^2 - m\Omega^2 y y_0 + \frac{m}{2}\Omega^2 y_0^2\right] f(y) = E_n f(y)$$

The equation above can be rewritten by completing the square (see Problem 9.2) and using the following definitions:

$$\Omega'^2 = \Omega^2 + \Omega_0^2 \qquad y_0' = \frac{y_0\Omega^2}{\Omega'^2} \qquad M = \frac{m\Omega'^2}{\Omega_0^2}$$

(*Continued*)

EXAMPLE 9.1.2 (*Continued*)

and we recall, by combining Eqs. 9.1.11 and 9.1.13, that

$$\Omega = \frac{qB}{mc}$$

The resulting Hamiltonian for the one-dimensional quantum wire then becomes

$$H = \frac{p_y^2}{2m} + \frac{m}{2}\Omega'^2(y - y_0')^2 + \frac{\hbar^2 k_x^2}{2M}$$

Notice that H has effectively two components, a harmonic oscillator part for motion along the y direction and a free-electron-like relationship for motion along the x direction but with mass M. Recall that the energy eigenvalues of the entire system will include the quantum levels from the z direction confinement as well as those from the harmonic oscillator in y and free motion in x.

we obtain

$$L = \frac{-cp_x}{qB} = -\frac{c\hbar k_x}{qB} = -\frac{chn_x}{qBL} \qquad 9.1.36$$

The maximum number of states per level, the degeneracy of the Landau level, is then found by solving Eq. 9.1.36 for n_x. The maximum number of states n_x is then

$$n_x = \frac{qB}{hc}L^2 \qquad 9.1.37$$

Dividing Eq. 9.1.37 through by L^2 yields the maximum number of states per unit area s,

$$s = \frac{n_x}{L^2} = \frac{qB}{hc} \qquad 9.1.38$$

9.2 CLASSICAL HALL EFFECT

The classical Hall effect is of great importance in semiconductors since it provides a convenient method by which the sign and magnitude of the carrier concentration can be measured experimentally. The basic experimental setup is sketched in Figure 9.2.1. An external magnetic field B is applied to a p-type

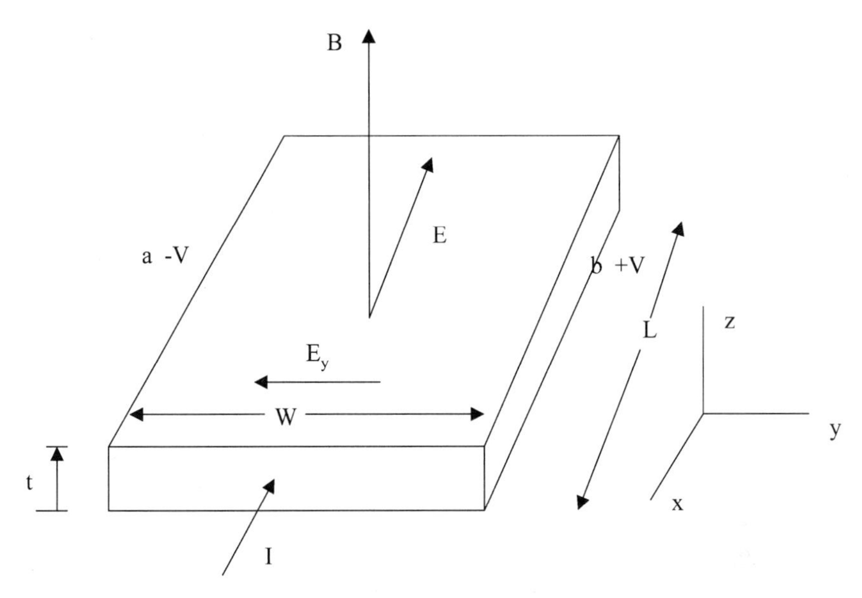

FIGURE 9.2.1 Classical Hall effect scheme. The magnetic field B is applied along the z axis perpendicular to the plane of the sample. The induced Hall field appears between a and b in the diagram.

semiconductor sample. For simplicity, we assume that the field is oriented along the z axis as shown in Figure 9.2.1. Let an electric field E be applied along the negative x direction. Thus the hole flow is also in the negative x direction. The Lorentz force law can be generalized with the addition of the electric field term from Eq. 9.1.19 as

$$\mathbf{F} = \frac{q}{c}\mathbf{v} \times \mathbf{B} + q\mathbf{E} \tag{9.2.1}$$

For holes the charge is positive. The magnetic field exerts a force on the holes directing them into the $+y$ direction. Therefore, the holes are deflected as they move through the sample toward the $+y$ side. This results in a disturbance in the space charge balance. More positive charge accumulates on the $+y$ side (right-hand side of the diagram) of the device than on the $-y$ side (left-hand side of the diagram). Similarly, the left-hand side becomes more negatively charged. This separation of charge results in the production of an electric field E_y that is directed from the right to the left sides of the sample (points a to b in Figure 9.2.1). Ultimately, steady-state conditions are obtained, in which the action of the magnetic field is balanced by the induced electric field E_y as

$$E_y = \frac{v_x B_z}{c} \tag{9.2.2}$$

The induced electric field E_y is called the *Hall field*. The resulting voltage, the Hall voltage, is given as

$$V_{ab} = E_y W \tag{9.2.3}$$

where W is the width of the sample. The magnitude of the current density is given as

$$J_x = qpv_x \tag{9.2.4}$$

The Hall field then becomes

$$E_y = \frac{J_x B_z}{qpc} = R_H J_x B_z \tag{9.2.5}$$

where R_H is called the *Hall coefficient*. Measurement of the Hall coefficient provides a direct measurement of the carrier concentration. This is easily seen from the definition of R_H, since

$$R_H = \frac{1}{qpc} \tag{9.2.6}$$

If an n-type material is used instead, the Hall coefficient is given as

$$R_H = -\frac{1}{qnc} \tag{9.2.7}$$

If mks units are used in place of cgs units, the factor c is removed from Eqs. 9.2.6 and 9.2.7.

In addition to the carrier concentration and carrier type, the Hall effect measurement can be used to determine the carrier mobility. The mobility can be determined as follows. Generally, the resistivity is a tensor quantity, and as we will see below, has off-diagonal terms. In the simple case where the current flow and the voltage drop are in the same direction, the resistivity can be taken as a scalar, ρ, and is defined as

$$\rho = \frac{R_x A}{L} \tag{9.2.8}$$

where R_x is the resistance in the direction of the current flow, A the area of the sample, and L its length. From Figure 9.2.1, the area of the sample is given as Wt. The resistance is equal to the ratio of the voltage along the x direction in the device, V_x, to the current I_x in that direction as

$$R_x = \frac{V_x}{I_x} \tag{9.2.9}$$

The conductivity σ is equal to $1/\rho$. In Brennan (1999, Chap. 6) the conductivity was determined to be

$$\sigma = q\mu_n n = \frac{1}{\rho} \quad 9.2.10$$

Solving for the magnitude of μ_n yields

$$|\mu_n| = \frac{\sigma}{qn} = \frac{c/\rho}{1/R_H} = \frac{cR_H}{\rho} \quad 9.2.11$$

where, again, cgs units are used. Substituting for ρ, the magnitude of the mobility becomes

$$|\mu_n| = \frac{cR_H I_x L}{VWt} \quad 9.2.12$$

which are all measurable quantities. The mobility determined from the Hall effect measurement is called the *Hall mobility*.

Typically, one uses a simple bridge to measure the Hall coefficient in a semiconductor as shown by Figure 9.2.2. Notice that the Hall voltage is measured across the sample between points iii and iv in the diagram. The voltage drop in the direction of the current flow, I, is measured between points ii and iv. However, Hall measurements can be made using an arbitrary form for the sample. It should be noted further that the Hall mobility is not necessarily the same as the mobility that is measured from a Haynes–Shockley time-of-flight measurement, which is usually referred to as the conductivity or drift mobility. Although the Hall and conductivity mobilities are usually similar, they are normally not precisely the same since nonlinear effects depending on the magnitude of the magnetic field can occur.

Using the Hall bridge of Figure 9.2.2, we can define the resistivity both along and perpendicular to the current path, making use of the more general tensor form of the resistivity. As mentioned above, the Hall voltage V_H is the voltage across the current path, while V_x is the longitudinal voltage along the current path. The Hall resistivity ρ_{xy} is defined as

$$\rho_{xy} = \frac{E_y}{J_x} = \frac{WE_y}{WJ_x} = \frac{V_H}{I} = \text{Hall resistance} \quad 9.2.13$$

where W is the width of the sample, E_y is the Hall field, and J_x is the current density along the x direction. The Hall field E_y is the field associated with the Hall voltage (i.e., it points from b to a as shown in Figure 9.2.1). It is clear from Eq. 9.2.13 that ρ_{xy} is the Hall resistance, which holds for a two-dimensional system (see below). The longitudinal resistivity ρ_{xx} is given as

$$\rho_{xx} = \frac{E_x}{J_x} = \frac{V_x W}{IL} \quad 9.2.14$$

EXAMPLE 9.2.1: Review of the Classical Hall Effect

Let us consider a simple classical Hall effect experiment. Assume that a Hall measurement is made on a silicon sample with the following data and that the setup is as discussed in the text. Length $L = 1.0$ cm; width $W = 0.2$ cm; thickness $t = 0.1$ cm; applied voltage $V = 0.245$ V; Hall voltage $V_H = 2.0$ mV; applied magnetic field $B = 1$ W/m^2; dc current $I = 5$ mA; Hall factor = 1.18.

a. Determine the Hall mobility.
b. Determine the drift mobility.

The Hall mobility is defined from Eq. 9.2.11 as

$$\mu_n = \frac{cR_H}{\rho} = cR_H\sigma$$

in cgs units. In mks units, the c is not present. Given that our data are in mks units, we find the Hall mobility from

$$\mu_n = R_H\sigma = \frac{1}{qn}\sigma$$

where we have used Eq. 9.2.6 in mks units for the absolute value of R_H. R_H can be found using Eq. 9.2.5 as

$$R_H = \frac{E_y}{J_x B_z} = \frac{V_H}{W}\frac{Wt}{I}\frac{1}{B_z} = \frac{V_H t}{IB_z}$$

The conductivity σ is equal to the ratio of the current density J to the electric field F as

$$\sigma = \frac{J}{F} = \frac{I}{Wt}\frac{L}{V}$$

Therefore, the Hall mobility can be found by taking the product of R_H and σ as

$$\mu_H = \frac{IL}{WtV}\frac{V_H t}{IB_z} = \frac{LV_H}{WVB_z}$$

Substituting the numbers, the Hall mobility is equal to 408 cm^2/V · s.

The ratio of the Hall mobility to the drift mobility is r_H, the Hall factor, equal to 1.18. Using this value, the drift mobility is calculated to be about 346 cm^2/V · s.

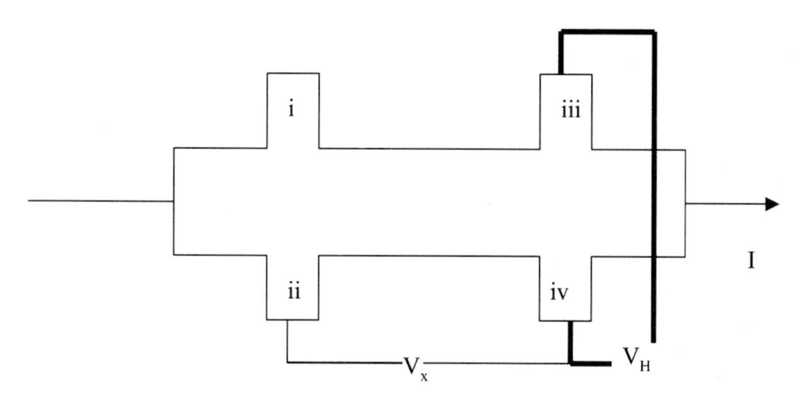

FIGURE 9.2.2 Bridge used in measuring the Hall effect. V_H is the Hall voltage and is measured across the sample between points iii and iv. V_x is the voltage drop in the direction of the current flow and is measured between points ii and iv.

which is the resistivity in the direction of the current flow.

An important relationship between the Hall resistance and the magnetic field can be determined as follows. The Lorentz force law for an electron is simply

$$\mathbf{F} = -q\mathbf{E} - \frac{q}{c}\mathbf{v} \times \mathbf{B} \qquad 9.2.15$$

In steady state, the force from the Hall field E_y balances the magnetic force. Therefore, the net forces balance and Eq. 9.2.15 becomes

$$|qE_y| = \left|\frac{q}{c}v_x B\right| \qquad 9.2.16$$

The current density J_x can be used to rewrite the velocity as

$$v_x = \frac{J_x}{qn} \qquad 9.2.17$$

Substituting Eq. 9.2.17 for v into Eq. 9.2.16 yields

$$E_y = \frac{J_x B}{qnc} \qquad 9.2.18$$

But the current density J_x is related to the Hall field as

$$J_x = \sigma_{xy} E_y = \frac{E_y}{\rho_{xy}} \qquad 9.2.19$$

Therefore, substituting Eq. 9.2.19 for J into Eq. 9.2.18 results in

$$\rho_{xy} = \frac{B}{qnc} \qquad 9.2.20$$

which is the Hall resistance. Therefore, the carrier concentration can be obtained from the measured Hall resistance.

9.3 INTEGER QUANTUM HALL EFFECT

Under certain conditions, the Hall resistance and conductivity behave unusually. When a high magnetic field is applied to a two-dimensional electron gas at very low temperatures, an exceptional situation occurs: The conductivity along the electric field direction vanishes for certain values of the magnetic field. Additionally, plateaus are observed in the Hall resistivity. It is important to recognize that these features do not occur when a magnetic field is applied to a three-dimensional system, like those discussed in Section 9.2, only in a two-dimensional system. As we discussed in Chapters 2, 3, and 8, a two-dimensional electron gas is formed at the interface of an inversion layer in a CMOS device or at the interface in a heterojunction structure. In both systems, the strong band bending at the interface results in spatial quantization in the direction normal to the interface. In the other two dimensions, no spatial quantization occurs, resulting in a two-dimensional system.

Experimental results for a GaAs–AlGaAs system are presented in Figure 9.3.1. The measurements are made at very low temperature and at relatively high magnetic field strengths, about 100 kilogauss. In the top of the figure, the Hall resistance is plotted as a function of the applied magnetic field strength: in the lower part of the figure, longitudinal resistance versus magnetic field strength is shown. As can be seen, the longitudinal resistance vanishes for particular values of the magnetic field. Correspondingly, at the same magnetic field strengths, the Hall resistance is constant, forming plateaus. Although finite at nonzero temperature, the resistivity extrapolates to zero at zero temperature. Therefore, a two-dimensional electron gas exposed to a sufficiently high magnetic field has no resistance.

A superconductor also does not exhibit any resistance. As discussed in Brennan (1999, Chap. 6.4), resistanceless current flow occurs in the absence of the possibility of scattering events. The possibility of scattering can vanish under certain conditions. This can be understood generally as follows. The allowed energy levels in a quantum system are separated in energy. As is well known, an electron can only be measured within a quantum state with a certain discrete energy. Let us consider a system that contains a number of degenerate states in each level. When all the degenerate states within a level are occupied, if there is a significant energy separation with the next-nearest energy level, the possibility of scattering can vanish. This can be understood easily since if all the states in the level are occupied, it is impossible for an electron to

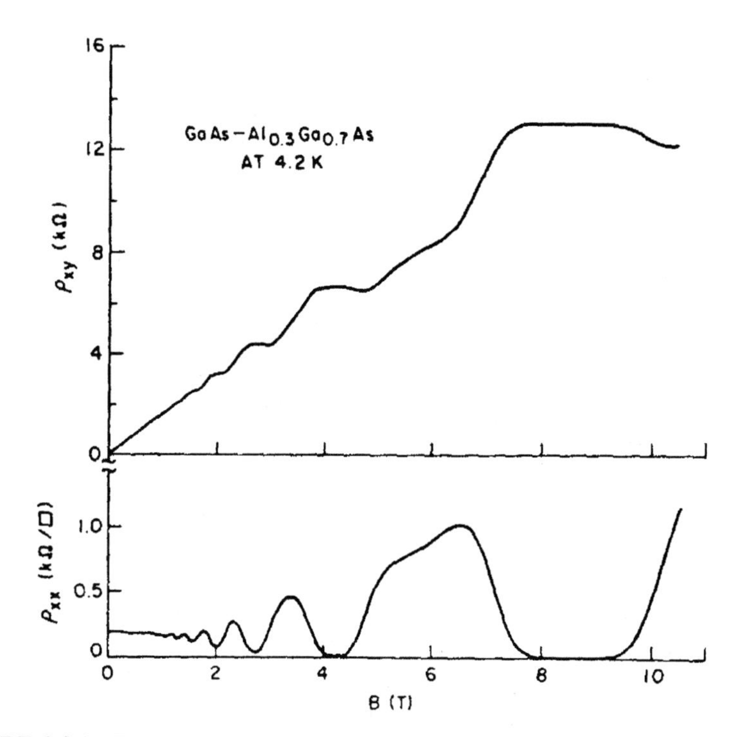

FIGURE 9.3.1 Experimental data of the Hall resistance and ρ_{xx} as a function of the applied magnetic field B. Plateaus in the Hall resistance are clearly shown that extend over a finite range of magnetic field strength. Notice that ρ_{xx} shows corresponding regions of zero resistivity. (Reprinted from Tsui and Gossard, 1981, with permission from D. C. Tsui and the American Physical Society.)

scatter into any of these states without violating the Pauli principle. Hence, in a completely filled level, no scattering events can occur. An electron must be transferred into a state within a different level in order to scatter. If the energy separation between the levels is greater than that available from the scattering event (i.e., a phonon energy, for example), scattering cannot occur. As a result, the possibility of scattering vanishes, leading to resistanceless current flow.

Another way of understanding resistanceless current flow involves the Boltzmann equation (Brennan, 1999, Sec. 6.1). The right-hand side of the Boltzmann equation is equal to the time rate of change of the distribution due to scatterings and is given by Brennan (1999, Eq. 6.1.7) as

$$\left(\frac{\partial f}{\partial t}\right)_{\text{scatterings}} = -\int \{f(x,k,t)[1-f(x,k't)]S(k,k') - f(x,k't)[1-f(x,k,t)]S(k',k)\}d^3k' \qquad 9.3.1$$

The terms $S(k,k')$ and $S(k',k)$ are the transition rates for scatterings from states k to k', and vice versa. The additional terms $f(x,k,t)$ and $1 - f(x,k',t)$ in the first part of Eq. 9.3.1 represent the probability that an electron is in the initial state (x,k) at time t and that a vacancy exists in the final state (x,k') at time t. If either the initial state is empty or the final state is filled, the scattering process cannot occur independent of the magnitude of $S(k,k')$. Obviously, if there is no electron to begin with, no scattering can occur. By the Pauli principle, if the final state is occupied, again no scattering can occur. It is important to notice that this is true even when the transition rate, $S(k,k')$ is not zero. The scattering rate can vanish if all the possible final states are occupied. As we will see below, this is the reason why the resistance vanishes in the integer quantum Hall effect (i.e., all possible final states of the system are occupied, leading to a zero scattering rate).

In a superconductor, electron pairing results in a new collective state that lies lower in energy than the free particle states on the Fermi surface. Therefore, an energy gap opens between the superconducting ground state and the normal conducting states (Brennan, 1999, Chap. 6). It takes a finite energy then to disturb the system from the superconducting state. If that amount of energy is not available, the superconducting state is undisturbed and the pairs move through the superconductor without resistance. In a superconductor the pairing formation produces a new state that lies lower in energy by a significant amount from the free particle states. Although the mechanism by which the resistance vanishes is very different in the quantum Hall effect, the lack of resistance is still related to the elimination of the possibility of scattering events.

The key to understanding the integer quantum Hall effect is as follows. As we discussed in Section 9.1, the application of an external magnetic field splits the allowed energy spectrum into Landau levels, each separated by a quantum $\hbar\Omega$. As shown in Figure 9.1.2, the allowed states are a series of cylinders each separated by energy $\hbar\Omega$. When all the states within a cylinder (Landau level) are occupied, the next available state lies at a significant energy above that level, in this case $\hbar\Omega$. Hence if the number of electrons present in the system exactly fills up the first Landau level, the system has a completely filled level and completely empty levels. As discussed above, there is no possibility of scattering in a completely filled level. Of course, no scatterings can occur in a completely empty level either since there are no electrons present to scatter. Therefore, within such a system, there is no possibility of scattering. The resistance of the system, ρ_{xx}, vanishes.

Why, though, is the Hall resistance, ρ_{xy}, quantized? To see the origin of the quantization of the Hall resistance it is important first to determine the degeneracy of a Landau level. When all the states within a Landau level are filled, an abrupt break occurs in the energy spectrum. So it is important to determine first how many electrons can be placed within each Landau level. In Section 9.1 we determined the maximum number of states per unit area s

to be (which is the degeneracy of the Landau level)

$$s = \frac{n_x}{L^2} = \frac{qB}{hc} \tag{9.3.2}$$

Let us consider the case where the total electron concentration n exactly fills the uppermost Landau level. The number of filled Landau levels is of course an integer, which we will refer to as ν. The total electron concentration n (recall that this is a two-dimensional system, so the concentration is given by the ratio of the total number of electrons to the unit area) is then given as

$$n = \nu s \tag{9.3.3}$$

The Hall resistance (recall that in two dimensions the Hall resistance and resistivity are the same) is given by Eq. 9.2.20 as

$$\rho_{xy} = \frac{B}{qnc} \tag{9.3.4}$$

Using Eq. 9.3.3, Eq. 9.3.4 becomes

$$\rho_{xy} = \frac{B}{q\nu sc} \tag{9.3.5}$$

But s is given by Eq. 9.3.2. Substituting Eq. 9.3.2 for s into Eq. 9.3.5 yields

$$\rho_{xy} = \frac{h}{\nu q^2} \tag{9.3.6}$$

where ν is an integer. Notice that Eq. 9.3.6 implies that the Hall resistance is quantized by an integral amount. This is the basic idea behind the integer quantum Hall effect.

As mentioned above, the resistance vanishes when all the states within the uppermost occupied Landau level are filled and all the states within the next Landau level are empty. Consequently, the Fermi level must lie in the energy gap between the uppermost occupied Landau level and the next Landau level. When this occurs, the resistance vanishes.

Although the theory above clearly explains the basic idea behind the integer quantum Hall effect, it does not completely explain the experimental observations. What is observed experimentally is that the plateaus in the Hall resistance (see Figure 9.3.1) persist over an extended range of applied magnetic field strengths. Additionally, there is a finite width of zero-resistance regions. This implies that the Landau levels must remain completely filled for the entire range of applied fields since the resistance vanishes throughout this region. However, as the field strength increases, according to Eq. 9.3.2, the degeneracy of the Landau levels increases. Hence, as the field is increased, more states would open up and the system would not show a finite width of the zero-resistance region. As a result, the resistance would no longer vanish since the Fermi level would now lie within a partially empty band. Additionally, if the electron density changes, the Fermi level will jump to the next Landau level, again resulting in dissipation during current flow since the Fermi level lies within a partially empty band. Hence there must be a mechanism by which the Fermi level does not move into the next available Landau level as the concentration increases. The presence of trap states formed between the Landau levels provides a means by which electrons are trapped, preventing their transfer into the next Landau level. As a result, the Fermi level remains within the gap between the Landau levels. The trap states arise from impurities or imperfections in the material. The traps behave as any localized defect: Once an electron becomes trapped in a defect state, it cannot move freely throughout the crystal (Brennan, 1999, Sec. 10.1). Hence electrons within the trap states cannot contribute to the current.

The question is, then: How can we explain the experimental observations using trap states? Provided that a variation in the concentration of electrons or applied field strength changes only the electron population of the trap states, by either adding or subtracting electrons, the Fermi energy remains between Landau levels. Notice that the Landau level below the Fermi energy remains completely filled, while the Landau level above the Fermi energy remains completely empty. Addition or subtraction of electrons from the trap states does not change the conductivity since these states cannot provide conduction anyway. Only the delocalized states can contribute to the current, and these are all either completely filled (Landau level below E_f) or completely empty (Landau level above E_f). Therefore, zero resistance is maintained over a finite field range.

It is informative to examine the density of states in a two-dimensional system in the presence of a magnetic field. In Section 9.1 we examined the density of states for a three-dimensional system and found that the three-dimensional density of states exhibits singularities that correspond to the emergence of each successive Landau level. From Eq. 9.1.33 it was observed that the denominator vanishes when the energy is equal to a Landau level. For a two-dimensional system with a magnetic field applied along the same axis as the confinement direction, assumed to be the z axis, the energy becomes quantized in all three directions. The applied magnetic field quantizes the energy in the x and y directions, as shown in Section 9.1, and the spatial confinement quantizes the energy in the z direction. The allowed energies of the system are completely quantized and the resulting density of states becomes a set of delta functions, as shown in Figure 9.3.2. If the system has impurities present, the impurities will produce localized states between the delta functions, of nonlocalized states. Such a situation is shown in Figure 9.3.3. Notice that localized states are formed between each Landau level. The localized states do not contribute any net current to the system. Hence there is a mobility gap formed, as shown in the diagram.

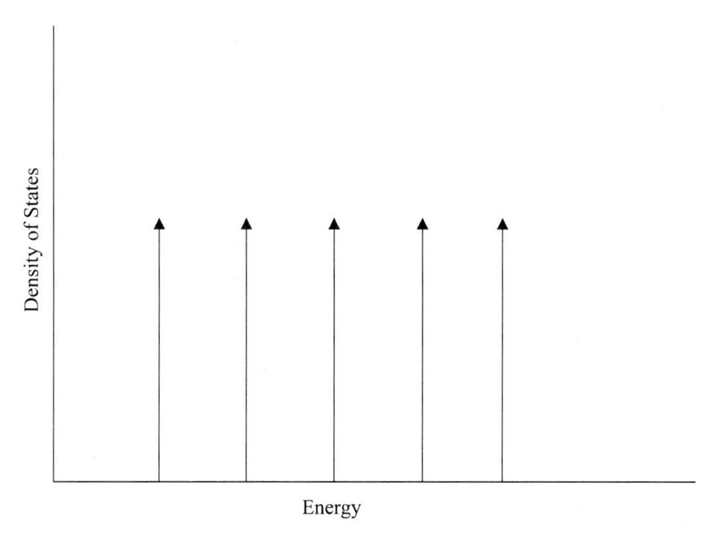

FIGURE 9.3.2 Density of states of a two-dimensional system in which an external magnetic field is applied parallel to the confinement direction. Each line represents an increasing integral multiple of $\hbar\Omega$.

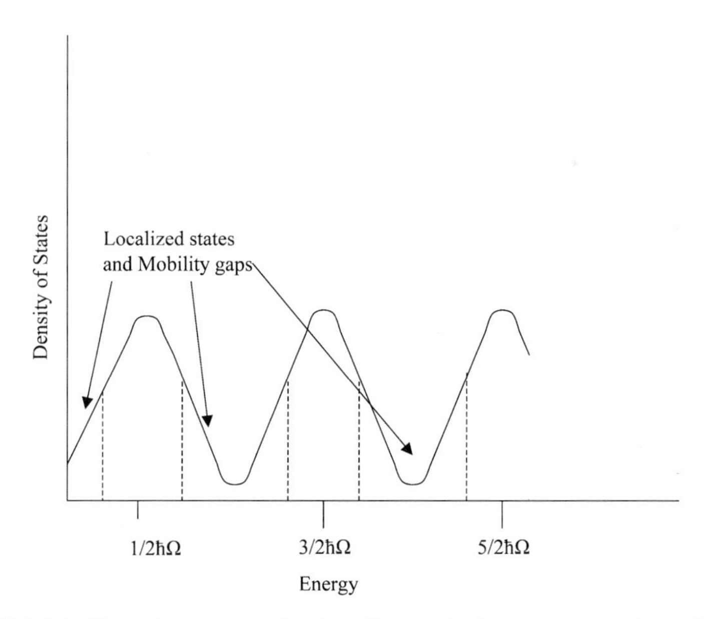

FIGURE 9.3.3 Two-dimensional density of states in the presence of impurities. The localized states lie between successive Landau levels, marked as multiples of $\hbar\Omega$ in the diagram.

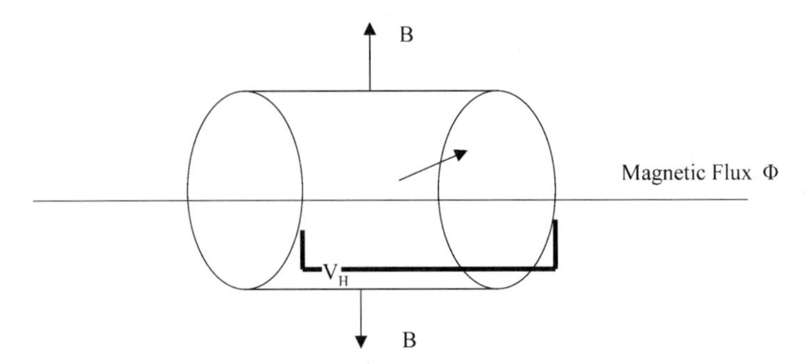

FIGURE 9.3.4 Geometry used in Laughlin's thought experiment. The arrows indicate the magnetic field that is directed radially outward from the surface. The current flows along the surface of the cylinder in a counterclockwise direction. The action of the magnetic field on the charge carriers in the current creates the voltage V_H.

Laughlin (1981) provided an interesting demonstration that ρ_{xy} remains quantized whereas ρ_{xx} tends toward zero when the Fermi level moves through regions of localized states within the gaps between Landau levels. Figure 9.3.4 is instrumental to understanding Lauglin's argument. The two-dimensional system is twisted so as to form a cylinder, as shown in Figure 9.3.4. A circulating current flows in the counterclockwise direction along the outside surface of the cylinder. A magnetic field is assumed to be directed radially outward from the cylinder. The action of the magnetic field on the charge carriers results in the voltage V_H shown in the diagram. The circulating current flow produces a magnetic flux Φ through the cylinder in accordance with Ampère's law. Let us consider the situation where the flux is adiabatically varied. From Faraday's law, the variation of the flux produces a voltage V given as

$$V = -\frac{1}{c}\frac{d\Phi}{dt} \tag{9.3.7}$$

The total power due to this adiabatic change in the flux is given as

$$\frac{dU}{dt} = IV \tag{9.3.8}$$

where U is the total energy of the system and I is the circulating current. Combining Eqs. 9.3.7 and 9.3.8 yields

$$\frac{dU}{dt} = -\frac{I}{c}\frac{d\Phi}{dt} \tag{9.3.9}$$

which becomes

$$I = -c\frac{dU}{d\Phi} \tag{9.3.10}$$

Equation 9.3.10 can be rewritten in a more convenient form using the vector potential A. The magnetic field B is equal to the curl of the vector potential, $\mathbf{A}$. The magnetic flux Φ is given as

$$\Phi = \int \mathbf{B} \cdot d\mathbf{a} \qquad 9.3.11$$

where $\mathbf{a}$ is the unit normal vector to the surface area. Using Stokes' theorem,

$$\int (\nabla \times \mathbf{A}) \cdot d\mathbf{a} = \int \mathbf{A} \cdot d\mathbf{l} \qquad 9.3.12$$

the flux can be written as

$$\Phi = \int \mathbf{B} \cdot d\mathbf{a} = \int (\nabla \times \mathbf{A}) \cdot d\mathbf{a} = \int \mathbf{A} \cdot d\mathbf{l} = AL \qquad 9.3.13$$

where $\mathbf{A}$ is the vector potential. Therefore, Eq. 9.3.10 can be rewritten as

$$I = -c\frac{dU}{d\Phi} = -c\frac{1}{L}\frac{dU}{dA} \qquad 9.3.14$$

where L is the circumference of the current loop. Now the current I can be determined from examining the change in the total energy with a change in the vector potential $\mathbf{A}$. Making the gauge transformation [see Brennan (1999, Sec. 1.2) for a discussion of gauge transformations]

$$\mathbf{A} \to \mathbf{A} + \delta\mathbf{A} = \mathbf{A} + \frac{\delta\Phi}{L} \qquad 9.3.15$$

the wavefunction gives a phase factor of

$$\psi \to \psi e^{(iq\delta A/\hbar c)x} \qquad 9.3.16$$

where x is the coordinate around the loop. If the electron is localized, the wavefunction is not extended in space. Hence a localized electron cannot respond to the flux and cannot carry any current, and the gauge transformation has no effect. Alternatively, if the electron is in a nonlocalized or extended state, gauge invariance requires that the wavefunctions be single-valued. In other words, the exponent in the phase term must be an integer times $2\pi i$ when x is equal to L (after the full circumference has been traversed). This implies that for $x = L$,

$$\frac{iq\delta AL}{\hbar c} = 2\pi i\nu \qquad 9.3.17$$

Therefore,

$$\nu = \frac{q\delta AL}{hc} \qquad \delta A = \frac{\nu hc}{qL} \tag{9.3.18}$$

where ν is an integer.

Consider the situation in which the magnetic flux in the cylinder is increased adiabatically. If the Fermi level lies within the mobility gap shown in Figure 9.3.2, the conductivity σ_{xx} vanishes. Under these conditions all the extended states within the Landau levels below the Fermi level are filled, and all localized states up to the Fermi level are filled. After the flux is changed by $\delta\Phi$, the delocalized states remain filled. Therefore, the summation over the energy levels remains the same, all the orbits before and after the flux change coincide. However, the energy of the system changes from Eq. 9.3.14. The only way this can happen is that an integral number of electrons are transferred through the system during the flux change. They enter the cylinder at one edge and leave the cylinder at the other edge.

The transfer of charge along the cylinder axis is easy to see from the affect of δA on the electrons. From the Schrödinger equation it is easy to see that the addition of δA simply shifts the centers of the wavefunctions from y_0 to $y_0 - \delta A/B$. Hence there is a transfer of charge from one edge of the cylinder to the other. The net change in energy is equal to the work done by transferring this charge,

$$\Delta U = -q\nu V_H \tag{9.3.19}$$

where ν is an integer. The unit flux quantum is hc/q. Therefore, $\delta\Phi = hc/q$. From Eq. 9.3.14 the current I becomes

$$I = -c\frac{\partial U}{\partial \Phi} = -c\frac{-q\nu V_H}{hc/q} = \frac{\nu q^2}{h}V_H \tag{9.3.20}$$

The resistance ρ_{xy} is

$$\rho_{xy} = \frac{V_H}{I} = \frac{h}{\nu q^2} \tag{9.3.21}$$

which is the result obtained in Eq. 9.3.5.

The result in Eq. 9.3.21 shows that one again obtains the integer quantum Hall effect if it is assumed that the Fermi level lies within the mobility gap and that the system is gauge invariant. The finite width of the Hall plateaus now immediately follows. Until the Fermi level moves through the mobility gap wherein the localized states lie, the resistance ρ_{xx} vanishes while the resistance ρ_{xy} plateaus. This forms the present understanding of the integer quantum Hall effect. In the next section we examine the main issues associated with the fractional quantum Hall effect.

9.4 FRACTIONAL QUANTUM HALL EFFECT

As we have seen above, the integer quantum Hall effect (IQHE) is traceable to the filling of various Landau levels. The resistance vanishes when all the states within the uppermost occupied Landau level are filled and all the states within the next Landau level are empty. The finite width of the plateaus in the Hall resistance arises from impurities. The Hall resistance is given by Eq. 9.3.6 as

$$\rho_{xy} = \frac{h}{\nu q^2} \tag{9.4.1}$$

Interestingly, ν can be an integer, as described in Section 9.3, or alternatively, a rational fraction, p/n, where p and n are, in turn, integers.

Fractional quantization was first observed by Tsui et al. (1982) using a GaAs–AlGaAs heterostructure. The heterostructure was of high uniformity such that very high carrier mobilities were present. The very high mobility of the sample is believed to result from the highly uniform interface and concomitant weak interface scattering. The experimental measurements were also performed at very low temperatures. Under these conditions, the electron correlations, necessary to produce the fractional quantum hall effect, are sufficiently greater than processes that disrupt correlated motion due to interface scattering and the thermal energy. Representative experimental results are shown in Figure 9.4.1. As can be seen from Figure 9.4.1, a plateau appears in ρ_{xy} at $p/n = \frac{1}{3}$, similar to those formed for the integers 1, 2, 3, and 4. Accompanying these plateaus the longitudinal resistance shows a pronounced minimum. Further experimental measurements, as shown in Figure 9.4.2, reveal that plateaus in ρ_{xy} and minimums of ρ_{xx} occur for many other rational fractions. Interestingly, virtually all the fractions are odd. However, there is a notable exception, that of $\frac{5}{2}$. It was determined empirically that the values of p/n are given by different progressions. These are

$$\nu = \frac{p}{2p+1} \qquad \nu = \frac{p}{4p+1} \tag{9.4.2}$$

where $p = 1, 2, 3, 4, \ldots$. Substitution of different integer values for p into the expressions given in Eq. 9.4.2 results in values for ν of

$$\begin{aligned} \nu &= \frac{p}{2p+1} = \frac{1}{3}; \frac{2}{5}; \frac{3}{7}; \frac{4}{9}; \ldots \\ \nu &= \frac{p}{4p+1} = \frac{1}{5}; \frac{2}{9}; \frac{3}{13}; \ldots \end{aligned} \tag{9.4.3}$$

The primary fractions, $\frac{1}{3}$ and $\frac{1}{5}$, are more readily observed experimentally than the higher-order fractions. The higher-order fractions require very high

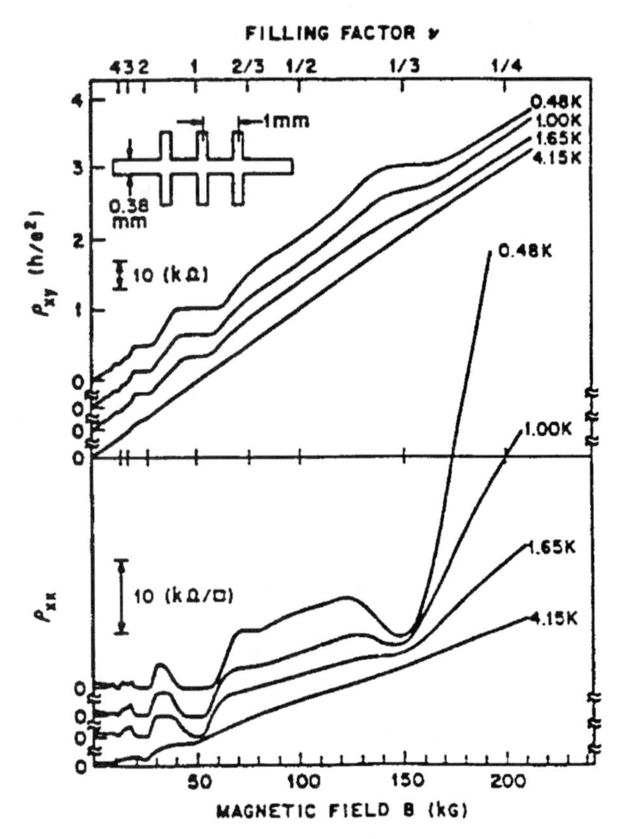

FIGURE 9.4.1 Experimental data for the Hall resistance and ρ_{xx} as a function of the applied magnetic field B and filling factor. As can be seen from the figure, at a fractional filling factor of $\frac{1}{3}$, a plateau appears in the Hall resistance clearly showing the fractional quantum Hall effect. (Reprinted from Tsui et al., 1982, with permission from D. C. Tsui and the American Physical Society.)

quality interfaces, with a concomitant lower interface scattering rate and lower temperatures to be observed. The higher-order fractions cannot be observed if the lower-order fractions are absent. In other words, each higher-order fraction is formed from the preceding lower-order fraction, forming a hierarchical set of states.

It is important to note that a fractional value for ν cannot arise physically from single-electron dynamics. As discussed in Section 9.3, the integer plateaus appear when the uppermost occupied Landau level is completely filled and the next-lying Landau level is completely empty. The presence of impurity states between the Landau levels ensures that the Fermi level will lie within the energy gap through a finite field range. As a result, the plateaus have a finite width and can thus be observed. The integer quantum Hall effect

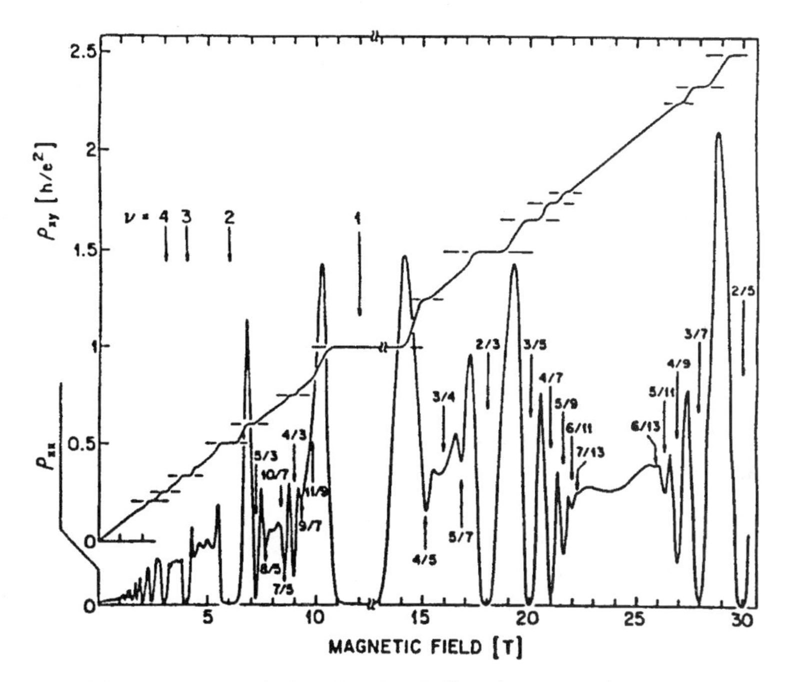

FIGURE 9.4.2 Experimental data for the Hall resistance and ρ_{xx} versus magnetic field showing both the integer and fractional quantum Hall effect. Plateaus in the Hall resistance at integral and fractional filling factors indicate the integer and fractional quantum Hall effects, respectively. (Reprinted from Tsui et al., 1987, with permission from D. C. Tsui and the American Physical Society.)

(IQHE) can then be thought to arise from the presence of an energy gap in the excitation spectrum that occurs at certain values of filling factors for the Landau levels. When a Landau level is completely filled and the next one is completely empty, the energy separation or gap between the Landau levels results in vanishing longitudinal resistance. It is important to recognize that the filling involves an integer number of electrons. Since a single electron is an indivisible quantity, fractional filling or emptying of states cannot occur by considering each electron individually. The explanation of the fractional quantum Hall effect (FQHE) then necessitates a collective state model for the electron system.

As mentioned above, the primary fractions, particularly $\frac{1}{3}$, are most readily observed experimentally. Higher-order fractions require increasingly higher mobility samples and lower-temperature operation for observation. Let us begin our discussion by examining how the $\frac{1}{3}$ state can be explained following the approach of Laughlin (1987), who recognized that the collective electron states could be represented using wavefunctions of the form

$$\psi_{1/m} = \prod_{j<k}^{N} (z_j - z_k)^m \exp\left(-\frac{1}{4}\sum_{i}^{N} |z_i|^2\right) \qquad 9.4.4$$

where m is an odd integer ranging from 1,3,5,...,N the total number of electrons present, z a complex number representing position in the complex plane, and i, j, and k are indices labeling each electron. The electrons are assumed to be spinless. Inspection of Eq. 9.4.4 shows that ψ obeys the Pauli principle; if two electrons are placed into the same quantum state, $z_j = z_k$ for example, the wavefunction vanishes, as it should. Additionally, the wavefunction given by Eq. 9.4.4 is antisymmetric under exchange of particle label if m is an odd integer. With this choice for m, the collective states are consistent with fermions. [For a discussion of symmetric and antisymmetric total wavefunctions, see Brennan (1999, Sec. 7.2)]. Notice that ψ would not change sign under particle exchange if m is chosen to be even instead. With an even value for m, ψ is symmetric and the sign is unchanged under particle exchange consistent with bosons. However, electrons are fermions, and thus should be described by a totally antisymmetric wavefunction as in Eq. 9.4.4. This requirement restricts m to odd integers.

Further inspection of Eq. 9.4.4 shows that the wavefunctions vanish rapidly when two electrons approach one another. The electrons tend to remain spatially apart from one another. Two or more electrons interact through direct Coulomb interaction, which is a function of the separation distance of the charges. This interaction is repulsive, and as such, as the two electrons approach one another, their interaction energy increases. Therefore, the wavefunctions of Eq. 9.4.4 ensure that the electrons remain spatially separated, thus lowering their interaction energy. It can further be demonstrated that the wavefunctions given by Eq. 9.4.4 are the unique, and thus nondegenerate, ground states for the system. The Laughlin wavefunctions given by Eq. 9.4.4 describe the ground state of the electron system and predict that the distribution of these electrons is optimally correlated to ensure that the repulsive Coulomb interaction is minimized. Thus the addition or removal of an electron from this system disturbs its inherent order, requiring a significant change in energy. The states corresponding to $\nu = 1/m$ are condensed many-particle ground states.

A question remains, though: How and why do these states form in the first place? Can we provide a physical picture of these fractional filling factor states? The $1/m$ states can be envisioned as follows. The quantum description of the magnetic field results in the quantization of magnetic flux, given as

$$\Phi = \frac{hc}{q} \qquad 9.4.5$$

The net magnetic field is comprised of a collection of flux quanta much as a net charge density is comprised of a collection of individual charges. We first consider the simplest system, that of a single electron in the Landau level. In the absence of an external magnetic field, the electron wavefunction is distributed over the two-dimensional plane with uniform probability. However, if an external magnetic field is applied to the system, the flux quanta associated with the external field alter the probability distribution for the electron. The resulting spatial probability distribution that describes the probability of localizing the electron at a given spatial point is no longer uniform. Instead, the spatial probability distribution resembles that of an egg carton (i.e., there are vortices that represent near-zero probability for finding the electron separated by regions of high probability for localizing the electron). In fact, at the center of each vortex the probability of localizing the initial electron is identically zero. Addition of another electron to the system requires that it be placed within a vortex, where the probability of localizing the initial electron is vanishingly small, such that the resulting total energy is minimized. The addition of more electrons follows a similar prescription except that interaction of the flux quanta and the total number of electrons initially present in the system determines the location of the vortices. Thus the behavior of the system is much like the Laughlin wavefunctions; the spatial separation of the electrons is optimized to ensure minimum energy of the collective system.

However, we still have not explained why the system has fractional quantization. In the argument above, each additional electron is added to the system until all the vortices arising from the flux quanta are filled. This corresponds to a filled Landau level leading to an integral filling factor. How, though, does a fractional filling factor occur? Consider the situation where the number of electrons is less than the number of vortices, a filling factor less than 1. Keep in mind that the energy of the collective system is reduced with increased spatial separation of the electrons. Ideally, the system will condense into a collective state in which the electrons are spatially separated as much as possible. The flux quanta and thus the vortices can be superimposed. In other words, unoccupied vortices can be combined to yield larger vortices. This is equivalent to saying that multiple flux quanta are bound together. Electrons placed at the centers (where their energy is minimized) of these larger vortices are necessarily more spatially separated. As a result, the repulsive interaction between the electrons is lowered and the collective energy of the system is reduced.

The fractional filling factors of $\nu = 1/m$ for odd m can now be understood. The case of $m = 3$ corresponds to the situation where three flux quanta are bound to each electron. For $m = 3$ the state is completely filled when only $\frac{1}{3}$ of the number of electrons otherwise needed to fill the Landau level are present. Hence the addition of another electron requires a substantial energy cost, due to the fact that the new electron has to be added to the next Landau level at significantly higher energy. As in the IQHE, when the highest occupied Landau level is completely filled and the next level is completely empty, the longitudinal resistance vanishes. Why, though, does the resistance persist away from the exact case of $\nu = \frac{1}{3}$? In the IQHE it was found that the presence of impurities is responsible for the persistence of the vanishing longitudinal resistance and the plateaus observed in the Hall resistance. Imperfections again play a similar role in the FQHE. Slight variation from $\nu = \frac{1}{3}$ results in the formation of quasiparticles, which are trapped by imperfections. Thus the vanishing longitudinal resistance persists.

The reader may wonder at this point why the fractional filling described above occurs at $\nu = \frac{1}{3}$. Energetically favorable configurations of the collective system occur for filling factors of $\frac{1}{3}, \frac{1}{5}, \frac{1}{7}$, and so on. Even filling factors are forbidden in the picture above since they would imply that the wavefunctions are symmetric, inconsistent with a fermion description. As mentioned above, there is a notable exception. One even denominator filling factor has been observed at $\nu = \frac{5}{2}$. It has been speculated that this case corresponds to a pairing action of electrons of opposite spins. Such a state would be a boson, and even filling factors would then be possible. Additional complexity has recently been discovered for filling factors of $\frac{1}{2}$ and $\frac{1}{4}$. Although not thought to be explainable using the theory above, vortex attachment is still thought to play a role in these even-denominator states. Basically, the system is viewed as a Fermi sea of dipoles instead of monopoles as in the theory above. To the author's knowledge, the aspects of this theory are still debated and have yet to be resolved. The interested reader is referred to the references, especially the article by Stormer et al. (1998).

Inspection of Eq. 9.4.3 shows that fractional filling factors different from $1/m$ occur. In addition, the features observed experimentally for these different fractional filling factors all appear alike. Given that the filling factors different from and the same as $1/m$ have similar measurable features, it is not unreasonable to expect that they have a common origin. Up to now the discussion has explained only how those states with filling factors of $1/m$ arise. How, then, do states with other fractional filling factors occur? Additionally, since it is reasonable to expect that all these states arise in the same manner, how can the theory above be generalized? The leading microscopic theory to explain the FQHE for factors other than $1/m$ is called the *composite fermion theory*. Basically, the composite fermion theory is an extension of the argument above wherein the system is comprised of composite fermions, the bound state of an electron, and an even number of vortices. In essence, a composite fermion is a new quasiparticle that consists of the union of vortices and electrons. The FQHE arises in this picture in much the same way as the IQHE except that composite fermions are the constitutive particles, as opposed to individual electrons. Another way of thinking about the system is that the FQHE of electrons is similar to the IQHE of composite fermions. Therefore, by reformulating the problem as a collection of composite fermions, the simple picture provided by the IQHE is retained. It should be noted that the interaction between charges is still important in this picture to stabilize the composite fermions, but once the composite fermions are created, their interaction is relatively weak. The actual explanation for why composite fermions form in the first place is not yet well understood. Nevertheless, postulating their existence yields a useful picture of the FQHE.

The number of states in the FQHE can be determined using the composite fermion picture as follows. The filling factor is determined from the inverse of the number of available flux quanta per electron. Let n represent the filling factor of the original electron system. To each electron an even number of flux quanta, $2m$, are added, producing a composite that behaves as a fermion—hence the name *composite fermions*. Composite fermions then consist of an electron and $2m$ flux quanta. What we want to do is relate the composite fermion system to the original electron system. The same external flux is available to the composite fermions in addition to the flux that is contained in each composite fermion (i.e., that available to each electron in the original electron system). Since n is the filling factor of the original electron system, the filling factor of the composite fermions is also n. The number of flux quanta available to an electron state is then given as $1/n$ plus $2m$. The inverse of $(1/n + 2m)$ is the filling factor. Thus the original electron system with a filling factor of

$$p = \frac{1}{1/n + 2m} = \frac{n}{2mn + 1} \tag{9.4.6}$$

is equivalent to a composite fermion system with filling factor n. It is interesting now to see what filling factors are allowed by Eq. 9.4.6. We recall that m can only be odd, $1, 3, 5, \ldots$. The variable n can be any integer. For $m = 1$, and substituting into Eq. 9.4.6 the successive integer values of n yields

$$p = \frac{1}{3}; \frac{2}{5}; \frac{3}{7}; \frac{4}{9}; \ldots \tag{9.4.7}$$

which corresponds with the first sequence given by Eq. 9.4.3. Choosing $m = 3$, and again substituting into Eq. 9.4.6 the successive integer values of n yields

$$p = \frac{1}{5}; \frac{2}{9}; \frac{3}{13}; \ldots \tag{9.4.8}$$

which agrees with the second sequence given by Eq. 9.4.3. Therefore, we see that all of the odd filling factors can be accounted for using the composite fermion theory. The full details of the theory are beyond the level of this book. Here we confine ourselves to an introductory account. The interested reader is referred to the references for further discussion.

9.5 SHUBNIKOV–DE HAAS OSCILLATIONS

Another interesting effect that is observed when an external magnetic field is applied to a two-dimensional electron gas is the presence of oscillations of longitudinal resistance. This effect, called the *Shubnikov–de Haas effect*, is highly useful in characterizing a two-dimensional electron system. Hence it is used extensively in studying the two-dimensional gas in MOSFETs, HEMTs, and general heterostructure systems.

The origin of the Shubnikov–de Haas effect is somewhat similar to that of the quantum Hall effect. The basic idea is that the electron occupancy and scattering rate is a strong function of the applied magnetic field. As the field is varied, the Landau levels move through the Fermi level, resulting in a change in the occupancy and hence the electron scattering rates. As shown in Figure 9.3.1, the density of states in a two-dimensional system in the presence of an applied magnetic field becomes a series of delta functions each separated by a quantum of energy $\hbar\Omega$. However, the presence of impurities and defects within the material acts to broaden the density of states as shown in Figure 9.3.2. Localized states form between each Landau level, due to traps and impurities. Electrons within these localized states cannot contribute to the conductivity of the sample. The origin of the Shubnikov–de Haas oscillations can be understood from the behavior of the Fermi level in this system. When the Fermi level lies within one of the mobility gaps shown in Figure 9.3.2, the conductivity of the sample is extremely small. This arises because all of the extended states within the lower Landau levels are filled, whereas the extended states in the higher-energy Landau levels are empty. A small change in the magnitude of the magnetic field will move the Fermi level only within the mobility gap where the states are all localized. Hence the conductivity does not change and remains vanishingly small. However, if the Fermi level lies within a partially filled Landau level, the conductivity of the sample is relatively high. Under these conditions, electrons can be readily scattered or accelerated into unfilled extended states. Thus the conductivity of the sample changes dramatically as the Fermi level moves through the Landau levels.

How, though, does the position of the Fermi level change? In Section 9.1 we found that the degeneracy of a Landau level is a function of the applied magnetic field strength. As the magnitude of the magnetic field increases, the number of states within a Landau level increases as shown by Eq. 9.1.37,

$$s = \frac{qB}{hc} \qquad 9.5.1$$

As the magnetic field strength increases, more states are added to the Landau levels. In other words, with increasing magnetic field strength the number of states that each Landau level can hold also increases. Consider a system with N electrons. At low temperature, all of the states, both extended and localized, below the Fermi level are filled. Let us assume that there are a few electrons in the uppermost Landau level. The Fermi level then lies within this Landau level. The electrons within the uppermost level reside in extended states. Since the uppermost level is not completely filled or empty, the electrons within it can readily be accelerated under application of an external electric field. They are thus free to conduct, yielding a high conductivity. If the magnetic field

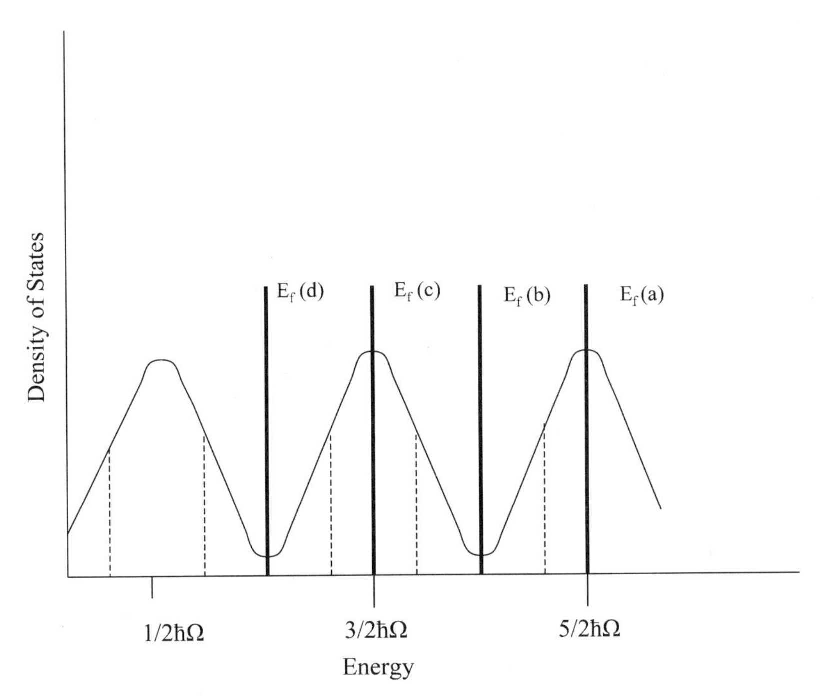

FIGURE 9.5.1 Density of states for a two-dimensional system in the presence of a magnetic field showing the position of the Fermi level for increasing magnitude of the magnetic field strength: *a* (lowest) *d* (highest) fields. The Fermi level moves lower in energy with increasing field.

strength is increased, the number of states within each Landau level increases. Therefore, the few electrons within the uppermost Landau level can now drop into these new, low-energy states that have been created by the higher magnetic field. If the number of new states equals the number of excess electrons, no electrons remain within the uppermost Landau level. The Fermi level moves into the mobility gap since the uppermost Landau level is now empty while all of the lower-energy Landau levels are filled. The electrons within the extended states cannot contribute to the conductivity since they have no place to be accelerated or scattered into, while the localized electrons also cannot conduct. Hence the system exhibits very low conductivity. Notice that the conductivity changes, then with the magnitude of the applied magnetic field.

From the discussion above the origin of the Shubnikov–de Haas oscillations can be understood. Experimentally, it is observed that the resistivity or, equivalently, the conductivity of the sample oscillates with magnetic field strength. As the magnetic field strength increases, the number of states within each Landau level increases. The lower-energy Landau levels can accommodate more electrons with increasing magnetic field. This implies that the Fermi level continues to go lower in energy in the sample as the magnetic field increases.

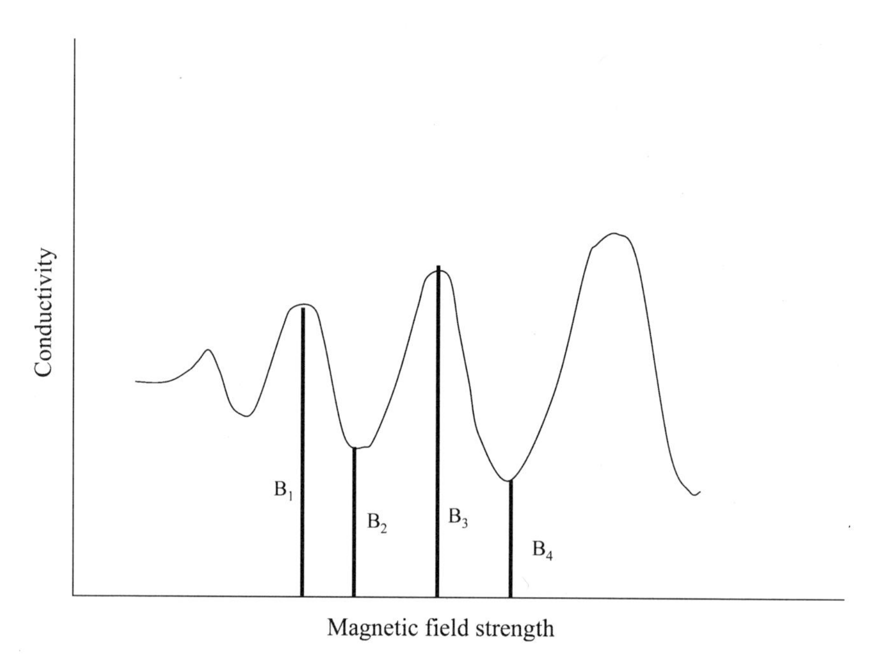

FIGURE 9.5.2 Conductivity as a function of applied magnetic field strength for a two-dimensional system. Notice that the mobility undergoes oscillations (Shubnikov–de Haas oscillations). As the magnetic field strength increases, the Fermi level moves through the various Landau levels and mobility gaps, resulting in an oscillation in the conductivity.

Assume at first that the Fermi level lies within a Landau level as shown in Figure 9.5.1 [marked as E_f (a)]. (In Figure 9.5.1 no allowance is made for the change in the density of states with magnetic field.) Since the Fermi level lies within a partially filled energy band, the mobility is relatively high, as shown in Figure 9.5.2 (line B_1). Figure 9.5.2 represents a sketch of the dependence of the conductivity on the magnetic field, not actual experimental data, and hence is only approximate. If the magnetic field strength is increased, the number of states below the Fermi level increases. As a result, the Fermi level drops into a mobility gap, as shown in Figure 9.5.1 [marked as E_f (b)], where the conductivity is low (see Figure 9.3.2 line marked as B_2). Further increase in the magnetic field results in a continued drop of the Fermi level into the next available Landau level, as shown as E_f (c) in Figure 9.5.1. Once the Fermi level drops into the next Landau level [case E_f (c) in Figure 9.5.1], extended state vacancies open up, producing a partially empty set of extended states within the Landau level. Under this condition, the conductivity increases, as can be seen in Figure 9.5.2, line B_3. As the magnetic field is increased further, more states are opened within the Landau levels. Once all the electrons within the uppermost Landau level drop into these new lower-energy states, the Fermi level moves once again into the mobility gap [case E_f (d) in Figure 9.5.1]. The conductivity drops again as shown in Figure 9.5.2, line B_4. Clearly, the conductivity undergoes oscillations with the magnetic field.

As shown in Figure 9.5.2, the amplitude of the oscillations is assumed to increase with increasing magnetic field strength. This is observed in experimental measurements of the Shubnikov–de Haas oscillations. From the magnitude of the oscillations, a value for the carrier effective mass can be determined. Additionally, by measuring the frequency of the Shubnikov–de Haas oscillations, a value for the two-dimensional carrier concentration can be obtained. Clearly, measurement of the Shubnikov–de Haas oscillations provides an important diagnostic tool characterizing carrier transport in two-dimensional electronic systems. The full physics of Shubnikov–de Haas oscillations is highly complicated by the presence of numerous Landau levels and is omitted here. The interested reader is referred to the book by Ridley (1993) for a thorough discussion.

PROBLEMS

9.1. Determine the form of kinetic energy operator for a system with a symmetric vector potential **A** given as

$$\mathbf{A} = \frac{(-By, Bx, 0)}{2}$$

9.2. Determine the expression for the Hamiltonian for the one-dimensional quantum wire structure discussed in Example 9.1.2. Use the definitions given in the example and complete the square.

9.3. Relate the Hall resistance for the quantum Hall effect to the fine structure constant α. The fine structure constant is defined as the ratio of the velocity in the first Bohr orbit to the velocity of light in mks units. To get the velocity of the first Bohr orbit, write the force equation for an electron in a circular orbit about a proton assuming classical motion. Assume that the orbital radii allowed are given as

$$r_n = \frac{4\pi\varepsilon_0\hbar^2}{mZq^2}n^2$$

where n is an integer that labels the orbit, Z the atomic number, and m the mass of the electron.

In this way, the quantum Hall effect can be used to determine fundamental physical constants.

REFERENCES

Numbers in parentheses at the end of sources indicate chapter(s) for which a source has been consulted.

Ambacher, O., Foutz, B., Smart, J., Shealy, J. R., Weimann, N. G., Chu, K., Murphy, M., Sierakowski, A. J., Schaff, W. J., Eastman, L. F., Dimitrov, R., Mitchell, A., and Stutzmann, M. (2000). Two dimensional electron gases induced by spontaneous and piezoelectric polarization in undoped and doped AlGaN/GaN heterostructures. *J. Appl. Phys.*, **87**, 334–44. (2)

Asai, S., and Wada, Y. (1997). Technology challenges for integration near and below 0.1 μm. *Proc. IEEE*, **85**, 505–20. (7)

Asbeck, Peter (2000). III–V HBTs for microwave applications: technology status and modeling challenges. *Proc. Bipolar/BiCMOS Circuits and Technology Meeting*, pp. 33–40. (4)

Asbeck, P. M., Yu, E. T., Lau, S. S., Sullivan, G. J., Van Hove, J., and Redwing, J. (1997). Piezoelectric charge densities in AlGaN/GaN HFETs. *Electron. Lett.*, **33**, 1230–31. (2)

Baccarani, G., Wordeman, M. R., and Dennard, R. H. (1984). Generalized scaling theory and its application to a $\frac{1}{4}$ micrometer MOSFET design. *IEEE Trans. Electron Devices*, **ED-31**, 452–62. (7)

Bakshi, P., Broido, D. A., and Kempa, K. (1991). Spontaneous polarization of electrons in quantum dashes. *J. Appl. Phys.*, **70**, 5150–52. (8)

Bandyopadhyay, S., Das, B., and Miller, A. E. (1994). Supercomputing with spin-polarized single electrons in a quantum coupled architecture. *Nanotechnology*, **5**, 113–33. (8)

Barnes, F. S., Su, W.-H., and Brennan, K. F. (1987). A potentially low-noise avalanche diode microwave amplifier. *IEEE Trans. Electron Devices*, **ED-34**, 966–72. (5)

Bean, J. C. (1990) Materials and technologies, in *High Speed Semiconductor Devices*, edited by S. M. Sze. New York: Wiley. (2)

Bhattacharya, P. (1997). *Semiconductor Optotelectronic Devices*, 2nd ed., Upper Saddle River, NJ: Prentice Hall. (2)

Birnbaum, J., and Williams, R. S. (2000). Physics and the information revolution. *Phys. Today*, Jan.

Brennan, K. F. (1999). *The Physics of Semiconductors with Applications to Optoelectronic Devices*, Cambridge: Cambridge University Press. (2–9)

Brennan, K. F., and Haralson, J., II (2000). Superlattice and multiquantum well avalanche photodetectors: physics, concepts and performance. *Superlattices Microstruct.* **28**, 77–104. (7)

Brennan, K., and Hess, K. (1983). Transient electronic transport in staircase heterostructures. *IEEE Electron Devices Lett.*, **EDL-4**, 419–21. (4)

Brennan, K., and Hess, K. (1984). High field transport in GaAs, InP and InAs. *Solid-State Electron.*, **27**, 347–57. (5)

Brennan, K., and Hess, K. (1986). A theory of enhanced impact ionization due to the gate field and mobility degradation in the inversion layer of MOSFETs. *IEEE Electron Devices Lett.*, **EDL-7**, 86–88. (7)

Brennan, K. F., and Mansour, N. S. (1991). Monte Carlo calculation of electron impact ionization in bulk InAs and HgCdTe. *J. Appl. Phys.*, **69**, 7844–47. (5)

Brennan, K. F., and Park, D. H. (1989). Theoretical comparison of electron real-space transfer in classical and quantum two-dimensional heterostructure systems. *J. Appl. Phys.*, **65**, 1156–63. (5)

Brennan, K. F., and Ruden, P. P., Eds. (2001). *Topics in High Field Transport in Semiconductors*. Singapore: World Scientific. (6)

Brennan, K., Hess, K., Tang, J. Y.-F., and Iafrate, G. J. (1983). Transient electronic transport in InP under the condition of high-energy electron injection. *IEEE Trans. Electron Devices*, **ED-30**, 1750–54. (4)

Brews, J. R. (1990). The submicron MOSFET, in *High Speed Semiconductor Devices*, edited by S. M. Sze. New York: Wiley. (7)

Bykhovski, A., Gelmont, B., and Shur, M. (1993). The influence of the strain-induced electric field on the charge distribution in GaN–AlN–GaN structure. *J. Appl. Phys.*, **74**, 6734–39. (2)

Bykhovski, A., Gelmont, B., and Shur, M. (1997). Elastic strain relaxation and piezoeffect in GaN–AlN, GaN–AlGaN, and GaN–InGaN superlattices. *J. Appl. Phys.*, **81**, 6332–38. (2)

Calecki, D., Palmier, J. F., and Chomette, A. (1984). Hopping conduction in multiquantum well structures. *J. Phys. C: Solid State Phys.*, **17**, 5017–30. (6)

Capasso, F., and Datta, S. (1990). Quantum electron devices. *Phys. Today*, **43**, 74–82. (6)

Capasso, F., and Kiehl, R. A. (1985). Resonant tunneling transistor with quantum well base and high-energy injection: a new negative differential resistance device. *J. Appl. Phys.*, **58**, 1366–72. (6)

Capasso, F., Tsang, W. T., and Williams, G. F. (1983). Staircase solid-state photomultipliers and avalanche photodiodes with enhanced ionization rates ratio. *IEEE Trans. Electron Devices*, **ED-30**, 381–90. (5)

Capasso, F., Mohammed, K., and Cho, A. Y. (1986a). Sequential resonant tunneling through a multiquantum well superlattice. *Appl. Phys. Lett.*, **48**, 478–80. (6)

Capasso, F., Sen. S., Gossard, A. C., Hutchinson, A. L., and English, J. H. (1986b). Quantum-well resonant tunneling bipolar transistor operating at room temperature. *IEEE Electron Devices Lett.*, **EDL-7**, 573–76. (6)

Capasso, F., Sen, S., Beltram, F., Lunardi, L. M., Vengurlekar, A. S., Smith, P. R., Shah, N. J., Malik, R. J., and Cho, A. Y. (1989). Quantum functional devices: resonant tunneling transistors, circuits with reduced complexity, and multiple-valued logic. *IEEE Trans. Electron Devices*, **ED-36**, 2065–81. (6)

Capasso, F., Sen, S., Lunardi, L. M., and Cho, A. Y. (1991). Quantum transistors and circuits break through the barriers. *Circuits Devices*, 18–25. (6)

Chang, M. F. (1996). *Current Trends in Heterojunction Bipolar Transistors*. Singapore: World Scientific. (App. C)

Chang, L. L., Esaki, L., and Tsu, R. (1974). Resonant tunneling in semiconductor double barriers. *Appl. Phys. Lett.*, **24**, 593–95. (6)

Chen, Y. C., Chin, P., Ingram, D., Lai, R., Grundbacher, R., Barsky, M., Block, T., Wojtowicz, M., Tran, L., Medvedev, V., Yen, H. C., Streit, D. C., and Brown, A. (1999). Composite-channel InP HEMT for W-band power amplifiers. *Proc. International Conference on Indium Phosphide and Related Materials.*

Collier, C. P., Wong, E. W., Belohradsky, M., Raymo, F. M., Stoddart, J. F., Kuekes, P. J., Williams, R. S., and Heath, J. R. (1999). Electronically configurable molecular-based logic gates. *Science*, **285**, 391–93. (8)

Conrad, M. (1986). The lure of molecular computing. *IEEE Spectrum*, 55–60. (8)

Cowles, J., Metzger, R. A., Guitierrez-Aitken, Brown, A. S., Streit, D., Oki, A., Kim, T., and Doolittle, A. (1997). Double heterojunction bipolar transistors with InP epitaxial layer grown by solid-source MBE. *Proc. International Conference on Indium Phosphide and Related Materials*, pp. 133–36. (4)

Critchlow, D. L. (1999). MOSFET scaling: the driver of VLSI technology. *Proc. IEEE*, **87**, 659–67. (7)

Dagnall, G. (2000). Solid source molecular beam epitaxial growth of 1.55-μm InAsP/InGaAsP edge-emitting lasers. Ph.D. dissertation, Georgia Institute of Technology. (2)

Datta, S. (1989). *Quantum Phenomena*. Reading, MA: Addison-Wesley. (6)

Datta, S. (1995). *Electronic Transport in Mesoscopic Systems*. Cambridge: Cambridge University Press. (6)

Del Alamo, J. A., and Somerville, M. A. (1999). Breakdown in millimeter wave power InP HEMT's: a comparision with GaAs PHEMTs. *IEEE J. Solid-State Circuits*, **34** (9), 1204–11. (3)

d'Espagnat, B. (1979). The quantum theory and reality. *Sci. Am.*, Nov., pp. 158–79. (8)

Dingle, R., Stormer, H. L., Gossard, A. C., and Wiegmann, W. (1978). Electron mobilities in modulation-doped semiconductor heterojunction superlattices. **33**, 6657. (2)

Dingle, R., Stormer, H. L., Gossard, A. C., and Wiegmann, W. (1980). Electronic properties of the GaAs–AlGaAs interface with applications to multi-interface heterojunction superlattices. **98**, 90–100. (2)

Doshi, B., Brennan, K. F., Bicknell-Tassius, R., and Grunthaner, F. (1998). The effect of strain-induced polarization fields on impact ionization in a multiquantum well structure. *Appl. Phys. Lett.*, **73**, 2784–86. (2)

Duncan, A., Ravaioli, U., and Jakumeit, J. (1998). Full-band Monte Carlo investigation of hot carrier trends in the scaling of metal-oxide-semiconductor field-effect transistors. *IEEE Trans. Electron Devices*, **ED-45**, 867–75. (7)

Eisenstein, J. P., and Stormer, H. L. (1990). The fractional quantum Hall effect. *Science*, **248**, 1510–16. (9)

Ellenbogen, J. C., and Love, J. C. (2000). Architectures for molecular electronic computers. 1. Logic structures and an adder designed from molecular electronic diodes. *Proc. IEEE*, **88**, 386–426. (8)

Fawcett, W., Boardman, A. D., and Swain, S. (1970). Monte Carlo determination of electron transport in gallium arsenide. *J. Phys. Chem. Solids*, **31**, 1963–90. (5)

Ferry, D. K. (1991). *Semiconductors*. New York: Macmillan. (6)

Ferry, D. K., and Barker, J. R. (1999). Issues in general quantum transport with complex potentials. *Appl Phys. Lett.*, **74**, 582–84. (9)

Ferry, D. K., and Goodnick, S. M. (1997). *Transport in Nanostructures*. Cambridge: Cambridge University Press. (8)

Fichtner, W., and Potzl, H. W. (1979). MOS modelling by analytical approximations. I. Subthreshold current and threshold voltage. *Int. J. Electron.*, **46**, 33–55.

Fiegna, C., Iwai, H., Wada, T., Saito, M., Sangiorgi, E., and Ricco, B. (1994). Scaling the MOS transistor below 0.1 μm: methodology, device structures, and technology requirements. *IEEE Trans. Electron Devices*, **ED-41**, 941–50. (7)

Frensley, W. R. (1987). Wigner-function model of a resonant-tunneling semiconductor device. *Phys. Rev. B*, **36**, 1570–80. (6)

Gaylord, T. K., and Brennan, K. F. (1988). Semiconductor superlattice electron wave interference filters. *Appl. Phys. Lett.*, **53**, 2047–49. (2)

Gaylord, T. K., and Brennan, K. F. (1989). Electron wave optics in semiconductors. *J. Appl. Phys.*, **65**, 814–20. (2)

Gelmont, B. L., Shur, M., and Stroscio, M. (1995). Polar optical-phonon scattering in three- and two-dimensional electron gases. *J. Appl. Phys.*, **77**, 657–60. (2)

Gershenfeld, N., and Chuang, I. L. (1998). Quantum computing with molecules. *Sci. Am.*, www.sciam.com/1998/0698issue/0698gershenfeld.html (8)

Gilden, M., and Hines, M. E. (1966). Electronic tuning effects in the Read microwave avalanche diode. *IEEE Trans. Electron Devices*, **ED-13**, 169–75. (5)

Golio, J. M. (1995). Commercial cellular phones and microwaves. *Compound Semicond.*, Sept./Oct. (1)

Golio, J. M. (2000). Low voltage/low power microwave electronics. IEEE Microwave Theory and Techniques Distinguished Lecture. (1)

Gribnikov, Z. S. (1972). *Fiz. Tekh. Poluprovodn.*, **3**, 1169 (in Russian). (5)

Grotjohn, T., and Hoefflinger, B. (1984). A parametric short-channel MOS transistor model for subthreshold and strong inversion current. *IEEE Trans. Electron Devices*, **ED-31**, 234–46. (7)

Grover, L. K. (1997). Quantum mechanics helps in searching for a needle in a haystack. *Phys. Rev. Lett.*, **79**, 325–28. (8)

Gunn, J. B. (1963). Microwave oscillations of current in III–V semiconductors. *Solid-State Commun.*, **1**, 88–91. (5)

Haddad, G. I., and Mazumder, P. (1997). Tunneling devices and applications in high functionality/speed digital circuits. *Solid-State Electron.*, **41**, 1515–24. (6)

Haddad, G. I., Greiling, P. T., and Schroeder, W. E. (1970). Basic principles and properties of avalanche transit-time diodes. *IEEE Trans. Microwave Theory Tech.*, **MTT-18**, 752–72. (5)

Halperin, B. I. (1982). Quantized Hall conductance, current-carrying edge states, and the existence of extended states in a two-dimensional disordered potential. *Phys. Rev. B*, **25**, 2185–90. (9)

Haroche, S., and Raimond, J.-M. (1996). Quantum computing: dream or nightmare? *Phys. Today*, Aug., pp. 51–52. (8)

Heath, J. R., Kuekes, P. J., Snider, G. S., and Williams, R. S. (1998). A defect-tolerant computer architecture: opportunities for nanotechnology. *Science*, **280**, 1716–21. (8)

Hess, K. (1988). *Advanced Theory of Semiconductor Devices*. Upper Saddle River, NJ: Prentice Hall. (2)

Hess, K., Morkoc, H., Shichijo, H., and Streetman, B. G. (1979). Negative differential resistance through real-space electron transfer. *Appl. Phys. Lett.*, **35**, 469–71. (5)

Hilsum, C. (1962). Transferred electron amplifiers and oscillators. *Proc. IRE*, **50**, 185–89. (5)

Hsu, L., and Walukiewicz, W. (1997). Electron mobility in $Al_xGa_{1-x}N/GaN$ heterostructures. *Phys. Rev. B*, **56**, 1520–28. (2)

Ibbetson, J. P., Fini, P. T., Ness, K. D., DenBaars, S. P., Speck, J. S., and Mishra, U. K. (2000). Polarization effects, surface states, and the source of electrons in AlGaN/GaN heterostructure field effect transistors. *Appl. Phys. Lett.*, **77**, 250–52. (2)

Ismail, K., Meyerson, B. S., and Wang, P. J. (1991). High electron mobility in modulation-doped Si/SiGe. *Appl. Phys. Lett.*, **58**, 2117–19. (8)

Ismail, K., Nelson, S. F., Chu, J. O., and Meyerson, B. S. (1993). Electron transport properties of Si/SiGe heterostructures: measurements and device implications. *Appl. Phys. Lett.*, **63**, 660–62. (8)

Jacoboni C., and Lugli, P. (1989). *The Monte Carlo Method for Semiconductor Device Simulation.* Vienna: Springer-Verlag. (5)

Jain, J. K. (1992). Microscopic theory of the fractional quantum Hall effect. *Adv. in Phys.*, **41**, 105–46. (9)

Kang, S., Doolittle, A., Lee, K. K., Dai, Z. R., Wang, Z. L, Stock, S. R., and Brown, A. S. (2000). Characterization of AlGaN/GaN structures on various substrates grown by radio frequency plasma assisted molecular beam epitaxy. *J. Electron. Mater.*, **30**, 156–61. (2)

Kastalsky, A., and Luryi, S. (1983). Novel real-space hot-electron transfer devices. *IEEE Electron Devices Lett.*, **EDL-4**, 334–36. (5)

Kastner, M. A. (1992). The single-electron transistor. *Rev. Mod. Phys.*, **64**, 849–58. (8)

Kawamura, T., and Das Sarma, S. (1992). Phonon-scattering-limited electron mobilities in $Al_xGa_{1-x}As/GaAs$ heterojunctions. *Phys. Rev. B*, **45**, 3612–27. (2)

Kim, T.-H. (2000). Growth and characterization of InP-based high electron mobility transistors. Ph.D. dissertation, Georgia Institute of Technology. (3)

Kluksdahl, N. C., Kriman, A. M., Ferry, D. K., and Ringhofer, C. (1989). Self-consistent study of the resonant-tunneling diode. *Phys. Rev. B*, **39**, 7720–35. (6)

Kohn, W., and Luttinger, J. M. (1957). Quantum theory of electrical transport phenomena. *Phys. Rev.*, **108**, 590–611. (6)

Krieger, J. B., and Iafrate, G. J. (1986). Time evolution of Bloch electrons in a homogeneous electric field. *Phys. Rev. B*, **33**, 5494–5500. (6)

Kroemer, H. (1964). The theory of the Gunn effect. *Proc. IEEE*, **52**, 1736. (5)

Kroemer, H. (1982). Heterostructure bipolar transistors and integrated circuits. *Proc. IEEE*, **70**, 13–25. (4)

Kuech, T. F., Collins, R. T., Smith, D. L., and Mailhiot, C. (1990). Field-effect transistor structure based on strain-induced polarization charges. *J. Appl. Phys.*, **67**, 2650–52. (2,3)

Landau, L. D., and Lifshitz, E. M. (1977). *Quantum Mechanics: Non-relativistic Theory*. Oxford: Pergamon Press. (9)

Larson, L. (1997). Integrated circuit technology options for RFIC's: present status and future directions. *Proc. IEEE 1997 Custom Integrated Circuits Conference*, pp. 169–75. (3)

Laughlin, R. B. (1981). Quantized Hall conductivity in two dimensions. *Phys. Rev. B*, **23**, 5632–33. (9)

Laughlin, R. B. (1987). Elementary theory: the incompressible quantum fluid, in *The Quantum Hall Effect*, edited by R. E. Prange and S. M. Girvin. Berlin: Springer-Verlag, pp. 233–302. (9)

Laux, S. E., and Fischetti, M. V. (1988). Monte-Carlo simulation of submicrometer Si n-MOSFETs at 77 and 300 K. *IEEE Electron Devices Lett.*, **EDL-9**, 467–69. (7)

Lent, C. S., and Tougaw, P. D. (1997). A device architecture for computing with quantum dots. *Proc. IEEE*, **85**, 541–57. (8)

Lew, A. Y., Zuo, S. L., Yu, E. T., and Miles, R. H. (1998). Correlation between atomic-scale structure and mobility anisotropy in InAs/GaInSb superlattices. *Phys. Rev. B*, 57, 6534–39. (2)

Liboff, R. L. (1992). *Introductory Quantum Mechanics*, 2nd ed., Reading MA: Addison-Wesley. (9)

Liu, W. (1999). *Fundamentals of III–V Devices, HBTs, MESFETs, and HFETs/HEMTs*. New York: Wiley-Interscience. (3,4)

Liu, H. C., and Sollner, T. C. L. G. (1994). High-frequency resonant-tunneling devices, in *Semiconductors and Semimetals*, edited by R. K. Willardson and A. C. Beer. San Diego, CA: Academic Press, pp. 359–419. (6)

Liu, R., Pai, C.-S., and Martinez, E. (1999). Interconnect technology trend for microelectronics. *Solid-State Electron.*, **43**, 1003–9. (7)

Lloyd, S. (1993). A potentially realizable quantum computer. *Science*, **261**, 1569–71. (8)

Lloyd, S. (2000). Ultimate physical limits to computation. *Nature*, **406**, 1047–54. (1)

Lundstrom, M. (2000). *Fundamentals of Carrier Transport*. Cambridge: Cambridge University Press. (5)

Luryi, S. (1985). Frequency limit of double-barrier resonant-tunneling oscillators. *App. Phys. Lett.*, **47**, 490–92.

Madelung, O., Ed. (1991). *Semiconductors: Group IV Elements and III–V Compounds*. Berlin: Springer-Verlag. (App. B)

Marsh, J. H., Houston, P. A., and Robson, P. N. (1981). In *Gallium Arsenide and Related Compounds*, edited by H. W. Thim. Bristol, Gloucestershire, England: Institute of Physics. (5)

Matthews, J. W., and Blakeslee, A. E. (1974). Defects in epitaxial multilayers. I. Misfit dislocations. *J. Cryst. Growth*, **27**, 118. (2)

Maziar, C. M., Klausmeier-Brown, M. E., and Lundstrom, M. S. (1986a). A proposed structure for collector transit-time reduction in AlGaAs/GaAs bipolar transistors. *IEEE Electron Devices Lett.*, **EDL-7**, 483–85. (4)

Maziar, C. M., Klausmeier-Brown, M. E., Bandyopadhyay, S., Lundstrom, M. S., and Datta, S. (1986b). Monte Carlo evaluation of electron transport in heterojunction bipolar transistor base structures. *IEEE Trans. Electron Devices*, **ED-33**, 881–87. (4)

Mazumder, P., Kulkarni, S., Bhattacharya, M., Sun, J. P., and Haddad, G. I. (1998). Digital circuit applications of resonant tunneling diodes. *Proc. IEEE*, **86**, 664–86. (6)

McDermott, B. T., Gertner, E. R., Pittman, S., Seabury, C. W., and Chang, M. F. (1996). Growth and doping GaAsSb via metalorganic chemical vapor deposition for InP heterojunction bipolar transistors. *Appl. Phys. Lett.*, **68**, 1386–88. (App. C)

Meindl, J. D. (1984). Ultra-large scale integration. *IEEE Trans. Electron Devices*, **ED-31**, 1555–61. (7)

Meyer, M. (1997). NEC's HBT philosophy. *Compound Semicond.*, May/June. (1)

Misawa, T. (1966a). Negative resistance in p-n junctions under avalanche breakdown conditions. Part I. *IEEE Trans. Electron Devices*, **ED-13**, 137–43. (5)

Misawa, T. (1966b). Negative resistance in p-n junctions under avalanche breakdown conditions. Part II. *IEEE Trans. Electron Devices*, **ED-13**, 143–51. (5)

Monemar, B., and Pozina, G. (2000). Group III-nitride based hetero and quantum structures. *Prog. Quantum Electron.*, **24**, 239–92. (App. C)

Mori, T., Ohnishi, H., Imamura, K., Muto, S., and Yokoyama, N. (1986). Resonant tunneling hot-electron transistor with current gain of 5. *Appl. Phys. Lett.*, **49**, 1779–80. (6)

Muraguchi, M. (1999). RF device trends for mobile communications. *Solid-State Electron.*, **43**, 1591–98. (1)

Nakajima, H., Tomizawa, M., and Ishibashi, T. (1992). Monte Carlo analysis of the space-charge effect in AlGaAs/GaAs ballistic collection transistors (BCTs) under high current injection. *IEEE Trans. Electron Devices*, **ED-39**, 1558–63. (4)

Nakazato, K., Blaikie, R. J., Cleaver, J. R. A., and Ahmed, H. (1993). Single-electron memory. *Electron. Lett.*, **29**, 384–85. (8)

Nguyen, L. D., Larson, L., and Mishra, U. (1992a). Ultra-high-speed modulation-doped field-effect transistors: a tutorial review. *Proc. IEEE*, **80**(4), 494–518. (3)

Nguyen, L. D., Brown, A. S., Thompson, M. A., and Jelloian, L. M. (1992b). 50 nm self-aligned gate pseudomorphic AlInAs/GaInAs high electron mobility transistors. *IEEE Trans. Electron Devices*, **ED-39**(9), 2007–14. (3)

Nye, J. F. (1985). *Physical Properties of Crystals: Their Representation by Tensors and Matrices*. New York: Oxford University Press. (2)

Pao, H. C., and Sah, C. T. (1966). Effects of diffusion current on characteristics of metal-oxide (insulator)–semiconductor transistors. *Solid-State Electron.*, **9**, 927–37. (7)

Park, D. H., and Brennan, K. F. (1989). Theoretical analysis of an AlGaAs/InGaAs pseudomorphic HEMT using an ensemble Monte Carlo simulation. *IEEE Trans. Electron Devices*, **ED-36**, 1254–63. (3)

Park, D. H., and Brennan, K. F. (1990). Monte Carlo simulation of 0.35 μm gate-length GaAs and InGaAs HEMTs. *IEEE Trans. Electron Devices*, **ED-37**, 618–28. (3)

Passlack, M., Abrokwah, J. K., and Lucero, R. (2000). Experimental observation of velocity overshoot in n-channel AlGaAs/InGaAs/GaAs enhancement mode MODFETs. *IEEE Electron Devices Lett.*, **EDL-21**, 518–20. (3)

Paul, D. J. (1998). Silicon germanium heterostructures in electronics: the present and the future. *Thin Solid Films*, **321**, 172–80. (4)

Pavlides, D. (1999). Reliability characteristics of GaAs and InP-based heterojunction bipolar transistors. *Microelectron. Reliab.*, **39**, 1801–8. (4)

Pierret, R. F. (1996). *Semiconductor Device Fundamentals*. Reading, MA: Addison-Wesley. (4)

Porod, W. (1997). Quantum-dot devices and quantum-dot cellular automata. *J. Franklin Inst.*, **334B**, 1147–75. (8)

Potter, R. C., Lakhani, A. A., Beyea, D., Hier, H., Hempfling, E., and Fathimulla, A. (1988). Three-dimensional integration of resonant tunneling structures for signal processing and three-state logic. *Appl. Phys. Lett.*, **52**, 2163–64. (6)

Preskill, J. (1998). Quantum computing: pro and con. *Proc. R. Soc. London Ser. A*, **454**, 469–86. (8)

Price, P. J. (1981). Two-dimensional electron transport in semiconductor layers. I. Phonon scattering. *Ann. Phys.*, **133**, 217–39. (6)

Quay, R., Hess, K., Reuter, R., Schlechtweg, M., Grave, T., Palankovski, V., and Selberherr, S. (2001). Nonlinear electronic transport and device performance of HEMTs. *IEEE Trans. Electron Devices*, **ED-48**, 210–17. (3)

Raghavan, G., Sokolich, M., and Stanchina, W. E. (2000). Indium phosphides ICs unleash the high frequency spectrum. *IEEE Spectrum*, **37**, 47–52. (4)

Read, W. T., Jr., (1958). A proposed high-frequency negative-resistance diode. *Bell Syst. Tech. J.*, **37**, 401–46. (5)

Reed, M. A., Zhou, C., Muller, C. J., Burgin, T. P., and Tour, J. M. (1997). Conductance of a molecular junction. *Science*, **278**, 252–53. (8)

Ridley, B. K. (1993). *Quantum Processes in Semiconductors*, 3rd ed., Oxford: Oxford University Press. (6)

Ridley, B. K. (1997). *Electrons and Phonons in Semiconductor Layers*. Cambridge: Cambridge University Press. (6)

Ridley, B. K., and Watkins, T. B. (1961). The possibility of negative resistance effects in semiconductors. *Proc. Phys. Soc.*, **78**, 293–304. (5)

Ridley, B. K., Foutz, B. E., and Eastman, L. F. (2000). Mobility of electrons in bulk GaN and $Al_xGa_{1-x}N$/GaN heterostructures. *Phys. Rev. B*, **61**, 16862–69. (2)

Rockett, P. I. (1988). Monte Carlo study of the influence of collector region velocity overshoot on the high-frequency performance of AlGaAs/GaAs heterojunction bipolar transistors. *IEEE Trans. Electron Devices*, **ED-35**, 1573–79. (4)

Rosenberg, J. J., Benlami, M., Kirchner, P. D., Woodall, J. M., and Pettit, G. D. (1985). *IEEE Electron Devices Lett.*, **EDL-6**, 491–93. (5)

Rossi, F., and Jacoboni, C. (1989). A quantum description of drift velocity overshoot at high electric fields in semiconductors. *Solid-State Electron.*, **32**, 1411–15. (6)

Roulston, D. J. (1990). *Bipolar Semiconductor Devices*. New York: McGraw-Hill. (4)

Ruch, J. G. (1972). Electron dynamics in short channel field-effect transistors. *IEEE Trans. Electron Devices*, **ED-19**, 652–54. (3)

Ruden, P. P. (1990). Heterostructure FET model including gate leakage. *IEEE Trans. Electron Devices*, **ED-37**, 2267–70. (3)

See, P., Paul, D. J., Hollander, B., Mantl, S., Zozuolenko, I. V., and Berggren, K.-F. (2001). High dc performance Si/SiGe resonant tunnelling diodes. *IEEE Electron Devices Lett.*, **EDL-22**, 182. (App. C)

Selberherr, S. (1984). *Analysis and Simulation of Semiconductor Devices*. Vienna: Springer-Verlag. (7)

Shen, J.-J., Brown, A. S., Metzger, R. A., Sievers, B., Bottomley, L., Eckert, P., and Carter, B. (1998). Modification of quantum dot properties via surface exchange and annealing: substrate temperature effects. *J. Vac. Sci. Technol. B: Microelectron. Nanometer Struct.*, **16**, 1326–29. (2)

Shimony, A. (1985). The reality of the quantum world. *Sci. Am.*, 46–53. (8)

Shor, P. W. (1999). Polynomial-time algorithms for prime factorization and discrete logarithms on a quantum computer. *SIAM Rev.*, **41**, 303–32. (8)

Shur, M. (1990). *Physics of Semiconductor Devices*. Upper Saddle River, NJ: Prentice Hall. (4)

Singh, J. (1993). *Physics of Semiconductors and Their Heterostructures*. New York: McGraw-Hill. (9)

Smith, D. L. (1986). Strain-generated electric fields in [111] growth axis strained-layer superlattices. *Solid-State Commun.*, **57**, 919–21. (2)

Smith, A. W., and Brennan, K. F. (1998). Hydrodynamic simulation of semiconductor devices. *Prog. Quantum Electron.*, **21**, 293–360. (3)

Smorchkova, I. P., Elsass, C. R., Ibbetson, J. P., Vetury, R., Heying, B., Fini, P., Haus, E., DenBaars, S. P., Speck, J. S., and Mishra, U. K. (1999). Polarization-induced charge and electron mobility in AlGaN/GaN heterostructures grown by plasma-assisted molecular-beam-epitaxy. *J. Appl. Phys.*, **86**, 4520–26. (2)

Snow, E. S., Shanabrook, B. V., and Gammon, D. (1990). Strain-induced two-dimensional electron gas in [111] growth-axis strained-layer structures. *Appl. Phys. Lett.*, **56**, 758–60. (2)

Snowden, C. M. (1988). *Semiconductor Device Modelling*. Exeter, Devonshire, England: Peter Pereginus. (3)

Steane, A. (1998). Quantum computing. *Rep. Prog. Phys.*, **61**, 117–73. (8)

Stone, A. D., and Lee, P. A. (1985). Effect of inelastic processes on resonant tunneling in one dimension. *Phys. Rev. Lett.*, **54**, 1196–99. (6)

Stone, N. J., Ahmed, H., and Nakazato, K. (1999). A high-speed silicon single-electron random access memory. *IEEE Electron Devices Lett.*, **EDL-20**, 583–85. (8)

Stormer, H. L., and Tsui, D. C. (1983). The quantized Hall effect. *Science*, **220**, 1241–46. (9)

Stormer, H. L., Yeh, A. S., Pan, W., Tsui, D. C., Pfeiffer, L. N., Baldwin, K. W., and West, K. W. (1998). Composite fermions at different levels. *Physica E*, **3**, 38–46. (9)

Streetman, B. G., and Banerjee, S. (2000). *Solid State Electronic Devices*, 5th ed., Upper Saddle River, NJ: Prentice Hall. (5)

Streit, D. (2001). Private communication. (4)

Su, L. T., Jacobs, J. B., Chung, J. E., and Antoniadis, D. A. (1994). Deep-submicrometer channel design in silicon-on-insulator (SOI) MOSFETs. *IEEE Electron Devices Lett.*, **EDL-15**, 366–69. (8)

Sun, J. P., Haddad, G. I., Mazumder, P., and Schulman, J. N. (1998). Resonant tunneling diodes: models and properties. *Proc. IEEE*, **86**, 641–60. (6)

Sze, S. M. (1981). *Physics of Semiconductor Devices*, 2nd ed., New York: Wiley. (5)

Sze, S. M. (1985). *Semiconductor Devices Physics and Technology*. New York: Wiley. (4)

Tang, J. Y., and Hess, K. (1982). Investigation of transient and electronic transport in GaAs following high energy injection. *IEEE Trans. Electron Devices*, **ED-29**, 1906–10. (4)

Tapuhi, E. (1991). Molecular electronics: a new interdisciplinary field of research. *Interdiscip. Sci. Rev.*, **16**, 45–60. (8)

Taur, Y., and Ning, T. H. (1998). *Fundamentals of Modern VLSI Devices*. Cambridge: Cambridge University Press. (7)

Taur, Y., Buchanan, D. A., Chen, W., Frank, D. J., Ismail, K. E., Lo, S.-H., Asi-Halasz, G. A., Viswanathan, R. G., Wann, H.-J. C., Wind, S. J., and Wong, H.-S. (1997). CMOS scaling into the nanometer regime. *Proc. IEEE*, **85**, 486–503. (7)

Teich, M. C., Matsuo, K., and Saleh, B. E. A. (1986). Excess noise factors for conventional and superlattice avalanche photodiodes and photomultiplier tubes. *IEEE J. Quantum Electron.*, **QE-22**, 1184–93. (5)

Thornber, K. K. (1991). Path integral method: use of Feynman path integrals in quantum transport theory, in *Quantum Transport in Semiconductors*, edited by D. K. Ferry and C. Jacoboni. New York: Plenum Press. (6)

Tiwari, S. (1992). *Compound Semiconductor Device Physics*. San Diego, CA: Academic Press. (3)

Tiwari, S., Rana, F., Hanafi, H., Harstein, A., Crabbe, E. F., and Chan, K. (1996). A silicon nanocrystals based memory. *Appl. Phys. Lett.*, **68**, 1377–79. (8)

Tsao, J. Y. (1993) *Materials Fundamentals of Molecular Beam Epitaxy*. San Diego, CA: Academic Press. (2)

Tsu, R., and Dohler, G. (1975). Hopping conduction in a "superlattice." *Phys. Rev. B*, **12**, 680–86. (6)

Tsu, R., and Esaki, L. (1973). Tunneling in a finite superlattice. *Appl. Phys. Lett.*, **22**, 562–64. (6)

Tsui, D. C., and Gossard, A. C. (1981). Resistance standard using quantization of the Hall resistance of GaAs–$Al_xGa_{1-x}As$ heterostructures. *Appl. Phys. Lett.*, **38**, 550–2. (9)

Tsui, D. C., Stormer, H. L., and Gossard, A. C. (1982). Two-dimensional magnetotransport in the extreme quantum limit. *Phys. Rev. Lett.*, **48**, 1559–62. (9)

Vaidyanathan M., and Pulfrey, D. L. (1999). Extrapolated f_{max} of heterojunction bipolar transistors. *IEEE Trans. Electron Devices*, **ED-46**, 301–9. (4)

Veeraraghavan, S., and Fossum, J. G. (1989). Short-channel effects in SOI MOSFETs. *IEEE Trans. Electron Devices*, **ED-36**, 522–28. (7)

Venables, J. A. (2000). *Introduction to Thin Film Processes*. Cambridge: Cambridge University Press. (2)

Villasenor, J., and Mangione-Smith, W. H. (1997). Configurable computing. *Sci. Am.*, **276**, 66–71. (8)

von Klitzing, K., Dorda, G., and Pepper, M. (1980). New method for high-accuracy determination of the fine-structure constant based on quantized Hall resistance. *Phys. Rev. Lett.*, **45**, 494–97. (9)

Weil, T., and Vinter, B. (1986). Calculation of phonon-assisted tunneling between two quantum wells. *J. Appl. Phys.*, **60**, 3227–31. (6)

Weisbuch, C., and Vinter, B. (1991). *Quantum Semiconductor Structures*. San Diego, CA: Academic Press. (9)

Whitesides, G. M., Mathias, J. P., and Seto, C. T. (1991). Molecular self-assembly and nanochemistry: a chemical strategy for the synthesis of nanostructures. *Science*, **254**, 1312–19. (8)

Willett, R. L. (1997). Experimental evidence for composite fermions. *Semicond. Sci. Technol.*, **12**, 495–524. (9)

Willett, R. L., Eisenstein, J. P., Stormer, H. L., Tsui, D. C., Gossard, A. C., and English, J. H. (1987). Observation of an even-denominator quantum number in the fractional quantum Hall effect. *Phys. Rev. Lett.*, **59**, Oct. 12.

Williams, R. S. (2000). Private communication. (8)

Windhorn, T. H., Cook, L. W., and Stillman, G. E. (1982). *IEEE Electron Devices Lett.*, **EDL-3**, 18–20. (5)

Wong, H.-S. P., Frank, D. J., Solomon, P. M., Wann, C. H. J., and Welser, J. J. (1999). Nanoscale CMOS. *Proc. IEEE*, **87**, 537–70. (7)

Woodward, T. K., McGill, T. C., and Burnham, R. D. (1987). Experimental realization of a resonant tunneling transistor. *Appl. Phys. Lett.*, **23**, 451–53. (6)

Yamada, T., Zhou, J.-R., Miyata, H., and Ferry, D. K. (1994). In-plane transport properties of $Si/Si_{1-x}Ge_x$ structure and its FET performance by computer simulation. *IEEE Trans. Electron Devices*, **ED-41**, 1513–22. (8)

Yano, K., Ishii, T., Hashimoto, T., Kobayashi, T., Murai, F., and Seki, K. (1994). Room-temperature single-electron memory. *IEEE Trans. Electron Devices*, **ED-41**, 1628–37. (8)

Yau, L. D. (1974). A simple theory to predict the threshold voltage of short-channel IGFETs. *Solid-State Electron.*, **17**, 1059–63. (7)

Yokoyama, K., and Hess, K. (1986). Monte Carlo study of electronic transport in $Al_{1-x}Ga_xAs/GaAs$ single-well heterostructures. *Phys. Rev. B*, **33**, 5595–606. (2)

Yokoyama, K., Tomizawa, M., and Yoshii, A. (1984). Accurate modeling of AlGaAs/GaAs heterostructure bipolar transistors by two-dimensional computer simulation. *IEEE Trans. Electron Devices*, **ED-31**, 1222–29. (4)

Yoon, K. S., Stringfellow, G. B., and Huber, R. J. (1987). Monte Carlo calculation of velocity-field characteristics in GaInAs/InP and GaInAs/AlInAs single-well heterostructures. *J. Appl. Phys.*, **62**, 1931–36. (6)

Yu, T.-H., and Brennan, K. F. (2001). Theoretical study of the two-dimensional electron mobility in strained III-nitride heterostructures. *J. Appl. Phys.*, **89**, 3827–34. (2)

Yu, E. T., McCaldin, J. O., and McGill, T. C. (1992). In *Solid State Physics: Advances in Research and Applications*. Vol. 46, edited by H. Ehrenreich and D. Turnbull. Boston: Academic Press, pp.1–146. (App. C)

Yu, T.-H., Yu, E. T., Sullivan, G. J., Asbeck, P. M., Wang, C. D., Qiao, D., and Lau, S. S. (1997). Measurement of piezoelectrically induced charge in GaN/AlGaN heterostructure field-effect transistors. *Appl. Phys. Lett.*, **71**, 2794–96. (2)

APPENDIX A

Physical Constants

Quantity	Symbol	Value	Units
Avogadro's constant	N_{AVO}	6.022×10^{23}	mol^{-1}
Boltzmann's constant	k_B	1.38×10^{-23}	J/K
		8.62×10^{-5}	eV/K
Electron charge	q	1.6×10^{-19}	C
Electron rest mass	m_0	0.511×10^{6}	eV/c^2
		9.11×10^{-31}	kg
Magnetic permeability	μ_0	1.2566×10^{-8}	H/cm
Permittivity—free space	ε_0	8.85×10^{-14}	F/cm
Planck's constant	h	4.14×10^{-15}	$eV \cdot s$
		6.63×10^{-34}	$J \cdot s$
Reduced Planck's constant	$\hbar$	6.58×10^{-16}	$eV \cdot s$
		1.055×10^{-34}	$J \cdot s$
Speed of light	c	3.0×10^{10}	cm/s
Thermal voltage (300 K)	k_bT/q	0.0259	V

APPENDIX B

Bulk Material Parameters

TABLE I Bulk Material Parameters for Silicon

Parameter	Value
Lattice constant (Å)	$a = 5.43$
Dielectric constant	11.9
Intrinsic carrier concentration (cm^{-3})	1.0×10^{10}
Energy bandgap (eV)	1.12
Sound velocity (cm/s)	9.04×10^{5}
Density (g/cm^3)	2.33
Effective mass at X (m^*/m_0) (transverse)	0.19
Effective mass at X (m^*/m_0) (longitudinal)	0.916
Effective mass at L (m^*/m_0) (transverse)	0.12
Effective mass at L (m^*/m_0) (longitudinal)	1.59
Heavy hole mass	0.537
Electron mobility at 300 K ($cm^2/V \cdot s$)	1450
Hole mobility at 300 K ($cm^2/V \cdot s$)	500
Nonparabolicity at X (eV^{-1})	0.5
Intravalley acoustic deformation potential (eV)	9.5
Optical phonon energy at Γ (eV)	0.062
Intervalley separation energy, X – L (eV)	1.17

TABLE II Bulk Material Parameters for Ge

Parameter	Value
Lattice constant (Å)	$a = 5.646$
Dielectric constant	16.0
Intrinsic carrier concentration (cm^{-3})	2.4×10^{13}
Energy bandgap (eV)	0.664
Sound velocity (cm/s)	3.63×10^5
Density (g/cm^3)	5.326
Effective mass at L (m^*/m_0) (transverse)	0.082
Effective mass at L (m^*/m_0) (longitudinal)	1.64
Averaged effective mass at X (m^*/m_0)	0.482
Effective mass at Γ (m^*/m_0)	0.038
Heavy hole mass	0.354
Electron mobility at 300 K ($cm^2/V \cdot s$)	3900
Hole mobility at 300 K ($cm^2/V \cdot s$)	1900
Optical phonon energy	0.037
Intervalley separation energy, L – X (eV)	0.18
Intervalley separation energy, L – Γ (eV)	0.14

TABLE III Bulk Material Parameters for GaAs

Parameter	Value
Lattice constant (Å)	$a = 5.65$
Low-frequency dielectric constant	12.90
High-frequency dielectric constant	10.92
Energy bandgap (eV)	1.425
Intrinsic carrier concentration (cm^{-3})	2.1×10^6
Electron mobility at 300 K ($cm^2/V \cdot s$)	8500
Hole mobility at 300 K ($cm^2/V \cdot s$)	400
Longitudinal sound velocity (cm/s) along (100) direction	4.73×10^5
Density (g/cm^3)	5.36
Effective mass at Γ (m^*/m_0)	0.067
Effective mass at L (m^*/m_0)	0.56
Effective mass at X (m^*/m_0)	0.85
Heavy hole mass	0.62
Nonparabolicity at Γ (eV^{-1})	0.690
Intravalley acoustic deformation potential (eV)	8.0
Optical phonon energy at Γ (eV)	0.035
Intervalley separation energy, Γ – L (eV)	0.284
Intervalley separation energy, Γ – X (eV)	0.476

TABLE IV Bulk Material Parameters for InP

Parameter	Value
Lattice constant (Å)	$a = 5.868$
Low-frequency dielectric constant	12.35
High-frequency dielectric constant	9.52
Intrinsic carrier concentration (cm^{-3})	1.2×10^8
Energy bandgap (eV)	1.35
Electron mobility at 300 K ($cm^2/V \cdot s$)	4600
Hole mobility at 300 K ($cm^2/V \cdot s$)	150
Longitudinal sound velocity (cm/s)	5.13×10^5
Transverse sound velocity (cm/s)	3.10×10^5
Density (g/cm^3)	4.787
Effective mass at Γ (m^*/m_0)	0.078
Effective mass at L (m^*/m_0)	0.26
Effective mass at X (m^*/m_0)	0.325
Heavy hole mass	0.45
Nonparabolicity at Γ (eV^{-1})	0.830
Intravalley acoustic deformation potential (eV)	8.0
Optical phonon energy at Γ (eV)	0.043
Intervalley separation energy, $\Gamma - L$ (eV)	0.54
Intervalley separation energy, $\Gamma - X$ (eV)	0.775

TABLE V Bulk Material Parameters for InAs

Parameter	Value
Lattice constant (Å)	$a = 6.0584$
Low-frequency dielectric constant	14.55
High-frequency dielectric constant	11.8
Intrinsic carrier concentration (cm^{-3})	1.3×10^{15}
Energy bandgap (eV) ($T = 300$ K)	0.354
Longitudinal sound velocity (cm/s)	4.35×10^5
Density (g/cm^3)	5.67
Effective mass at Γ (m^*/m_0)	0.023
Effective mass at L (m^*/m_0)	0.286
Effective mass at X (m^*/m_0)	0.640
Heavy hole mass	0.43
Nonparabolicity at Γ (eV^{-1})	2.33
Electron mobility at 300 K ($cm^2/V \cdot s$)	3.3×10^4
Hole mobility at 300 K (cm2/V · s)	450
Intravalley acoustic deformation potential (eV)	8.0
Optical phonon energy at Γ (eV)	0.0302
Intervalley separation energy, $\Gamma - L$ (eV)	0.79
Intervalley separation energy, $\Gamma - X$ (eV)	1.85

TABLE VI Bulk Material Parameters for Wurtzite Phase InN

Parameter	Value
Lattice constant (Å)	$a = 3.54$, $c = 5.7$
Low-frequency dielectric constant	15.4
High-frequency dielectric constant	8.4
Energy bandgap (eV)	1.86
Longitudinal sound velocity (cm/s)	6.24×10^5
Transverse sound velocity (cm/s)	2.55×10^5
Density (g/cm^3)	6.81
Effective mass at Γ (m^*/m_0)	0.11
Nonparabolicity at Γ (eV^{-1})	0.419
Electron mobility at 300 K (cm^2/V $\cdot$ s)	3000
Intravalley acoustic deformation potential (eV)	7.1
Optical phonon energy at Γ (eV)	0.089
Piezoelectric coupling constant, K_{av}^2	0.0652

TABLE VII Bulk Material Parameters for GaN

Parameter	Zincblende Phase Value	Wurtzite Phase Value
Lattice constant (Å)	$a = 4.50$	$a = 3.189$, $c = 5.185$
Low-frequency dielectric constant	9.5	9.5
High-frequency dielectric constant	5.35	5.35
Energy bandgap (eV) ($T = 300$ K)	3.279	3.44
Longitudinal sound velocity (cm/s)	4.57×10^5	4.33×10^5
Density (g/cm^3)	6.095	6.095
Average effective mass at Γ (m^*/m_0)	0.15	0.19
Average heavy hole mass	1.37	2.53
Nonparabolicity at Γ (eV^{-1})	0.19	In-plane = 0.22, c-axis = 4.45
Intravalley acoustic deformation potential (eV)	8.3	7.8
Optical phonon energy at Γ (eV)	0.0909	0.0909
Piezoelectric constant (C/m^2)	0.375	0.375

TABLE VIII Bulk Material Parameters for SiC

Parameter	Cubic Phase Value	4H Phase Value
Lattice constant (Å)	$a = 4.35$	$a = 3.073$, $c = 10.053$
Low-frequency dielectric constant	9.72	10.0
High-frequency dielectric constant	6.52	6.7
Energy bandgap (eV) (T = 300 K)	2.39	3.26
Longitudinal sound velocity (cm/s)	1.12×10^6	1.35×10^6
Density (g/cm^3)	3.166	3.12
Average effective mass at X (m^*/m_0)	0.345	—
Intravalley acoustic deformation potential (eV)	6.5	6.5
Optical phonon energy at Γ (eV)	0.12	0.12
Piezoelectric constant (C/m^2)	0.375	0.375

TABLE IX Bulk Material Parameters for ZnS

Parameter	Zincblende Phase Value	Wurtzite Phase Value
Lattice constant (Å)	$a = 5.411$	$a = 3.814$, $c = 6.257$
Low-frequency dielectric constant	8.32	9.6
High-frequency dielectric constant	5.15	5.70
Energy bandgap (eV) (T = 300 K)	3.6	3.8
Longitudinal sound velocity (cm/s)	5.2×10^5	5.868×10^5
Density (g/cm^3)	4.075	4.075
Average effective mass at Γ (m^*/m_0)	0.34	0.28
Nonparabolicity at Γ (eV^{-1})	0.69	Averaged = 0.69
Intravalley acoustic deformation potential (eV)	4.9	4.9
Optical phonon energy at Γ (eV)	0.043	0.0426
Piezoelectric constant (C/m^2)	0.375	0.375

TABLE X Bulk Material Parameters for ZnSe

Parameter	Value
Lattice constant (Å)	$a = 5.66$
Low-frequency dielectric constant	9.20
High-frequency dielectric constant	6.20
Energy bandgap (eV) (T = 300 K)	2.70
Longitudinal sound velocity (cm/s)	4.58×10^5
Density (g/cm^3)	5.42
Effective mass at Γ (m^*/m_0)	0.17
Effective mass at L (m^*/m_0)	0.51
Effective mass at X (m^*/m_0)	0.316
Heavy hole mass	0.60
Nonparabolicity at Γ (eV^{-1})	0.69
Optical phonon energy at Γ (eV)	0.0314
Intervalley separation energy, Γ – L (eV)	1.58
Intervalley separation energy, Γ – X (eV)	1.49

TABLE XI Bulk Material Parameters for $Al_xGa_{1-x}As$

Parameter	Value
Lattice constant (Å)	$a = 5.65 + 0.0078x$
Low-frequency dielectric constant	$13.18 - 3.12x$
High-frequency dielectric constant	$10.89 - 2.73x$
Energy bandgap (eV) (T = 300 K)	$1.425 + 1.247x$
Longitudinal sound velocity (cm/s) along (100) direction	$4.7 \times 10^5 + 0.9 \times 10^5 x$
Density (g/cm^3)	$5.36 - 1.6x$
Effective mass at Γ (m^*/m_0)	$0.067 + 0.083x$
Effective mass at L (m^*/m_0)	$0.56 + 0.1x$
Effective mass at X (m^*/m_0)	$0.85 - 0.14x$
Heavy hole mass	$0.62 + 0.14x$
Electron mobility at 300 K (cm^2/V · s) ($x < 0.45$)	$8500 - 22000x + 10000x^2$
Hole mobility at 300 K (cm^2/V · s)	$400 - 970x + 740x^2$
Acoustic deformation potential (eV)	$6.7 - 1.2x$
Optical phonon energy at Γ (eV)	$0.036 - 6.55x + 1.79x^2$
Intervalley separation energy, Γ – L (eV)	$0.284 - 0.605x$
Intervalley separation energy, Γ – X (eV)	$0.476 - 1.122x + 0.143x^2$

TABLE XII Bulk Material Parameters for $Ga_{0.47}In_{0.53}As$

Parameter	Value
Lattice constant (Å)	$a = 5.867$
Low-frequency dielectric constant	13.85
High-frequency dielectric constant	11.09
Intrinsic carrier concentration (cm^{-3})	9.04×10^{11}
Energy bandgap (eV) ($T = 300$ K)	0.75
Longitudinal sound velocity (cm/s)	4.55×10^5
Density (g/cm^3)	5.48
Effective mass at Γ (m^*/m_0)	0.0463
Effective mass at L (m^*/m_0)	0.256
Effective mass at X (m^*/m_0)	0.529
Heavy hole mass	0.61
Nonparabolicity at Γ (eV^{-1})	1.18
Optical phonon energy at Γ (eV)	0.0327
Intervalley separation energy, Γ − L (eV)	0.58
Intervalley separation energy, Γ − X (eV)	1.02

TABLE XIII Bulk Material Parameters for $Al_{0.48}In_{0.52}As$

Parameter	Value
Lattice constant (Å)	$a = 5.867$
Low-frequency dielectric constant	12.42
High-frequency dielectric constant	10.28
Intrinsic carrier concentration (cm^{-3})	9.54×10^5
Energy bandgap (eV) ($T = 300$ K)	1.49
Longitudinal sound velocity (cm/s)	4.97×10^5
Density (g/cm^3)	4.75
Effective mass at Γ (m^*/m_0)	0.084
Effective mass at L (m^*/m_0)	0.274
Effective mass at X (m^*/m_0)	0.496
Heavy hole mass	0.677
Nonparabolicity at Γ (eV^{-1})	0.571
Optical phonon energy at Γ (eV)	0.041
Intervalley separation energy, Γ − L (eV)	0.16
Intervalley separation energy, Γ − X (eV)	0.22

TABLE XIV Bulk Material Parameters for $Ga_{0.5}In_{0.5}P$

Parameter	Value
Lattice constant (Å)	$a = 5.65$
Low-frequency dielectric constant	11.75
High-frequency dielectric constant	9.34
Intrinsic carrier concentration (cm^{-3})	220
Energy bandgap (eV) (T = 300 K)	1.92
Longitudinal sound velocity (cm/s)	5.49×10^5
Density (g/cm^3)	4.47
Effective mass at Γ (m^*/m_0)	0.105
Effective mass at L (m^*/m_0)	0.242
Effective mass at X (m^*/m_0)	0.61
Heavy hole mass	0.48
Nonparabolicity at Γ (eV^{-1})	0.52
Optical phonon energy at Γ (eV)	0.0464
Intervalley separation energy, Γ – L (eV)	0.125
Intervalley separation energy, Γ – X (eV)	0.217

TABLE XV Bulk Material Parameters for $Hg_{0.70}Cd_{0.30}Te$

Parameter	Value
Lattice constant (Å)	$a = 6.464$
Low-frequency dielectric constant	16.24
High-frequency dielectric constant	11.73
Intrinsic carrier concentration (cm^{-3}) (T = 77 K)	6.75×10^8
Energy bandgap (eV) (T = 77 K)	0.25
Longitudinal sound velocity (cm/s)	1.96×10^5
Density (g/cm^3)	7.374
Effective mass at Γ (m^*/m_0)	0.021
Effective mass at L (m^*/m_0)	0.23
Effective mass at X (m^*/m_0)	0.43
Heavy hole mass	0.46
Nonparabolicity at Γ (eV^{-1})	3.83
Optical phonon energy at Γ (eV)	0.018
Intervalley separation energy, Γ – L (eV)	1.31
Intervalley separation energy, Γ – X (eV)	1.95

APPENDIX C

Heterojunction Properties

Heterojunctions	ΔE_C(eV)	ΔE_v(eV)	Ref[a]
$Al_{0.3}Ga_{0.7}As$–GaAs	0.24	0.13	[1]
$Ga_{0.51}In_{0.49}P$–GaAs: ordered	0.03	0.4	[1]
$Ga_{0.51}In_{0.49}P$–GaAs: disordered	0.22	0.24	[1]
InP–$Ga_{0.47}In_{0.53}As$	0.	0.37	[1]
$Al_{0.48}In_{0.52}As$–$Ga_{0.47}In_{0.53}As$	0.5	0.21	[1]
$GaAs_{0.51}Sb_{0.49}$–InP	0.01	0.75	[2]
InAs–GaSb	0.88	0.51	[3]
GaSb–AlSb	0.5	0.35	[3]
InAs–AlSb	1.35	0.13	[3]
AlN–GaN [0 0 0 1]	2.05	0.8	[4]
	2.4	0.5	[4]
	1.5	1.36	[4]
AlN–GaN [1 1 0] cubic T	2.2	0.7	[4]
AlN–GaN [0 0 0 1] T	2.3	0.57	[4]
	2.2	0.68	[4]
AlN–GaN [0 0 0 1]	2.3	0.6	[4]
	0.6	0.93	[4]
InN–GaN [0 0 0 1]	0.9	0.59	[4]
GaN–InN [0 0 0 1]	0.7	0.7	[4]
GaN–InN [1 1 0] cubic T	1.2	0.3	[4]
AlN–InN [0 0 0 1]	3.0	1.32	[4]
InN–AlN [0 0 0 1]	2.6	1.71	[4]
AlN–SiC [0 0 1] cubic T		2	[4]
AlN–SiC [1 1 0] cubic T		1.6	[4]
GaN–SiC [1 0 0] cubic T		1.3	[4]
GaN–SiC [1 1 0] cubic T		0.4	[4]
Si–$Si_{0.6}Ge_{0.4}$	0.11	0.3	[5]
Si–$Si_{0.8}Ge_{0.2}$	0.02	0.15	[5]
$Si_{0.7}Ge_{0.3}$–Si	0.29	0.07	[5]

[a] [1] Chang (1996), [2] McDermott et al. (1996), [3] Yu et al. (1992), [4] Monemar and Pozina (2000), [5] See et al. (2001).

INDEX

ROWAN UNIVERSITY LIBRARY
3 3001 00858 4707